Physiology

Commissioning Editor: Timothy Horne
Development Editor: Barbara Simmons
Project Manager: Emma Riley
Designer: Stewart Larking
Illustration Manager: Merlyn Harvey
Illustrator: Cactus

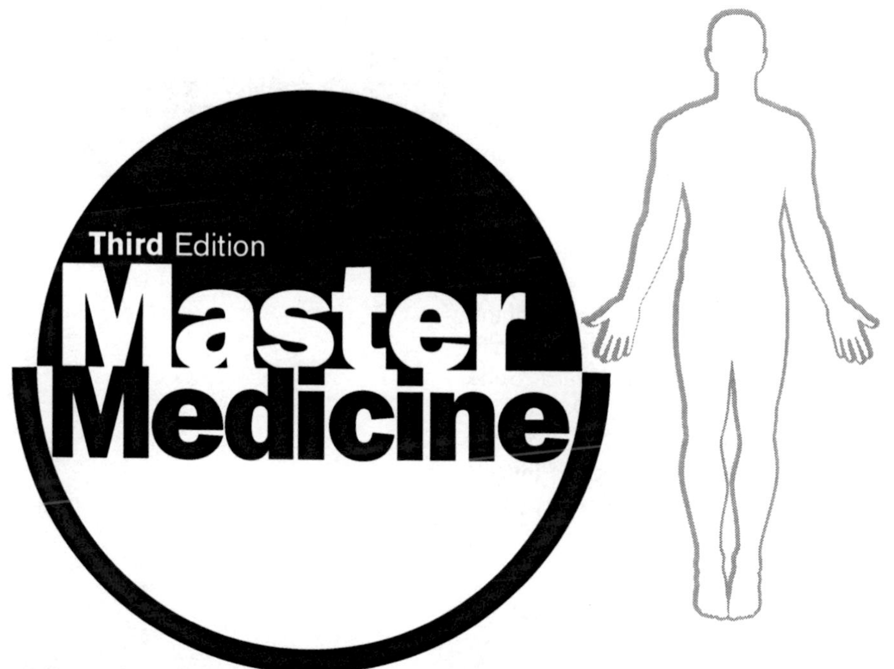

Physiology

A clinical core text of human physiology with self-assessment

J G McGEOWN
BSc MB BCh BAO PhD
Reader in Physiology
School of Medicine and Dentistry
Queen's University
Belfast

Edinburgh London New York Oxford Philadelphia St Louis Sydney Toronto 2007

CHURCHILL
LIVINGSTONE
ELSEVIER

An imprint of Elsevier Limited

First edition 1997
Second edition 2002
Third edition: 2007

ISBN-13: 978 0443 102929

British Library Cataloguing in Publication Data
A catalogue record for this book is available from the British Library

Library of Congress Cataloging in Publication Data
A catalog record for this book is available from the Library of Congress

Note
Neither the Publisher nor the Author assumes any responsibility for any loss or injury and/or damage to persons or property arising out of or related to any use of the material contained in this book. It is the responsibility of the treating practitioner, relying on independent expertise and knowledge of the patient, to determine the best treatment and method of application for the patient.

Transferred to Digital Printing in 2010

Contents

Using this book

Philosophy of the book

This section aims to help you:

- understand how self-assessment can make learning easier and more enjoyable
- use the book to increase your understanding as well as your knowledge
- plan your learning.

How much do you know about the control of blood pressure? Are they the right things? Can you answer examination problems on physiological control systems? This book aims to help you with these questions. Principles are illustrated and mechanisms explained while essential information is presented in a concise and ordered fashion. Do not think, though, that this book offers a 'syllabus'. It is impossible to draw boundaries around medical knowledge and learning is a continuous process carried out throughout your career. With this in mind, we want you to develop the ability to discriminate material which must be understood from areas which you need to know about, and topics which you might simply be aware of.

You are probably working towards one or more examinations and one aim is to show you how to overcome this barrier. Learning is not simply for the purpose of passing exams, however, so the book aims both to help you to pass and to develop useful knowledge and understanding.

Layout and content of the book

In order to use this book fully, you need to know why it is laid out as it is. The overview to each chapter sets out the scope of material covered. This has been chosen to reflect those things which anyone starting a medical career needs to know and which an examiner expects them to know. If you have already studied a given system, you might start by testing yourself using the self-assessment section at the end of each chapter. This will help to steer you towards areas that you need to work on. Alternatively, you may go straight into the main body of the chapter and check at the end that you have grasped the material; if not, then you will need to do further work and perhaps read about the topics elsewhere.

The main part of the text describes the important physiology for the different systems of the body. Within each section, the essential information is presented. These essentials are summarized at the start of each subsection in the form of learning objectives, which provide a guide as to what you should know when you have studied that subject. One useful study technique is to check whether you have achieved these objectives or not. Relevant structural information is given first, both anatomical and histological, followed by a description of the main functions carried out. This will generally involve an outline of what a system does and how its performance is measured. Consideration will then be given to the mechanisms which explain these observed functions. Particular emphasis is placed on the controls which regulate function to meet body needs. Finally, important examples will be given of the consequences of abnormal function in each system to provide a link with your clinical study and training.

You have to be sure that you are reaching the required standards, so the final section of each chapter is there to help you to check your knowledge and understanding. The self-assessment is in the form of multiple choice questions, single best answer questions, matching items questions, short note questions, modified essay questions, data interpretation, clinical scenarios and sample viva questions. Some of these formats may be new to you, but all are used in testing aspiring clinicians. Questions are mainly centred on aspects of physiology which are generally regarded as important in clinical practice and so are emphasized in examinations. Detailed answers are given with reference to relevant sections of the text; the answers also contain information and explanations which you will not find elsewhere, so you have to do the assessments to get the most out of this book.

Studying the subject

Use of the self-assessment sections will encourage you to store information and to test your ongoing improvement. You must keep testing your current level of knowledge, both strengths and weaknesses.

Overall, the aim is to help you to learn through the use of interlinked steps. Initially you must decide:

- What do you already know about the subject?
- What do you want to learn about the subject?

Knowledge is acquired more easily if it can be put into a framework, so:

- Ask yourself what things need explaining—what do you not understand?
- Try to expand on these things—explain as much as you can and be as specific as possible about what confuses you.
- Use resources such as textbooks, teachers and your peers to address these confusing issues.
- Check that you are no longer confused using relevant self-assessment materials.

If you can, discuss problems with colleagues/friends. The areas which you understand least well will become apparent when you try to explain them to someone else. You will also benefit from hearing a different perspective on a problem.

Approach to examinations

Your first task is to map out on a sheet of paper an assessment of your strong, reasonable and weak areas. This gives you a rough outline of your revision schedule, which you must then fit in with the time available. Clearly, if your exams are looming large you will have to be ruthless in the time allocated to your strong areas. The discipline of learning is closely linked to preparation for examinations. Many of us opt for a process of superficial learning which is directed towards retention of facts and recall under exam conditions. Ideally, you should acquire a deeper understanding but, recognizing constraints on your time, a pragmatic approach to learning which combines the necessity of passing exams with longer term needs is important.

Next, you need to know how you will be examined. For written examinations you need to know the length of the exam, the types of question (for example, MCQ, modified essay, etc.), the number of questions to be answered and what choice you will have. The hardest step is to determine the likely scope of the exam; medicine does not draw boundaries around knowledge either in breadth or depth and different teachers may have slightly different emphases. Most medical and dental courses now emphasize a core curriculum and this text seeks to identify aspects of physiology which are clinically relevant. The best approach is to combine lecture notes, textbooks (appropriate to the level of study) and past examination papers. The last of these also helps indicate the depth of knowledge expected in a given area. Trying to 'spot' specific questions, however, is a dangerous practice.

The next stage is to map out the time available for preparation. You must be realistic in this, allowing time for breaks and working steadily, not cramming. If you do attempt to cram, you have to realize that if the examination requires understanding you may be in trouble. An approach based on active learning through the use of key steps, learning objectives and self-assessment has much to commend it. For a subject such as cardiovascular physiology, you might set out the topics to be covered and then attempt to summarize your knowledge about each in note form. In this way, prior knowledge will be brought to the front of your mind and gaps in your knowledge/understanding become apparent. It is much more efficient to go to textbooks having been through this exercise as you are then 'looking' for information and explanations.

As the examination draws near, you should start to attempt practice questions and complete papers. It is not sufficient to have the necessary knowledge and understanding; you need to demonstrate these to the examiners. Many people pay insufficient attention to the type of question they are going to encounter. Moving the focus away from books and lecture notes to actual questions helps in identifying where knowledge is still lacking and what work is still to be done. Additionally, attempting complete exams allows concentration to be built up.

Examination methods

Multiple choice questions or multiple True/False questions

Most multiple choice questions test recall of information. The aim is to gain the maximum marks from what you can remember. The common form consists of a stem with several different phrases which complete the statement. Each statement is to be considered in isolation from the rest and you have to decide whether it is 'True' or 'False'. There is no need for 'Trues' and 'Falses' to balance out for statements based on the same stem; they may all be 'True' or all 'False'. The stem must be read with great care and if it is long, with several lines of text or data, then you should try and summarize it by extracting the essential elements. Make sure you look out for the 'little' words in the stem such as 'only', 'rarely', 'usually', 'never' and 'always'. Negatives such as 'not', 'unusual' and 'unsuccessful' often cause marks to be lost. 'May occur' has entirely different connotations to 'characteristic'. The latter generally indicates a feature which is normally observed, and the absence of which would represent an exception to a general rule, e.g., regular elections are a characteristic of a democratic society. Regular (if dubious) elections may occur in a dictatorship, but they are not characteristic.

Remember to check the marking method before starting. Most employ a negative system in which marks are lost for incorrect answers. The temptation is to adopt a cautious approach answering a relatively small number of questions. This can lead to problems, however, as we all make simple mistakes or even disagree vehemently with the answer favoured by the examiner! Caution may lead you to answer too few questions to pass after the marks have been deducted for incorrect answers.

Distracters are the technical term for parts of questions which sound as though they are correct but are definitely incorrect in the context of the complete statement. A good example would be to take the stem 'Parasympathetic nervous activity' and have as a completion statement 'leads to a rise in heart rate during exercise'. This is false since, although heart rate does rise in exercise, this is not caused by parasympathetic activity, which slows the heart. Another type of distracter is to give what appears to be the numerically correct value for a variable but to change the units, e.g., systemic arterial oxygen tension is normally about 95 mmHg; 95 kPa would be incorrect but could catch out the unwary. This is the commonest area where students lose marks even though they know the answer.

Single best answer questions

This is another form of multiple choice question which is sometimes favoured over the True/False format discussed above because the increased number of options means that negative marking is not required. Only one of the possible answers is to be chosen as being the 'most correct'. Care should be exercised, as several of the possible answers may be partially correct and none of them may appear to be absolutely correct. It is the answer which you judge to be the best of the options available which should be chosen. (This reflects some aspects of clinical decision making, in which the best solution has to be chosen among several options, all of which have both benefits and possible drawbacks.) Since there is no negative marking all questions should be attempted.

Essays

Although essays are increasingly rare in examinations, you may be asked to write an essay to test your ability to integrate information. Relevant facts will receive marks, as will a logical development of the argument or theme. Conversely, good marks will not be obtained for an essay which is a set of unconnected statements. Length matters little if there is no cohesion. Relevant graphs and diagrams should also be included but must be properly labelled.

Most people are aware of the need to 'plan' their answer yet few do this. Make sure that what you put in your plan is pertinent to the question asked as irrelevant material is, at best, a waste of valuable time and, at worst, causes the examiner to doubt your understanding. It is especially important in an examination based on essays that time is managed and all questions are given equal weight, unless guided otherwise in the instructions. A brilliant answer in one essay will not compensate for not attempting another because of lack of time. Nobody can get more than 100% (usually 70–80%, tops) on a single answer! It may even be useful to begin with the questions about which you feel you have least to say so that any time left over can be safely devoted to your areas of strength at the end.

Matching item questions

This form of question is another variant of the multiple choice question which is sometimes favoured over the True/False format discussed above because the increased number of options means that negative marking is not required. The main problem that can arise with these questions (other than simply not knowing the answer) is that more than one of the options may seem appropriate to a single statement. All you can do under these circumstances is decide which answer you think is best. Check the instructions carefully to see whether each answer can be used several times (as in my examples) or once only.

Short notes

Short notes are usually marked from a 'marking template' or 'model answer' which gives a mark(s) for each important fact (also called criterion marking). Nothing is gained for style or superfluous information. The aim is to set out your knowledge in an ordered, concise manner. The major faults of students are, firstly, devoting too much time to a single question thereby neglecting the rest and, secondly, not limiting their answer to the question asked. For example, in a question about the control of erythrocyte production, all facts about erythropoiesis should not be listed, only those relevant to its regulation.

Modified essay questions and clinical scenarios

These provide for the development of a theme using a series of descriptive paragraphs which set the scene interlinked with specific questions which lead the examinee on as the 'story' unfolds. They are widely used in clinical subjects, where they are often known as 'Patient management problems', but they are sometimes used to test knowledge in basic medical sciences as well. MEQs and clinical scenarios are usually criterion marked so mentioning all the relevant information is vital and a short note format may be used. There is also often a problem-solving element which requires application of basic principles to the specific situation described.

One of the main faults in dealing with this sort of question is for students to attempt to use material from near the end of the problem to help them answer earlier questions. The MEQ is commonly designed to suggest a range of equally likely interpretations based on the initial information. Full marks depend on mentioning all of these. Some explanations may then be excluded, and new possibilities introduced, by providing further information. Your answers to a question should reflect the information given up to that point. Using material which comes later may lead you to exclude some of the most relevant responses. Also, you should attempt all sections of the question (unless the exam instructions indicate negative marking). There is usually some degree of independence of the individual sections of the problem so you should be able to attempt later sections even if you cannot answer the initial questions.

Data interpretation

This involves the application of knowledge to solve a problem. In your revision, you should aim for an understanding of principles since it is impossible to memorize all the different data combinations. In the exam, a helpful approach is often to translate numbers into a description; for example an arterial blood pH below 7.4 represents an acidosis and the ECG tracing of a heart rate of 120 beats min^{-1} shows a tachycardia. Pattern recognition can then be attempted. This type of question is usually not negatively marked so put down an answer even if you are far from sure that it is right. Conversely, there is no point in listing four possibilities, if the question asks for one response. The examiner will not choose from your answers; the first response is likely to be taken!

Vivas

A few points on technique may be useful. The viva examination can be a nerve-wracking experience. You are normally faced with two examiners who may react with irritation, boredom or indifference to what you say. You may feel that the viva has gone well and yet you failed, or, more commonly, think that the exam has gone badly, simply because of the apparent attitude of the examiners.

Your main aim during the viva should be to control the questioning of the examiners so that they are constantly asking you about things you know about. It is worth running mock vivas with your colleagues to practise this skill. The 'examiner' also learns much in this situation since asking sensible questions requires deep understanding. Major points relating to any of the systems may be asked but if you are aware of having performed badly in an exam essay make sure you cover related material before a 'pass–fail' viva. Always give the simplest possible answer to the question asked; there are very few intentional traps in vivas but many students confuse themselves by looking for them. The examiners are likely to want to explore the limits of your knowledge so do not be upset if they push you hard. It is all right to say you do not know something and this allows the examiners to change tack to see what you do know about.

Normal values

A student is expected to know the values of certain physiological variables as they are essential to any discussion of normal physiology as well as for making decisions in emergency medicine. Values in the text have been limited to those which you are likely to be expected to know. Make sure you are also clear what the units of measurement are in each case.

Conclusions

I have set out a framework for using this book, but you should amend this according to your own needs and the examinations you are facing. Whatever approach you adopt, your aim should be for an understanding of the principles involved as well as the memorization of facts.

Reference table of normal values Many variables show a considerable range of normal values but mid-values have been quoted in most cases

Variable	Value	Units
Body fluid volumes		
Blood volume	70	$ml\,kg^{-1}$ body wt
Plasma volume	40	$ml\,kg^{-1}$ body wt
Total body fluid volume	60 (males) 50 (females)	% body wt
Intracellular fluid volume	2/3 total body fluid volume	–
Extracellular fluid volume	1/3 total body fluid volume	–
Blood cells/platelets		
Haematocrit (packed cell volume)	45% (0.45)	–
Haemoglobin concentration	14	$g\,dl^{-1}$
Red cell (erythrocyte) count	5	$\times 10^{12}\,L^{-1}$
Reticulocyte count	2	%
Erythrocyte sedimentation rate (ESR)	<5 (males) <7 (females)	$mm\,h^{-1}$
White cell (leucocyte) count	6	$\times 10^{9}\,L^{-1}$
Neutrophils	3.5	$\times 10^{9}\,L^{-1}$
Lymphocytes	2	$\times 10^{9}\,L^{-1}$
Eosinophils	0.2	$\times 10^{9}\,L^{-1}$
Monocytes	0.5	$\times 10^{9}\,L^{-1}$
Basophils	<0.1	$\times 10^{9}\,L^{-1}$
Platelet count	250	$\times 10^{9}\,L^{-1}$
Plasma		
Plasma protein concentration	60	$g\,L^{-1}$
Plasma oncotic (colloid osmotic) pressure	25	mmHg
Plasma osmolality	300	$mosmol\,kg^{-1}$
Na^+	140	$mmol\,L^{-1}$
K^+	4.5	$mmol\,L^{-1}$
Ca^{2+}	2.5	$mmol\,L^{-1}$
Cl^-	105	$mmol\,L^{-1}$
HCO_3^-	25–30	$mmol\,L^{-1}$
Glucose	5	$mmol\,L^{-1}$
Urea	5	$mmol\,L^{-1}$

Variable	Value	Units
Cardiovascular function		
Resting cardiac output	5	$L\,min^{-1}$
Resting heart rate	70	min^{-1}
Resting stroke volume	70	ml
Systemic arterial pressure	120/80	mmHg
Pulmonary arterial pressure	25/10	mmHg
Central venous pressure	3–8	cmH_2O
	2–6	mmHg
Respiratory function		
Compliance (lung + chest wall)	100	$ml\,cmH_2O^{-1}$
Intrapleural pressure during quiet breathing	–4 (expiration) –9 (inspiration)	cmH_2O
Tidal volume	500	ml
Respiratory rate	12	min^{-1}
Respiratory minute volume	6	$L\,min^{-1}$
Dead space	150	ml
Alveolar ventilation rate	4.2	$L\,min^{-1}$
FEV_1/FVC	>75	%
Blood gases and acid–base balance (also HCO_3^- above)		
Systemic arterial pH	7.38–7.42	–
Systemic arterial P_{O_2}	13 98	kPa mmHg
Systemic arterial P_{CO_2}	5.3 40	kPa mmHg
Base excess	–2–+2	$mmol\,L^{-1}$
Renal function		
Renal blood flow	1.2	$L\,min^{-1}$
Glomerular filtration rate	120	$ml\,min^{-1}$
Average urine output (normal hydration)	1	$ml\,min^{-1}$
Renal glucose threshold	11	$mmol\,L^{-1}$
Nutritional status		
Body mass index	20–30	$kg\,m^{-2}$
Hormones		
Thyroxine (T_4)	100	$nmol\,L^{-1}$
Triiodothyronine (T_3)	2	$nmol\,L^{-1}$
TSH	2.5	$mU\,L^{-1}$
Cortisol	400 (peak) 150 (trough)	$nmol\,L^{-1}$ $nmol\,L^{-1}$
Urinary free cortisol	<300	$nmol\,24\,h^{-1}$

Basic principles and cell physiology

Chapter 1

Overview

The physiology of the body is often described in terms of the homeostatic mechanisms which control the cellular environment. Cell function also relies on an adequate energy supply and the appropriate integration of the activities of the different cells within the body through efficient signalling mechanisms, both electrical and chemical. The cell membrane is very important in many of these functions since it forms a barrier between the intracellular and extracellular environments across which important molecules must be transported. The properties of the membranes of excitable cells also account for the generation of action potentials, and many of the receptors which recognize specific chemical messengers are membrane bound.

1.1 Cells, systems and homeostasis

Learning objectives

At the end of this section you should be able to:

- describe the main components of the cell
- identify the main body fluid compartments and estimate their likely volumes
- explain the principles of indicator dilution methods and suggest appropriate indicators for specific compartments
- outline the principles of homeostasis and describe how these apply to temperature regulation.

Physiology is the study of normal biological function. Although the ultimate goal is to understand the intact human, function is often considered in terms of individual physiological systems, each of which consists of different organs and structural elements. The different components of any *system* may be spread widely throughout the body but they serve a *common functional* purpose. Therefore, we talk of the cardiovascular system, consisting of the heart, arteries, capillaries, veins, lymphatics and related control mechanisms, all of which are involved in the circulation of the blood. The fundamental biological unit is the cell, however, and each organ's function reflects the integrated activity of the various specialized cells within it.

Cell structure

Although cells from different organs show considerable differences in shape (morphology) and function, all cells share some general characteristics. Each is bounded by a cell membrane (or *plasma membrane*) which separates the aqueous solution inside the cell (the *cytoplasm* or *intracellular fluid*) from the aqueous solution outside (the *extracellular fluid*). These two fluids have very different ionic compositions (see Section 5.1) and these differences are more easily maintained because ions cannot readily cross the double layer (*bilayer*) of phospholipids in the plasma membrane (Fig. 1). These are oriented with their hydrophilic heads adjacent to the aqueous solutions on either side, while their hydrophobic tails form a fatty core. Membrane proteins move about within the two-dimensional confines of the lipid layer (the fluid mosaic model of the membrane).

Membranes also delimit a number of subcellular units known as *organelles*. These include:

- the *nucleus*, which contains the genetic code for protein synthesis in the form of deoxyribonucleic acid (DNA). Since proteins are key structural and functional molecules in the cell, changes in gene expression regulate how a cell looks and behaves.

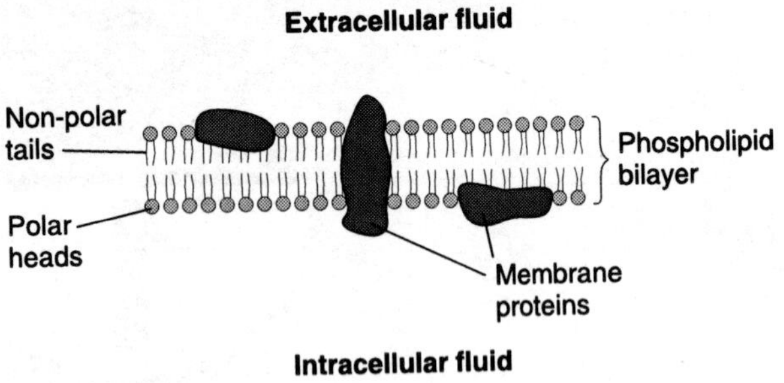

Fig. 1 The plasma membrane. Membrane proteins may be exposed on either side or may cross the whole thickness of the lipid bilayer.

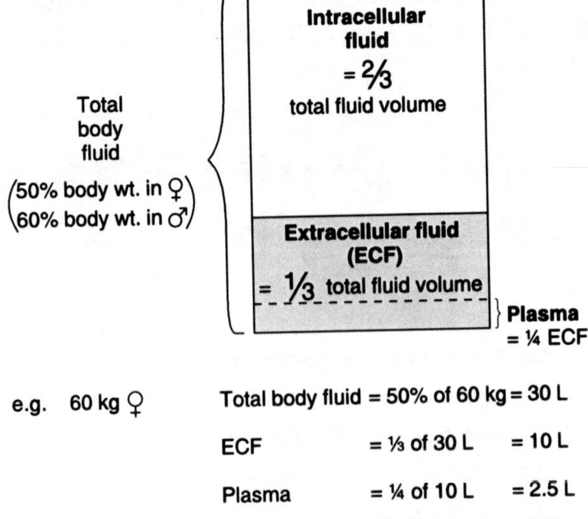

Fig. 2 Fluid compartments within the body showing the relative volumes occupied by intracellular fluid, extracellular fluid and plasma (part of the extracellular space).

- *mitochondria*, which are surrounded by both an outer membrane and a highly folded inner membrane. Mitochondria are responsible for oxidative breakdown of food molecules, providing the cell with usable energy in the form of ATP.
- the *endoplasmic reticulum* (or *sarcoplasmic reticulum* in muscle cells), which is an interconnecting series of membranous channels and vesicles. In some regions this has a large number of *ribosomes* attached to the surface, thus leading to the distinction between rough and smooth endoplasmic reticulum. Ribosomes play a key role in protein synthesis.
- the *Golgi apparatus*, a system of stacked vesicles found close to the nucleus. This is involved in the formation of secretory vesicles and lysosomes.

Body fluid compartments

The body contains many different aqueous solutions, which may be classified into a series of fluid compartments depending on their location (Fig. 2). The main subdivision is between intracellular and extracellular fluids. The extracellular compartment represents all fluid not inside cells, including the plasma component of blood, aqueous humour in the eye, synovial fluid in joints and cerebrospinal fluid within the central nervous system. The largest single extracellular subcompartment, however, consists of the *interstitial fluid*, which lies in the connective tissue matrix surrounding most body cells. It is normally this fluid which is in direct contact with the cell membrane and controlling interstitial conditions is vital for normal cell function. Much of this control is achieved by the continuous circulation of blood through the cardiovascular system. The rate of blood flow to any region and the plasma concentrations of different solutes greatly influence the cellular environment since, with the exception of plasma proteins, water and solutes move freely between plasma and the interstitium.

Volume measurement by indicator (dye) dilution methods

The volume of a given fluid compartment, e.g., the extracellular fluid volume, is usually estimated indirectly using an indicator substance. If a known quantity of indicator becomes evenly distributed through that compartment, measurement of the final concentration of indicator allows the volume of fluid to be calculated using Equation 2 below.

$$\text{Concentration} = \frac{\text{Amount of indicator}}{\text{Fluid volume}} \qquad \text{(Eq. 1)}$$

$$\therefore \text{Fluid volume} = \frac{\text{Amount of indicator}}{\text{Concentration}} \qquad \text{(Eq. 2)}$$

This technique is called the *indicator*, or *dye dilution method*, and it relies on finding an indicator which fulfils two main criteria.

- It should become evenly distributed through the whole of the fluid compartment of interest, otherwise the concentration will be artificially high leading to an underestimate of the volume concerned.
- It should not leak out into other fluid compartments or be rapidly metabolized or cleared from the body, since the concentration of indicator would be reduced under such circumstances, leading to an overestimate of the relevant volume.

For these reasons different indicators are appropriate when estimating the volume of different fluid spaces within the body.

Blood and plasma volume

Blood volume can be measured by taking a sample of an individual's blood and radioactively labelling the red cells, e.g., with chromium-51. These cells are reinjected, given time to mix evenly through the blood volume, and another sample of blood is taken. The radioactivity per unit volume is used to calculate the total blood volume (normal value is 4–5 L).

Plasma volume is usually estimated using a dye which binds to plasma protein, such as Evans' blue dye. This is restricted to the plasma space because the protein does not readily escape from the circulation. Plasma proteins radiolabelled with iodine-131 can also be used. These indicators do not enter the red cells so the calculated volume is that of plasma only (2.5–3.0 L), and not the total blood volume.

Extracellular fluid volume

Extracellular fluid volume is not easily measured and different indicators give different results in a given subject. Substances which have been used include radioactive sodium, radioactive bromide, the polysaccharide inulin, and sucrose. Mean values are about 10–15 L. The plasma volume represents about 25% of the extracellular space and, since there is free exchange of fluid across the capillary wall, any increase or decrease in extracellular fluid volume will be reflected by a parallel change in plasma volume. This means that mechanisms which affect the total extracellular fluid volume will also alter the plasma and, therefore, blood volumes, e.g., renal fluid reabsorption (Section 5.4).

Homeostasis

Normal cell function relies on appropriate environmental conditions since the temperature, pH, ionic concentrations, O_2 and CO_2 levels in the extracellular fluid all influence biochemical activity inside the cell. The mechanisms which control the body's internal environment are called *homeostatic mechanisms*. These keep conditions inside the body relatively constant despite considerable changes in the external environment.

The following general elements are present in all homeostatic control systems (Fig. 3).

- *Detectors*, or receptors, sensitive to the variable to be controlled.

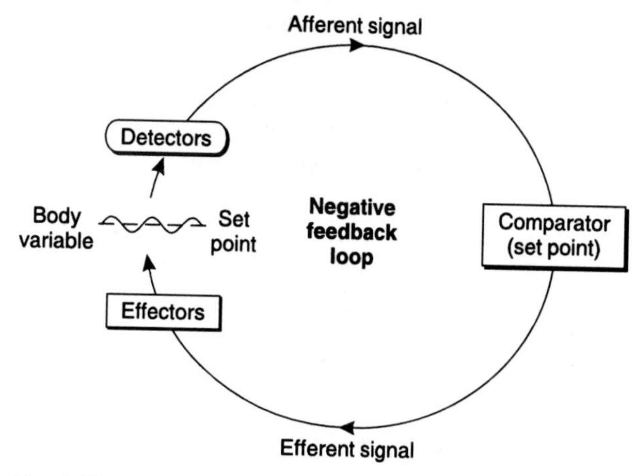

Fig. 3 Elements of a homeostatic control system. The controlled variable may oscillate, or hunt, around a mean value (the set point).

- A comparator, or *integrating centre*, which receives the afferent signal from the detectors and compares this measured value of the variable with a preferred value, or *set point*, for the system. If the actual value differs from the set point the integrator generates an efferent (outgoing) signal.
- *Effector* systems which can change the measured variable are activated by the signals from the integrator. Their effect is to bring the variable back towards the set point.

In such a system, any deviation of the variable away from its normal (set) value stimulates responses which reduce that deviation. This is referred to as *negative feedback* control and is an important element of homeostasis. The overall effect is to maintain a constant environment but often a phenomenon known as hunting is observed in which the controlled variable oscillates around a fixed mean value rather than remaining exactly at the set point (Fig. 3).

Biological systems may also demonstrate *positive feedback* in which deviations from the steady state are actually amplified, rather than diminished, by a feedback loop. Such mechanisms, however, have no part to play in homeostasis.

Homeostatic regulation of body temperature

Regulation of body temperature provides a good example of homeostasis. The body core is normally held close to 37°C, even though the environmental temperature (which affects the rate of heat loss from the body) and rate of metabolic heat production may vary widely.

Temperature detectors (*thermoreceptors*) monitor both the core temperature (the regulated variable) and the peripheral temperature (a separate but relevant variable). The afferent inputs go to an integrating centre (the *thermoregulatory centre*) in the *hypothalamus* of the brain, which compares core temperature with the set point of 37°C. If these differ, outgoing nerve signals activate a number of effector systems which alter the rates of heat production and heat loss.

In cold conditions:

- reflex increases in muscle tone and shivering liberate metabolic energy as heat
- in infants, there may also be increased heat production in brown fat (Section 9.6)
- heat loss is reduced by sympathetic constriction of peripheral blood vessels, which decreases skin blood flow; erection of skin hair (piloerection) also occurs but has limited significance in humans.

In hot conditions the rate of heat loss can be greatly accelerated by:

- increased sweating stimulated by sympathetic nerves which in this specific case release the neurotransmitter acetylcholine
- dilatation of cutaneous blood vessels caused by reduced activity in vasoconstrictor sympathetic nerves releasing the more typical sympathetic neurotransmitter noradrenaline (norepinephrine). There may also be active dilatation of arterioles (Section 3.7).

Behavioural aspects are important too, e.g., we dress up warmly, exercise and eat hot food when we are cold, while preferring light clothes and cool drinks in hot weather.

Box 1 Clinical note: Pyrexia

Pyrexia or fever is a temporary elevation of the core temperature commonly caused by infection. It is not due to any breakdown of the detector or effector systems responsible for temperature control but results from a change in the set point. Toxins released from bacteria and immune cells cause the hypothalamus to respond to a core temperature of 37°C as if it were abnormally low and so it increases heat production, e.g., through shivering (rigors). This raises body temperature.

1.2 Energy sources in the cell

Learning objectives

At the end of this section you should be able to:

- outline the main steps in cellular energy exchange.

Body cells require energy to carry out mechanical work (e.g., in muscle cells), to transport ions or molecules against concentration gradients (e.g., ion pumps) and for the synthesis of complex molecules. These active processes are ultimately fuelled by the energy released through the breakdown of carbohydrate, fat and protein from our diet. The initial series of reactions in the breakdown of glucose is referred to as glycolysis and does not require O_2 (anaerobic metabolism). This leads to the formation of lactic acid and produces small amounts of *adenosine triphosphate (ATP)*, the main source of usable energy within the cell. In the presence of O_2 (aerobic conditions), however, intramitochondrial enzymes catalyse the complete catabolism of the products of glycolysis, releasing much greater amounts of ATP and generating CO_2 and water as byproducts (Section 4.7). Whereas glycolysis leads to a net production of two molecules of ATP per glucose molecule, oxidative metabolism can produce an additional 34–36 ATP molecules. Subsequent hydrolysis of the high-energy phosphate bonds in ATP by *ATPase enzymes* makes energy available for the active processes of the cell.

1.3 Transport across cell membranes

Learning objectives

At the end of this section you should be able to:

- list the factors affecting solute diffusion across cell membranes
- identify the conditions necessary for osmosis and understand the differences between osmolality, tonicity and osmotic pressure
- describe the characteristics of carrier-mediated transport and differentiate between facilitated diffusion, primary active transport and secondary active transport
- describe what is meant by endocytosis and exocytosis.

There is a constant traffic across cell membranes, supplying O_2 and substrate molecules for intracellular metabolism and removing CO_2, waste substances and active products. A variety of transport mechanisms are involved.

Diffusion

Diffusion can occur whenever a substance is present at a higher concentration on one side of the cell membrane than the other. It results in net movement from high to low concentration, i.e., a net flux of the ion or molecule. No energy source is required so this is referred to as a *passive transport* mechanism. Diffusion of materials into and out of cells is affected by the following factors.

Solute concentration gradients, i.e., not the absolute concentrations but the difference in concentration across the cell membrane.

Membrane permeability to the solute. The plasma membrane is selectively permeable to fatty and small nonpolar molecules which dissolve in the membrane lipid. Thus, fatty acids, steroid hormones, O_2 and CO_2 all diffuse readily into cells. The permeability to water-soluble (lipid-insoluble) ions and large polar molecules such as proteins, however, is generally low. Certain ions can diffuse across the cell membrane much more readily (i.e., they are more permeant) than their lipid solubility would predict. This is because of membrane proteins which bridge the lipid barrier and provide an easier route for ion diffusion. These may take the form of carrier molecules, which bind to the ion and then move it across the membrane by changing conformation, or they may provide fluid-filled channels through which the ions can pass (Section 1.4). The roles of both these types of molecule are considered in more detail later in the chapter. Specific carriers and channels are selective for different kinds of ion and so membrane permeability to a given ion may differ widely from cell to cell, depending on which proteins are present.

Transmembrane voltage gradients affect the movement of ions. If the inside of the membrane is negative with respect to the outside, cations (positively charged) will be electrostatically attracted into the cell and anions (negatively charged) will be repelled outwards. The net transmembrane flux of an ion is proportional to the combined effect of the electrical and concentration gradients acting on it, i.e., on the *electrochemical gradient* for that ion.

Molecular weight of the diffusing substance. Small molecules diffuse more rapidly.

Diffusion distance. Diffusion is too slow to allow effective exchange over distances of more than about 100 μm.

Membrane surface area. For a given set of conditions rate of diffusion is proportional to the surface area of membrane.

Osmosis

Osmosis depends on the passive diffusion of water across a membrane from a region of low solute concentration (effectively high water concentration) to a region of high solute concentration (low water concentration). Osmosis requires:

- a solute concentration gradient across a membrane
- a *semipermeable membrane*, i.e., permeable to the solvent (water) but not the solute. If the membrane is highly permeable to the solute, conditions on either side of the membrane will rapidly equilibrate by solute diffusion, thus removing the osmotic driving force.

Cell membranes are permeable to water molecules (because of their small size), so any solute which cannot cross the membrane (an impermeant solute) can generate an osmotic gradient. Normally cells exist in osmotic equilibrium, the osmotically active particles inside the cell being balanced by those in the extracellular fluid. Any disturbance of this balance will lead to a net movement of water across the cell membrane and a change in cell volume.

The osmotic properties of a solution can be described in several ways.

Osmolality is defined as the total number of dissolved particles per kg of solvent (H_2O), and has units of mosmol kg^{-1}. Osmolality will be used throughout this text since this determines osmotic effects and is usually reported in biochemistry tests. Since H_2O has a density of 1 kg L^{-1}, however, osmolality for dilute solutions is very similar to *osmolarity,* which refers to the number of particles per litre of solution (mosmol L^{-1}), and physiology and medical texts often use the terms osmolarity and osmolality interchangeably. Substances which dissociate in solution increase the number of dissolved particles, and this raises the osmolality above the molar concentration of solute. For example, 1 mmol L^{-1} of glucose has an osmolality of 1 mosmol kg^{-1} but a 1 mmol L^{-1} NaCl solution has an osmolality of 2 mosmol kg^{-1} since each NaCl molecule dissociates to produce two ions.

Tonicity is a biological term relating to the actual effect of a solution on living cells, specifically erythrocytes. A solution may be:

- *Isotonic*, i.e., in osmotic equilibrium with the osmotically active solutes in intracellular fluid under normal conditions. Plasma and interstitial fluid are, by definition, isotonic in healthy individuals, and intravenous fluid supplements should generally be isotonic to avoid red cell damage.
- *Hypertonic*, i.e., contains a higher concentration of osmotically active particles than the intracellular fluid, leading to osmotic water loss and cell shrinking (crenation). Plasma can become hypertonic in certain disease states due to an increased concentration of solutes, e.g., increased glucose in diabetes mellitus (see Section 8.7).
- *Hypotonic*, i.e., contains a lower concentration of osmotically active particles than the intracellular fluid, leading to osmotic cell swelling and possibly cell lysis. Body fluids become hypotonic if their relative water content is increased. This dilution may occur because of electrolyte (usually Na^+) loss or water retention.

Tonicity depends both on the osmolality of a solution and the ease with which the solute in question can pass through the cell membrane. Readily diffusible substances have no osmotic effect on a cell even at high osmolality. Thus a 300 mosmol kg^{-1} solution of NaCl (an effectively impermeant solute) is isotonic while a 300 mosmol kg^{-1} solution of urea (which crosses cell membranes readily) is extremely hypotonic, leading to water absorption and almost immediate cell lysis caused by the unbalanced osmotic effect of trapped intracellular solutes (mainly K^+ and inorganic anions).

Osmotic pressure is the hydrostatic pressure that would be necessary to exactly oppose the osmotic effect of a solution and prevent any net water movement. It is usually expressed in mmHg or kPa (kPa = 1000 N m^{-2} = mmHg × 0.133). The osmotic pressure exerted on the cell membrane by isotonic fluids is over 770 kPa, i.e., over 7.5 × atmospheric pressure. Normally this is exactly balanced by the osmotic pressure resulting from impermeant intracellular solutes, i.e., the osmotic pressure gradient across the cell membrane is zero.

Carrier-mediated transport

Carrier proteins in the cell membrane bind to a specific substrate and then undergo some conformational change. As a result, the substrate is transported across the membrane and released on the other side. The rate of transport increases as the substrate concentration increases, but the maximum rate of transport (V_{max}) is dependent on the density of carriers in a given cell, since all transport sites will be occupied above a certain substrate concentration (Fig. 4). This is referred to as saturation. *Substrate specificity* and *saturation* are two hallmarks of carrier-mediated transport, whether it is passive (facilitated diffusion) or active.

Facilitated diffusion

The energy driving facilitated diffusion is the substrate concentration gradient, so this is a passive process, i.e., no additional energy input is required. Movement across the membrane is, nevertheless, dependent on the availability of transport or carrier proteins (Fig. 5). Cellular absorption of glucose from the extracellular fluid is one example.

Active transport

In active transport systems carrier molecules transport molecules or ions against concentration or electrical gradients. Such carriers must use energy from some other source to do the necessary work. The energy can be provided in two ways.

Primary active transport uses the energy released by hydrolysis of ATP, e.g., to transport Na^+ out of the cell and K^+ into the cell against their electrochemical gradients (the Na^+/K^+ ATPase, or Na^+/K^+ pump: Fig. 5).

Secondary active transport uses the energy released during the passive movement of one substance down its electrochemical gradient to transport another substance against a concentration gradient (Fig. 5). For example, sodium cotransport systems couple Na^+ diffusion into intestinal cells

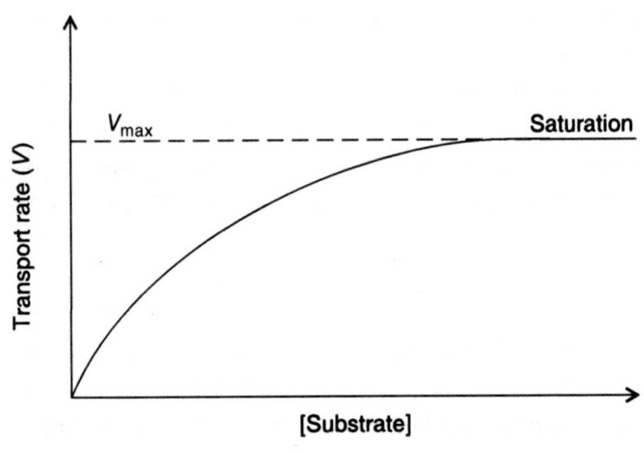

Fig. 4 The rate of transport as a function of substrate concentration for carrier-mediated transport.

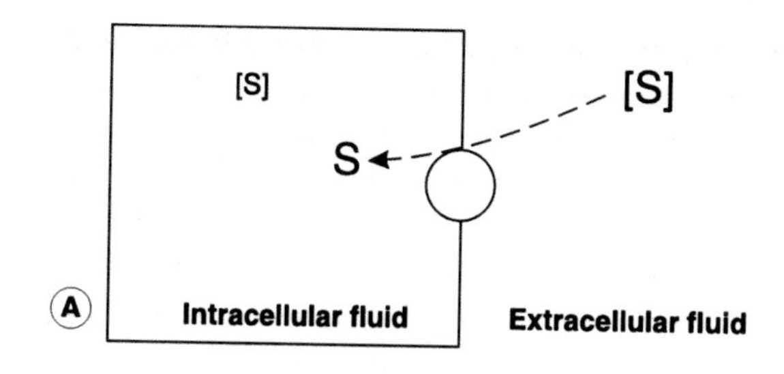

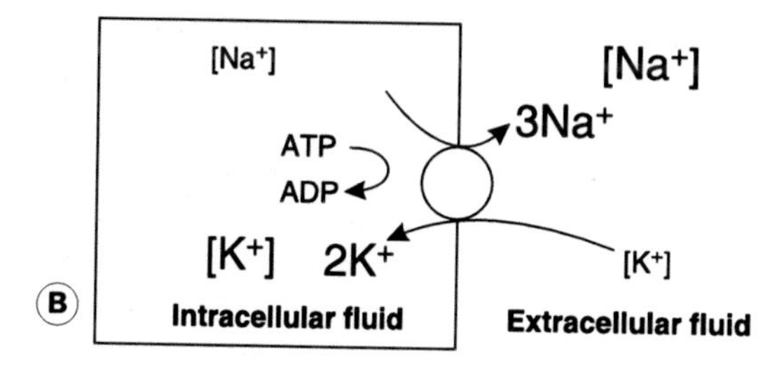

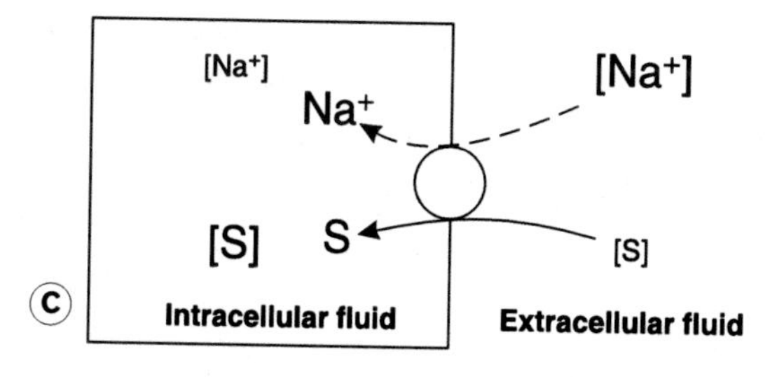

[S] ≡ High concentration

[S] ≡ Low concentration

Fig. 5 Carrier-mediated transport mechanisms. (A) In facilitated diffusion the substrate (S) moves down a concentration gradient. (B) Primary active transport links ATP breakdown to the transport of a substrate against a concentration gradient. (C) Secondary active transport uses passive movement of one substance to drive a substrate against a concentration gradient.

with the absorption of glucose from the gut lumen (Section 7.4). Absorption can continue even if the concentration of glucose is higher inside the cell than outside. Although the cotransport step itself does not require an external energy source, such systems ultimately depend on active pumping of Na^+ back into the extracellular space by Na^+/K^+ ATPase to maintain the normal Na^+ diffusion gradient into the cell.

Vesicular transport

Vesicular transport depends on the transport of substances within membrane-bound vesicles. Limitations on transport resulting from large molecular size and low membrane permeability can be circumvented since the vesicle contents never physically

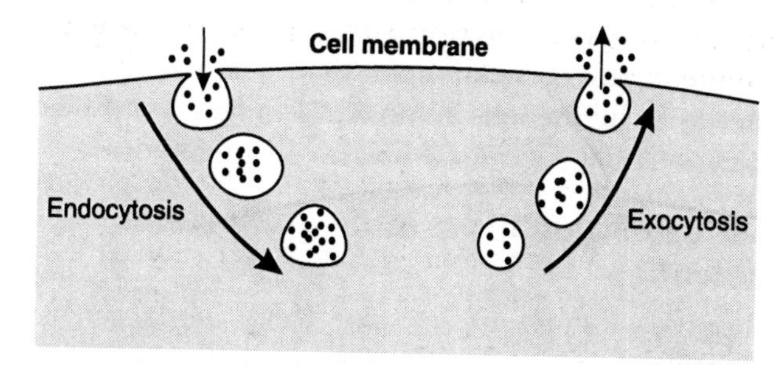

Fig. 6 Fluid and large membrane-impermeant molecules can be moved in and out of cells using vesicular transport.

cross the membrane. *Endocytosis* involves the invagination of a portion of plasma membrane into the cell, which forms a vesicle containing extracellular fluid and near membrane molecules. In *exocytosis* a vesicle from within the cell fuses with the membrane, releasing its contents to the extracellular fluid (Fig. 6). Exocytosis is particularly important in the release of glandular secretions and chemical transmitters.

1.4 Electrical signals and excitable cells

Learning objectives

At the end of this section you should be able to:

- explain the term resting membrane potential (RMP)
- outline the role of diffusion potentials and electrogenic pumps in generating the RMP
- draw the typical membrane potential changes during a nerve action potential (AP) and list the main AP properties
- describe how changes in cell ionic conductances produce an AP
- explain how voltage controlled ion channels regulate ionic conductance
- explain how local circuits lead to unidirectional AP propagation.

Electrical signalling within the nerves and muscles of the body is a vital aspect of body function. These cells are said to be *excitable* because they are capable of generating self-propagating electrical signals known as action potentials.

Resting membrane potential

Electrical recordings from nerves show that there is a potential difference of about 70 mV across the cell membrane with the inside negative with respect to the outside. We say that the resting membrane potential is –70 mV. This can be explained by the fact that the cell membrane separates two solutions with different ionic concentrations and is not equally permeable to all the ions involved. As a result, a diffusion potential is generated across the membrane.

Diffusion potentials and equilibrium potentials

Suppose we start with zero electrical potential across the cell membrane (Fig. 7). The concentration of K^+ is higher inside the cell than outside and the membrane is permeable to K^+, so K^+ diffuses outwards. Permeability is selective, however, and the large intracellular anions cannot follow K^+. Consequently, an imbalance of charge builds up across the membrane producing a potential difference, with the inside negative with respect to the outside. This is a diffusion potential. The voltage gradient opposes further diffusion of the positive K^+ ions, making it more and more difficult for them to leave the cell. The diffusion potential will increase until an equilibrium state is achieved in which the concentration gradient is exactly balanced by the opposing voltage gradient. The potential difference under these conditions is known as the equilibrium potential for K^+ (E_K). This depends on the ratio of $[K^+]$ on either side of the membrane and can be calculated using the *Nernst equation*:

$$E_K \frac{RT}{zF} \log_e \frac{[K^+]_o}{[K^+]_i} \qquad \text{(Eq. 3)}$$

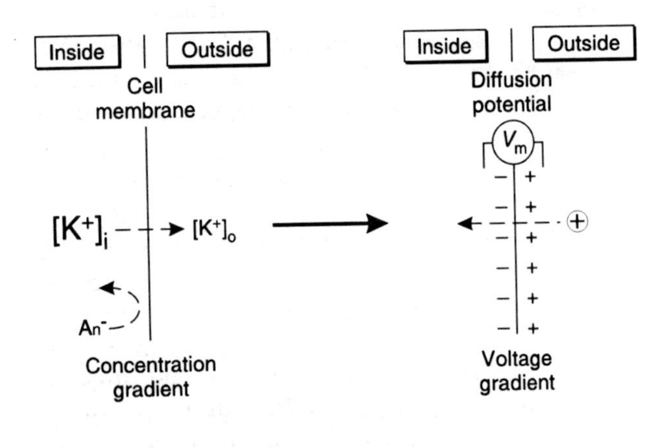

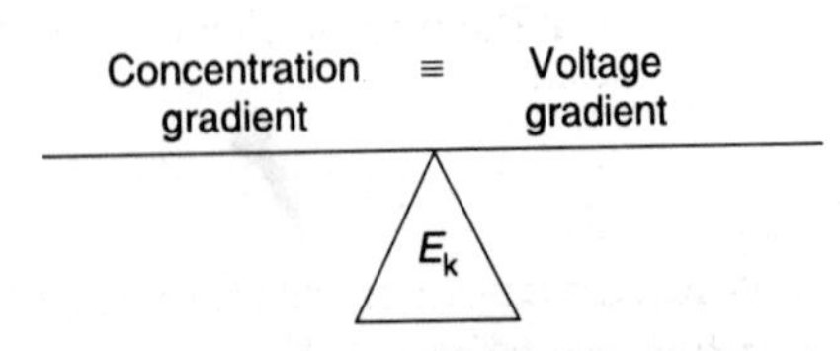

Fig. 7 A diffusion potential results from movement of K^+ down its concentration gradient across a membrane that is impermeable to anions. At E_K, there is no net movement of K^+.

where R = ideal gas constant, T = absolute (or thermodynamic) temperature (°C + 273K), F = Faraday's constant and z = ionic valency (+1 for K^+).

If we apply this equation to a nerve cell, the concentration of K^+ is much higher inside the cell than outside and the calculated value for E_K is approximately –90 mV. The measured value of the resting membrane potential (–70 mV) is more positive than this, so it cannot be explained solely in terms of the diffusion potential generated by K^+. In fact other ions can also cross the membrane, particularly Na^+. The concentration gradient for Na^+ is in the opposite direction to K^+ (Fig. 9, below) and the calculated value for E_{Na} is approximately +65 mV. The Na^+ gradient tends to make the membrane potential more positive than it would otherwise be, but because the resting membrane is much more permeable to K^+ than Na^+, the resting potential is much closer to E_K than E_{Na}. One equation which takes account of the involvement of both K^+ and Na^+ in determining the resting membrane potential (RMP) is:

$$\text{RMP} = \frac{RT}{F} \log_e \frac{[K^+]_o + \alpha[Na^+]_o}{[K^+]_i + \alpha[Na^+]_i} \qquad \text{(Eq. 4)}$$

$$\text{where } \alpha = \frac{\text{Permeability to } Na^+}{\text{Permeability to } K^+}$$

This equation works well using $\alpha = 0.01$, i.e., assuming the membrane permeability to K^+ is 100 × permeability to Na^+ at rest.

Electrogenic ion pumps

Any current flowing across the cell membrane will affect its potential, and one possible source of such currents is the ionic pump which maintains the normal transmembrane concentration gradients. The Na^+/K^+ ATPase pumps $3Na^+$ out of the cell for every $2K^+$ transported in (Fig. 5). This imbalance means that current flows from the inside to the outside of the membrane during active pumping. Such systems are said to be electrogenic and can affect the resting membrane potential. In this case, the net loss of positive charge from inside the cell makes the membrane potential more negative than it would otherwise be. Although this is probably not an important mechanism in determining the resting membrane potential in nerve and striated muscle, it may make a significant contribution in some smooth muscles.

Action potentials

If a nerve cell is stimulated by injecting (positive) electric current, the membrane potential becomes less negative (Fig. 8). We say the membrane potential

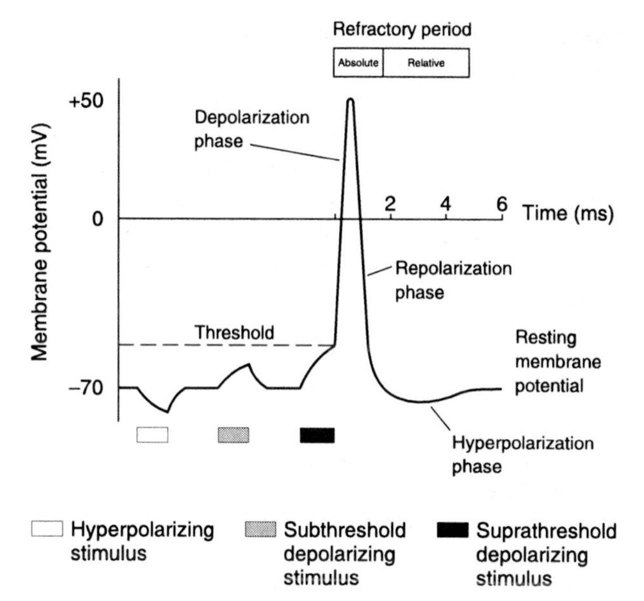

Fig. 8 Intracellular recordings of membrane potential in a nerve. If depolarization reaches threshold, the nerve produces an action potential. Further stimulation is inhibited during the refractory period.

has reduced (because the magnitude of the potential difference is reduced even though it is less negative) or that the membrane has been *depolarized*. With small stimuli (subthreshold), the membrane potential simply returns to normal after the stimulus ceases. If, however, the membrane is depolarized to a certain level, known as the *threshold potential*, the nerve itself generates a series of changes in the potential, known as an action potential. Action potentials are a feature of nerves and muscles, and it is the ability to generate these characteristic electrical signals which typifies excitable tissues.

The action potential in nerve has an initial phase of *rapid depolarization* which reverses the potential difference across the membrane, reaching a peak at about +50 mV within a few tenths of a millisecond. The membrane then begins to *repolarize*, falling back to the normal resting potential about 1 ms after initiation of the action potential. The potential may actually become more negative than normal for a time (*hyperpolarization*), but eventually returns to the resting potential after a further 2–3 ms.

Action potential properties

Action potentials demonstrate a number of important properties.

- Action potentials obey an *all-or-none law*. Subthreshold stimuli do not elicit any active voltage changes, while any stimulus which exceeds threshold (suprathreshold stimuli) will lead to a full action potential (Fig. 8). Increasing the stimulus further has no effect on action potential shape or size.
- During an action potential it is either impossible or more difficult than normal to stimulate a further action potential and the cell is said to be *refractory* (Fig. 8). During the early part of the action potential there is an absolute refractory period when no stimulus is effective. This is followed by the relative refractory period during which an action potential can be stimulated but a larger than normal stimulus is required.
- Action potentials are *self-propagating*, i.e., once an action potential has been generated by an external stimulus at one site, it will automatically be conducted over the whole of the excitable membrane.

Generation of an action potential

The mechanisms responsible for action potential production depend on two main features:

- transmembrane ionic concentration gradients
- voltage-dependent changes in the membrane permeability to Na^+ and K^+.

Changes in ionic permeability may also be expressed as changes in the electrical resistance (r) of the membrane to the flow of ionic current. Usually these changes are described in terms of the conductance (g) of the membrane for a given ion, where conductance is defined as the inverse of the resistance ($g = 1/r$). Therefore, when permeability to a particular ion increases, the resistance to current carried by that ion decreases, while conductance increases.

Applying these ideas to nerve action potentials, the resting K^+ permeability of the membrane is greater than that to Na^+, but any depolarizing stimulus leads to a rapid increase in Na^+ permeability. In electrical terms there is an increase in the *sodium conductance* (g_{Na}). Sodium ions will diffuse into the cell, driven by the high Na^+ concentration outside the cell and drawn by the negative charge inside (i.e., by the electrochemical gradient for Na^+ which is directed into the cell). The resulting *inward* sodium current (I_{Na}) depolarizes the membrane further setting up a *positive feedback loop* (Fig. 9). This accounts for the rapid depolarization phase of the action potential. At its peak, the potential comes quite close to E_{Na} because the high Na^+ permeability makes it the dominant ion affecting the membrane potential at this time (Fig. 10).

Sodium conductance does not remain high but rapidly falls back to a low level, even though the membrane is depolarized. At the same time the *potassium conductance* (g_K) starts to rise (Fig. 9). This

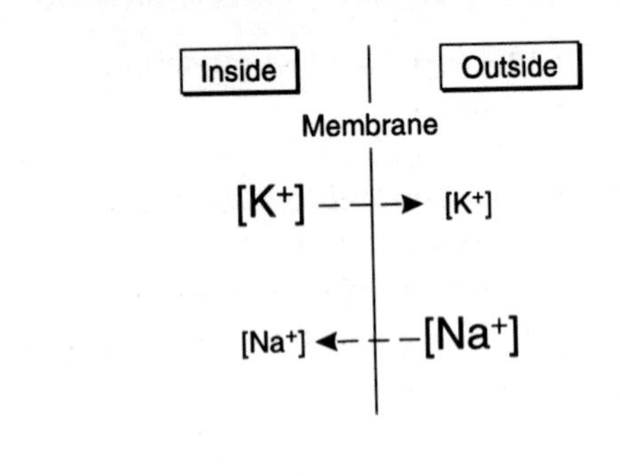

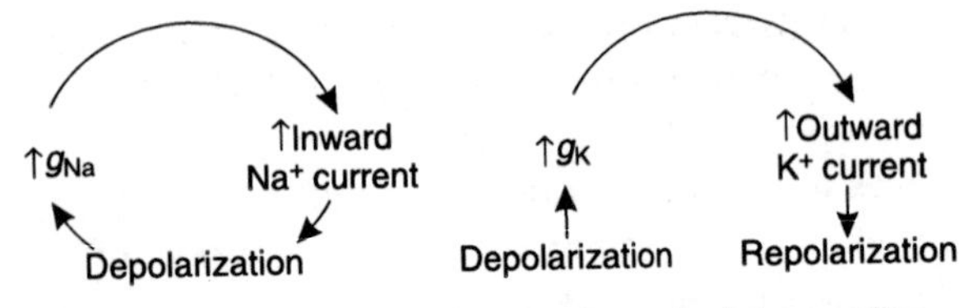

Fig. 9 The mechanisms underlying action potential generation. Increased Na^+ conductance leads to the initial depolarization, while increased K^+ conductance accounts for repolarization.

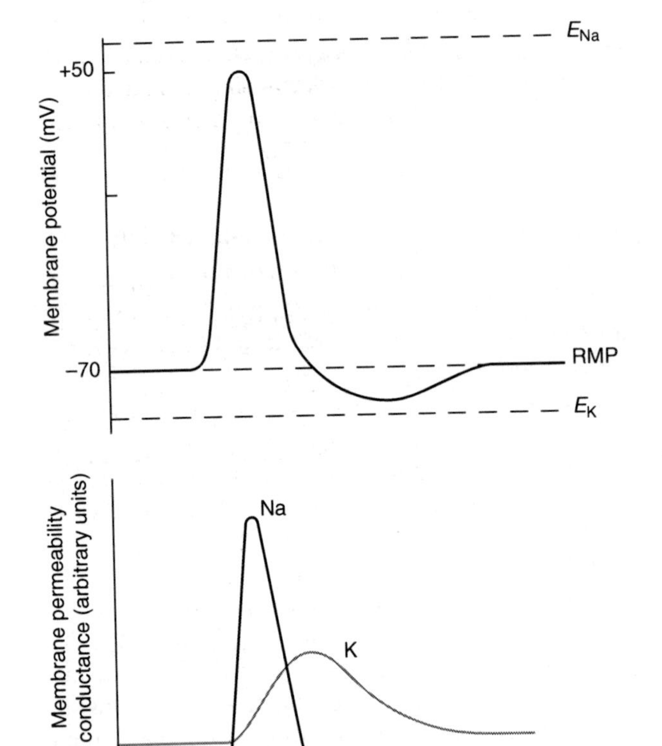

Fig. 10 Changes in membrane conductance/permeability to Na^+ and K^+, and the action potential that they produce. At maximum permeability for an ion, the membrane potential approaches the ion's equilibrium potential.

is also a voltage-dependent event induced by membrane depolarization, but it develops more slowly than the changes in g_{Na}. The result is an increase in *outward* potassium current (I_K) driven by the large electrochemical gradient for K^+ which exists near the peak of the action potential, when the membrane is very positive relative to E_K. This current repolarizes the membrane (Fig. 10). Potassium conductance remains high for some time after the membrane potential has returned to resting levels, causing hyperpolarization as the membrane is driven closer than normal to E_K (Fig. 10). Conductance then falls back to normal, and the membrane returns to the resting potential.

Voltage-controlled ion channels

In order to understand how voltage-dependent changes in ion conductances come about, we must consider how Na^+ and K^+ cross the membrane. They pass through membrane-bound ion channels, specialized proteins each of which is selectively permeable to a given type of ion. In polarized conditions, these channels mainly exist in a closed, non-conducting state so that no ionic current can flow and membrane conductance and permeability are low for that ion. Depolarizing the membrane causes a change in the channel conformation so that more of the channels open (Fig. 11). This is true for both the Na^+ and K^+ channels in nerve and explains the voltage-dependent conductance increases (g_{Na} and g_K) during an action potential (Fig. 10). The sodium channels open or activate more quickly, however, so that g_{Na} rises much faster than g_K.

Sodium channels also demonstrate a second voltage-dependent property, known as *inactivation*. This mechanism closes the channels again during membrane depolarization, causing g_{Na} to fall. Because inactivation is slower than activation, membrane depolarization causes the Na^+ conductance to rise rapidly to a peak (activation) before falling again (inactivation), as observed during an action potential. Inactivation also explains the absolute refractory period. The depolarizing phase of the action potential relies on an inward Na^+ current through open channels, but inactivated Na^+ channels cannot be reopened unless they first return to their normal, closed state. This only occurs after the membrane has been repolarized, so it is impossible to generate enough Na^+ current to produce another action potential until repolarization is almost complete.

Potassium channels do not inactivate. The fall in g_K following repolarization is simply caused by reduced activation at negative membrane potentials. Throughout the period for which g_K is higher than normal, the outward K^+ current will be increased. This hyperpolarizes the membrane and also opposes the depolarizing effects of any stimulating inward current, giving rise to the relative refractory period.

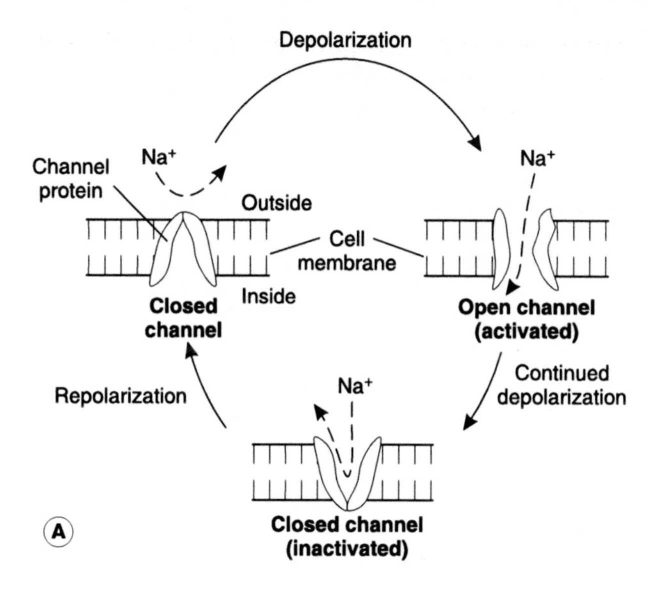

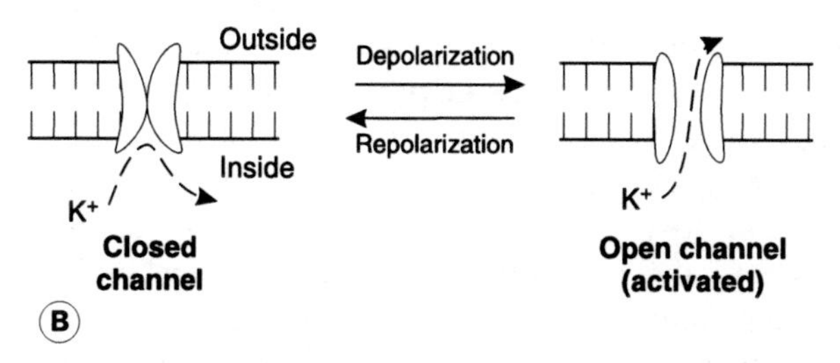

Fig. 11 Voltage-induced changes in ion channel state. (A) Sodium channels. (B) Potassium channels.

During this time larger than normal stimuli are required to depolarize the membrane to the threshold for generation of an action potential.

Action potential conduction or propagation

At the peak of an action potential, the inside of the membrane is positive with respect to the outside. This potential gradient is opposite to that for the resting membrane on either side, so current flows from positive to negative both inside and outside the cell, setting up *local* currents or local circuits (Fig. 12). The transfer of positive charge depolarizes the membrane adjacent to the action potential and, once the threshold is reached, an action potential is generated in that region also. This acts as a further source of depolarizing current and so the process is continued. Once an action potential has been initiated at one point, it will be conducted across the rest of the membrane in this way, i.e., it is self-propagating. The refractory properties of the membrane ensure that conduction is normally *unidirectional*. This is because, although local currents will spread in both directions from a site of depolarization, the membrane on the side from which the action potential has just come will be refractory and cannot easily be excited again. The velocity of propagation is increased with increased axon diameter (decreasing resistance to internal current flow) and with myelination (saltatory conduction; see Section 7.2).

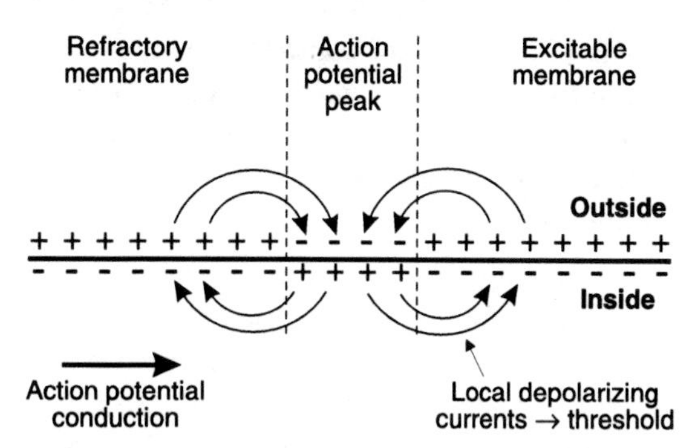

Fig. 12 Unidirectional propagation of an action potential caused by local currents generated by the reversal of membrane potential during an action potential.

Excitable cells: pumping of Na^+ and K^+

Ion pumping plays no direct role in the generation of an action potential but is important over longer periods of time for the maintenance of normal transmembrane concentration gradients which are essential both for the maintenance of the resting potential and for the generation of action potentials. Although it should be emphasized that the total number of ions crossing the membrane during a single action potential is very small indeed, prolonged activity could eventually alter the intracellular ionic conditions significantly. The Na^+/K^+ ATPase pump opposes the tendency for Na^+ accumulation and K^+ depletion inside the cell during repeated generation of action potentials (Fig. 5).

1.5 Chemical signals

Learning objectives

At the end of this section you should be able to:

- list and define the main types of chemical transmission
- explain the role of receptors and G proteins in chemical signalling
- outline the main signal transduction mechanisms.

Many aspects of function are controlled by chemical signals in the form of messenger molecules secreted into the extracellular fluid. Chemical signals may be classified in terms of the source of the messenger molecules and the route by which they reach their target.

- **Endocrine** hormones are transported round the body in the circulation (see Section 8). They may be secreted by specialized ductless endocrine glands, e.g., the adrenal and thyroid glands, or by individual cells within an organ which also has other non-endocrine functions, e.g., the hormone-secreting cells in the gastrointestinal mucosa.
- **Neurotransmitters** are released from nerves and reach their target by diffusing across a narrow synaptic cleft (see Section 7.2).
- **Paracrine** control involves secretion into the interstitial space to affect adjacent cells. **Autocrine** control is a variant in which the messenger acts on the secretory cell itself.

Cell receptors

Chemical messengers produce their effects by first binding to specific cell receptors (see Fig. 14, below). The receptor is usually a membrane protein, but some hormones actually cross the cell membrane and interact with a receptor inside the cell, e.g., steroid hormones. Binding of messenger to receptor leads to a functional response, e.g., when the neuro–transmitter acetylcholine binds to receptors in the membrane of skeletal muscle cells, those cells contract. Molecules which bind to and activate a given receptor are referred to as receptor agonists.

The requirement for receptor binding helps explain how an endocrine hormone, which circulates throughout the body, can selectively stimulate relevant target cells. These cells express the necessary receptor protein while other cells, which do not carry the receptor, are not affected. It should be noted that activation of a given receptor may produce different types of response in different cells, while different receptors may exist for a given messenger.

Box 2 Clinical note: Drugs and receptors

Many prescribed drugs act by binding to specific receptors on cells. This allows them either to activate the receptor directly (agonists) or to prevent their activation by the normal chemical messengers within the body (receptor antagonists or blockers). The hunt for new, selective, receptor-binding drugs is an important aspect of pharmacological research.

G proteins and receptor-dependent responses

The conversion of a chemical signal to a functional response is called *signal transduction*. This often involves receptor-associated proteins on the inner aspect of the cell membrane known as GTP-binding proteins, or simply as G proteins (Fig. 13). In the inactivated, or 'off' state, G proteins bind preferentially to *guanosine diphosphate* (GDP), and have little effect on their intracellular targets. When an agonist binds to the receptor, however, the associated G protein shows an increased affinity for *guanosine triphosphate* (*GTP*). Binding to GTP activates the G protein and causes it to dissociate from the receptor inside the cell. It can now associate with other membrane-bound signal transduction molecules, affecting their activity in turn (see below). The G protein does not remain permanently switched 'on', however, as it also has GTP-ase activity, which hydrolyses the GTP to GDP. This inactivates the G protein again, which reassociates with the receptor until agonist molecule is bound once more.

Although the details of signal-transduction are specific to the agonist, receptor and cell type involved,

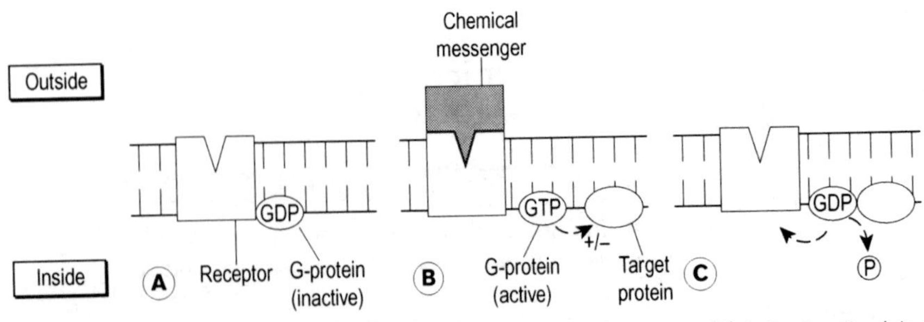

Fig. 13 Cell receptors for chemical messengers often act via G proteins. (A) In the 'unstimulated' state the G protein binds GDP and is inactive. (B) Receptor binding of an agonist leads to an increased affinity for GTP, which activates the G protein, altering the activity in other membrane-bound target proteins (e.g., enzymes for second messengers or ion channels). (C) The G protein breaks down the GTP to GDP, and becomes inactive again.

receptor-dependent signal transduction pathways generally involve one of the following mechanisms (Fig. 14):

Ion channels. Receptors may open or close membrane ion channels, leading to changes in the membrane potential of a cell. This is particularly a feature of neurotransmitter receptors in excitable tissues such as nerve and muscle. The receptor may act directly on the channel, or via G proteins or second messengers (see below). Ion channel dependent responses can be very rapid (milliseconds).

Second messengers. Receptors may produce a change in the concentration of intracellular chemicals known as second messengers. G proteins often act as intermediates between a receptor and a membrane-bound enzyme responsible for synthesis of the second messenger. Some G proteins are excitatory (causing increases in the levels of second messenger) and some inhibitory (causing reductions in second messenger levels). Important second messengers include:

- *cyclic adenosine monophosphate* (*cAMP*), manufactured from ATP by the action of the enzyme adenylate cyclase. It is broken down by phosphodiesterases.
- *cyclic guanosine monophosphate* (*cGMP*) manufactured from GTP by the enzyme guanylate cyclase. It is broken down by phosphodiesterases.
- *inositol 1,4,5-trisphosphate* (IP_3) and diacylglycerol manufactured from the membrane lipid phosphatidylinositol by the enzyme phospholipase C.
- free Ca^{2+}, which may be released from intracellular Ca^{2+} stores or enter the cell from the extracellular fluid via Ca^{2+} channels in the plasma membrane.

Second messenger responses are usually rapid (milliseconds–seconds).

Kinase activity. Second messengers often exert their actions through kinases, enzymes which phosphorylate intracellular proteins (usually on serine or threonine residues), altering their activity. Some important kinases are:

- protein kinase A (PKA), activated by cAMP
- protein kinase G (PKG), activated by cGMP
- protein kinase C (PKC), activated by diacylglycerol and Ca^{2+}
- calcium–calmodulin dependent kinases, activated when Ca^{2+} binds to an intracellular binding protein known as calmodulin.
- receptor protein-tyrosine kinases, in which binding of an agonist activates the kinase activity of the receptor itself, which phosphorylates tyrosine, rather than serine or threonine, residues on intracellular target proteins. Insulin and other growth factors activate receptors of this type, thus promoting cell growth and differentiation. This ultimately results in alteration in the rate of DNA *transcription* to mRNA, and thus controls the levels of *gene expression* as protein.

The downstream targets of the different kinases listed above are often other intracellular enzymes, allowing for control of a wide range of cellular

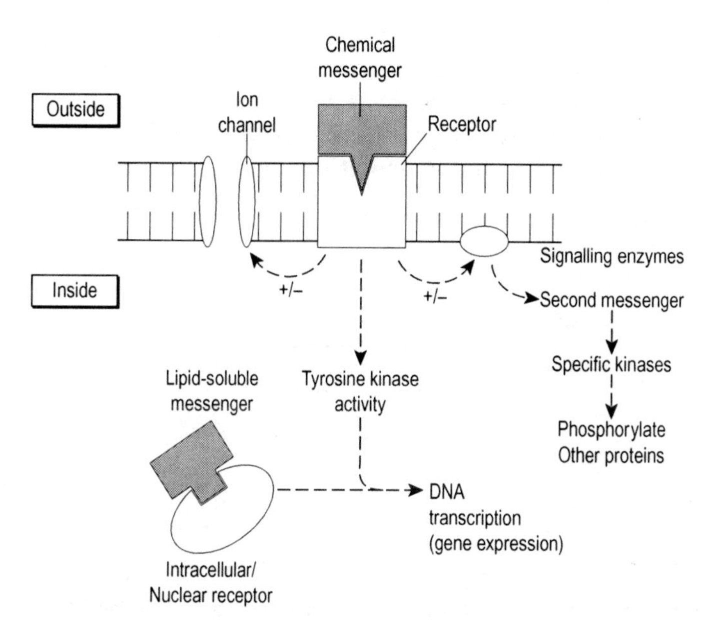

Fig. 14 Some of the signal transduction mechanisms which may be activated by cell receptors.

One

reactions. The intracellular effects of kinases are reversed by specific phosphatases, which dephosphorylate the relevant proteins. Response times for kinase-dependent transduction pathways are intermediate (seconds to hours).

Intracellular receptors. Activation of intracellular receptors by thyroid or steroid hormones, for example, can also modify the rate of *transcription* of DNA into mRNA, thereby altering the production and cell content of functionally relevant proteins, e.g., specific enzymes or transport molecules. Responses are relatively slow (minutes to hours).

Box 3 Clinical note: Cancer and abnormalities of cell signalling

The receptor tyrosine kinases normally activated by growth factors stimulate powerful signalling pathways which promote cell growth and division. Mutations of these kinases, or of the downstream molecules on which they act, can lead to permanent, uncontrolled cell replication. Many types of human cancer are now known to involve cell signalling abnormalities of this type.

1.6 Contractile mechanisms in muscle

Learning objectives

At the end of this section you should be able to:

- describe the structure of skeletal muscle from the level of the cell down to the contractile proteins
- draw a labelled diagram of a sarcomere
- explain the sliding-filament hypothesis of contraction and outline the main events in the cross bridge cycle
- relate the sliding-filament hypothesis to the length–tension relationship in muscle
- describe the events responsible for excitation–contraction coupling in striated and smooth muscle
- explain the functional classification of skeletal muscle fibres
- outline the similarities and differences between skeletal, cardiac and smooth muscles.

There are three major types of muscle in the body, *skeletal, cardiac* and *smooth* muscle, but in each case the ability actively to contract, shortening and developing tension in the process, is crucial. In the following description, the contractile process will be considered at a cellular level using skeletal muscle as a model. Differences between this and other muscle types will then be discussed.

Skeletal muscle structure

Each muscle contains muscle cells or fibres arranged in parallel with one another and with the long axis of the muscle. A single muscle fibre may be tens of centimetres long and contains multiple nuclei arranged around the periphery of the cell. The cells are rich in mitochondria, which supply the ATP necessary for contraction, and have an extremely well developed sarcoplasmic reticulum (a Ca^{2+} store). Light microscopy shows up a series of regular alternating light and dark bands running across the width of each fibre and this has led to the term *striated* muscle being applied to both skeletal and cardiac muscle (which also demonstrates this phenomenon).

Myofibrils

At higher magnification, a skeletal muscle fibre can be seen to consist of a series of parallel subunits known as myofibrils. These contain the *contractile proteins* of the cell and are alternately banded light and dark. The light bands are termed the *I bands* and the dark bands the *A bands* (Fig. 15). It is the parallel alignment of the banding pattern in adjacent myofibrils across the width of each fibre which accounts for the striated appearance of the whole cell. The Z *line* (or disc) bisects each I band and the interval between two adjacent Z lines is called a *sarcomere.*

Myofilaments

The A and I bands of the myofibrils are formed by parallel elements known as the *thick* myofilaments (these show up as the A bands) and the *thin* myofilaments (these are present in the I bands and are attached to the Z line at one end). These myofila-

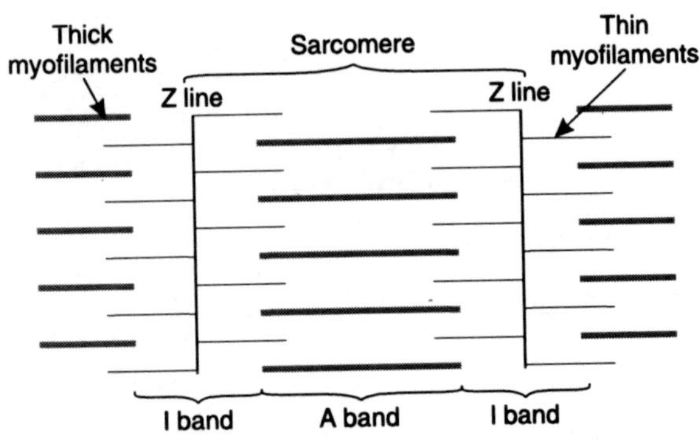

Fig. 15 A myofibril showing the arrangement of the thick and thin myofilaments within each sarcomere.

ments are formed from the contractile and regulatory protein molecules responsible for tension production and shortening in the muscle. The myofilaments overlap each other, with the thin myofilaments interdigitated between adjacent thick myofilaments (Fig. 15).

The thick myofilaments consist of molecules of the contractile protein *myosin*, each with a long body attached to a double head (Fig. 16A). Myosin molecules are arranged along the long axis of the myofilament with the heads projecting laterally towards the thin myofilaments. Each head has binding sites for actin and for ATP, which is broken down to release the chemical energy necessary for muscle contraction.

Thin myofilaments contain at least three different proteins (Fig. 16B). *Actin*, the second contractile protein, is arranged in a double-stranded helix rather like a double rope of pearls twisted about its long axis. A molecule of *tropomyosin* lies along the groove between the two actin strands and at regular intervals there are also molecules of *troponin*, which are attached to both actin and tropomyosin. Troponin also has binding sites for Ca^{2+} and is important in the regulation of contraction.

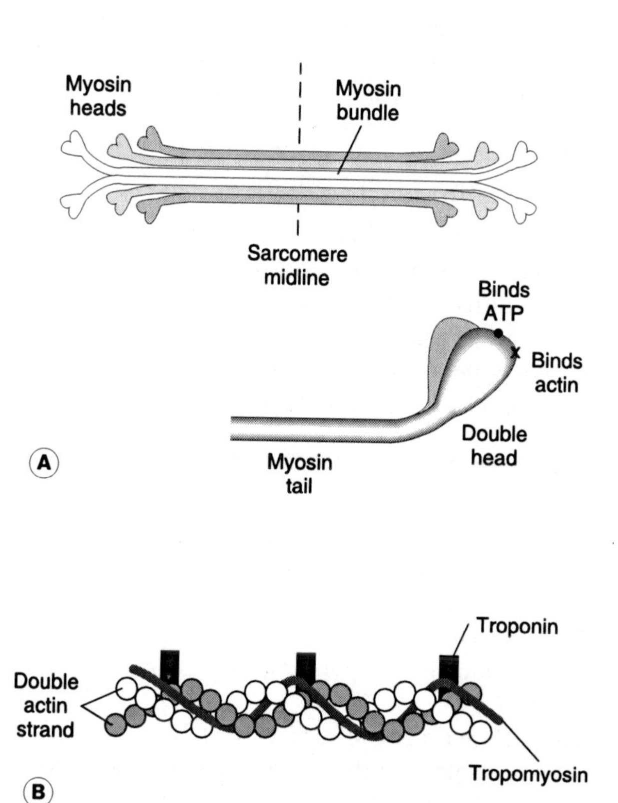

Fig. 16 The molecular structure of thick (A) and thin (B) myofilaments.

The sliding-filament model of contraction

The sliding-filament hypothesis proposes that the thick and thin myofilaments slide past one another during cell contraction. This is supported by the observation that neither thick nor thin myofilaments change length during active muscle contraction. Instead, the thin myofilaments slide between the thick myofilaments as they are drawn towards the midline of the A band. This shortens the sarcomere and contracts the muscle. Electron micrographs show the presence of *cross bridges* projecting from the thick myofilaments. These are believed to be the myosin heads which bind to the thin myofilaments, attaching to the actin molecules. They then move towards the centre of the sarcomere, pulling the thin filament inwards, before releasing the actin and moving back to their original position where they bind to a new site on the actin (Fig. 17). Since many different cross bridges are formed and broken asynchronously at different sites along the myofilaments, smooth contraction is possible despite the fact that any individual cross bridge is continually detaching and then reattaching to actin at a different point along the thin myofilament. This is termed the *cross bridge cycle*. It converts the energy released from ATP by the myosin ATPase to mechanical energy, with substantial production of heat as a byproduct.

Length–tension relationships in skeletal muscle

The sliding-filament hypothesis helps explain one of the functional characteristics of skeletal muscle contraction, i.e., the fact that the amount of active tension developed during stimulation of a muscle depends

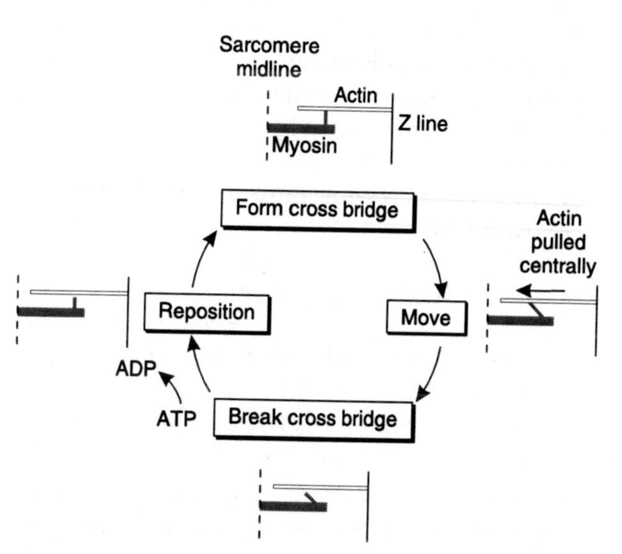

Fig. 17 Summary of the proposed sequence of events in the cross bridge cycle.

One

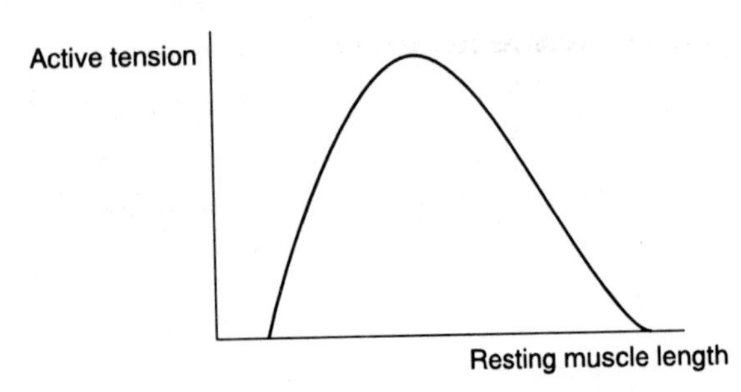

Fig. 18 The active tension developed during contraction of a skeletal muscle is dependent on its resting length prior to stimulation (see text).

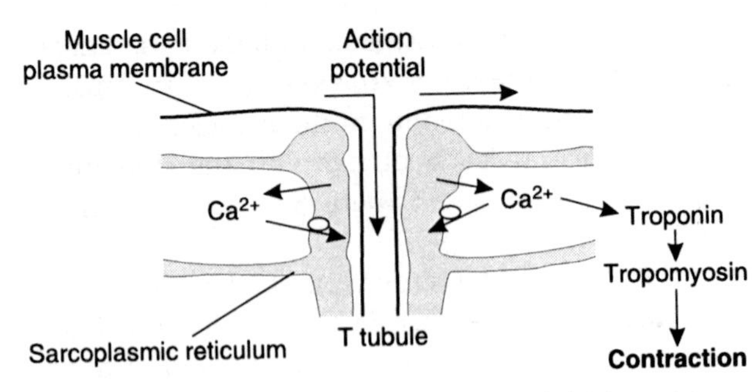

Fig. 19 Excitation–contraction coupling within skeletal muscle.

on its resting length (Fig. 18). If the muscle is stretched beyond the point at which there is no overlap of the myofilaments then no force can be generated. As the resting length of the muscle is decreased, the degree of overlap of the filaments increases and there is increasing tension developed in response to stimulation. This is to be expected if force production depends on the presence of cross bridges between the thick and thin myofilaments, since the number of such links would increase with increasing filament overlap. Eventually, however, all the available cross bridges will have formed and tension cannot increase any further. As the resting length of the muscle is decreased beyond this point, the thin myofilaments start to interfere with each other and active force production decreases again.

Excitation–contraction coupling

Skeletal muscle cells remain relaxed until stimulated by an excitatory impulse in a motor nerve. This results in the firing of an action potential on the muscle membrane, and it is this depolarization of the muscle fibre that signals contraction. The cellular events linking depolarization with the contractile response constitute the excitation–contraction coupling mechanisms of the cell.

The action potential is first conducted across the plasma membrane of the muscle fibre. At regular intervals, the membrane penetrates deeply into the muscle fibre forming structures known as *transverse*, or *T tubules*. Any action potential is rapidly conducted along the tubular membrane into the centre of the fibre, rather than simply passing over the outer surface of the cell. This helps to synchronize the contractile response across the width of the fibre.

Depolarization results in the release of large amounts of Ca^{2+} from the *sarcoplasmic reticulum*, which is in close contact with the plasma membrane in the region of the T tubules. Cytoplasmic Ca^{2+} levels rise and this acts as an intracellular signal which activates contraction (Fig. 19). The mechanism responsible for this involves binding of Ca^{2+} to troponin on the thin myofilaments, which causes troponin to change its conformation, or shape. At rest, troponin and tropomyosin block any binding between myosin and actin, but this inhibition is removed following the formation of Ca^{2+}–troponin complexes. Cross bridge cycling can then proceed, producing contraction. After repolarization of the cell, however, the sarcoplasmic reticulum actively pumps Ca^{2+} back out of the cytoplasm using energy released by breakdown of ATP, thus reducing intracellular $[Ca^{2+}]$ once more. Troponin is no longer Ca^{2+} bound so tropomyosin moves back into its normal position, where it prevents further actin–myosin interaction, and relaxation occurs.

Skeletal muscle: classification of fibres by speed and metabolism

Skeletal muscles may be divided into two groups based on their speed of contraction.

Slow twitch, or ***type I***, fibres are found in high proportions in the postural muscles of the back, for example, where slow, sustained contractions are necessary. Type I fibres rely on *oxidative metabolism* to generate the necessary ATP and contain large numbers of *mitochondria*. They resist fatigue well and appear red in colour because of the presence of intracellular *myoglobin*, an O_2-carrying molecule which favours O_2 uptake from the blood and acts as an O_2 store inside the cell.

Fast twitch, or ***type II***, fibres may be subdivided into IIA and IIB depending on their metabolic profiles.

Type IIA, or *fast oxidative fibres* are found in muscles such as the soleus in the calf and also rely on oxidative metabolism. They contain myoglobin and are moderately resistant to fatigue.

Type IIB, or *fast glycolytic fibres*, by comparison, appear white because of the absence of myoglobin. These cells contain high concentrations of glycogen and rely almost exclusively on anaerobic glycolysis, so ATP production is not limited by O_2 and glucose delivery from the circulation. This allows for rapid, powerful contractions, but these are of necessity brief because of the limited glycogen stores. Such fibres are found in high proportion in the extraocular muscles, which produce the rapid movements of the eyes.

Cardiac muscle

Like skeletal muscle, cardiac muscle is striated and the contractile process is identical to that outlined above. Excitation–contraction coupling is also similar in that the necessary rise in intracellular [Ca^{2+}] is achieved by release from the sarcoplasmic reticulum Ca^{2+} stores. Unlike skeletal muscle, however, this release is not triggered directly by depolarization of the cell membrane, but by entry of Ca^{2+} from the extracellular fluid through voltage-dependent membrane channels, which open during the cardiac action potential (Section 3.3). This process is referred to as Ca^{2+}-induced Ca^{2+} release.

Another major difference between cardiac and skeletal muscle is seen in their electrical activity, with *spontaneous* (*myogenic*) activity and prolonged action potentials in cardiac cells (Section 3.2). This activity may be modified by the autonomic innervation of the heart but is not dependent upon it, so that a denervated heart will still beat rhythmically. Electrical activity in skeletal muscle, by contrast, is *neurogenic*, and is totally dependent on stimulation by somatic motor neurones (Section 7.5).

Smooth muscle

So named because of the absence of striations, smooth muscle is found in the walls of the blood vessels and viscera of the body. It demonstrates spontaneous electrical and mechanical activity, whether *phasic*, as in the rhythmic activity of the gastrointestinal tract, or *tonic*, as in the sustained contractions of the vasculature. This activity may be modified by circulating hormones or autonomic nerves (Section 7.6). Although the velocity of contraction is much slower than in striated muscle, smooth muscles can maintain prolonged contractions at a small fraction of the energy cost necessary in other muscle types.

Contractile mechanism

Smooth muscle contraction relies on the interaction between actin and myosin. These are not organized into parallel myofilament arrays as they are in skeletal and cardiac myofibrils, however, so no striations result. In the absence of Z plates, the actin filaments attach to anchoring points on the cell membrane (*dense plates*) and within the cell (*dense bodies*) and are interconnected by myosin filaments (Fig. 20A). Cross bridge formation pulls the actin filaments together and the cell shortens. In some types of smooth muscle, adjacent cells are linked in series by *tight junctions* through which force can be transmitted. There may also be *gap junctions*, providing low-resistance pathways for the communication of electrical signals and thus allowing a layer of smooth muscle cells to act as a coordinated functional unit within the body (Fig. 20B). This differs from skeletal muscle, where cells act in parallel and are electrically insulated from each other, but is rather similar to the functional syncytium seen in cardiac muscle (Section 3.2).

Excitation–contraction coupling

Some of the most marked differences between smooth and striated muscles are observed in their

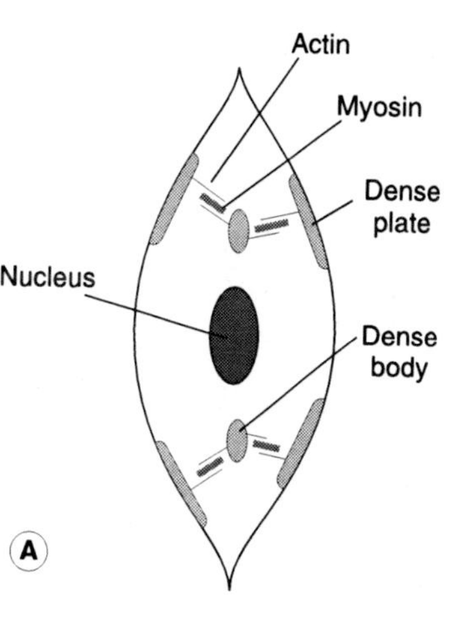

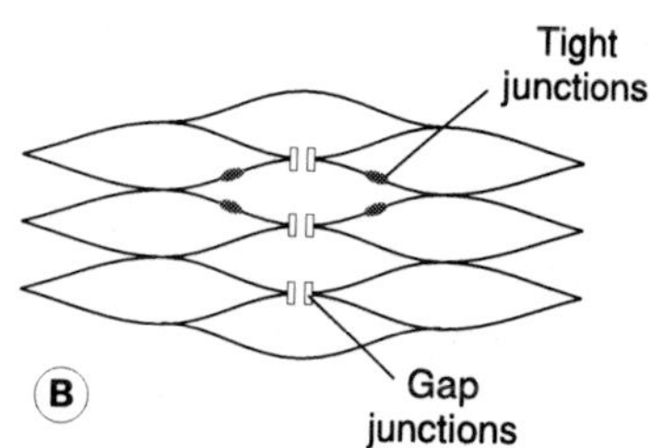

Fig. 20 Smooth muscle structure. (A) A single cell showing actin filaments anchored to dense bodies and plates and interconnected by myosin filaments. (B) A sheet of smooth muscle cells mechanically attached by tight junctions and electrically linked through low-resistance gap junctions.

respective excitation–contraction coupling mechanisms (Fig. 21). Contraction is initiated by a rise in intracellular [Ca^{2+}], either as a result of Ca^{2+} entry across the plasma membrane through voltage-dependent Ca^{2+} channels following cell depolarization, or from release of intracellular Ca^{2+} stores. Unlike skeletal muscle, however, membrane depolarization does not directly stimulate release of Ca^{2+} from the sarcoplasmic reticulum in smooth muscle. Stored Ca^{2+} may be released, however, in response to second messengers such as IP_3 generated following receptor binding of chemical messengers (*agonist-induced Ca^{2+} release*; Section 1.5). Calcium itself can also activate release, so Ca^{2+} entry following cell depolarization may be amplified by Ca^{2+}-induced Ca^{2+} release, similar to that seen in cardiac muscle.

The mechanisms whereby increases in cytoplasmic [Ca^{2+}] promote contraction are different in smooth muscle compared with striated muscle (Fig. 21). Smooth muscle does not contain troponin and so regulation through a troponin–tropomyosin mechanism, analogous to that in skeletal and cardiac muscle, cannot occur. Instead, Ca^{2+} binds to a regulatory protein known as calmodulin. The Ca^{2+}–calmodulin complex activates an enzyme called *myosin light chain kinase (MLCK)*, which phosphorylates smooth muscle myosin. This activates the myosin ATPase system and cross bridge cycling commences. When [Ca^{2+}] falls again, MLCK is deactivated and myosin is dephosphorylated by a phosphatase, terminating the contraction.

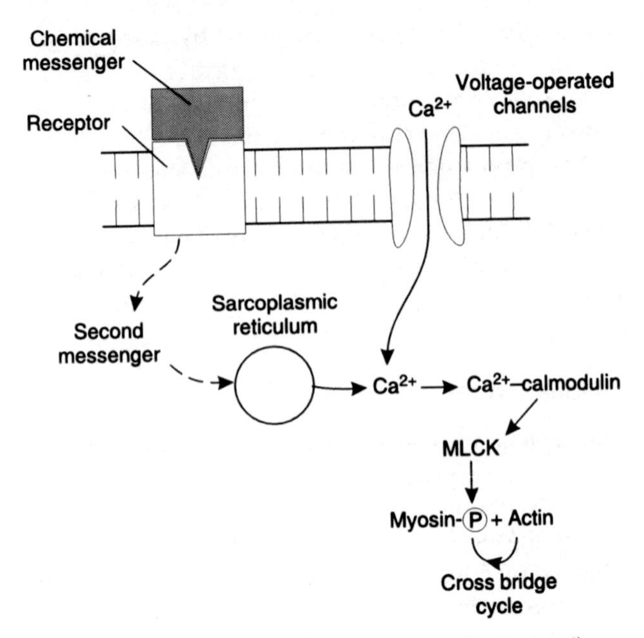

Fig. 21 Summary of excitation–contraction coupling in smooth muscle.

Self-assessment: questions

Multiple true/false questions

Each of the following statements consists of a stem followed by a number of possible endings. State whether each statement is True or False. For each stem, all, several or none of the statements may be True.

1. The plasma membrane:
   a. is permeable to lipophilic molecules
   b. is made up of a lipid core sandwiched between two layers of protein
   c. may contain proteins which confer ion permeability
   d. may burst in hypotonic extracellular solutions
   e. can generate action potentials in excitable cells

2. Homeostatic control:
   a. tends to keep extracellular conditions constant
   b. relies on positive feedback mechanisms
   c. of body temperature uses parasympathetic nerves to control heat loss
   d. requires detectors, comparators and effectors
   e. is more likely to keep a variable close to the set point if the feedback gain is high

3. Diffusion of a substance across the cell membrane:
   a. is an active process
   b. may require a carrier molecule
   c. is generally faster for polar than nonpolar molecules of similar molecular weight
   d. may provide energy for transport of another molecule
   e. increases in rate at higher temperatures

4. In an experiment, erythrocytes are suspended in a variety of different solutions. Cell volume would be expected to:
   a. stay constant in a 300 mmol L^{-1} NaCl solution
   b. increase in a 150 mmol L^{-1} NaCl solution
   c. increase in a hypotonic solution
   d. stay constant in a 300 mosmol kg^{-1} urea solution
   e. stay constant in a 300 mosmol kg^{-1} NaCl solution

5. Carrier-mediated transport:
   a. is always active
   b. is normally faster than simple diffusion of the transported substrate
   c. may be electrogenic
   d. may be inhibited by a molecule structurally similar to the normal substrate
   e. may require ATP

6. The resting membrane potential in nerves:
   a. is likely to be close to E_K if the resting membrane permeability to K^+ is high
   b. is about –70 mV
   c. can be explained solely in terms of a K^+ diffusion potential
   d. would be expected to decrease (become less negative) if the extracellular K^+ concentration ($[K^+]_o$) increased
   e. is such that ions carry current into and out of the cell at the same rate

7. The action potential in nerves:
   a. has a rapid depolarizing phase owing to an inward sodium current
   b. reaches a peak much closer to E_{Na} than E_K
   c. depends on voltage-dependent activation and inactivation of both Na^+ and K^+ channels
   d. would rise to a more positive peak if the extracellular Na^+ concentration ($[Na^+]_o$) were increased
   e. repolarizes because of an increase in inward K^+ current
   f. is conducted along an axon as the result of the internal release of chemical messengers

8. Hormone receptors:
   a. of a given type always activate the same responses in different cells
   b. are evenly distributed over all body cells
   c. may influence DNA transcription
   d. have specific binding properties
   e. may inhibit production of an intracellular messenger

9. During skeletal muscle contraction:
   - **a.** the myofilaments do not change length
   - **b.** the A and the I bands do not change length
   - **c.** calcium is released from the T tubule
   - **d.** cross bridges are formed between myosin and actin
   - **e.** maximal tension can be generated when the muscle is close to its normal resting length

10. Muscle contraction:
    - **a.** in smooth muscle develops less force per unit cross section than in skeletal muscle
    - **b.** in smooth muscle uses ATP less rapidly than in skeletal muscle
    - **c.** in type I skeletal muscle fibres relies on aerobic metabolism
    - **d.** in all type II skeletal muscle fibres is anaerobic
    - **e.** leads to fatigue more rapidly in type II than in type I skeletal muscle fibres

Single best answer questions

For each of the following questions, choose the single best answer.

1. In a cell most ATP is normally generated by:
   - **a.** the nucleus
   - **b.** the Golgi apparatus
   - **c.** the endoplasmic reticulum
   - **d.** the mitochondria
   - **e.** the lysosomes

2. Substance X is charged, readily soluble in water and diffuses freely across capillary membranes. For which fluid volume is X most likely to be a useful dilution indicator?
   - **a.** total body water volume
   - **b.** plasma volume
   - **c.** blood volume
   - **d.** interstitial fluid volume
   - **e.** extracellular fluid volume

3. Sweating is a useful mechanism for lowering body temperature because:
   - **a.** it reduces heat production in the skin
   - **b.** it increases heat loss by evaporation
   - **c.** it increases the thermal conductivity of the skin
   - **d.** it increases the electrical conductivity of the skin
   - **e.** it increases heat loss by convection

4. During diffusion, the net flux of a cation into a cell through open ion channels is likely to be decreased by:
   - **a.** increasing the temperature
   - **b.** decreasing the membrane potential
   - **c.** increasing the intracellular concentration of that cation
   - **d.** decreasing the extracellular concentration of that cation
   - **e.** increasing the membrane potential

5. In an experiment, you discover that chemicals which poison the mitochondria slow the rate of cellular uptake of substance Y. From this you may reasonably conclude that:
   - **a.** Y is normally metabolized by the mitochondria
   - **b.** Y is taken up by facilitated diffusion
   - **c.** Y is taken up by primary active transport
   - **d.** Y is taken up by secondary active transport
   - **e.** Y is taken up by an active process

6. During a nerve action potential the inward Na^+ current increases to a peak and then declines again. This is best explained by saying that:
   - **a.** voltage-sensitive Na^+ channels are activated by depolarization and then deactivated by repolarization
   - **b.** voltage-sensitive Na^+ channels are activated by depolarization and then inactivated by repolarization
   - **c.** the driving force promoting Na^+ entry gets smaller as the membrane depolarizes
   - **d.** voltage-sensitive Na^+ channels are first activated and then inactivated by depolarization
   - **e.** voltage-sensitive Na^+ channels are first activated and then deactivated by depolarization

7. Which of the following changes are most likely to increase the speed of conduction of action potentials along an axon?
   - **a.** increasing the electrical resistance of the extracellular solution

b. increasing the resistance of intracellular solution
c. increasing the diameter of the axon
d. increasing the threshold for action potential generation
e. reducing the temperature

8. There is a form of GTP known as GTP-γS which can bind to G proteins but which cannot be broken down to GDP by them. Intracellular application of this substance would be expected to:
a. cause persistent activation of receptor-activated G proteins
b. cause persistent inhibition of receptor-activated G proteins
c. increase the intracellular levels of second messengers
d. decrease the intracellular levels of second messengers
e. promote receptor tyrosine kinase activity

9. Complete removal of extracellular Ca^{2+} from a skeletal muscle fibre would be expected to:
a. immediately inhibit contraction completely
b. immediately increase the force of contraction
c. reduce the force of contraction in a time-dependent manner
d. reduce the force of contraction in a use-dependent manner
e. reduce the force of contraction in a time- and use-dependent manner

10. Which of the following is unlikely to inhibit contraction in smooth muscle?
a. reduction in calmodulin levels
b. inhibition of myosin light chain phosphatase
c. inhibition of myosin light chain kinase
d. intracellular chelation of Ca^{2+}
e. blockade of Ca^{2+} channels

Matching item questions

Theme: Cell transport

Options

A. Diffusion
B. Osmosis
C. Carrier-mediated transport
D. Facilitated diffusion
E. Primary active transport
F. Secondary active transport
G. Endocytosis
H. Exocytosis

For each of the descriptions below choose the most appropriate option from the list above. Each option may be used once, more than once or not at all.

1. This is a passive transport process which can be saturated at high substrate concentrations.
2. This relies on a solute concentration gradient to drive the movement of solvent across the plasma membrane.
3. This allows particles and large extracellular molecules (e.g., intact proteins) to be brought into the cell.
4. This uses ATP breakdown to move substrates from low to high concentration.
5. This process is passive and tends to be slow in the case of lipophobic substrates.
6. This uses a concentration gradient for one substance to drive the transport of another.

Theme: Excitable properties of nerves

Options

A. Voltage-gated ion channels
B. Ligand-gated ion channels
C. Activation
D. Inactivation
E. Deactivation
F. Depolarization
G. Repolarization
H. Hyperpolarization

For each of the descriptions below choose the most appropriate option from the list above. Each option may be used once, more than once or not at all.

1. Closure of ion channels due to repolarization of the cell membrane.
2. This phase of the action potential is generated by opening of Na^+ channels and demonstrates positive feedback.
3. Results from an increase in the g_K:g_{Na} above that normally seen at the resting membrane potential.

Questions

4. Reduction in g_{Na} due to membrane depolarization.
5. Increase in ion channel open probability due to membrane depolarization.
6. Ion channels which are activated by neurotransmitters at synapses.

Short notes

Write short notes on the following:

a. homeostasis
b. body fluid compartments
c. fluid tonicity
d. diffusion potentials
e. action potential propagation in axons
f. second messengers
g. contractile proteins

Modified essay

The cell membrane separates intracellular fluid from interstitial fluid. The ionic concentrations ([ion]) are very different on either side of the membrane but the cell normally exists in a steady state in which the intracellular concentrations ($[ion]_i$) remain constant.

Questions

1. What two conditions must be fulfilled if a substance is to diffuse across a cell membrane?
2. What other factors affect the rate of diffusion?
3. What mechanisms limit the effects of extracellular ions on $[ion]_i$ in cells?
4. How does membrane potential affect the rate of diffusion of anions and cations?

Body fluids are aqueous solutions and water permeates freely across cell membranes. Diffusion of water is referred to as osmosis and is important in the control of cell volume.

Questions

5. What two conditions must be fulfilled if osmosis is to take place across the cell membrane?
6. Which ions are most important in the osmotic control of cell volume? Why are they important?
7. If a sample of plasma was diluted with isotonic saline, what effect would this have on the volume of red cells suspended in the plasma? What difference would it make if distilled water were used instead of saline?

Data interpretation

An experiment is carried out with a view to determining which ions might be most important in controlling the resting membrane potential in frog skeletal muscle. The resting membrane potential was first measured using an intracellular electrode and the value obtained was −87 mV. The intracellular ($[ion]_i$) and extracellular ($[ion]_o$) concentrations of a number of ions were then measured and the results are given in Table 1.

Questions

1. Use the Nernst equation (Eq. 5 below) to calculate the transmembrane equilibrium potential for each of these ions.

$$E_x = \frac{RT}{zF}\log_e\frac{[X^z]_o}{[X^z]_i} \quad \text{(Eq. 5)}$$

(Assume $\frac{RT}{F} = 25\,\text{mV}$ under these conditions.)

2. Let us assume that the resting membrane potential in frog skeletal muscle is chiefly determined by diffusion potentials generated by permeant (diffusible) ions. Based on the experimental results above, which ion or ions would appear to diffuse through the resting membrane most easily? Explain your answer.

Table 1 Ion concentrations inside ($[ion]_i$) and outside ($[ion]_o$) a frog skeletal muscle cell

Ion	$[Ion]_i$ (mmol L^{-1})	$[Ion]_o$ (mmol L^{-1})
K^+	125	2.5
Na^+	10	110
Ca^{2+}	10^{-4} (0.1 µmol L^{-1})	1.1
Cl^-	1.5	78

Resting membrane potential is −87 mV.

3. In which direction will each of the four ions listed move across the membrane at the measured resting potential? Will the resulting current be inward or outward in each case?
4. Using the results from this experiment, what is the net electrical driving force (in mV) acting on each of these ions at the resting potential? What is the significance of the sign (+ or –) of this driving potential?

Clinical scenario

A 67-year-old man is brought into accident and emergency in a semi-conscious state in the early hours of the morning. He was found by passers by in the roadside and appeared to have been living rough recently. There was an empty bottle of gin in his pocket and he was obviously drunk. They had called for an ambulance as they were worried about him, especially because it was a very cold January night and there was a severe frost.

On examination the man has cold, cyanosed peripheries. His peripheral pulse is weak and difficult to palpate. The aural temperature (i.e., measured with an ear thermometer) is 33.2°C.

Questions

1. What is the likely diagnosis?
2. Why is aural temperature preferred to the use of an oral or skin thermometer?
3. What are the main methods of heat loss from the body? What relevance might the history of living rough and alcohol consumption have in this context?
4. Might reduced heat production have played a role, given the history in this case?
5. What is the primary goal in treating this person?
6. How might you achieve this?
7. What is the main fatal consequence of hypothermia?

Viva questions

1. What are the main fluid compartments within the body?
2. How might you estimate blood volume?
3. Tell me about the role of receptors in cell signalling.

Multiple true/false answers

1. a. **True.** These dissolve into the lipid membrane.
   b. **False.** A bilipid layer with proteins inserted into it.
   c. **True.** Some membrane proteins are ion channels.
   d. **True.** Because of osmotic swelling of the cell.
   e. **True.** Depends on voltage-controlled ion channels.

2. a. **True.** Controls the internal environment of the body, particularly the interstitium.
   b. **False.** Negative feedback.
   c. **False.** Sympathetic nerves control heat loss by regulating skin blood flow and sweating.
   d. **True.**
   e. **True.** High-gain systems give large corrective responses.

3. a. **False.** Diffusion is passive.
   b. **True.** Facilitated diffusion.
   c. **False.** Nonpolar molecules are more lipid soluble and diffuse across the membrane more easily.
   d. **True.** In secondary active transport systems.
   e. **True.** Because of increased thermal solute motion.

4. a. **False.** 300 mmol L^{-1} NaCl equals 600 mosmol kg^{-1}; this is a strongly hypertonic NaCl solution and would cause cell shrinkage.
   b. **False.** 150 mmol L^{-1} NaCl equals 300 mosmol kg^{-1}, i.e., isotonic saline.
   c. **True.** Osmotic swelling.
   d. **False.** Urea diffuses freely into the cell and has almost no osmotic effect on the cells.
   e. **True.** This is isotonic saline.

5. a. **False.** Facilitated diffusion is passive.
   b. **True.** This is one of its benefits.
   c. **True.** For example, the $3Na^+/2K^+$ exchange pump.
   d. **True.** The inhibitor competes for the substrate-binding site. Competitive inhibition is characteristic of carrier-mediated transport.
   e. **True.** Active transport.

6. a. **True.** Resting K^+ permeability greatly exceeds Na^+ permeability in nerves and skeletal muscle.
   b. **True.** It is larger, or more negative, in skeletal muscle, about −90 mV.
   c. **False.** Resting potential is less negative than E_K because of the contribution from Na^+ and other ions.
   d. **True.** The concentration gradient driving K^+ out of the cell is decreased so the voltage gradient necessary to balance it, E_K, is less negative. The membrane potential decreases because K^+ is the dominant permeant ion.
   e. **True.** For membrane potential to remain constant the rates of charge movement must be equal and opposite. The resting potential is not identical to any one ion's equilibrium potential, so there is a driving force favouring ion diffusion. If we assume that only Na^+ and K^+ are involved, then the outward movement of K^+ must be balanced by the inward movement of Na^+.

7. a. **True.** Positive feedback accelerates this.
   b. **True.** Because the membrane permeability to Na^+ is high at the peak of the action potential.
   c. **False.** K^+ channels do not inactivate.
   d. **True.** It would increase the $[Na^+]$ gradient into the cell giving an increase in E_{Na}.
   e. **False.** The repolarizing K^+ current is outward.
   f. **False.** Conduction relies on generation of local depolarizing currents along the axon.

8. a. **False.** The response depends on the cell type as well as the receptor and the hormone.
   b. **False.** Variations in receptor expression help regulate hormone sensitivity in different tissues.
   c. **True.** For example, intracellular steroid hormone receptors.
   d. **True.**
   e. **True.** Both promotion and inhibition of second messenger production may be mediated by different types of G protein.

9. a. **True.** They slide past each other.
   b. **False.** The A bands stay constant in length but the I bands shorten as the thin myofilaments slide between the thick.
   c. **False.** T tubule depolarization triggers Ca^{2+} release from the sarcoplasmic reticulum.
   d. **True.** Contraction depends on the cross bridge cycle.
   e. **True.** Active tension depends on the resting length, and the normal length in the body seems close to optimal.

10. a. **False.** Contraction is slower, but force per unit cross section is higher in smooth muscle.
    b. **True.** Slower rate of myosin ATPase activity.
    c. **True.**
    d. **False.** Type IIA fibres are aerobic.
    e. **True.** Fatigue resistance declines in the order type I > type IIA > type IIB.

Single best answers

1. d. Mitochondria are the site of aerobic ATP production. Smaller amounts of ATP can be generated anaerobically within the cell cytoplasm.
2. e. Such a substance would spread throughout the extracellular space but is unlikely to enter body cells due to its charge. It should spread throughout the interstitial and plasma volumes and so would not give a correct estimate of either of these volumes on their own.
3. b.
4. b. Decreasing the membrane potential, i.e., making the inside of the cell less negative, decreases the electrical gradient favouring movement of positively charged cations into the cell. All the other changes suggested would either increase the electrochemical gradient favouring influx or, in the case of an increase in temperature, increase the kinetic energy of the ions, leading to an increased rate of flux. NB: these effects should be carefully distinguished from the effects of membrane potential on channel opening, which may be promoted by membrane depolarization. The question excludes these possibilities by referring to flux through open channels.
5. e. The implication is that reduced ATP production slows uptake, i.e., chemical energy is necessary for the process. One cannot distinguish between primary and secondary active transport on this basis, however, as ATP is necessary in both cases, although it plays an indirect role in secondary active transport.
6. d. Both activation and inactivation of these channels result from membrane depolarization, causing a rapid increase in g_{Na} followed by a decrease. The driving force for Na^+ influx actually decreases during depolarization, as the membrane comes closer to E_{Na}, and then increases again during repolarization. This cannot explain an increase in current followed by a decrease. Also, note the specific meaning of activation (channel opening due to depolarization), inactivation (channel closure due to depolarization) and deactivation (channel closure due to repolarization; i.e., the reversal of activation).
7. c. This reduces the resistance to the flow of the intracellular local currents which depolarize the adjacent membranes. All the other changes would be expected to slow conduction by decreasing the extracellular or intracellular component of the local currents, or by increasing the amount of current flow necessary to get the membrane to threshold.
8. a. G proteins are activated, or switched on, when they are bound to GTP. Binding of the GTP-γS will persistently activate the

protein because breakdown to GDP, which normally reverses activation, cannot occur.

9. e. Excitation–contraction in skeletal muscle relies on voltage dependent Ca^{2+} release from the sarcoplasmic reticulum. Most of the Ca^{2+} released is recycled back into the sarcoplasmic reticulum and so is available to be released again, even if there is no external Ca^{2+}. A fraction of the Ca^{2+} released is pumped out across the cell membrane, however, and so the total intracellular Ca^{2+} declines over time (time-dependent). The rate of this decline increases the more often the cell is stimulated (use-dependent). Removal of extracellular Ca^{2+} would immediately inhibit contraction in cardiac muscle, however, since release from the sarcoplasmic reticulum is triggered by Ca^{2+} entry in that case.

10. b. All the other changes would inhibit excitation–contraction coupling pathways in smooth muscle. Inhibiting the phosphatase would actually promote contraction by slowing the rate of dephosphorylation of MLCK.

Matching item answers

Theme: Cell transport

1. D. Facilitated diffusion can only occur down a concentration gradient but the number of available carriers determines the maximum rate of transport, i.e., transport can be saturated.
2. B. In osmosis, a concentration gradient across the plasma membrane for a nondiffusible solute leads to movement of the solvent, water, in the reverse direction.
3. G. Relatively large structures which are unlikely to cross the plasma membrane can be taken up in endocytotic vesicles.
4. E. Transport against a concentration gradient—or an electrochemical gradient in the case of ions—has to be active, i.e., requires energy. Primary active transport uses an ATPase to break down ATP, releasing the energy required.
5. A. Lipophobic substrates do not cross the plasma membrane readily in the absence of an appropriate carrier system.
6. F. This form of active transport is secondary in the sense that the energy does not come directly from ATP. ATP is required in the longer term, however, since primary active transport maintains the concentration gradient for diffusion of the driving substrate.

Theme: Excitable properties of nerves

1. E. Reversal of activation.
2. F. Depolarization causes increased channel opening leading to increased I_{Na} leading to further depolarization.
3. H. Under these circumstances the membrane will be more polarized than normal.
4. D. Voltage dependent inactivation.
5. C. The total number of channels open at any time can be described in terms of the open probability, i.e., the probability that any given channel will be open at any given time. Activation = an increase in open probability.
6. B. The ion channels are activated by binding of the transmitter, rather than by depolarization.

Short note answers

a. Homeostasis is the maintenance of a constant internal environment in the body. This requires receptors and effectors to detect and control the relevant variable and an integrating centre to determine the normal value (set point). Variations from normal produce changes which return the variable towards normal, i.e., there is negative feedback.

b. Body fluid (50–60% of total mass) may be divided into intracellular (2/3 total fluid) and extracellular (1/3 total) compartments. The ionic make-up of the fluid in these regions is different, e.g. [Na^+] is high extracellularly while [K^+] is high intracellularly. The extracellular space contains a number of subcompartments, of which the interstitial space (fluid within tissues) and the plasma volume (1/4 of extracellular fluid) are most important, although cerebrospinal fluid, synovial fluid and aqueous humour also contribute.

c. Fluid tonicity refers to the osmotic effect of a fluid at the cell membrane, particularly at the erythrocyte membrane. This is determined by solute concentration and the permeability of the cell membrane to the solute. Isotonic solutions produce osmotic equilibrium at the cell membrane. If solute concentration is less than that of plasma (300 mosmol kg^{-1}) or if the membrane is permeable to the solute (e.g., urea solutions) the solution is hypotonic and water will enter the cells causing swelling (possibly lysis). If the concentration of impermeant solute is higher than normal, the solution is hypertonic and cells will shrink (crenate).

d. Diffusion potentials are caused by selective ion diffusion across membranes, e.g., K^+ can diffuse out of cells but large anions cannot, so the gradient from high $[K^+]$ inside the cell to low $[K^+]$ outside leads to a negative diffusion potential inside the cell. At the equilibrium potential, the electrical and concentration gradients balance and no net ion movement occurs.

e. Action potentials depolarize a region of membrane, which acts as a source for local currents. These depolarize adjacent, resting membrane until an action potential is generated once threshold is reached. This propagation is normally unidirectional because of the refractoriness of axon which has recently fired an action potential. The velocity of propagation or conduction increases with increased axon diameter (decreased resistance to internal current flow) and with myelination (saltatory conduction).

f. Second messengers act as intracellular signalling molecules which play a key role in the conversion, or transduction, of an extracellular message into a cellular response. Their levels rise and fall in response to activation of cell receptors by their specific agonists, and receptor-associated, heterotrimeric GTP-binding proteins (G proteins) often link this activation to increases or decreases in the rate of synthesis of the relevant G protein. The second messengers, in turn, often activate intracellular kinase enzymes. These phosphorylate other enzymes and functionally important proteins, modifying their activity. Some important second messengers are listed below, along with their means of synthesis, intracellular targets and major removal mechanisms. Second messengers also have specific mechanisms which reduce the concentration of the messenger once receptor activation ceases, e.g., breakdown of cAMP and cGMP by phosphodiesterases, and removal of cytoplasmic Ca^{2+} by reuptake into stores or removal across the plasma membrane.

Messenger	Synthesis	Target
Cyclic AMP (cAMP)	*From* ATP *by* adenylate cyclase	Protein kinase A
Cyclic GMP (cGMP)	*From* GTP *by* guanylate cyclase	Protein kinase G
Inositol 1,4,5-trisphosphate + diacylglycerol	*From* phosphatidylinositol 4,5-bisphosphate *by* phospholipase C	Intracellular Ca^{2+} release Protein kinase C
Ca^{2+}	*Released from* stores *or entry from* extracellular fluid	Troponin Ca–calmodulin sensitive kinases

g. The contractile proteins actin and myosin are incorporated into the myofilaments in muscle and are responsible for contraction. Myosin heads form cross bridges with actin, pulling the thin myofilaments towards the middle of the thick myofilaments and shortening the sarcomere. This is controlled by the troponin–tropomyosin system in striated muscle and by myosin light chain kinase in smooth muscle.

Modified essay answers

1. (i) There must be a concentration gradient driving diffusion and (ii) the membrane must be permeable to the substance.
2. Available surface area, temperature, molecular size, diffusion distance, membrane voltage for ions.

3. (i) The cell membrane has limited permeability to ions, (ii) active transport systems (e.g., Na^+/K^+ ATPase, Ca^{2+} ATPase) tend to reverse any changes in $[\text{ion}]_i$ caused by diffusion.
4. Electrical potentials produce an electrostatic force, which will affect charged species, e.g., a positive charge inside the membrane will attract anions into the cells and repel cations. Net movement of an ion will depend on the balance of the chemical and electrical gradients acting on it, i.e., the electrochemical gradient.
5. (i) There must be a concentration gradient of solute across the cell membrane, and (ii) the membrane must be permeable to water but relatively impermeable to the solute, i.e., semipermeable for that solute.
6. Sodium ions and Cl^- in the extracellular fluid and K^+ and inorganic anions (phosphate and sulphate) inside the cell. These ions are important because: (i) they are osmotically active at cell membranes, which have only limited permeability to them, (ii) they make the largest contributions to the total osmolality of the relevant solutions, and (iii) there is a concentration gradient across the membrane for each. This is maintained both by low membrane permeability and by active transport systems (for Na^+ and K^+).
7. Dilution of plasma with isotonic saline has no effect on the total osmolality or tonicity and, therefore, there would be no change in red cell volume. Dilution with water leads to proportionate reduction in total plasma osmolality. The resulting hypotonic solution would cause red cell swelling and possibly cell lysis.

Data interpretation answers

1. $E_K = -98\,\text{mV}$; $E_{Na} = +60\,\text{mV}$; $E_{Cl} = -99\,\text{mV}$ ($z = -1$ for Cl^-); $E_{Ca} = +126\,\text{mV}$ ($z = +2$ for Ca^{2+}).
2. The resting potential of $-87\,\text{mV}$ is much closer to the equilibrium potentials for K^+ and Cl^- than to those of the other two ions. This suggests that the membrane is most permeable to these ions, since the effect of any ion on the membrane potential largely depends on how easily that ion can diffuse across the membrane (diffusion potentials depend on diffusion). Taken to its extreme, if the membrane were infinitely permeable to an ion, then the membrane potential would always equal the equilibrium potential for that ion, while if there was zero permeability, the ion could have no effect on the membrane potential at all.
3. At voltages positive to the equilibrium potential, cations (positively charged) move outward and anions (negative charge) move inwards. The current (which is named for the direction of positive charge movement) is in the same direction as ion movement for cations but in the opposite direction for anions. In this case K^+ will move outward, while Na^+, Ca^{2+} and Cl^- move inwards. K^+ and Cl^- both produce an outward current while Na^+ and Ca^{2+} produce inward currents.
4. The driving force at any membrane potential (V_m) is given by the difference between V_m and E_X for any given ion ($= -87 - Ex$ mV in this case). This gives values of $+11\,\text{mV}$ for K^+, $-147\,\text{mV}$ for Na^+, $+12\,\text{mV}$ for Cl^- and $-213\,\text{mV}$ for Ca^{2+}. If the membrane potential is positive with respect to E_X ($V_m - E_X$ is positive), cations will be driven out of the cell and anions attracted in (outward currents in both cases), and vice versa for negative differences.

Clinical scenario answers

1. The likely diagnosis is hypothermia; a reduction in core body temperature.
2. The aurally recorded temperature more closely reflects core temperature. The oral temperature can also be affected by recent hot or cold drinks which produce local heating or cooling. Skin temperature is less informative still. It reflects peripheral rather than core temperature, and is very variable across the body. It is also affected by local blood flow; indeed, local skin temperature can be used as an indirect measure of skin blood flow.
3. Heat is lost by conduction, convection and evaporation. This last still plays a role in cold conditions as there is fluid evaporation during breathing and from insensible loss from the skin. Heat loss is increased when clothing is damp, as may well be the case in someone living out of doors, mainly due to reduced heat insulation against conductive losses. Torn or damaged clothes are also less efficient in preventing convection, so warm air can escape from the body. Alcohol greatly exaggerates these effects, since it acts as a peripheral

vasodilator, opposing the normal sympathetic vasoconstrictor tone which helps restrict loss in cold conditions, and increasing the rate of peripheral heat loss.

4. Body temperature reflects the balance between heat production and heat loss. Inadequate nutrition will reduce metabolic heat generation.
5. To get the core temperature back towards normal (37°C).
6. The core temperature will rise if the rate of heat loss is slower than the rate of metabolic heat production. This can be further accelerated if the gradient is reversed so that the body actually absorbs heat from its environment. This can be done most quickly by using a warm bath at or above body temperature, the method of choice in acute hypothermia. A slower method is to insulate the body with warm, dry blankets or clothing, thus reducing heat loss and allowing metabolic heat to bring the core temperature back up to normal.
7. If the core temperature falls too low (below 30°C), fatal cardiac arrhythmias, such as ventricular fibrillation, may result.

Viva answers

1. You would need to mention total body water and how this is divided into the intracellular and extracellular compartments. The extracellular fluid has several subcompartments, most importantly the interstitial fluid and plasma. Questioning is likely to go on to consider the probable volumes of these compartments.
2. It is best to begin with a brief explanation of the indicator dilution principle and the criteria for a suitable indicator before suggesting a specific technique. This allows you to get across foundation material, which you should know with confidence, before going on to detail, where the chance of error is higher. Radiolabelled red cells are suitable markers in this case. It is also possible to use a plasma volume indicator, e.g., labelled plasma protein, providing the haematocrit is used to scale the answer up to total blood volume (see Section 2.1). The examiners will expect you to be able to suggest normal values for blood and plasma volumes.
3. Explain what is meant by a receptor, emphasizing their usual site in the plasma membrane and the presence of specific binding sites for chemical messenger molecules. You should be ready to discuss the main signal transduction pathways. Be prepared for follow-up questions dealing with G proteins, second messengers and intracellular receptors to cell-permeant signalling molecules, e.g., steroid and thyroid hormones.

Chapter 2

Blood and related physiology

Overview

Blood is circulated around the body within the cardiovascular system, transporting O_2, necessary metabolic substrates and hormones to the cells of the body, while removing CO_2 and waste products. Plasma, the liquid phase of blood, has many functions, involving colloid osmotic effects, transport, signalling, immunity and clotting. The cellular elements of blood are responsible for gas transport, immunity and some aspects of haemostasis.

2.1 Blood volume and constituents

Learning objectives

At the end of this section you should be able to:
- state the normal values for blood volume, plasma volume and haematocrit.

Blood volume averages about 70 ml kg^{-1} body weight, giving a total of approximately 5 L in an adult. This consists of a suspension of red cells, white cells and platelets (the *formed elements*) in *plasma*. Centrifuging a sample of blood separates the formed elements from the plasma and the ratio of the volume of the packed red cells to the total blood volume is referred to as the *haematocrit* or *packed cell volume* (Fig. 22), which is normally about 0.45, or 45%. A thin layer of white cells and platelets can also be identified at the interface between the red cells and the plasma (the *buffy coat*). Plasma comprises 55% of total blood volume, i.e. about 3 litres in an adult.

2.2 Plasma constituents

Learning objectives

At the end of this section you should be able to:
- classify the plasma proteins and outline their functions.

Plasma has the same ionic composition as the rest of the extracellular fluid (Section 5.1) but also contains plasma proteins (total concentration of about 60 g L^{-1}), with a range of important functions.

- *Albumin* (synthesized by the liver) is the protein in highest concentration and so it makes the greatest contribution to the colloid osmotic pressure of plasma (Section 3.8). It also acts as a nonspecific transport protein for a number of substances with a low solubility in water, e.g., bilirubin.
- *Globulins* include specific transport proteins (e.g., transferrin for iron), clotting factors, the complement system and inactive precursors of certain hormones, e.g., angiotensinogen. One subtype, known as gamma globulins, or immunoglobulins, acts as circulating antibodies important in specific immune responses.
- *Fibrinogen* is converted to fibrin during clotting.

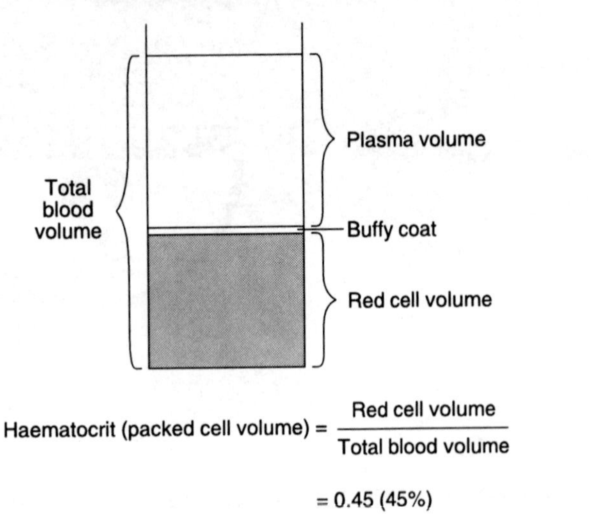

$$\text{Haematocrit (packed cell volume)} = \frac{\text{Red cell volume}}{\text{Total blood volume}}$$

$$= 0.45\ (45\%)$$

Fig. 22 Determination of the haematocrit from a centrifuged blood sample.

2.3 Erythrocytes

Learning objectives

At the end of this section you should be able to:

- state normal values for [haemoglobin], red cell count, red cell life span and erythrocyte sedimentation rate (ESR)
- outline the structure of haemoglobin and its role in O_2 transport
- outline how red cells are manufactured and destroyed
- explain how nutritional deficiencies lead to anaemia
- describe how ABO and Rh blood groups are determined and the consequences of blood group mismatch.

Red blood cells, or erythrocytes, have several functions but are particularly important as O_2 delivery systems. This reflects the O_2 transport characteristics of *haemoglobin*, which is packaged inside erythrocytes so that it does not leak out of capillaries. Haemoglobin consists of four peptide chains (the globin structure) each linked to a haem molecule consisting of a porphyrin ring structure encircling a ferrous iron ion (Fe^{2+}). The reversible binding of O_2 to these ions accounts for 97% of the normal O_2-carrying capacity of blood, so haemoglobin concentration determines O_2 transport capacity. Normal haemoglobin values are in the range of 14–16 $g\,dl^{-1}$ in men and 12–14 $g\,dl^{-1}$ in women. A low blood haemoglobin concentration is referred to as *anaemia*.

Erythrocyte development

The process of red cell production, or *erythropoiesis*, begins in the embryonic yolk sac and is continued in the liver, spleen and lymph nodes in the maturing fetus. By the end of pregnancy and after birth, however, the process is restricted to bone marrow. As time progresses, the contribution from long bones decreases and in adult life only the marrow of membranous bones, such as the vertebrae, ribs and pelvis, is involved.

Stages in erythrocyte development

Pluripotential, or uncommitted, *stem* cells, which have the potential to produce any type of blood cell, divide and develop into erythroid stem cells committed to form erythrocytes (Fig. 23). These divide further and mature, synthesizing haemoglobin and eventually forming *normoblasts*. Nuclear material is extruded and the endoplasmic reticulum resorbed, producing first a *reticulocyte*, containing a few remnants of endoplasmic reticulum, and then an erythrocyte. Normally only these last two cell types are found in the circulation, with reticulocytes making up less than 2% of the total. This percentage rises during periods of rapid erythrocyte synthesis, when more immature cells enter the circulation. Mature red cells take the form of biconcave discs which deform easily within the narrow capillaries. The normal red cell count in blood is 4×10^{12} to $6 \times 10^{12}\,L^{-1}$ ($4\text{–}6 \times 10^{6}\,mm^{-3}$).

Erythrocyte destruction

Ageing erythrocytes are destroyed, often in the *spleen*, after an average life span of 120 days. The phagocytic cells of the reticuloendothelial system degrade the haemoglobin released, with iron from the haem and amino acids from the globin molecules being recycled. The porphyrin ring is converted to *bilirubin*, which is further metabolized by the liver and then excreted in bile (Section 6.3).

Control of erythrocyte production

Erythropoiesis is controlled by the kidney, which releases a hormone known as *erythropoietin* if the delivery of O_2 to renal cells falls below normal. This will occur if the concentration of circulating haemoglobin is reduced, i.e., during anaemia. The bone marrow responds by increasing red cell production, thus increasing the haemoglobin content back to normal. Since this control loop is sensitive to tissue O_2 levels rather than the actual haemoglobin concentration, other conditions which reduce the O_2 content

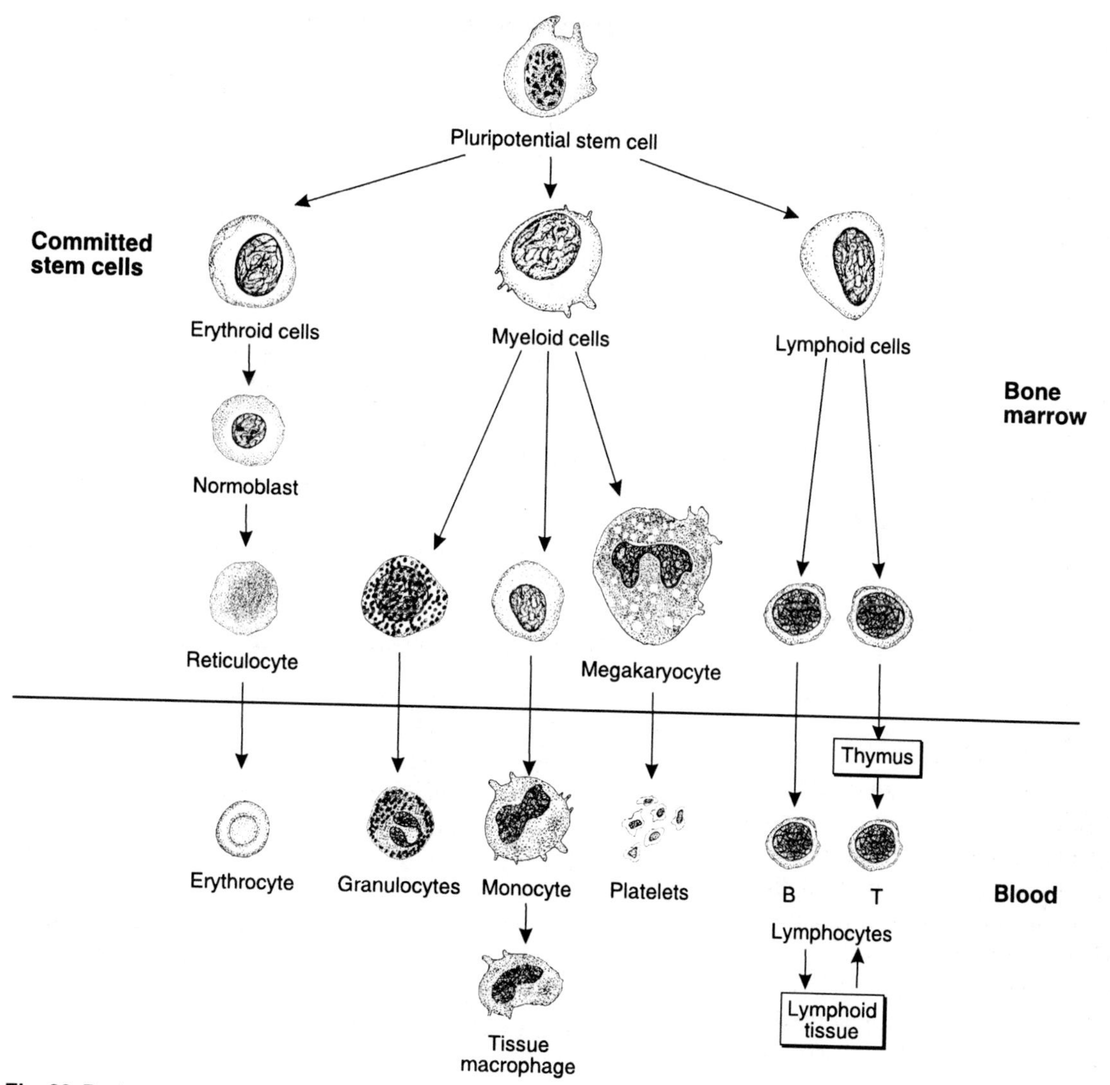

Fig. 23 Production of blood cells.

of blood will also stimulate erythropoiesis, even if the haemoglobin concentration is normal. This is seen at high altitudes, where the partial pressures of O_2 in the lungs and blood are reduced (Section 4.6). Over a period of weeks at high altitudes, erythropoietin stimulates an increase in the haemoglobin concentration, with a rise in haematocrit and red cell count (*compensatory polycythaemia*). It is for this reason that athletes wishing to increase the O_2-carrying capacity of their blood often train at altitude.

Box 4 Clinical note: Use and abuse of erythropoietin

One of the problems associated with chronic renal failure is anaemia. This can be treated with injections of erythropoietin which substitute for normal renal production of the hormone, stimulating marrow production of red cells. Synthetic erythropoietin (EPO) is also used by athletes, however, in order to increase haemoglobin levels and improve performance without having to train at altitude. This practice can lead to dangerously high haematocrits, with risk of blood clotting in arteries due to slow blood flow.

Nutritional requirements for red cell production

Erythropoiesis and haemoglobin synthesis require adequate supplies of the vitamins B_{12} (cyanocobalamin) and *folic acid*, as well as the mineral *iron*. Deficiencies of these may cause anaemia.

Vitamin B_{12} and folate

If B_{12} or folate levels are reduced, cell division and maturation are adversely affected. This is particularly important at sites of rapid cell turnover, such as the bone marrow. There is a reduction in the red cell count so that the overall haemoglobin concentration falls. The erythrocytes which do form are abnormally large (*macrocytes*), so this is known as a *macrocytic anaemia*. Abnormal erythrocyte precursors called *megaloblasts* are found in the marrow, so the term *megaloblastic anaemia* is also used. It should be appreciated that there is no problem with haemoglobin synthesis within the developing cells; there are just too few red cells produced.

Deficiency of B_{12} or folate can arise in two ways. The diet itself may include inadequate amounts of the normal source foods for these vitamins, e.g., animal products (for both B_{12} and folate) and green vegetables (rich in folate). Vitamin B_{12} deficiency can also occur with a normal diet as a result of reduced B_{12} absorption. Parietal cells in the stomach normally secrete *intrinsic factor*, which binds to B_{12}, and it is the resulting complex that is absorbed from the ileum (Section 6.4). The B_{12} is then transported to the liver, which normally stores 1–2 years' supply (liver is an excellent dietary source of B_{12}). In *pernicious anaemia*, there is reduced secretion of intrinsic factor. This leads to malabsorption of B_{12} and a megaloblastic, macrocytic anaemia results. These patients often also appear mildly jaundiced because of increased bilirubin production following haemolysis of the abnormally fragile red cells. Treatment involves regular intramuscular injections with B_{12}, thereby bypassing the normal intestinal absorption mechanisms.

Iron

If the supply of iron is inadequate, haemoglobin synthesis is restricted. This leads to an anaemia in which erythrocytes contain less haemoglobin than normal (they are *hypochromic*) and are, as a result, smaller than normal (*microcytic*). Iron-deficiency anaemia can occur whenever iron demand exceeds supply. That may be because of reduced iron intake or increased iron loss, e.g., because of chronic bleeding. Men normally lose approximately 1 mg of iron daily, chiefly through the shedding of intestinal epithelium. In menstruating women, this can rise to 2 mg or more. Since the average diet provides 10–15 mg of iron each day, of which about 10% is normally absorbed, iron balance is usually maintained in men, but women run an appreciable risk of becoming iron depleted. Pregnancy increases iron demand further and iron supplements may be necessary.

Iron is mainly absorbed in the ferrous form by means of intestinal *transferrin* and an iron–transferrin receptor (Section 6.4). The rate of absorption increases when iron demand is increased within the body. Plasma transferrin transports iron within the circulation and excess iron is stored as *ferritin* or *haemosiderin*, particularly in the liver, spleen and bone marrow. Mean iron stores are 2–3 times larger in men than women, presumably as a consequence of menstrual losses.

Erythrocyte sedimentation rate

In an undisturbed vertical column of anticoagulated blood, erythrocytes slowly settle out, leaving a clear column of plasma above them. This rate of sedimentation increases in certain disease states and the erythrocyte sedimentation rate (ESR) is a widely used clinical investigation. Normal values lie in the range 5–10 $mm\,h^{-1}$ but may be higher during pregnancy. Abnormally high ESR values are often associated with an increase in immunoglobulins. This favours aggregation of red cells into stacks called *rouleaux*, which sediment more rapidly than single cells.

Blood groups and transfusion

The ability to replace blood lost following trauma or surgery is a vital aspect of modern medicine. This relies on an understanding of immune reactions against red blood cells, since these may be fatal and must be avoided if blood from one person (the donor) is to be safely transfused into another (the recipient). Whenever antigens on the surface of erythrocytes (*agglutinogens*) come into contact with specific antibodies directed against them (*agglutinins*), the cells clump together or agglutinate. By testing for *agglutination* of red cells with known antibodies, the erythrocyte antigens can be identified and this defines the blood group. A variety of different antigen types have been identified but those of the ABO and Rhesus systems are the most common.

ABO system

There are two main antigens in this system, A and B, and these give rise to four different blood groups

Table 2 Summary of the ABO blood group system. Inheritance of the A and B antigens is autosomal codominant so there are four possible combinations of red cell antigens. The plasma contains antibodies to any antigen not present on the red cells

Genotype	Antigens on cells	Blood group	Antibodies in plasma
OO	None	O	Anti-A and anti-B
OA or *AA*	A	A	Anti-B
OB or *BB*	B	B	Anti-A
AB	A and B	AB	None

as shown in Table 2. Of these, blood groups O and A are almost equally common and together account for over 85% of the population in Western Europe. It should be noted that plasma always contains preformed antibodies (IgM class) against A or B antigens which are not already present on our own erythrocytes, whether we have been sensitized by exposure to foreign red cells or not. This breaks the general rule in immune responses, since antibodies against all other foreign antigens are only secreted in appreciable amounts after exposure to that antigen (Section 2.5). It may be that we are all exposed to A and B antigens from another source, e.g., intestinal bacteria, and only become tolerant to the antigens also present on our erythrocytes. Whatever the mechanism, the consequence is that a major immune reaction can be expected on the first exposure to blood of the wrong ABO group.

Rhesus (Rh) factor

Blood is either Rh positive or Rh negative depending on whether red cells carry one of the Rh antigens or not. There are three main Rh antigens, C, D and E, but D is the most common. Inheritance is dominant so that the genotypes *Dd* and *DD* both result in D positive blood. Over 85% of Western Europeans are Rh positive.

A Rh-negative recipient will mount an immune response against Rh-positive blood but, unlike ABO agglutinins, there are normally no anti-Rh antibodies in plasma from a Rh-negative individual. Therefore, there is unlikely to be any Rh-dependent agglutination following an initial transfusion with Rh-positive blood. This exposure sensitizes the immune system to the Rh antigen, however, so that subsequent mismatched transfusions can lead to prompt agglutination and haemolysis of the donor cells. Rhesus sensitization can also occur when a Rh-negative mother gives birth to a Rh-positive baby. Fetal red cells, normally separated from the maternal circulation by the placenta, may enter the mother's blood during delivery as the placenta is sheared off the uterine wall. This stimulates production of anti-Rh antibodies by the mother and, if there is a subsequent Rh-positive pregnancy, these antibodies (IgG class) cross into the fetal circulation, leading to haemolysis. The resulting jaundice, anaemia and heart failure may threaten the baby's life. These problems of the 'Rhesus baby' have largely been eradicated by injecting all D-negative mothers with anti-D antibodies shortly after the birth of each baby. Any circulating D-positive cells become coated with exogenous antibody and are destroyed before they can stimulate the maternal immune system. This is an example of temporary passive immunity (conferred by the injected immunoglobulins) preventing the development of permanent active immunity. It should be noted that fetal/maternal ABO incompatibility is common but has no damaging consequences because the IgM class anti-A and anti-B antibodies are too large to cross the placenta.

Transfusions and cross matching

The terms *universal donor* (blood group O, Rh negative) and *universal recipient* (blood group AB, Rh positive) are sometimes used to indicate conditions in which transfusions may be attempted without knowing the blood group of both donor and recipient. The reasoning is that the red cells of the universal donor carry no antigens and so cannot be agglutinated, while the plasma of the universal recipient contains no antibodies and could not agglutinate donor cells, regardless of their group. Possible agglutination of recipient cells is regarded as unlikely, since agglutinins in the donor plasma will be diluted following transfusion. Use of O, Rh negative blood for a patient of unknown blood group is reserved for life-threatening emergencies, however, and donor and recipient blood groups should normally match. Indeed, the existence of a wide range of rarer erythrocyte antigens means that blood of the appropriate group should actually be tested against samples of the patient's cells and plasma before being transfused. Donor cells are mixed with recipient plasma, while recipient cells are mixed with donor plasma; there should be no agglutination in either case. This is referred to as *cross matching*.

2.4 Leucocytes

Learning objectives

At the end of this section you should be able to:

- state normal values for white cell count and percentages of each leucocyte type (the differential white cell count)
- explain how leucocytes are classified morphologically and outline the function of each type
- explain the term reticuloendothelial system
- outline how leucocytes are manufactured.

White blood cells, or leucocytes, are vitally important in the disposal of damaged and ageing tissue and in the immune responses which protect us from infections and cancer cell proliferation. The total blood white cell count is normally in the range 4×10^9 to $10 \times 10^9 L^{-1}$ ($4–10 \times 10^3 mm^{-3}$), but this may increase markedly during infection or inflammation.

Leucocyte types

Based on their histological appearances, five main types of leucocyte may be identified. These fall into two morphological groups. *Polymorphonuclear granulocytes* have irregular, multilobed nuclei and a high density of cytoplasmic granules (Fig. 23). Neutrophils, eosinophils and basophils all belong in this group. Lymphocytes and monocytes, by comparison, lack granules and have large, regular nuclei, and so they are classified as *mononuclear agranulocytes*. The basic characteristics of individual leucocyte types are listed below.

Neutrophils comprise 60–70% of circulating leucocytes. They are highly mobile and can engulf debris or foreign organisms through the process of *phagocytosis*, trapping the target in a vesicle which then fuses with a lysosome (Fig. 24). Organic material is digested by lysosomal enzymes, although inorganic material may remain within the cytoplasm indefinitely.

Eosinophils make up 1–4% of circulating leucocytes. They are phagocytic and are particularly involved in the destruction of parasitic worms but may also contribute to allergic responses.

Basophils generally account for under 0.5% of leucocytes. These phagocytes release histamine and heparin and are involved in allergic responses.

Lymphocytes are the only nonphagocytic white cells and represent 25–30% of blood leucocytes. They are central to specific immune defences within the body (Section 2.5) and can be subdivided into B and T lymphocytes.

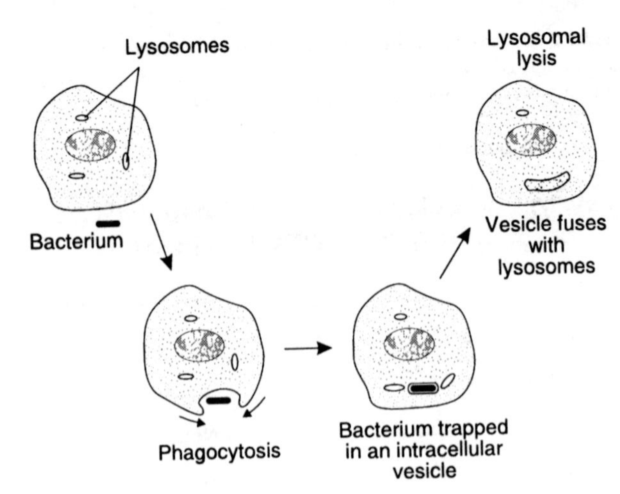

Fig. 24 Phagocytosis of a bacterium.

Monocytes constitute 2–5% of leucocytes. They have the greatest phagocytic potential of all body cells. *Tissue macrophages* are believed to be monocytes which have migrated into the connective tissues. The monocyte/macrophage system forms the core of the reticuloendothelial system.

Reticuloendothelial system

This phagocytic system is almost synonymous with the tissue macrophage system. Mobile macrophages rove freely through connective tissues, but many more remain relatively fixed in a given region. When activated, however, these will also become mobile, being chemotactically attracted to local sites of infection or damage. There are dense aggregations of macrophage-type cells within the reticular tissue of the lymph nodes, spleen and bone marrow. Alveolar macrophages in the lung and the phagocytic Küpffer cells in the liver, along with microglial cells in nervous tissue, are also regarded as part of the reticuloendothelial system, as are the blood monocytes.

Leucocyte production

Leucocytes originate from the pluripotential stem cells in the bone marrow, which divide and mature giving two separate leucocyte stem cell lines (Fig. 23).

The myeloid line. This gives rise to the three types of granulocyte as well as monocytes and macrophages. These cells all have important phagocytic roles. The myeloid stem cell line also produces large multinucleate *megakaryocytes* from which *platelets* are derived.

The lymphoid line. From this stem cell line the lymphocytes are produced (Fig. 23). B cells mature in the marrow before being distributed to the lymphoid tissues of the body, i.e. the lymph nodes, spleen, thymus and Peyer's patches in the intestinal submucosa. T lymphocyte precursor cells are believed to migrate initially to the thymus, where they mature fully before being redistributed to other lymphoid sites. Lymphocytes can replicate and develop further within the lymphoid tissues of the body and there is a continuous recirculation of lymphocytes from blood to lymph and back again.

2.5 Immune responses

Learning objectives

At the end of this section you should be able to:

- list and classify immune mechanisms as nonspecific or specific (innate or acquired)
- describe the mechanisms of nonspecific (innate) immunity, paying particular attention to the inflammatory response and complement functions
- describe the mechanisms of specific (acquired) immunity, differentiating clearly between antibody-mediated and cell-mediated responses
- explain with examples the differences between active and passive immunity.

The immune defences of the body are classically subdivided into those which are nonspecific and innate, and those which are specific and acquired. This is a useful division, but it should be remembered that there are many points of interaction between the two systems. For example, lymphocytes only produce specific antibodies against foreign molecules when these antigens are first processed by nonspecific phagocytic cells such as macrophages. At the same time, antibodies lead to antigen removal by amplifying pre-existing, nonspecific responses. These two elements of immunity are, therefore, highly interdependent.

Nonspecific or innate immunity

This depends on interrelated defence mechanisms which act against any foreign or abnormal cell, i.e., they are nonspecific. They are also said to be innate since they do not depend on previous exposure to a particular organism. Nonspecific immune mechanisms include physical barriers to infection, inflammation, complement activation and natural killer cell activity.

Mechanical and chemical barriers against infection

The epithelia covering skin and lining the gastrointestinal, genitourinary and respiratory tracts provide mechanical barriers which help exclude damaging organisms. This is aided by mucus secretion in the respiratory tract, which traps dirt and pathogens and allows them to be swept out by the action of epithelial cilia. Chemical secretions, such as acid in the stomach and vagina, also play a role and may defend against colonization of the epithelial surface, e.g., preventing vaginal yeast infections.

The inflammatory response and phagocytosis

Inflammation is a set of local cellular and vascular responses to tissue damage or infection which accelerates the destruction and phagocytic removal of invading organisms and debris.

Phagocytic responses in inflammation Tissue macrophages adjacent to a site of bacterial invasion, for example, become mobile and phagocytically active. Chemicals released from injured and infected cells (chemotaxins) act as attractants for these cells, directing their movements towards the damaged area in a process called *chemotaxis*. Local macrophages are soon reinforced by the migration of blood neutrophils and monocytes into the region (Fig. 25). These stick to the endothelium of capillaries in the affected area (*leucocyte margination*) and then invaginate themselves through the clefts between the endothelial cells using active amoeboid movements (*diapedesis*).

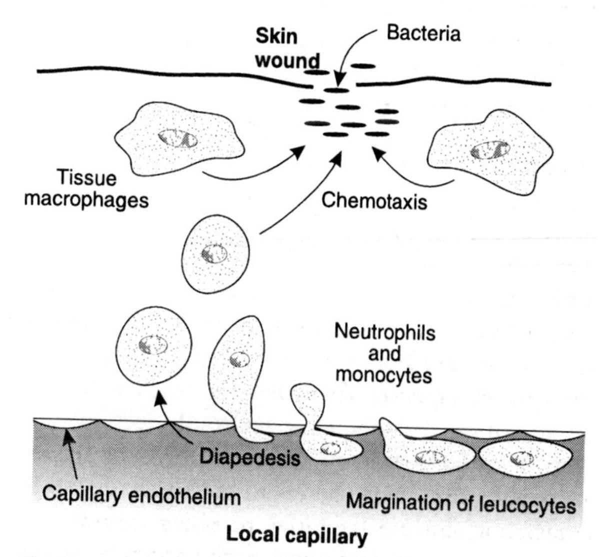

Fig. 25 Diapedesis of blood granulocytes and monocytes and chemotaxis of local macrophages increase tissue phagocyte density during inflammation.

The blood-borne phagocytes escape into the tissues where they are chemotactically attracted to assist in removal of infectious or toxic agents and tissue debris. The resulting congregation of large numbers of neutrophils and macrophages within a tissue is the histological hallmark of acute inflammation.

Vascular responses in inflammation Leucocyte aggregation in inflamed tissues is accelerated by increased blood flow caused by dilatation of local blood vessels. This transports more white cells into the region, while an increase in capillary permeability promotes diapedesis. These vascular changes are stimulated by the generation of a variety of vasodilator substances within damaged tissues.

Kinins. Activation of a cascade of proteolytic enzymes known as the *kinin–kallikrein* system produces vasoactive kinins, particularly bradykinin.

Vasodilators. Basophils and mast cells release the vasodilators bradykinin, 5-hydroxytryptamine (serotonin) and histamine, as well as the anticoagulant heparin.

Prostaglandins. Activated phagocytic cells also stimulate prostaglandin synthesis and this may amplify the mechanisms of inflammation. Drugs which inhibit prostaglandin production, such as aspirin and glucocorticoid steroid hormones, can be used as antiinflammatory agents.

Local and systemic effects of inflammation The inflammatory response leads to a number of characteristic effects at the site of injury or infection:

- Increased blood flow causes redness and raises the temperature locally.
- Increased capillary permeability allows fluid to leak into the tissues, causing local swelling (oedema). Plasma proteins, including clotting factors, also leak out so that a meshwork of fibrin clot is laid down in the interstitium. This provides a mechanical barrier to the spread of infection.
- The release of tissue-damage products stimulates nociceptors, causing pain (Section 7.4).

Acute inflammation (particularly if caused by infection) may also be associated with generalized (systemic) responses.

Body temperature. This may rise as a result of resetting of the hypothalamic set point (Section 1.1). Fever (*pyrexia*) seems to increase the activity of phagocytic cells and is promoted both by exogenous pyrogens, including bacterial products known as endotoxins, and by endogenous pyrogens released by phagocytes. Endogenous pyrogens stimulate prostaglandin synthesis and this is blocked by drugs such as aspirin. This probably explains their ability to reduce fever (their antipyretic action).

Blood white cell count. There is an increase in the blood white cell count (a *leucocytosis*), most of the increase being the result of extra neutrophils (a *neutrophilia*). This reflects both the rapid mobilization of neutrophils already present in the bone marrow and an increased rate of production in the marrow.

Activation of the complement system

Foreign cells, especially certain bacteria, carry surface molecules which activate the complement system of plasma proteins. These normally exist as a family of inactive precursors which are activated by proteolytic cleavage. The system is organized as a *cascade* so that each activated component activates the next in the sequence (Fig. 26). Thus, activating one component can lead to activation of the entire complement system. Nonspecific activation is referred to as activation by the *alternate pathway*, distinguishing it from the *classical pathway* to activation, which requires antibody.

Complement function Complement activation defends against infection both by killing cells directly and by promoting phagocytosis (Fig. 26).

Direct cell killing. The activated forms of the last five components in the system (C_5–C_9) combine to form a protein called the *membrane attack complex*. This becomes inserted in the plasma membrane of the invading cell, where it forms a large pore. When the density of such pores is high, cell lysis results.

Phagocytosis. Complement activation produces a series of biologically active complement fragments which increase the efficiency of phagocytosis. This involves several mechanisms:

- *Opsonization* makes target cells more susceptible to phagocytosis. Complement fragments, which act as opsonins, bind to the surface of bacteria. Phagocytic cells, which carry receptors for these fragments, become attached to them and this increases the efficiency of bacterial phagocytosis.
- *Chemotaxis* by complement fragments attracts more phagocytic cells to an infected or damaged region.
- *Vasodilatation* occurs and capillary permeability is increased following complement activation, thus amplifying the inflammatory response.

Natural killer cells

These are lymphocyte-like cells with granular cytoplasm. They can identify and destroy tumour cells or cells infected with viruses. The mechanism is not

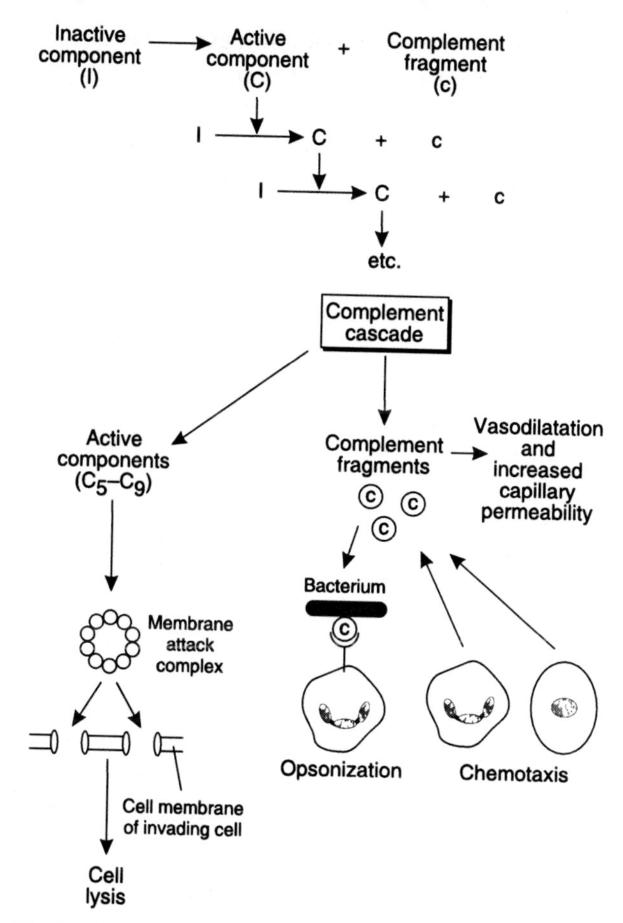

Fig. 26 Nonspecific activation of the complement cascade by the alternate pathway.

specific to any particular virus or tumour type and previous antigen exposure is not necessary.

Nonspecific regulators of immunity

There are a large number of molecules, other than specific antibodies, which are important in regulating immune defences. *Interferons* are proteins which can be released by any virus-infected cell and they inhibit viral replication in other cells. Activated leucocytes and macrophages also produce a variety of nonantibody proteins which participate in the regulation of immunity. Phagocytic cells (granulocytes, monocytes and macrophages) secrete *cytokines*, while lymphocytes produce *lymphokines*. These lymphokines can amplify nonspecific immune mechanisms whenever a lymphocyte is activated by an antigen. Any product of an immune cell (other than an antibody) which signals to other immune cells may also be termed an *interleukin*.

Specific or acquired immunity

Specific immunity refers to a number of mechanisms whereby susceptibility to infection by a particular organism is greatly reduced following initial exposure to it. This is clearly demonstrated with measles, mumps, rubella and other childhood infections, in which the immunity acquired following a first infection usually protects us if we are exposed to the same organism later in life. Indeed, clinical infection is not necessary; vaccination uses weakened or killed pathogens to stimulate immunity without producing illness. In both infection and vaccination the protection acquired is highly specific; having had a bout of measles does not reduce one's chances of contracting rubella or mumps, for example.

Specific immunity relies on the ability of the immune system to respond to foreign molecules, which are called *antigens*, and the defence mechanisms thus activated are targeted against the relevant antigen. Antigens are usually large (molecular weight >10 000), complex molecules. Proteins are particularly antigenic, although polysaccharides, lipoproteins and glycolipids can also act as antigens. Much smaller molecules, known as *haptens*, may also stimulate an immune response, but only if attached to a protein for initial presentation to the immune system. Following sensitization, however, subsequent responses can be stimulated by the hapten molecule alone. This mechanism accounts for allergic reactions to small organic molecules, including antibiotics like penicillin, and even to inorganic species such as nickel.

Specific immune responses rely on lymphocytes and may be mediated either by antibodies (humoral responses) or cells. Antibody-mediated immunity depends on B lymphocytes while T lymphocytes are responsible for cell-mediated immunity.

Antibody-mediated responses

These are particularly important in protection against bacterial infections, although they play a role in immunity against some viruses.

B lymphocyte activation This is the first step in antibody production and it occurs when foreign antigens come into contact with B lymphocytes. These cells carry surface immunoglobulins, or antibodies, which act as antigen-specific membrane receptors. When these bind to the relevant antigen, the antigen-receptor complex is endocytosed by the lymphocyte, which is then activated. Each B cell will only bind with one type of antigen and this leads to production of circulating antibodies with the same specificity. It is important to appreciate that although B lymphocytes specific for all possible antigens appear to be present in the body, they do not normally secrete antibody. Specific antibody only appears in body fluids once the relevant B cell has been activated by antigen binding, after which it multiplies and the daughter cells become transformed into *plasma* cells and *memory* cells (Fig. 27). Plasma cells act as factories for the relevant antibody, which they secrete in large amounts. Memory cells are identical with the parent lymphocyte and they greatly expand the reserve of cells capable of being activated during subsequent antigen exposure. Activated B lymphocytes also carry processed antigen in association with normal surface antigens known as histocompatibility antigens on their cell membrane. This allows them to act as antigen presenting cells which can in turn activate T lymphocytes (see below).

Primary and secondary responses On first exposure to an antigen, the immune response, as indicated by the plasma antibody concentration (or antibody titre), takes some weeks to develop fully (Fig. 28). This is referred to as the primary response and, although it can limit the duration of an infection, it is generally too slow to prevent it altogether. Antibody levels fall again as time passes, but a second exposure to the same antigen produces a much more rapid antibody response which peaks at a higher concentration and declines more slowly. This secondary response depends on the memory cells formed following initial B lymphocyte activation. These provide a reservoir of lymphocytes which can be rapidly activated during future exposure to the relevant antigen. This amplified antibody response may overwhelm a potential pathogen before it can cause the symptoms of infection, i.e., it provides acquired, antibody-mediated immunity.

Antibody structure Antibodies belong to the class of proteins known as gamma globulins or *immunoglobulins* (Ig). The basic monomer unit consists of two heavy and two light peptide chains linked together by disulphide bridges to produce a Y-shaped structure (Fig. 29). Each monomer has two antigen-binding sites, the specificity of which is determined by their amino acid sequence. This region is often referred to as the variable region of the antibody, since the seemingly limitless range of antigen specificities is produced by equally numerous variations in amino acid sequence. The make-up of the rest of the antibody, however, does not change with changing antigen specificity. Differences in the structure of this constant region allow immunoglobulins to be divided into five different antibody classes (see below). Enzyme digestion can split antibodies into two fragments, an antigen-binding fragment (F_{ab}) and a crystallizable fragment (F_c) (Fig. 29). Variations in the F_c region account for the different properties of different antibody types:

- IgG is the most abundant Ig in blood and is produced in large quantities during the secondary response. It promotes phagocytosis and cell lysis.
- IgM molecules are pentamers constructed from five basic monomer units (Fig. 29B) and are secreted early on in antibody responses. IgM promotes agglutination, phagocytosis and cell lysis.
- IgD is a surface antibody on B lymphocytes but its role is unclear.

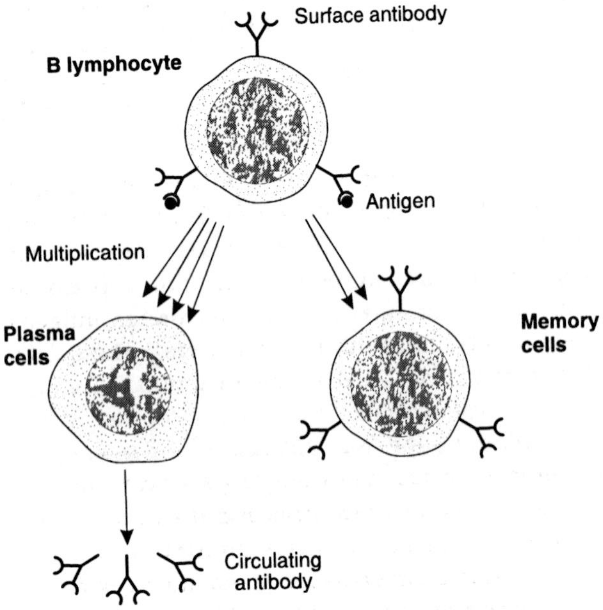

Fig. 27 B lymphocyte activation by antigen binding leads to formation of antibody-secreting plasma cells and antigen-sensitive memory cells.

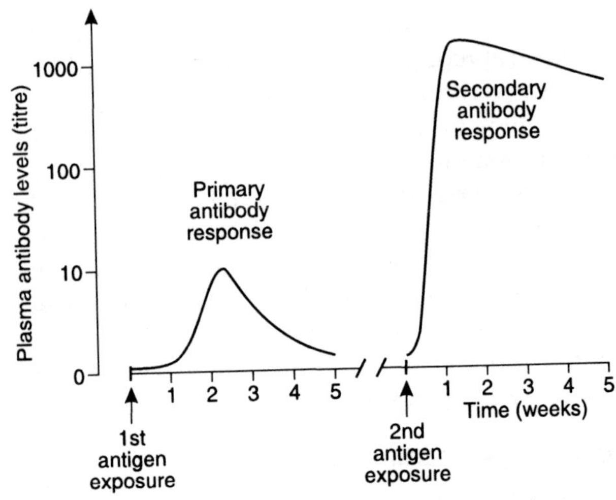

Fig. 28 Plasma antibody levels following first and second exposures to a given antigen.

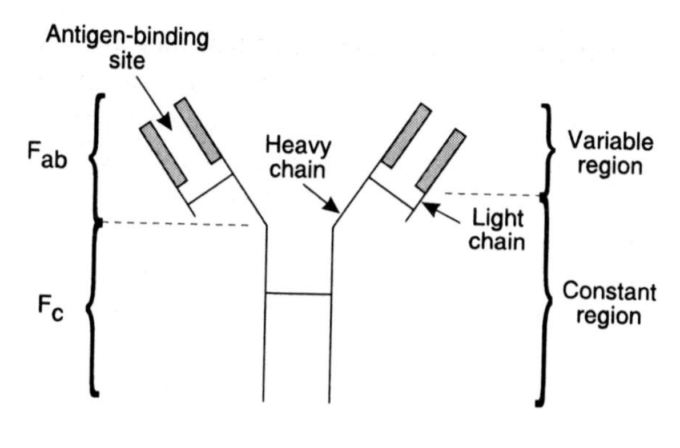

A Antibody monomer (≡ Y)

B Antibody pentamer (IgM)

C Antibody dimer (IgA)

Fig. 29 Immunoglobulin structure.

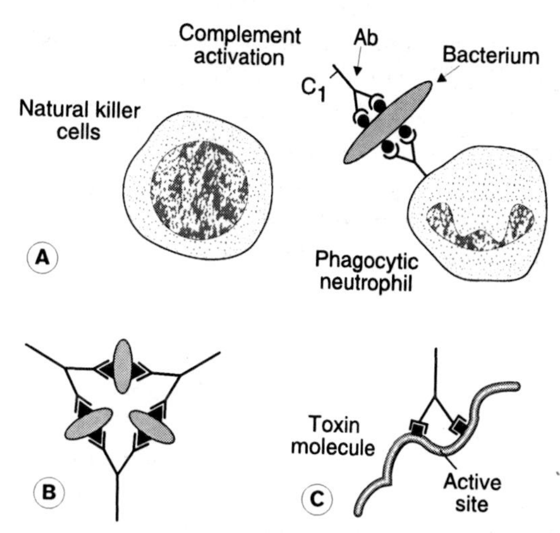

Fig. 30 Mechanisms of antibody (Ab) function. (A) Targeting of nonspecific immunity. (B) Agglutination reactions. (C) Neutralization of toxins.

- IgA molecules are dimers formed from two monomer units (Fig. 29C). IgA is present in secretions including saliva, tears and breast milk.
- IgE is important in parasitic infections and some allergic responses. It stimulates release of bradykinin, histamine, 5-hydroxytryptamine and other inflammatory mediators from mast cells and basophils.

Antibody function Antibodies bind to antigens, thus promoting their destruction through a range of different mechanisms:

1. Targeting and amplification of nonspecific immunity is the major strategy employed by antibody-mediated immunity to provide protection against specific pathogens. IgG and IgM class antibodies are particularly important in promoting bacterial lysis and phagocytosis in this way (Fig. 30A).
   - Antibodies activate the complement system by the classical pathway. The F_c region of antigen-bound antibodies can bind and cleave the C_1 complement component. This leads to activation of the entire complement cascade with the effects described above.
   - Antibodies can act as opsonins, using receptors for the F_c region to bind phagocytes to antigens.
   - Antibody-coated cells are more likely to be attacked by natural killer cells.
2. Agglutination refers to the clumping together of bacteria or foreign cells into large-scale lattices held together by antibody linkages (Fig. 30B). This is possible because all antibodies have more than one antigen-binding site and so can interconnect target cells, physically hindering the spread of infectious agents and increasing the likelihood of phagocytosis. IgM antibodies are the most effective because their pentameric structure has 10 binding sites per molecule (Fig. 29). Agglutination reactions are used in blood group testing (Section 2.3).
3. Neutralization of toxins and inactivation of some viruses may result from immunoglobulins cloaking biologically active sites (Fig. 30C). This is how antisera to snake and insect bites work.

Cell-mediated immunity

This relies on T lymphocytes, which do not manufacture circulating antibodies. It is particularly important in combating viral and fungal infections as well as in immune responses against potential cancer cells.

T lymphocyte activation T cells are activated by exposure to a foreign antigen which is identified and bound by specific surface receptors. This recognition step only occurs, however, if the antigen is closely associated with other, normal, cell surface antigens, known as histocompatibility antigens (Fig. 31). This can come about in one of two main ways:

1. Cytotoxic T lymphocytes can be activated directly by body cells carrying abnormal antigens on their surface. This occurs during viral illnesses, since foreign viral antigens are expressed on the plasma membrane of any infected cells. Cancer cells also carry abnormal surface antigens, which can lead to T lymphocyte activation, and this may prevent the

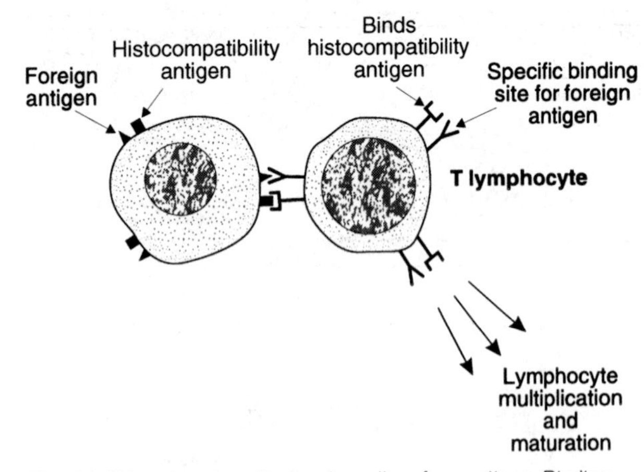

Fig. 31 T lymphocyte activation by cell surface antigen. Binding requires the presence of both foreign and normal histocompatibility antigens. The activating cell may be any body cell carrying an abnormal antigen or a specialized antigen presenting cell which has phagocytosed antigenic material.

development of clinical disease. Cell-mediated immunity is also important in organ transplant rejection when the donor antigens do not exactly match those of the recipient. As their name implies, cytotoxic T cells can kill the abnormal cells directly.

2. Helper T lymphocytes can be activated by exogenous antigens, such as those derived from bacteria and environmental allergens. These antigens must first be processed by antigen presenting cells (sometimes referred to as dendritic cells), however, which engulf the foreign agent and then break it down using lysosomal enzymes. Individual antigens can then be presented on the surface of the antigen presenting cell in association with the normal histocompatibility antigens those cells carry. Macrophages and B lymphocytes both act as antigen presenting cells.

T lymphocyte function As with B lymphocytes, T cell activation leads to multiplication and maturation. There are three main types of T lymphocyte, however, each with a different function.

Cytotoxic T lymphocytes can lyse cells carrying the antigen to which they are sensitive and are particularly important in the immune response to viruses and cancer cells.

Helper T lymphocytes are vital in both antibody- and cell-mediated immunity, but they have no direct effects on foreign antigens or cells. They are activated by antigen presenting cells such as macrophages or activated B lymphocytes, rather than cell surface antigens on infected or mutated body cells as is the case for other T lymphocytes. Once stimulated by the appropriate antigen, they release a number of lymphokines, which stimulate other immune cells. This increases macrophage activity as well as promoting multiplication and activation of T and B lymphocytes. Helper T lymphocytes are thus vital to all immune responses.

Suppressor T lymphocytes inhibit lymphocyte function. This may provide a mechanism for the control of the immune response which reduces the risk of immune damage to normal body cells.

Cell-mediated responses persist longer than antibody responses since activated T lymphocytes survive longer than plasma cells. T lymphocyte immunity also demonstrates memory, with more rapid and vigorous responses to a given antigen on repeated exposure.

Active and passive immunity

Activation of lymphocytes provides active defence against infection. If antibodies are injected or absorbed into the body these will also protect us from a given antigen. This is called passive immunity but, unlike active immunity, it only provides short term protection (a few months) since there is no production of memory cells. A physiological example of passive immunity is the protection a newborn baby gains from the maternal immunoglobulins in its body (IgG crosses the placenta and IgA is present in breast milk). As these antibodies are degraded over the first 2–3 months of life, the level of passive protection wanes, but by this time the infant's immune system is mature enough to mount active responses. In adult life, injections of immunoglobulin may be used to give short term passive protection, e.g., against some forms of viral hepatitis.

Box 5 Clinical note: Immunosuppression and organ transplantation

One of the major barriers to successful organ transplantation is the difficulty in matching donor organ antigens to those of the recipient. An exact match is rarely possible because of the shortage of donor organs and the wide range of tissue antigens, and so there is always a risk that an immune response will be mounted against the transplanted organ (graft rejection). Immunosuppressant drugs are used to reduce this risk, but these inhibit immune responses in a nonspecific fashion and leave the patient susceptible to infections. Nevertheless, use of well-matched organs and careful management of drug doses can give excellent graft survival rates.

2.6 Platelets and haemostasis

Learning objectives

At the end of this section you should be able to:
- state the normal platelet count
- describe the mechanisms of haemostasis
- explain how the clotting cascade works, differentiate intrinsic from extrinsic activation and name the factors involved in the last three (common) steps
- outline the mechanisms of clot removal.

Platelets (or thrombocytes) are blood-borne cell fragments which contain organelles and enzymes but no nuclear material. They bud off from *megakaryocytes*, giant multinucleate bone marrow cells derived from the myeloid stem cell line (Fig. 23). The normal platelet count in blood is 150×10^9 to $300 \times 10^9 L^{-1}$ ($150–300 \times 10^3 mm^{-3}$). The role of platelets is to help prevent or stop bleeding, a process called haemostasis. Haemostasis involves vascular spasm, platelet plug formation and clotting (coagulation).

Vascular spasm

Damage to the wall of a blood vessel leads to rapid smooth muscle contraction. This may help to narrow the vascular defect and reduce blood flow. Vascular spasm, however, can only reduce or arrest blood loss in the short term, as the vessels eventually relax again. Other haemostatic mechanisms are required if bleeding is to be arrested permanently.

Platelet plugging

Platelets normally circulate freely, but when they are exposed to tissue collagen following vascular damage they become adherent, sticking to the vessel wall and to each other (Fig. 32). At the same time, they release a number of chemicals, including *adenosine diphosphate (ADP)* and a product of arachidonic acid metabolism called *thromboxane A_2*. These increase platelet stickiness, resulting in positive feedback favouring the formation of an ever larger platelet plug, which can stop bleeding by occluding the opening. The bleeding time (the time taken for visible blood loss from a small puncture wound to stop) is usually less than 5 minutes, but this may increase considerably if platelet function is defective. Platelet adherence to undamaged blood vessels is normally inhibited by *prostacyclin* (another arachidonic acid derivative) released from the endothelium.

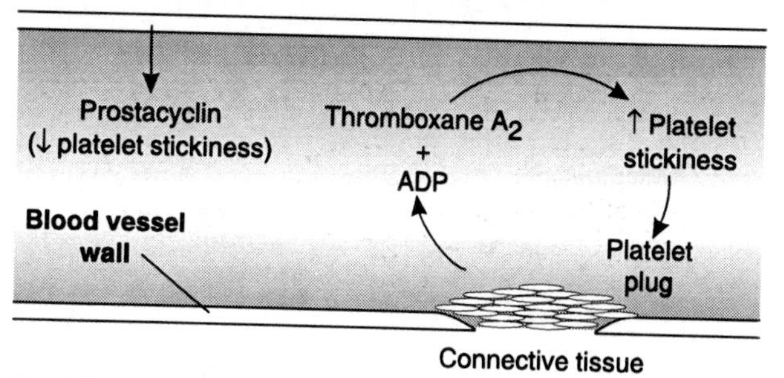

Fig. 32 Platelets form plugs in regions of vascular damage.

Clotting or coagulation

Platelet plugs are insecure, temporary structures but they are quickly anchored in place by the formation of a fibrin clot. Clotting, or coagulation, depends on a family of plasma proteins known as *clotting factors*, several of which are manufactured in the liver in vitamin K-dependent reactions. They are normally present as inactive proenzymes but, once activated, each factor activates the next by proteolytic cleavage. This continues in a sequential fashion down the coagulation cascade and finally results in the conversion of prothrombin (factor II) to active *thrombin*. Thrombin catalyses first the conversion of fibrinogen (factor I) to *fibrin* monomers, and then their polymerization into cross-linked fibrin strands. The resulting clot, which traps erythrocytes and other blood cells, bridges the opening in the blood vessel wall and secures the platelets in position. After some time, clot retraction pulls the damaged edges of the blood vessel together, squeezing out clear serum (serum is the fluid left after plasma has clotted).

Intrinsic and extrinsic clotting pathways

Clotting can be activated through two different mechanisms, referred to as the intrinsic and extrinsic mechanisms or pathways (Fig. 33). Whenever blood is exposed to an abnormal surface, such as collagen in the wall of a torn blood vessel, clotting occurs by the intrinsic pathway. Aggregated platelets play an important role, releasing a phospholipid cofactor called platelet factor 3. The extrinsic pathway is initiated when plasma is mixed with specific products of tissue damage, collectively known as tissue thromboplastin. This bypasses some of the initial steps in the intrinsic pathway. The two pathways converge, however, at the activation of factor X, the step prior to thrombin formation. Several steps in both

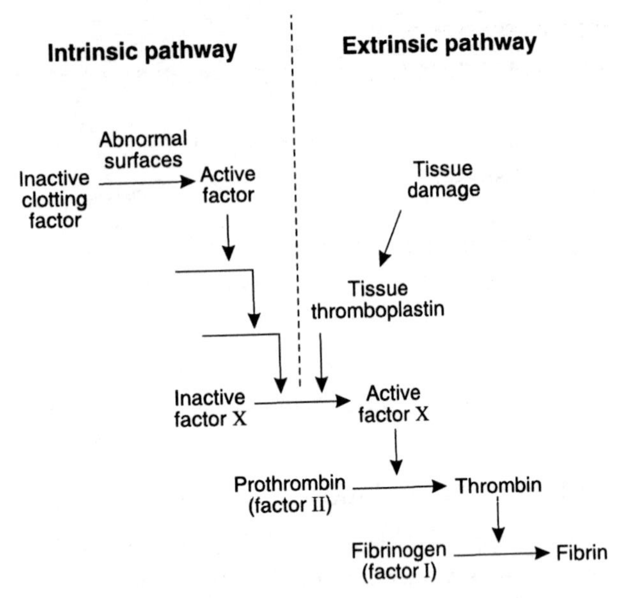

Fig. 33 Activation of the clotting cascade by the intrinsic or extrinsic pathways. Both pathways require Ca^{2+} and the intrinsic pathway is accelerated by platelet factor 3 (PF_3).

pathways require Ca^{2+} which explains why Ca^{2+}-chelating agents such as citrate are effective anticoagulants.

Fibrinolysis and clot removal

Blood clotting cannot be allowed to continue uninhibited or blood vessels would soon become permanently blocked. A number of natural substances act as anticoagulants, inhibiting clot formation in normal vessels. There is also a mechanism for removing clots after damage to a blood vessel has been repaired. This depends on the proteolytic enzyme *plasmin*, which is formed from the inactive plasma protein plasminogen following activation of the coagulation cascade. Plasmin catalyses the breakdown of fibrin, a process called *fibrinolysis*, and phagocytic cells then remove the clot debris.

Box 6 Clinical note: Clotting defects

Deficiencies of clotting factors can lead to severe bleeding problems. This may arise because of liver failure or vitamin K deficiency, both of which inhibit synthesis of several clotting factors. There may also be an inherited deficiency of nearly any clotting factor. The most common is ***haemophilia***, the classic form of which results from factor VIII deficiency caused by an X chromosome-linked genetic defect. Affected individuals are nearly always male, while the condition is inherited through carrier females. Haemophiliacs bleed spontaneously from a variety of sites and this may require treatment with concentrated extracts of factor VIII from normal plasma.

Self-assessment: questions

Multiple true/false questions

Each of the following statements consists of a stem followed a number of possible endings. State whether each statement is True or False. For each stem, all, several or none of the statements may be true.

1. Plasma:
   a. proteins account for at least 70% of total plasma osmolality
   b. volume is proportional to 1/haematocrit at constant blood volume
   c. staining with haemoglobin may result from haemolysis
   d. contains inactive hormone precursors
   e. proteins influence the erythrocyte sedimentation rate

2. Erythrocyte formation:
   a. takes place mainly in the marrow of long bones during adult life
   b. may be stimulated by a reduction in arterial O_2 content
   c. may be reduced in chronic renal failure
   d. normally produces new erythrocytes at a rate closer to 2×10^{11} day^{-1} than 2×10^{10} day^{-1} in a 70 kg adult male
   e. may slow down following gastrectomy

3. ABO blood group status:
   a. is defined by the antibodies in our plasma
   b. is most commonly group AB in Western Europeans
   c. may be different for a mother and her baby
   d. is tested using erythrocyte coagulation reactions
   e. is autosomally inherited

4. Circulating leucocytes:
   a. are all derived from pluripotential stem cells in bone marrow
   b. chiefly consist of lymphocytes
   c. may leave the circulation at sites of inflammation
   d. are derived from two separate committed cell lines in bone marrow
   e. known as monocytes form part of the reticuloendothelial system

5. Complement activation:
   a. involves proteolytic cleavage of inactive proenzymes
   b. may be initiated through the F_c portion of an antibody
   c. is an important element of both nonspecific and specific immunity
   d. can occur by intrinsic and extrinsic pathways
   e. promotes phagocytosis through formation of the membrane attack complex

6. B lymphocytes:
   a. secrete circulating antibodies
   b. carry antigen receptors on their surface
   c. require helper T lymphocytes during activation
   d. secrete cytokines
   e. mature fully in bone marrow

7. Immunoglobulins:
   a. of the IgM class have 10 antigen-binding sites per molecule
   b. of the IgA class can cross the placenta
   c. of the IgE class can stimulate degranulation of mast cells and basophils
   d. of the IgG class are dominant in secondary antibody responses
   e. consist of light and heavy peptide chains linked by hydrogen bonds
   f. of the IgM class act as anti-ABO agglutinins

8. Blood clotting:
   a. requires Ca^{2+}
   b. is promoted by platelet plugging
   c. defects usually prolong the bleeding time
   d. is initiated by tissue thromboplastin in the intrinsic coagulation pathway
   e. stimulates fibrinolysis

Single best answer questions

For each of the following questions choose the single best answer.

1. A sample of blood is repeatedly ejected from a syringe through a small bore needle. The haematocrit for this sample is reduced from 0.47 to 0.35 after this treatment. What percentage of the red cells have been mechanically destroyed during this procedure?
   - **a.** 12%
   - **b.** 45%
   - **c.** 26%
   - **d.** 0.12%
   - **e.** 0.26%

2. At physiological pH, plasma proteins:
   - **a.** are anions
   - **b.** move towards the cathode during electrophoresis
   - **c.** are all globulins
   - **d.** are only found in the vascular space
   - **e.** cannot bind to Ca^{2+} ions

3. With regard to blood groups, a Rhesus –ve baby:
   - **a.** must have had parents both of whom were also Rhesus –ve
   - **b.** will have anti-Rhesus antibodies in its circulation at birth
   - **c.** may be affected in utero by the presence of Rhesus antigen in the maternal circulation
   - **d.** poses a risk to the well-being of the mother if she is Rhesus +ve
   - **e.** may be sensitized against the Rhesus antigen by exposure to Rhesus +ve blood

4. The presence of polymorphonuclear granulocytes in peripheral tissues:
   - **a.** results from specific, acquired immune responses
   - **b.** is promoted by local vasoconstriction
   - **c.** results in high tissue levels of antibody
   - **d.** is a hallmark of acute inflammation
   - **e.** inhibits chemotaxis

5. T lymphocyte activation:
   - **a.** always depends on the function of antigen presenting cells
   - **b.** always depends on the presence of appropriate histocompatibility antigens
   - **c.** is only important in cell-mediated immunity, and not in antibody-mediated immunity
   - **d.** does not occur on first exposure to an antigen
   - **e.** leads to short term responses lasting days rather than weeks

6. With regard to blood transfusion:
   - **a.** people of blood group AB, Rhesus +ve are referred to as 'universal donors' because they carry no antibodies against relevant red cell antigens in their blood
   - **b.** people of blood group AB, Rhesus +ve are referred to as 'universal donors' because they carry all the relevant blood group antigens on their red cells
   - **c.** people of blood group O, Rhesus –ve are referred to as universal recipients because they have neither blood group antigens nor antibodies in their blood
   - **d.** people of blood group AB, Rhesus +ve are referred to as 'universal recipients' because they carry all the relevant blood group antigens on their red cells
   - **e.** people of blood group AB, Rhesus +ve are referred to as 'universal recipients' because they carry no antibodies against relevant red cell antigens in their blood

Matching item questions

Theme: Leucocytes

Options

A. Neutrophils
B. Eosinophils
C. Basophils
D. B lymphocytes
E. T lymphocytes
F. Monocytes
G. Plasma cells
H. Memory cells
I. Killer cells

For each of the descriptions below choose the most appropriate option from the list above. Each option may be used once, more than once or not at all.

1. These cells promote the rapid secondary immune response.
2. These are highly phagocytic agranular cells.
3. These cells secrete antibody into the circulation.

4. These cells contribute to nonspecific immunity by destroying abnormal body cells.
5. These cells mature within the thymus.
6. These cells promote inflammation by releasing vasodilator substances.

Theme: Haemostasis

Options

A. Platelets
B. Vascular spasm
C. Prostacyclin
D. Thromboxane A_2
E. Ca^{2+}
F. Factor X
G. Factor II
H. Fibrinogen
I. Plasmin

For each of the descriptions below choose the most appropriate option from the list above. Each option may be used once, more than once or not at all.

1. An anti-thrombotic product of arachadonic acid metabolism.
2. Is chelated by EDTA, preventing coagulation.
3. Might be used as a 'clot-busting' agent to remove thrombi within coronary arteries.
4. Is activated by thrombin in the coagulation cascade.
5. Promotes platelet plugging.
6. Is also known as prothrombin in its inactive form.

Short notes

Write short notes on the following:

a. plasma composition
b. red cell destruction
c. phagocytic cells
d. chemotaxins

Modified essay

Many blood vessels are relatively fragile structures, and so there is a constant risk of blood loss through damage to vascular tissue, even in the absence of trauma.

Questions

1. Which mechanisms normally prevent or limit bleeding and what stimuli activate them?
2. What role does positive feedback play in haemostasis?

Two simple tests of haemostatic function are the bleeding time (time taken for bleeding to stop, normally 1–5 minutes) and the clotting time (time taken for a blood sample to clot in a glass tube, normally 4–8 minutes).

Questions

3. Is the intrinsic or extrinsic coagulation mechanism activated when blood clots in a glass tube?
4. If the clotting time is prolonged regardless of whether the intrinsic or extrinsic mechanism is activated, what does this suggest about the site of the defect in the coagulation pathways?
5. If an individual has an abnormally long bleeding time but a normal clotting time, what is the likely cause of the problem?

The risk of blood loss from the circulation is paralleled by a risk of blockade of the blood vessels by unwanted clots. When a clot develops within an intact vessel, obstructing flow, this is referred to as thrombosis. This is a major cause of ischaemic heart disease and cerebrovascular problems (e.g., strokes). Clots may detach from their site of origin and circulate with the blood to become lodged in some distant blood vessel, a process called embolism. Various mechanisms help prevent these unwanted events.

Questions

6. What mechanisms help prevent blood clots forming in normal vessels? What vascular abnormalities might promote thrombosis?
7. What mechanisms promote removal of unwanted clot?

Data interpretation

The following results were obtained on analysis of a blood sample from an anaemic patient:

Haemoglobin concentration:	$8.3\,g\,dl^{-1}$
Red cell count:	$2.2 \times 10^{12} L^{-1}$
Haematocrit (PCV):	0.25

Questions

1. Comment on these results.
2. Use the values above (the primary indices) to calculate the following values: mean cell volume, mean cell haemoglobin, mean cell haemoglobin concentration (the secondary indices). Give the appropriate units in each case.
3. How would you classify this anaemia?
4. Suggest a deficiency state which may have caused these abnormalities?

Clinical scenario

A young male adult is admitted to hospital. He complains that he feels weak and tires easily. On questioning, it becomes clear that he has been bleeding into his gastrointestinal tract, probably from a gastric ulcer. The patient is also found to have a mild fever (38.6°C).

The medical house officer sends a sample of the patient's blood to the haematology laboratory in a bottle containing a Ca^{2+}-chelating agent. The requested investigations include measurement of the haemoglobin concentration and a blood cell count. The results are given in Table 3.

Questions

1. What is the purpose of the Ca^{2+}-chelating agent in the sample bottle?
2. How might the patient's symptoms of weakness and tiredness be explained on the basis of the haematology results?
3. Comment on the reticulocyte count in the light of the other blood results.
4. Which of the blood results is most consistent with the patient's fever and what may be the cause of both changes?

It is decided that blood transfusion may be necessary and so a further sample of blood is sent for grouping and cross matching. The patient's red cells and plasma are separated by centrifugation and the red cells tested for agglutination in the presence of different agglutinins (antibodies) known to be specific for the A or B antigens. The results are shown in Table 4.

Questions

5. What is the patient's ABO blood group? Which of the ABO agglutinins (antibodies) would you expect to find in his plasma?
6. Which observations in the agglutination table act as positive controls (i.e., demonstrate that a positive result is obtained when expected) and which provide negative controls (i.e., demonstrate that a positive result is not obtained when not expected)?

Viva questions

1. How is erythrocyte production controlled?
2. What different types of immune mechanisms do you know?
3. What is meant by the term 'bleeding time'? Why might the bleeding time be prolonged?

Table 3 Haematological analysis for a patient's blood sample

Variable	Measured value	Normal value
Haemoglobin	$9.6\,g\,dl^{-1}$	$14–16\,g\,dl^{-1}$
Red cell count	$3.3 \times 10^{12}\,L^{-1}$	$4–6 \times 10^{12}\,L^{-1}$
Reticulocytes	9%	0–2%
White cell count	$15.6 \times 10^{9}\,L^{-1}$	$4–10 \times 10^{9}\,L^{-1}$
Platelet count	$190 \times 10^{9}\,L^{-1}$	$150–400 \times 10^{9}\,L^{-1}$

Table 4 Results from ABO blood group testing

Red cell group	Saline	Anti-A antibody	Anti-B antibody
O	–	–	–
A	–	++	–
B	–	–	++
Unknown	–	++	++ (patient's cells)

++, agglutination

Self-assessment: answers

Multiple true/false answers

1. a. **False.** Total osmolality is dictated by dissolved ions. Plasma proteins are responsible for the colloid osmotic pressure at the capillary wall, however.
   b. **False.** Plasma volume is proportional to [1 – haematocrit] at constant blood volume.
   c. **True.** Haemoglobin is released from lysed erythrocytes.
   d. **True.** For example, angiotensinogen.
   e. **True.** By promoting rouleaux formation.

2. a. **False.** Although the marrow of long bones contributes to erythropoiesis in children, membranous bones are almost exclusively responsible in adults.
   b. **True.** This stimulates renal erythropoietin.
   c. **True.** Because of a lack of erythropoietin.
   d. **True.** You can estimate this from other values in the text: blood volume = $70\,ml\,kg^{-1}$, so total blood volume = 4.9 L in 70 kg individual; erythrocyte count = $5 \times 10^{12}\,L^{-1}$, so total number of erythrocytes = 24.5×10^{12}. Erythrocyte life span = 120 days, so replacement needs = $24.5 \times 10^{12}/120 = 2.0 \times 10^{11}$ cells day^{-1}.
   e. **True.** Because of a lack of intrinsic factor and consequent vitamin B_{12} deficiency. Anaemia may take months to appear because of liver stores of B_{12}.

3. a. **False.** Defined by the antigens on erythrocytes. There is, however, a one-to-one correspondence between ABO blood group and plasma antibodies; see Table 2.
   b. **False.** AB is rarest; A and O are most common.
   c. **True.** For example, a group O mother and group A father may have a baby with either group O or group A.
   d. **False.** Agglutination reactions.
   e. **True.**

4. a. **True.** As are erythrocytes and platelets.
   b. **False.** Neutrophils normally account for two-thirds of all leucocytes.
   c. **True.** This process is called diapedesis.
   d. **True.** The myeloid and lymphoid lines (Fig. 23).
   e. **True.**

5. a. **True.** Each complement component is activated in this way.
   b. **True.** Important in antibody-mediated immunity.
   c. **True.**
   d. **False.** Classical and alternate pathways.
   e. **False.** Membrane attack complex lyses cells directly. It is complement fragments, cleaved during activation, which act as opsonins and enhance phagocytosis.

6. a. **False.** Plasma cells produced by activated B lymphocytes secrete antibody.
   b. **True.** These have the same antigen specificity as the antibody secreted after lymphocyte activation.
   c. **True.** B lymphocyte activation does not take place in the absence of helper T lymphocyte activation.
   d. **False.** Cytokines are secreted by granulocytes and monocytes; lymphocytes secrete lymphokines.
   e. **True.** Unlike T lymphocytes which complete their development in the thymus.

7. a. **True.** IgM has five monomer units per molecule with two binding sites per monomer.
   b. **False.** Only IgG can cross the placenta; IgA is present in secretions and breast milk.
   c. **True.** Releases vasoactive substances like bradykinin and histamine.
   d. **True.**
   e. **False.** The peptide chains are linked by disulphide bridges.
   f. **True.** The strong agglutinating action of IgM reflects its multiple binding site structure.

8. a. **True.** In both intrinsic and extrinsic pathways.
   b. **True.** Because of platelet release of a phospholipid called platelet factor 3.

c. **False.** Bleeding time is more dependent on platelet function.
d. **False.** Tissue thromboplastin initiates the extrinsic pathway.
e. **True.** Through conversion of plasminogen to plasmin.

Single best answers

1. c. 26%. This is calculated by expressing the reduction in haematocrit as a % of the original haematocrit, since the haematocrit gives us a measure of the red cell volume.
2. a. The negative charge promotes binding to cations such as H^+ and Ca^{2+}. Although capillaries are relatively impermeable to proteins, some do diffuse into the interstitial fluid.
3. e. A Rhesus –ve baby may have parents one or both of whom may be heterozygous for the relevant gene. Since the Rhesus gene is autosomal dominant, they will still be Rhesus +ve. Rhesus –ve individuals only produce measurable levels of antibody following sensitization by exposure to the Rhesus antigen.
4. d. The presence of a large number of granulocytes, usually mainly neutrophils, is a histopathological hallmark of acute inflammation in that tissue. This is promoted and targeted by specific immune responses but does not require them, as it is part of the nonspecific, or innate immune response. The movement of granulocytes into the tissues is promoted by vasodilatation and by the presence of chemotaxic agents in the damaged/infected tissue.
5. b. The T lymphocytes can only recognize relevant foreign antigens when they are presented in the context of normal histocompatibility antigens, either on the surface of the damaged cell itself (cytotoxic T lymphocytes) or by an antigen presenting. Helper and suppressor T lymphocytes help modulate antibody-mediated, as well as cell-mediated responses.
6. e. AB, +ve is the 'universal recipient' because such individuals do not produce antibodies against antigens A, B or the Rhesus antigen, all of which they carry on their own red cells. Answer e. is favoured over d. as the pathological consequences of a mismatched transfusion result from the recipient's antibodies binding to the donor antigens, triggering a generalized immune response in the donor.

Matching item answers

Theme: Leucocytes

1. H. The production of large numbers of memory cells following the first exposure to an antigen leads to a rapid and amplified response on subsequent exposure.
2. F. Monocytes form part of the reticuloendothelial system.
3. G. Plasma cells actively secrete antibody; B lymphocytes, from which they are derived, do not.
4. I. Killer cells are nonspecific in the sense that they do not respond to any particular antigen, but destroy any cell with abnormal antigens.
5. E. The T in T lymphocyte stands for thymus.
6. C. Vasodilators released include bradykinin, histamine and 5-hydroxytryptamine.

Theme: Haemostasis

1. C.
2. E. This is the basis of some anticoagulants used to prevent laboratory blood samples from clotting.
3. I.
4. H. Fibrinogen is activated by thrombin (active factor II) to form fibrin, the molecular skeleton of a clot.
5. D. This is another arachadonic acid metabolite.
6. G.

Short note answers

a. Plasma consists of an aqueous solution of electrolytes, proteins and nutrients (glucose, amino acids, lipids and fatty acids). The dominant cation is Na^+ (140 mmol L^{-1}), with smaller concentrations of K^+ (4.5 mmol L^{-1}) and Ca^{2+} (2.5 mmol L^{-1}). The main anions are Cl^- (105 mmol L^{-1}) and HCO_3^- (25 mmol L^{-1}).

Proteins include albumin (osmotically active and transport functions), globulins (transport, signalling molecules and immunoglobulins) and fibrinogen (clotting).

b. Red cell destruction mainly occurs within the spleen and becomes more likely as a cell ages, mean life span being 120 days. Haemoglobin is degraded in reticuloendothelial cells, releasing iron and globin molecules, which are broken into amino acids for future protein synthesis. The porphyrin ring forms bilirubin, which is excreted in bile.

c. Phagocytic cells engulf foreign material and debris and destroy them using lysosomal enzymes. They are mobile and can be attracted to sites of inflammation by chemotaxic agents. Phagocytic cells include monocytes and all three types of polymorphonuclear granulocyte (neutrophils, basophils and eosinophils) in blood, as well as tissue macrophages, which are believed to be monocytes which have escaped from the circulation. There are also phagocytic Küpffer cells in the liver and microglial cells in the brain. These, along with monocytes and macrophages (especially in the lymph nodes, spleen and bone marrow), constitute the phagocytic reticuloendothelial system.

d. Chemotaxins are chemicals which mobilize and attract phagocytes. Important examples include tissue damage products and complement fragments. In acute inflammation, these agents lead to the aggregation of a large concentration of neutrophils and macrophages in the damaged tissues as part of the nonspecific immune response.

Modified essay answers

1. The haemostatic mechanisms are vascular spasm (in response to damage to vascular smooth muscle or the surrounding tissues), plugging of vascular openings by aggregated platelets (activated by exposure of platelets to collagen) and clotting (activated by collagen exposure, or contact with tissue thromboplastin).
2. Positive feedback is particularly important in platelet plugging. Adherent platelets release thromboxane A_2 and ADP, both of which further enhance platelet stickiness thus increasing the size of the platelet plug. There are also positive feedback elements within the coagulation process since some activated clotting factors promote earlier steps in the cascade, as well as directly activating the next factor in the chain.
3. This is the intrinsic coagulation mechanism. No extrinsic agent is added to the blood but contact with the abnormal surface (glass) activates clotting.
4. This suggests a defect in those stages of the coagulation cascade which are common to both pathways, i.e., factor X, prothrombin or fibrinogen (Fig. 33). It could also indicate a reduced plasma concentration of Ca^{2+}, since Ca^{2+} is necessary for both mechanisms.
5. This suggests an abnormality of platelet function, possibly caused by a reduced platelet count. Bleeding time is normally shorter than clotting time and is largely determined by the rate of platelet plug formation.
6. Rapid blood flow is an important mechanical factor, making it difficult for platelets to attach to the endothelium. Normal endothelial cells also release prostacyclin, which reduces platelet stickiness, and plasma contains a number of anticoagulants, such as antithrombin III, which inhibit fibrin formation. Sluggish blood flow (particularly likely in veins) and damage to the endothelium favours platelet adhesion and local coagulation leading to clot formation.
7. Activation of the clotting cascade also leads to conversion of plasminogen to plasmin, a fibrinolytic enzyme. Clot debris is removed by phagocytic cells.

Data interpretation answers

1. All three parameters are lower than normal.
2. Mean cell volume (MCV) = (haematocrit)/(red cell count) = $0.25/(2.2 \times 10^{12})$ L = 0.114×10^{-12} L = 114 fl (1 fl = 10^{-15} L).
3. Mean cell haemoglobin (MCH) = (haemoglobin concentration)/(red cell count)
4. The haemoglobin concentration needs to be expressed L^{-1}, not dl^{-1}, i.e., it has to be scaled up by a factor of 10. Therefore, MCH = $83/(2.2 \times 10^{12})$ g = 37.7×10^{-12} g = 37.7 pg (1 pg = 10^{-12} g).

5. Mean cell haemoglobin concentration (MCHC) = MCH/MCV = (haemoglobin concentration)/(haematocrit) = $8.3/0.25\,g\,dl^{-1} = 33\,g\,dl^{-1}$.
6. This is a macrocytic anaemia, i.e., the red cells are larger than normal (increased MCV). The MCH is also increased, because of the increased cell size, but the MCHC is not altered. This suggests there is a problem with cell division within the bone marrow but that haemoglobin production can occur as normal.
7. Macrocytic anaemia has a range of causes but deficiencies of vitamins B_{12} and/or folate are among the relatively common causes. Microcytic anaemia, on the other hand, typically results from iron deficiency.

Clinical scenario answers

1. It acts as an anticoagulant by removing the free Ca^{2+} necessary for the coagulation cascade.
2. The patient is anaemic, i.e., he has a reduced concentration of haemoglobin. Fatigue is a common symptom of anaemia, presumably because of the reduced O_2-carrying capacity of the blood.
3. An elevated reticulocyte count suggests an increased rate of erythrocyte production in the marrow. This is a response to the anaemia, mediated by renal release of erythropoietin.
4. An elevated white cell count and fever are both systemic features of inflammation, often caused by bacterial or viral infection.
5. Blood group is AB. There should be no ABO antibodies in his plasma (otherwise they would agglutinate his own red cells).
6. All the tests using saline are negative controls. This excludes nonspecific agglutination. The tests using known cells provide both positive and negative controls. For example, group O cells should not agglutinate with any of the antibodies (all negative controls), while group A cells should agglutinate with anti-A (positive control) but not with anti-B (negative control).

Viva answers

1. It is important to emphasize the role of renal erythropoietin as the physiological stimulus to red cell production in the marrow. You should make it clear that erythropoietin is produced in response to the low tissue O_2 levels which result from anaemia rather than being a direct response to low haemoglobin. This explains why haemoglobin levels can actually be raised following chronic arterial hypoxia (decreased arterial O_2 content), e.g., due to disease or high altitude (Section 4.6). Causes of anaemia may well be asked about and the discussion may go on to consider sites of red cell production at different ages or the significance of a reticulocytosis.
2. This is a huge topic with great scope for getting lost. Start with some general principles of classification, e.g., nonspecific and specific immunity. Build on this by listing the main mechanisms involved under each heading. It is logical to discuss nonspecific mechanisms first because many specific mechanisms act to target and amplify these. It is likely that the examiner will direct you with some subsidiary questioning at this stage. You should expect to be asked about inflammation, antibody-mediated immunity or cell-mediated immunity in more detail.
3. Bleeding time is defined as the time taken for a small wound (usually inflicted with a lancet into the ear lobe) to stop bleeding. You should be able to state a normal value. It is important to identify the three mechanisms of haemostasis (vascular spasm, platelet plugging and coagulation) and then you should comment that platelet defects/deficiencies are the most likely to affect bleeding time, although any of the three may cause bleeding to be prolonged. Questioning is likely to go on to deal with factors affecting platelet stickiness but the discussion is also likely to take in activation of the coagulation cascade.

Cardiovascular physiology

Overview

Cardiovascular function reflects the properties both of the different blood vessels which make up the circulation and of the heart, which pumps blood through those vessels. Cardiac physiology depends on the electrical and mechanical properties of cardiac cells, and on their integrated activity in the intact heart. This is under a number of controls; some are intrinsic to the heart itself while other, extrinsic systems depend on nerves and hormones. As a result, the output from the heart can be varied to meet changing tissue needs.

Peripheral blood flow depends both on the arterial pressure generated by the heart and on local resistance to flow. These factors are again under both local and extrinsic controls. Nutrient and gas exchange, the main goal of the cardiovascular system, occurs within the microcirculation. Blood is then returned through the veins to the heart, while excess capillary filtrate is returned to the blood from the interstitium by the lymphatics.

3.1 Relevant structure

Learning objectives

At the end of this section you should be able to:

- describe the systemic and pulmonary cardiovascular circuits
- describe and draw diagrams to illustrate the main structural features of the heart at both whole organ and cellular levels.

Components of the circulation

The cardiovascular system consists of a muscular pump, the heart, and two functionally distinct circulations, systemic and pulmonary (Fig. 34). Blood flow through the *systemic* circulation depends on contraction of the left ventricle of the heart, which forces blood into the aorta. Large arteries branch off this to supply different areas of the body, forming regional circulations which are effectively in parallel with one another. Within each tissue, smaller and smaller arterial vessels are arranged in series, terminating in the arterioles, which supply the tissue capillaries. These act as the sites of gas and nutrient exchange between blood and the interstitial fluid. Capillaries drain through venules into veins, which empty in turn into the superior and inferior venae cavae. It is these large venous vessels which return systemic blood to the right atrium of the heart.

The right side of the heart drives blood through the *pulmonary* circulation. The right ventricle fills from the right atrium and then contracts, driving blood into the pulmonary trunk, which supplies the pulmonary arteries. Pulmonary capillaries lie in close contact with the air-filled alveoli of the lung and are involved in the gaseous diffusion which oxygenates the blood (Section 4.3). The pulmonary veins return this oxygenated blood to the left atrium, which empties into the left ventricle, thus completing the double circulation.

Cardiac structure

The heart is a double pump arranged as a muscular cone with its apex directed downwards and to the left and its base behind the upper sternum. Each pumping unit consists of two chambers, with a thin-walled *atrium* opening into a more muscular ventricle. Endocardium lines the inner surface of the muscular myocardium, which is covered in turn by the epicardium. The whole heart is surrounded by a tough, fibrous sac, the pericardium. *Atrioventricular valves* prevent back flow of blood from ventricles to atria (Fig. 35). On the right side, this valve has three cusps and is known as the tricuspid valve, while on the left it has two cusps and is called the mitral valve. There are also one-way *semilunar valves* at the ventricular outlets, where blood flows into the pulmonary trunk (*pulmonary valve*) or aorta (*aortic valve*). The chambers on the right and left sides of the heart are separated by shared dividing walls known as the atrial and ventricular septa.

Pulmonary circulation
Pulmonary trunk
Pulmonary vein
V
A
Right heart
Left heart
A
V
Vena cava
Aorta
Deoxygenated
Oxygenated
Systemic circulation

A = Atrium
V = Ventricle

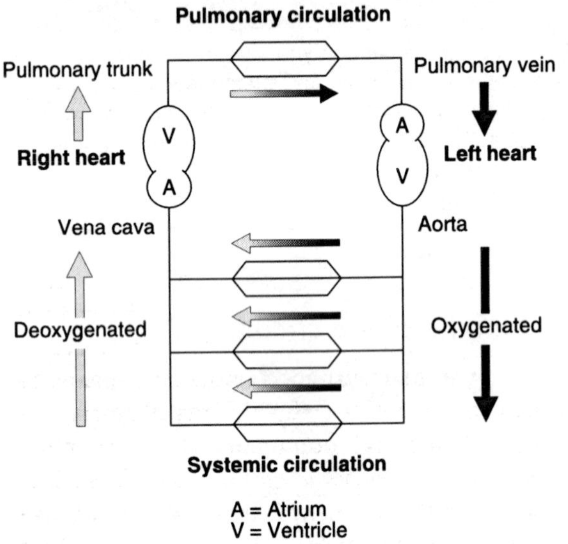

Fig. 34 Diagrammatic representation of the cardiovascular system (see text).

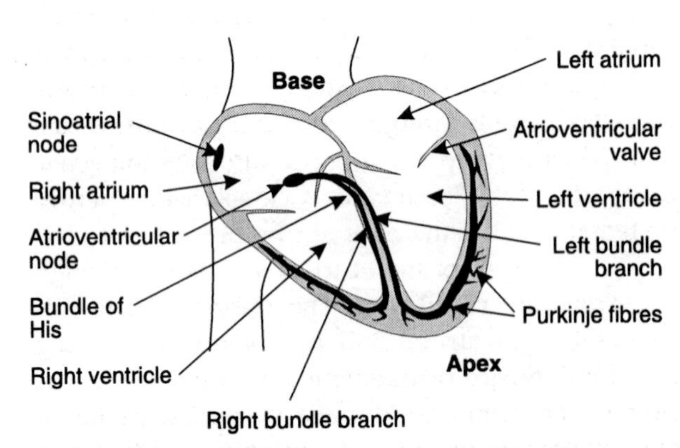

Fig. 35 The functional anatomy of the heart including the main electrical conducting pathways.

Cardiac muscle cells

The muscle cells (also known as cardiac myocytes or cardiac fibres) of the heart contain contractile proteins, as described in Section 1.6. Cardiac muscle is rich in mitochondria, reflecting its dependence on aerobic metabolism, and the cells contain only one nucleus. The fibres may branch at either end and they connect with the next cell in series through a region of close cellular association known as an *intercalated disc* (Fig. 36). Desmosomes link the membranes of adjacent cells at these sites. There are also special transmembrane channel proteins, which form gap junctions connecting the cytoplasm of the cells on either side of the intercalated discs. This provides electrical continuity from cell to cell allowing easy transmission of action potentials. Because of this the heart behaves as a functional syncytium, with rapid conduction of electrical signals leading to well-coordinated contraction.

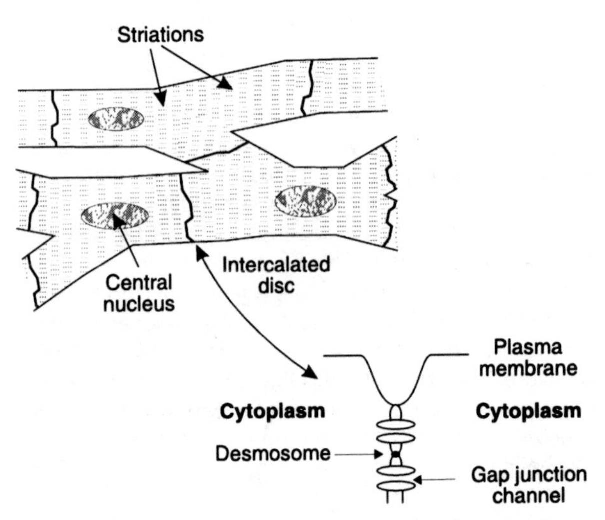

Fig. 36 Uninucleate, branching cardiac muscle cells make close contact with each other via the intercalated discs. Desmosomes link adjacent cell membranes at these sites while gap junction channels provide low electrical resistance pathways from cell to cell.

3.2 Electrical properties of the heart

Learning objectives

At the end of this section you should be able to:

- describe, and draw a diagram of, a ventricular action potential
- explain the significance of the action potential shape in terms of the mechanical function of the heart
- outline the ionic mechanisms which produce the cardiac action potential
- explain how electrical activity is spontaneously generated and conducted within the heart
- explain in principle how the ECG is generated and recorded
- draw a labelled diagram of a typical ECG and identify the electrical activity responsible for each waveform.

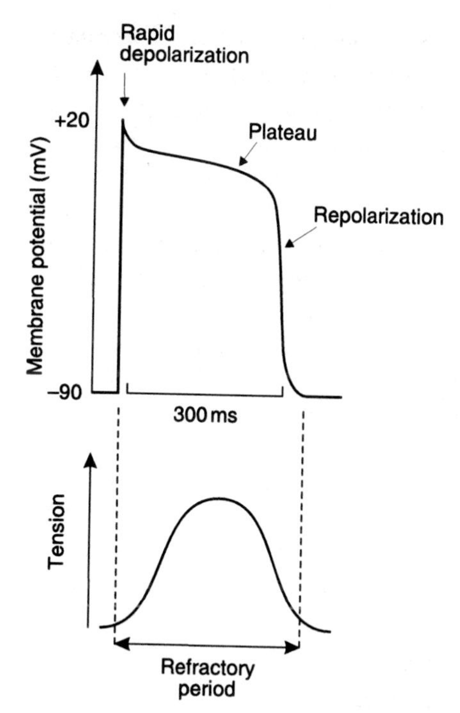

Fig. 37 Action potential and contractile response for a ventricular cell. The prolonged refractory period ensures that relaxation occurs before further contraction can be generated.

Cardiac muscle cells are excitable, and the signal which leads to each contraction is the generation of an action potential at the cell membrane. It is important to understand the cellular mechanisms which generate these action potentials. Equally important, however, is an understanding of how this electrical activity is conducted from cell to cell so as to trigger coordinated mechanical activity and efficient, rhythmical pumping in the intact heart.

Cardiac action potentials

Intracellular recordings from ventricular cells show that the resting membrane potential is similar to that in skeletal muscle, but more negative than that in nerve, at about −90 mV. As with any excitable cell, depolarization of the membrane to threshold leads to the production of an 'all or none' action potential with a characteristic shape (Fig. 37). There is an initial rapid depolarization to about +20 mV followed by an initial, partial repolarization of some 5–10 mV. Further repolarization is very slow indeed and this produces an *action potential plateau* in which membrane potential remains close to 0 mV for some 150–300 ms. After this, the membrane repolarizes rapidly, returning to the resting potential.

It is the plateau which distinguishes cardiac action potentials from those in skeletal muscle and nerve, as it greatly prolongs the action potential. This has important functional consequences for the mechanical activity of the heart. Cardiac cells are absolutely refractory to stimulation for the whole duration of the action potential and show a high degree of relative refractoriness for an additional 50 ms (Section 1.4). Therefore, a second action potential cannot be generated for a period of up to 350 ms after stimulation in ventricular cells. Contraction and relaxation are complete within this time (Fig. 37) and so it is impossible to get summation of contractions, or continuous tetanic contractions, as are commonly seen in skeletal muscle at higher frequencies of stimulation (Section 1.6). Since effective cardiac pumping depends on cyclical contraction and relaxation, the prolonged action potential protects against pump failure caused by sustained contraction, which would prevent the heart from refilling. It also sets an upper limit on the rate of contraction, which cannot exceed about 3–4 beats s^{-1} in the ventricle. Atrial rates can be considerably higher, however, because the atrial action potential, and, therefore, the refractory period, is shorter (less than 200 ms).

Mechanism of the cardiac action potential

The mechanisms underlying the cardiac action potential are similar to those in nerve, i.e., they depend on transmembrane ion gradients and voltage-sensitive changes in membrane permeability, or conductance, to those ions (Section 1.4). Three ions, Na^+, Ca^{2+} and K^+, are involved (Fig. 38).

Sodium channels. Depolarization first opens (activates) Na^+ channels, increasing Na^+ conductance. This leads to an *inward* Na^+ current which causes further depolarization. The resulting positive feedback accounts for the initial rapid depolarization phase, as in nerve. Sodium conductance then declines again because of depolarization-induced inactivation of Na^+ channels. The cell remains refractory to stimulation until these have returned to their resting, closed state following repolarization.

Calcium channels. There are also Ca^{2+} channels in the membrane, which open in response to depolarization. They activate more slowly than the Na^+ channels but once open they allow Ca^{2+} to flow into the cell (Fig. 38). This *inward* Ca^{2+} current keeps the membrane depolarized and thus maintains the plateau in the action potential.

Potassium channels. The conductance changes to K^+ in cardiac cells are more complicated than those described for nerve. Potassium conductance first *decreases* following depolarization, so that during the plateau there is actually less outward K^+ current than normal. This makes it easier for the inward Ca^{2+} current to maintain depolarization, since there is very little current flowing in the opposite direction. After 200 ms or so, however, K^+ conductance *rises*, increasing the outward current. This K^+ current repolarizes the membrane. Repolarization is assisted by the reduction in the opposing, inward Ca^{2+} current which also occurs in the latter part of the action potential (Fig. 38). The fact that K^+ conductance first decreases and then increases in response to the initial depolarization reflects the fact that there is more than one type of K^+ channel in these cells and each type responds differently to voltage changes.

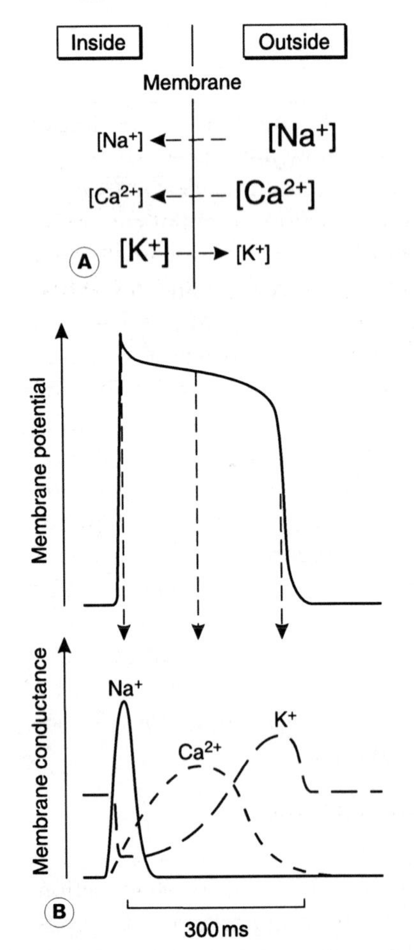

Fig. 38 The flow of ions across the cardiac cell membrane (A) and the changes in ion conductance during the action potential (B).

Spontaneous electrical activity (automaticity)

An isolated heart beats regularly without any extrinsic stimulation from nerves or hormones. This mechanical automaticity reflects the fact that the action potentials which are conducted throughout the heart (the signals which lead to contraction) are generated spontaneously within the cardiac muscle itself. Such activity is said to be *myogenic* (unlike action potentials in skeletal muscle which are neurogenic, i.e., they are only produced in response to nervous stimuli). The cells responsible for spontaneous action potential production are often referred to as *pacemaker cells* because they determine the rate at which the heart beats. Electrical recordings from such cells show that, instead of a constant resting membrane potential between action potentials, there is a steadily depolarizing potential known as a *pacemaker potential* (Fig. 39). When threshold is reached an action potential fires, and then the cycle of events is repeated.

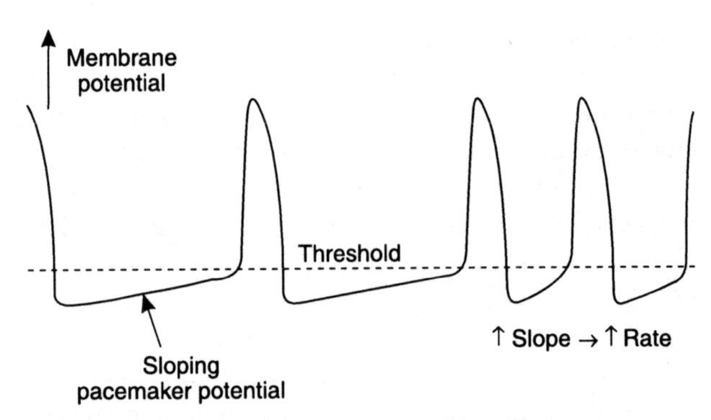

Fig. 39 Electrical activity in cardiac pacemaker cells from the sino-atrial node. The rate of spontaneous action potential production increases as the slope of the depolarizing pacemaker potential increases.

Several cell types in the heart are capable of pacemaker activity, but in the intact organ it is the fastest pacemaker which drives the rest of the heart. Once an action potential is generated at one site, it is rapidly conducted throughout the cardiac muscle and will trigger an action potential in slower pacemaker regions before they can reach the threshold potential spontaneously. Normally the pacemaking frequency is highest in a group of specialized cells called the *sino-atrial node*, which also have a different shape of action potential (Fig. 39). These cells dictate the rate of electrical events in the rest of the heart. This produces a resting heart rate of 60–70 beats min^{-1}, and the regular pattern of excitation and contraction which results is called *sinus rhythm*.

Conducting pathways in the heart

Cardiac action potentials normally originate in the pacemaker cells of the sino-atrial node, which is located in the wall of the right atrium close to the superior vena cava (Fig. 35). Action potentials are conducted away from the sino-atrial node through the normal atrial fibres. This is made possible by the interbranching structure of cardiac cells and the easy transmission of action potentials from one cell to another via the low-resistance gap junctions in the intercalated discs (Fig. 36). The conduction velocity through the atrial muscle is $0.3\,m\,s^{-1}$ and this is rapid enough to produce a coordinated atrial contraction which forces blood into the ventricles.

The *atrioventricular node* is located on the atrial septum in the lower half of the right atrium. It consists of specialized conducting tissue which is capable of pacemaker activity but is normally driven by conducted action potentials from the sino-atrial node. Conduction through the atrioventricular node is slow (about $0.05\,m\,s^{-1}$) and this delays transmission of the action potential to the ventricle, ensuring that ventricular contraction will not commence until atrial contraction is completed. Action potentials are conducted from the atria to the ventricles by the *atrioventricular bundle*, or bundle of His. This divides into right and left *bundle branches*, and these carry the impulses down either side of the ventricular septum towards the apex of the heart. These branches are continuous with the *Purkinje fibres*, which ramify through the ventricular muscle itself. Conduction through the atrioventricular bundle, the bundle branches and the Purkinje fibres is rapid (2–4 $m\,s^{-1}$) and this promotes synchronized contraction throughout the muscle of the ventricles. Local conduction from one ventricular cell to the next is much slower ($0.5\,m\,s^{-1}$).

Electrocardiogram (ECG)

Because action potentials do not fire simultaneously in all cardiac cells but are conducted across the myocardium from one region to another, extracellular potential differences are generated between one area of the heart and the next. These can be recorded from the skin as very small potential differences (approximately 1 mV) and such a record is known as an electrocardiogram (ECG). This is an important clinical tool, used both in the diagnosis of abnormal cardiac rhythms (*arrhythmias*) or defects in the conduction pathways and when investigating possible damage to the bulk of the myocardium, e.g., caused by ischaemia.

To understand the principle underlying the ECG, let us consider an area of myocardium with two surface electrodes being used to record any potential differences from the overlying skin (Fig. 40A). At rest there will be zero potential difference between the electrodes, but as the depolarizing phase of an action potential spreads across the heart, the transmembrane potential will be reversed in one area while other areas remain polarized. This generates a

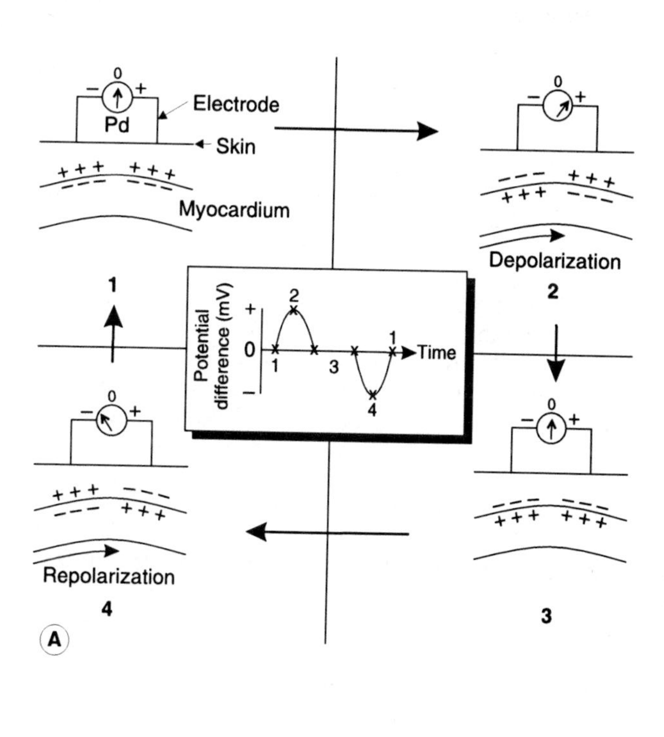

Fig. 40 (A) The principles underlying the electrocardiogram (ECG). (B) Sample trace showing the form of real ECGs.

potential difference which is recorded as a positive deflection, assuming the electrodes are attached to the positive and negative poles of the voltmeter as shown (reversing this connection would reverse the sign of the recorded potential difference but would not alter the shape of the ECG). Eventually, all the local myocardium will be depolarized, so the potentials at the two surface electrodes will be identical, giving zero potential difference again. Once repolarization starts to spread, however, a negative potential will be recorded until the whole tissue returns to the resting potential again.

A typical ECG recording consists of three different waves (Fig. 40B). There is an initial positive deflection known as the *P wave*. This is generated by the spread of depolarization across the atria. It is followed by the *QRS complex*, which is produced by the spread of depolarization across the ventricles. These contain more muscle and so generate larger surface potentials. The final *T wave* is the result of repolarization of the ventricles. The T wave is usually smaller than the QRS wave because repolarization spreads over the surface of the ventricles more slowly than does depolarization. The same is true for the atria, and this explains why no ECG wave attributable to repolarization of the atria is seen. The exact shape of the ECG depends on which points on the body surface are used for recording. Different electrode combinations (different recording leads) can give information about different parts of the heart.

The simplest piece of information which can be determined from the ECG is heart rate. The beat to beat interval can be measured between equivalent points on consecutive waves, e.g., from the peak of one P wave to the next. The total time for 10 beats is often measured from the ECG and divided to give an average beat to beat interval. This is then used to calculate an average heart rate, e.g., distance from first P wave to 11th P wave (10 × P–P intervals) = 208 mm; average = 20.8 mm = 0.83 s (standard ECG chart speed = 25 mm s^{-1}). Heart rate = 60 ÷ 0.83 = 72 beats min^{-1}.

Box 7 Clinical note: Cardiac arrhythmias

Any abnormal or irregular pattern of cardiac contraction is referred to as an arrhythmia. These arise from defects of pacemaking activity or abnormal electrical conduction through the heart. The normal rhythm which results from the activity of the sino-atrial node is referred to as sinus rhythm. The heart rate is not actually constant in normal individuals but varies during the respiratory cycle, speeding up during inspiration and slowing again during expiration. This is referred to as sinus arrhythmia and is particularly marked in fit young individuals. Pathological arrhythmias include atrial fibrillation, in which there is no coordinated atrial contraction. Action potentials are conducted randomly to the ventricles, producing a very irregular pulse, and there are no P waves on the ECG. This condition may lead to heart failure, but some cardiac function is maintained since atrial contraction is not necessary for passive filling of the ventricles (Fig. 41). Ventricular fibrillation, by comparison, is fatal, since there is no effective cardiac output in the absence of coordinated ventricular contraction. This is a common arrhythmia in those who have suffered from myocardial infarction but it may be reversed by the immediate application of DC electrical shocks from a defibrillator which synchronize depolarization of the myocardium.

3.3 Cardiac contractility

Learning objectives

At the end of this section you should be able to:

- describe the cellular mechanisms of contraction and excitation–contraction coupling in the heart.

The cellular basis of contraction is essentially the same as in skeletal muscle, with thick and thin myofilaments sliding past one another because of the formation of mobile cross bridges between myosin and actin (Section 1.6). Cross bridge formation is regulated by intracellular Ca^{2+} so that increases in the cytoplasmic [Ca^{2+}] lead to contraction through changes in the shape of the troponin–tropomyosin system on the thin myofilaments. The event which triggers contraction is the action potential. In cardiac muscle this stimulates Ca^{2+} release from the sarcoplasmic reticulum but the mechanism differs from skeletal muscle, in which Ca^{2+} release is activated directly by membrane depolarization (Section 1.6). In the heart, Ca^{2+} entry during the plateau phase of the action potential is the important stimulus (Section 3.2), but the effect of this influx on the intracellular [Ca^{2+}] is amplified by Ca^{2+}-induced Ca^{2+} release from the cardiac sarcoplasmic reticulum. Once the plasma membrane has repolarized, Ca^{2+} is rapidly removed from the cytoplasm (both by transport across the plasma membrane and reuptake into the sarcoplasmic reticulum), and the cell relaxes.

3.4 The cardiac cycle

Learning objectives

At the end of this section you should be able to:

- draw a diagram showing electrical and mechanical events of the (left) cardiac cycle; this should be labelled with values for pressure, volume and time
- explain how pressure changes are generated
- explain how pressure gradients open and close valves and the generation of heart sounds
- name and explain the phases of ventricular volume change.

The cardiac cycle refers to the mechanical and electrical events during a single cycle of contraction and relaxation. It describes both the patterns of change observed in individual measurements of mechanical and electrical function and how the timings of events in different chambers of the heart relate to each other (Fig. 41).

Mechanical events

Ventricular pressure

This is the simplest of the three relevant pressure waves. During most of *ventricular diastole*, when the ventricle is relaxed, intraventricular pressure is low (less than 1 mmHg). This rises to about 5 mmHg at the end of diastole because of atrial systole, which forces additional blood into the ventricle. As the atrium relaxes, pressure falls but *ventricular systole*, or contraction, then commences and pressure rises rapidly, reaching a peak of about 120 mmHg in the left ventricle. (The wave form in the right ventricle is identical in shape to that in the left but the peak pressure during systole is much lower, at about 25 mmHg.) Pressure falls back to its original low value as the ventricle relaxes (diastole). In Figure 41

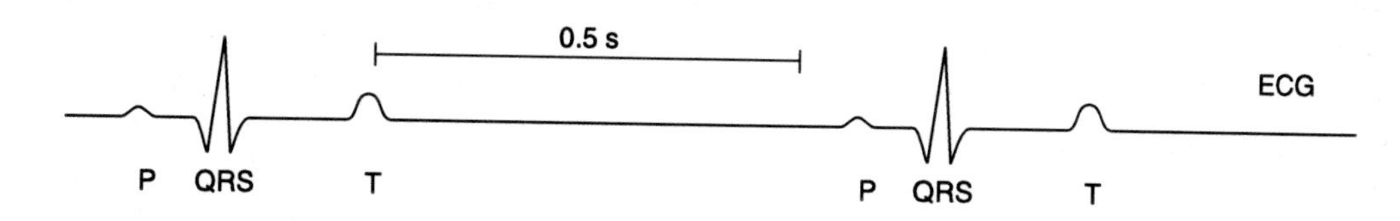

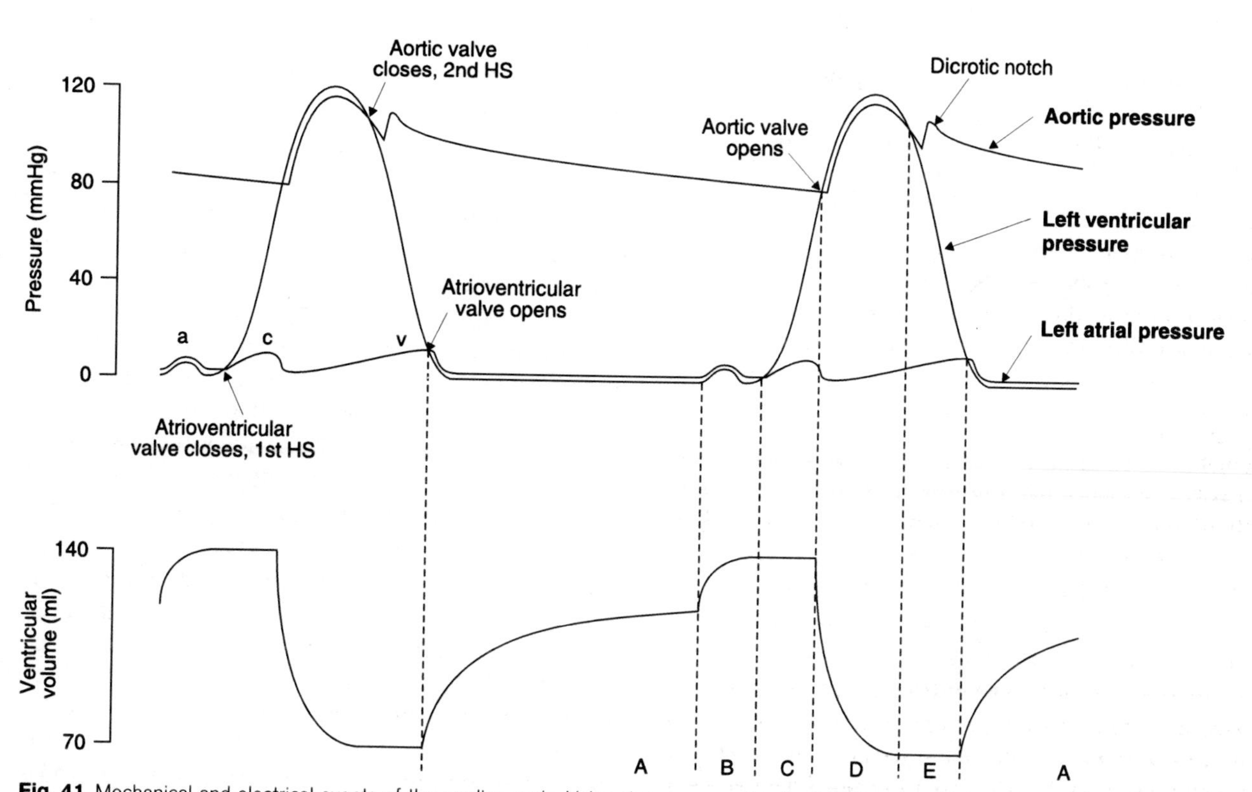

Fig. 41 Mechanical and electrical events of the cardiac cycle. Valve closure generates the heart sounds (HS). The ventricular volume record (bottom graph) can be divided into five phases: passive filling (A); atrial ejection (B); isovolumetric contraction (C); ventricular ejection (D); and isovolumetric relaxation (E).

the cardiac cycle lasts 0.9 s (i.e., resting heart rate equals 66 beats min^{-1}), systole lasting 0.3 s and diastole 0.6 s. As heart rate increases, it is mainly diastole which shortens, and if the heart rate rises too high, cardiac pumping becomes very inefficient because of inadequate ventricular filling during diastole.

Atrial pressure

There are three peaks in atrial pressure, known as the 'a', 'c' and 'v' waves. Throughout ventricular diastole there is a pressure gradient favouring blood flow through the open atrioventricular valve into the ventricle. Atrial pressure remains constant, at about 1 mmHg, until atrial systole, when active atrial contraction raises it to a peak of about 6 mmHg. This is the *'a' wave*. The atrium then relaxes and ventricular pressure starts to rise so the atrial and ventricular pressure curves cross each other. This reverses the pressure gradient and, as a result, the *atrioventricular valve* (which only permits flow from atrium to ventricle) *closes*. The back pressure on the valve cusps during early ventricular systole, when both the inflow and outflow valves of the ventricles are closed, causes a secondary rise in atrial pressure, the *'c' wave*. As soon as the aortic or pulmonary valves open, atrial pressure falls rapidly to almost zero (the 'x' descent). Blood continues to enter the atria from the venous system, however, and since the atrioventricular valves are closed this produces a steady increase in pressure, the *'v' wave*. This peaks at 3–4 mmHg just before the rapidly falling pressure in the relaxing ventricle drops below that in the atrium (the 'y' descent). The *atrioventricular valves reopen* and the atrial pressure drops back to about 1 mmHg. The 'a' and 'v' waves, or perhaps more correctly, the 'x' and 'y' descents, can be observed during careful examination of the jugular veins of the neck (the 'c' wave is rarely visible; see Section 3.9).

Aortic pressure

During ventricular diastole, there is a gradual decline in aortic pressure to a minimum of about 80 mmHg. Throughout this period, aortic pressure is higher than left ventricular pressure so the *aortic valve remains closed*. During systole, the ventricular pressure rises above aortic pressure, *opening* the valve and allowing blood to be ejected into the aorta. Aortic pressure follows ventricular pressure closely during systole, peaking at about 120 mmHg, but as the pressure in the ventricles starts to decline there is a much slower drop in aortic pressure. This reflects the elasticity of the aorta, which is stretched by the rapid inflow of blood during ventricular systole. The stored elastic energy is then released during diastole, as the walls of the aorta passively recoil, thus maintaining aortic pressure. As a result, the aortic and ventricular pressure waves cross once more, and the *aortic valve closes*. This halts the back flow of blood towards the ventricle and the force generated as the momentum of this blood is dissipated shows up as a brief increase in aortic pressure, resulting in the *incisura* or *dicrotic notch* (Fig. 41).

Pressures in the pulmonary trunk follow a similar pattern but are much lower at 25/10 mmHg as compared with 120/80 mmHg for the aorta (systolic/diastolic pressure). This reflects the lower resistance of the pulmonary circulation, pulmonary and systemic blood flow being almost equal.

Heart sounds

Closure of the valves of the heart produces mechanical vibrations which are audible at the chest wall as the heart sounds. The *first heart sound* is caused by the closure of the atrioventricular valves and so marks the beginning of ventricular systole. The *second heart sound* is caused by closure of the aortic and pulmonary valves. Clinically, ventricular systole is regarded as the period between the first and second heart sounds, while diastole is the time between the second heart sound and the first heart sound of the next cycle. It should be noted that valve opening makes no detectable sound.

Ventricular volume

At the end of systole, there is about 70 ml of blood in the ventricle. This increases to about 125 ml during diastole as a result of *passive* filling from the atrium (Fig. 41). *Active* filling during atrial systole at the end of ventricular diastole only increases filling by a further 25%, to about 140 ml. The dominance of passive filling explains why ventricular function is still possible in the absence of coordinated atrial contraction, e.g., during atrial fibrillation.

Following closure of the atrioventricular valve, but before the aortic valve opens, the ventricles are effectively closed boxes whose volume cannot change. There is, therefore, a short period of *isovolumetric contraction*, with constant volume but rising ventricular pressure. Opening of the aortic valve then leads to the *ejection phase* during which ventricular volume drops from 140 ml to 70 ml. Therefore, about 70 ml of blood are ejected in each cycle, i.e., the *stroke volume* is 70 ml. The normal *ejection fraction* (a useful measure of ventricular function) is, therefore, around 50% (stroke volume/end diastolic volume). Ejection is followed by a period of *isovolu-*

metric relaxation during which atrioventricular and aortic valves are again closed. Passive filling then recommences with opening of the atrioventricular valve.

Electrical events

The electrical signals in the heart act as stimuli for the mechanical responses and so each wave on the ECG is associated with the onset of the relevant pressure change (Fig. 41). The P wave indicates depolarization of the atria and comes at the start of the pressure wave caused by atrial contraction. Similarly, the spread of depolarization over the ventricles, recorded as the QRS complex, marks the onset of ventricular contraction. This is followed by ventricular repolarization and the resulting T wave overlaps the drop in pressure caused by ventricular relaxation.

3.5 Cardiac output

Learning objectives

At the end of this section you should be able to:
- define cardiac output and state normal values
- explain in principle how cardiac output may be measured
- explain Starling's law of the heart using appropriate graphs
- describe how nerves and hormones modify cardiac function
- outline the mechanisms which increase cardiac output during exercise
- describe the physiological consequences of cardiac failure.

The cardiac output is a measure of the heart's ability to pump blood and is defined as the *volume of blood expelled by either ventricle in 1 minute*. The outputs of the two sides of the heart are normally equal so we do not distinguish between the right and left ventricles in this definition.

Cardiac output is determined by two features of cardiac function, the heart rate and the volume of blood ejected during a single contraction of the ventricle (the stroke volume).

$$\text{Cardiac output} = \text{Heart rate} \times \text{stroke volume} \quad \text{(Eq. 6)}$$

At rest:

cardiac output = 70 beats $\text{min}^{-1} \times 70\,\text{ml beat}^{-1}$ = $4900\,\text{ml min}^{-1}$ (approximately $5\,\text{L min}^{-1}$).

Measurement of cardiac output

Fick's method

The Fick technique for the estimation of cardiac output is a specific application of Fick's principle, which states that the rate of addition or removal of a substance in any part of the circulation equals the blood flow multiplied by the resulting change in concentration. We can apply this to the pulmonary circulation where there is a continuous uptake of O_2 from the alveoli and, as a result, $[O_2]$ in blood increases (Fig. 42). The relationship between the magnitude of the change in $[O_2]$ and the pulmonary blood flow can be derived from the considerations outlined in Box 8.

In practice, the rate of O_2 absorption is measured using a device called a spirometer and is about $200\,\text{ml min}^{-1}$ at rest. Systemic arterial $[O_2]$ can be measured from any peripheral artery and usually equals $200\,\text{ml L}^{-1}$. Systemic venous $[O_2]$ cannot be measured from a peripheral vein, however, since venous $[O_2]$ varies for different organs. A mixed venous sample must be obtained instead, either from the right ventricle or the pulmonary trunk. Such samples give an $[O_2]$ of $160\,\text{ml L}^{-1}$. Inserting these values into Fick's equation (Eq. 7) gives a value of $5\,\text{L min}^{-1}$ for the resting cardiac output (Fig. 42).

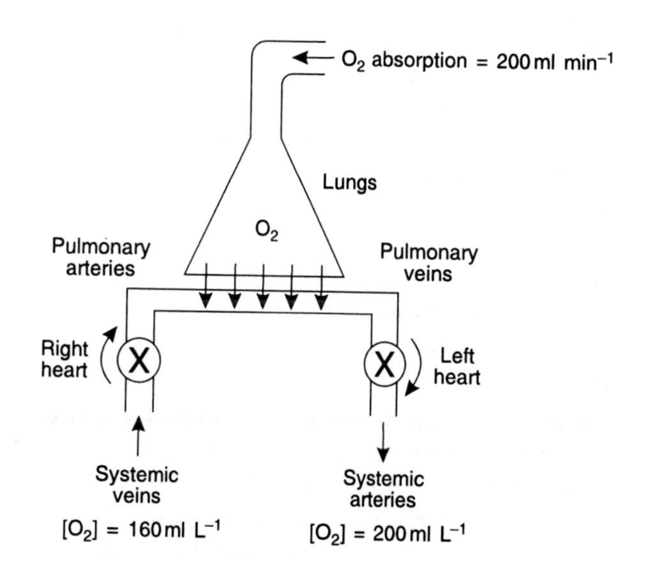

$$\text{Cardiac output} = \frac{\text{Rate } O_2 \text{ absorption}}{[O_2]\text{ systemic arteries} - [O_2]\text{ systemic veins}}$$

$$= \frac{200\,\text{ml min}^{-1}}{(200-160)\,\text{ml L}^{-1}} = 5\,\text{L min}^{-1}$$

Fig. 42 Application of Fick's principle to the measurement of cardiac output.

Three

Box 8 Background mathematics: Cardiac output from Fick's method

Each minute, the amount of O_2 entering the lungs in pulmonary arterial blood = (Pulmonary blood flow) × (Pulmonary arterial [O_2]).

Each minute the amount of O_2 leaving the lungs in the pulmonary venous blood = (Pulmonary blood flow) × (Pulmonary venous [O_2]).

The difference between these two values must equal the rate at which O_2 is absorbed from the alveoli into the pulmonary capillaries.

∴ (Pulmonary blood flow × Pulmonary venous [O_2])
 − (Pulmonary blood flow × Pulmonary arterial [O_2])
 = Rate of alveolar O_2 absorption.

Since pulmonary blood flow equals cardiac output we can simplify this to:

Cardiac output × (Pulmonary venous [O_2] − Pulmonary arterial [O_2]) = Rate of O_2 absorption.

Or:

$$\text{Cardiac output} = \frac{\text{Rate of } O_2 \text{ absorption}}{\text{Pulmonary venous } [O_2] - \text{Pulmonary arterial } [O_2]}$$

Since it is easier to take systemic than pulmonary blood samples, we can rewrite this expression in the usual form of Fick's equation:

$$\text{Cardiac output} = \frac{\text{Rate of } O_2 \text{ absorption}}{\text{Systemic arterial } [O_2] - \text{Systemic venous } [O_2]} \quad \text{(Eq. 7)}$$

Box 9 Background mathematics: Cardiac output using indicator dilution

Figure 43 shows the changes in dye concentration (*c*). The area under the curve (*A*) produced by extrapolating the initial decline can be used to calculate the average concentration of indicator for the time of one circulation (*t*).

$$\text{Average } c\,(\text{mg L}^{-1}) = \frac{A\,(\text{mg L}^{-1}\text{min})}{t\,(\text{min})} \quad \text{(Eq. 8)}$$

In theory, we could also calculate the average indicator concentration using the expression:

$$\text{Average } c = \frac{Y\,(\text{mg})}{\text{Relevant blood volume (L)}}$$

The indicator is dissolved in the total volume of blood flowing through the heart in time *t*, i.e., cardiac output × *t*

$$\therefore \text{Average } c = \frac{Y}{\text{Cardiac output} \times t} \quad \text{(Eq. 9)}$$

By equating Eq. 8 and Eq. 9 we get:

$$\frac{A}{t} = \frac{Y}{\text{Cardic output} \times t}$$

$$\therefore \text{Cardiac output } (\text{L min}^{-1}) = \frac{Y\,(\text{mg})}{A\,(\text{mg L}^{-1}\text{min})} \quad \text{(Eq. 10)}$$

Indicator dilution technique

Following injection of a known amount (*Y* mg) of a measurable dye (the indicator) into the central venous system close to the heart, indicator concentration (*c*) in systemic arterial blood rapidly rises to a peak and then declines again (Fig. 43). Recirculation of the indicator leads to secondary peaks on the concentration curve, but extrapolation of the initial decline gives an estimate of what the concentration curve would have been like in the absence of such recirculation. Box 9 shows how cardiac output can be calculated from these studies.

In choosing an indicator, one requires a physiologically inert substance which is retained in the circulation and is not metabolized. An adaptation of the method which fulfils all of these criteria uses

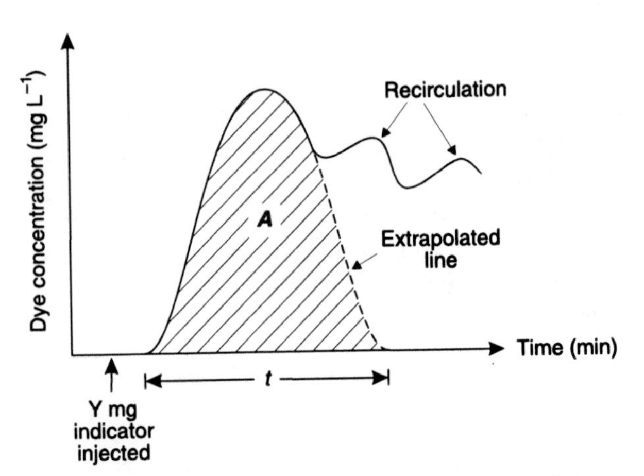

Fig. 43 Measurement of cardiac output by the indicator dilution technique. The area *A* is measured and used to calculate cardiac output (see Box 9).

injections of warm or cold saline into the pulmonary trunk and the resulting change in blood temperature is measured a few centimetres further on. This is referred to as the *thermodilution* technique.

Control of cardiac output

Since cardiac output is determined by heart rate and stroke volume, relevant control mechanisms must regulate one or both of these variables. These mechanisms may be divided into two groups; those which depend on *intrinsic* properties of the heart itself and *extrinsic* controls which modify cardiac performance.

Intrinsic control of cardiac function

The most important intrinsic mechanism involved in the control of cardiac output is usually referred to as *Starling's law of the heart*, or the Frank–Starling mechanism, after the two physiologists who first described it. They observed an increase in the force of contraction as the resting length of the cardiac muscle fibres was increased. This resulted in increases in stroke volume and, therefore, in cardiac output, as the volume of the ventricle immediately before contraction (the *end diastolic volume*) was increased (Fig. 44A). This is independent of any extrinsic nervous and hormonal controls. Similar results are obtained if some other measure of the degree of ventricular filling is used, e.g., *venous return* (the total rate of blood flow into the atrium from the veins) or atrial pressure. One way of summarizing the overall effect of the Frank–Starling mechanism on cardiac function is as a plot of cardiac output against atrial pressure, which demonstrates clearly that increased ventricular filling leads to increased output across the normal range of atrial pressures (Fig. 44B).

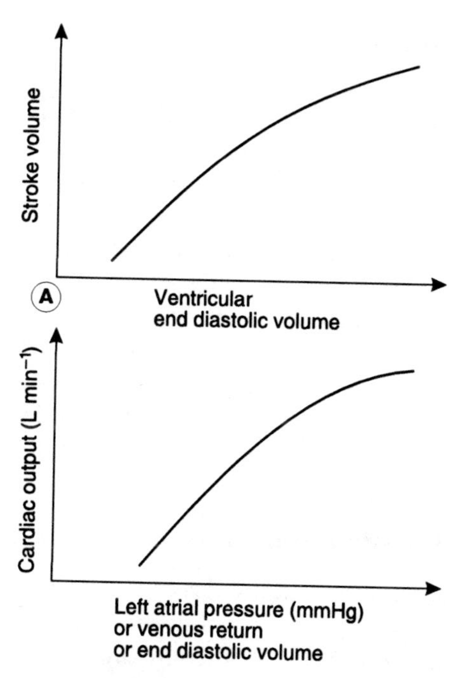

Fig. 44 Starling's law of the heart. (A) As ventricular volume increases so does the force of contraction, increasing stroke volume. (B) The effect of this on cardiac output can be plotted using atrial pressure, venous return, or end diastolic volume on the *x* axis, since these variables are all interrelated.

Starling's law helps explain two important features of cardiac function, namely that cardiac output equals venous return and that the average outputs from the two ventricles are equal. If venous return suddenly rises above ventricular output, blood will accumulate in the ventricle, increasing the end diastolic volume. Starling's law predicts that this will lead to an increase in both stroke volume and cardiac output until a new steady state is reached in which cardiac output equals venous return again. Since the output from one ventricle is responsible for the venous return to the other side of the heart in the intact circulation, this mechanism will also ensure that the cardiac output from the two ventricles remains equal. For example, if the cardiac output from the left ventricle increases, this will increase right venous return and right ventricular output will rise as a consequence.

Preload and afterload As in any muscle, the contractile performance of the heart is influenced by the mechanical forces, or loads, which it experiences. Preload refers to the level of stretch in the relaxed muscle immediately before it contracts. In the heart this is largely dictated by the venous return. Increases in venous return increase the preload, i.e., they increase the stretch in the muscle fibres before contraction. This increases contractile force and, therefore, cardiac output, as described by Starling's law (Fig. 44).

Afterload refers to the force that the muscle must generate during contraction and this is most obviously affected by changes in arterial pressure. In a normal heart, however, increasing the systemic arterial pressure (left ventricular afterload) has little effect on cardiac output, at least over the range of pressures likely to be experienced. Therefore, it is the preload, effectively the rate of venous return, which is most important in determining cardiac output.

Extrinsic control of cardiac function

Extrinsic control is that exerted by the autonomic nerves and circulating hormones which can affect the activity of the heart.

Nervous control

Both the *sympathetic* and *parasympathetic* divisions of the autonomic nervous system help regulate the heart's activity (Section 7.6). They can alter the heart rate (*chronotropic* effects) or the strength of ventricular contraction (*inotropic* effects) but are not under conscious control. Instead, they are reflexly regulated by inputs from cardiovascular and other receptors, e.g., from the baroreceptors, which help regulate arterial pressure, or through connections with higher centres in the brain, e.g., in response to emotions such as anger or fear.

Sympathetic nerves Sympathetic nervous activity to the heart is controlled from a number of regions in the central nervous system (CNS), including the medulla oblongata. The postganglionic nerves release the neurotransmitter *noradrenaline* (norepinephrine) and supply both pacemaking and contractile cells throughout the atria and ventricles. Stimulation of these nerves leads to increases both in heart rate (positive chronotropic effect) and in myocardial contractility (positive inotropic effect). This increases the cardiac output at any given atrial filling pressure, pushing the relationship between cardiac output and atrial pressure upwards and to the left (Fig. 45A). There is a limit to the benefits of increasing heart rate, however, since the time available for diastolic filling between contractions decreases, and rates over 200 beats min^{-1} may lead to a decline in cardiac output.

Parasympathetic nerves These come from the medulla oblongata in the brain and reach the heart via the *vagus nerve*. They supply the sino-atrial and atrioventricular nodes, releasing *acetylcholine* when stimulated, and this slows the heart (a negative chronotropic effect) through its influence on pacemaker activity. Maximal parasympathetic activity can reduce the heart rate to 30–40 beats min^{-1} (vagal brake). The result is a reduction in cardiac output, pushing the cardiac function curve down and to the right (Fig. 45A). Parasympathetic nerves are not distributed to ventricular muscle and so they have no appreciable effect on contractility (i.e., no inotropic effect).

Resting nervous tone Under resting conditions, there is activity in both the sympathetic and parasympathetic nerves supplying the heart. This is referred to as resting nervous tone and cardiac function can be altered by either decreasing or increasing the frequency of stimulation by either set of nerves. For example, heart rate and cardiac output may be doubled by completely blocking the resting parasympathetic tone (Fig. 45B).

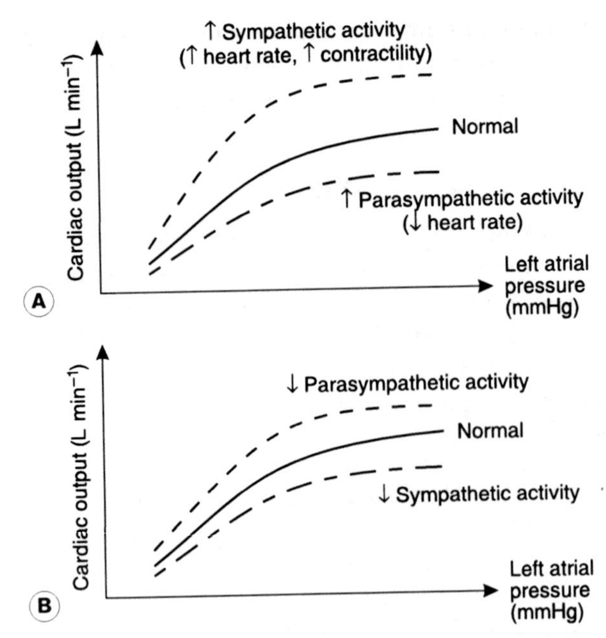

Fig. 45 Effects of autonomic nervous control on cardiac function. (A) Increased stimulation. (B) Reduced activity also alters cardiac output because there is a basal level of tonic activity in these nerves.

Hormonal control

The most important circulating hormones affecting the heart are the *catecholamines*, i.e., noradrenaline (also called norepinephrine) and adrenaline (or epinephrine). These are released by adrenal medullary cells in response to increased preganglionic sympathetic nervous stimulation of that gland (Section 8.6). They have the same effects on the heart as noradrenaline (norepinephrine) from sympathetic nerves, increasing both heart rate and myocardial contractility, and thus increasing cardiac output. Circulating catecholamines prolong the period of cardiac stimulation for some minutes after sympathetic nervous activity has returned to normal.

Exercise and cardiac function

During exercise, cardiac output rises to meet the increased blood flow requirements of skeletal muscle. The maximal output is about 25–30 L min^{-1} in normal individuals. Several factors contribute to this.

Venous return. This rises as a result of compression of the veins by contracting skeletal muscle and, therefore, increases cardiac output through the Starling mechanism. Unidirectional venous valves ensure that blood is forced towards the right atrium (see Fig. 56).

Sympathetic nervous activity increases both in response to higher centres involved in initiating exercise and reflexly by receptors in the exercising muscles themselves. Heart rate and myocardial contractility are increased as a result, helping boost cardiac output. In practice, the rise in heart rate is the dominant effect and there is usually little change in stroke volume. Sympathetic stimulation also leads to active constriction of the smooth muscle in the walls of veins (*venoconstriction*), further enhancing venous return.

Parasympathetic (vagal) activity to the heart. This is inhibited during exercise and this also contributes to the increase in heart rate.

Cardiac hypertrophy. This is seen as a response to prolonged periods of exercise over months or years in trained athletes. This increases the muscle bulk of the left ventricle through enlargement of the individual cardiac fibres, thus raising the stroke volume. As a result, the maximum cardiac output achieved during strenuous activity is considerably higher than in untrained individuals, reaching values up to 35–40 L min^{-1}. Resting cardiac output is normal, however, since athletes also have a lower than normal resting heart rate (40–50 min^{-1}) due to increased parasympathetic (vagal) tone.

Box 10 Clinical note: Cardiac failure

If the muscle of the heart becomes damaged, e.g., because of decreased myocardial blood supply (myocardial ischaemia), or if the demands placed on the heart are abnormally large, e.g., in hypertension where the arterial pressure is abnormally high, the heart may fail to perform its pump functions adequately. This failure affects the body in two main ways. Firstly, the cardiac output may fall below the required level, so that tissue perfusion is reduced (forward failure). This leads to fatigue and poor exercise tolerance. Secondly, blood accumulates in the failing ventricle, which becomes dilated, and this leads to an increase in atrial pressure (backward failure). The pressure in the veins emptying into the affected atrium will also rise, and this increases capillary pressure (Section 3.8). The result is an accumulation of fluid (oedema) either in the pulmonary tissue (in the case of left ventricular failure) or in the skin and systemic organs (right ventricular failure). Pulmonary oedema is particularly serious since it interferes with gas exchange in the alveoli (Section 4.3), causing breathlessness.

3.6 Control of arterial pressure

Learning objectives

At the end of this section you should be able to:

- describe how blood pressure is measured and state normal values
- explain how cardiac output and peripheral resistance affect arterial blood pressure
- provide an integrated description of how nerves and hormones regulate blood pressure
- differentiate between reflexes responsible for short term pressure control and possible long term control mechanisms.

The unqualified term blood pressure usually means arterial blood pressure. Arterial pressure acts as the driving force for blood flow through the tissues of the body. If blood pressure drops (*hypotension*), perfusion of the organs may fail, and this becomes an acute, life-threatening situation if blood flow through the brain or heart is impaired. An abnormally high blood pressure (*hypertension*) can damage the heart and vasculature over the long term and may require prolonged drug treatment. Reliable measurement of blood pressure is, therefore, of both physiological and clinical significance.

Measurement of blood pressure

Pressure may be measured directly by inserting a cannula into an artery, but this is only appropriate for experimental or intensive care situations. Indirect measurement makes use of an inflatable cuff which can be wrapped around the upper arm, over the biceps area. Cuff pressure is recorded using a device known as a *sphygmomanometer*. The pressure is raised well above the normal arterial pressure, to approximately 200 mmHg, and then the cuff is slowly deflated. At the same time a stethoscope is used to listen, or auscultate, over the brachial artery distal to the cuff. This method is sometimes referred to as the *auscultatory method*. Initially there is silence because the high pressure in the cuff completely collapses the artery, but as cuff pressure falls, a short tapping sound, known as the first *Korotkoff sound*, is heard. This occurs when the highest pressure achieved in the artery (the systolic pressure) is slightly greater than the cuff pressure, and so a brief jet of blood is forced through in each cardiac cycle. This turbulent flow vibrates the arterial walls and produces the Korotkoff sounds. As the cuff deflates further,

Three

arterial pressure exceeds cuff pressure for longer and longer in each cycle, allowing blood to flow through for longer and with less turbulence. As a result, the Korotkoff sounds become longer and quieter. Eventually the sound disappears completely (the fifth and last Korotkoff sound) as blood flow becomes continuous and nonturbulent because the cuff pressure is below arterial diastolic pressure. The sphygmomanometer pressures corresponding with the first and last Korotkoff sounds are recorded as the systolic and diastolic arterial pressures.

Arterial pressures are normally reported in terms of the *systolic/diastolic* pressures, with normal values of *120/80 mmHg* in systemic arteries. (Equivalent values in pulmonary arteries are considerably lower; about 25/10 mmHg under resting conditions.) Sometimes *mean arterial pressure* (MAP) is calculated:

$$\text{MAP} = \text{Diastolic pressure} + \frac{\text{Pulse pressure}}{3} \qquad \text{(Eq. 11)}$$

where Pulse pressure = (Systolic pressure – Diastolic pressure).

A simple (arithmetic) average of systolic and diastolic values is not used for the mean pressure because diastole normally lasts twice as long as systole (Fig. 41).

Blood pressure and peripheral resistance

For fluid to flow through a pipe, there must be a pressure gradient between the two ends of the pipe (Fig. 46). The size of that gradient (ΔP) equals the rate of fluid flow (Q) times the resistance to flow (R), i.e.

$$\Delta P = Q \times R \qquad \text{(Eq. 12)}$$

Applying this to the systemic circulation, ΔP is the pressure gradient between the aorta and the right atrium. Since atrial pressure (approximately 1 mmHg) is negligible in comparison with arterial pressure (normally 120/80 mmHg), ΔP effectively equals arterial blood pressure. The total blood flow through the systemic circulation, i.e., the cardiac output, is Q. Thus:

$$\text{Arterial pressure } (\Delta P) = \text{Cardiac output } (Q) \times \text{Peripheral resistance } (R) \qquad \text{(Eq. 13)}$$

where *peripheral resistance* is the total resistance to blood flow through the systemic circulation.

This is an important relationship because it indicates that blood pressure may be regulated through changes in either cardiac output *or* peripheral resistance.

Q = Flow rate

R = Resistance

P = Pressure

$P_1 \longrightarrow Q \qquad P_2$

$P_1 - P_2 = \Delta P = Q \times R$

∴ Arterial pressure = Cardiac output × Peripheral resistance

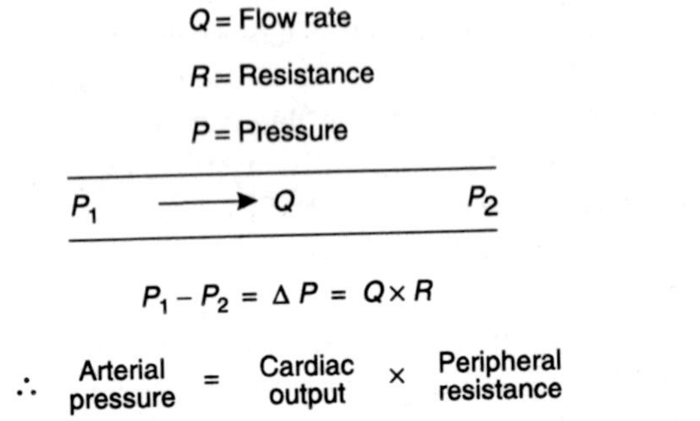

Fig. 46 The main physical variables controlling flow in the circulation.

Pressures and resistances in different types of blood vessel

Using Equation 12 we can see that if Q is constant, then the pressure difference between two points in the circulation will be proportional to the resistance of the blood vessels between those two points. The total blood flow through any class of systemic blood vessel (e.g., all arteries or all veins) must be equal to the cardiac output, so the drop in pressure along any type of blood vessel is proportional to the total resistance to blood flow offered by all vessels of that type. Blood pressure measurements can be made at different sites around the systemic circulation and the overall pressure profile determined (Fig. 47). The pressure in the aorta and large arteries is high and pulsatile (about 120/80 mmHg), and there is only a small drop in pressure along their length. Both the mean pressure and the amplitude of the pulse pressure (systolic pressure – diastolic pressure) decrease somewhat more in the small arteries but the largest

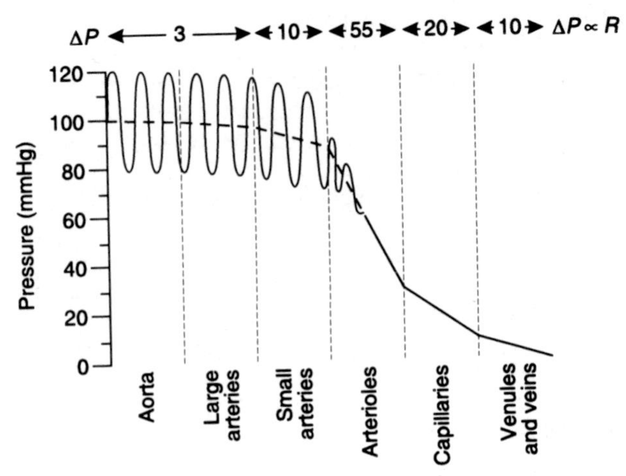

Fig. 47 Pressure profile of the systemic circulation. The pressure drop (ΔP) along each vascular element reflects its resistance since the total flow through each class of vessel equals the cardiac output.

pressure drop occurs within the arterioles. This tells us that the single largest contribution to the *peripheral resistance* comes from the *arterioles* (about 55% of total resistance). Arterioles have a considerable amount of smooth muscle in their walls and peripheral resistance can be actively controlled by constriction or dilatation of these vessels (*vasoconstriction* and *vasodilatation*, respectively). The resulting changes in arteriolar diameter dramatically affect their resistance since the resistance (R) of a single vessel is inversely proportional to the fourth power of the radius (r), i.e.

$$R \propto \frac{1}{r^4} \qquad \text{(Eq. 14)}$$

Physical factors affecting vascular resistance

The total resistance offered by all the blood vessels in the body of any given type represents the balance between the resistance of a single vessel and the total number of such vessels in parallel with each other. The resistance of a single blood vessel to fluid flow is mainly dictated by its radius (Eq. 14), so an individual capillary (diameter of 5–10 µm) offers a much greater resistance to blood flow than a single arteriole (diameter of 10–100 µm). Within the circulation as a whole, however, many blood vessels of each type are arranged in parallel, and this reduces their total resistance. The total resistance of all capillaries taken together is less than that of all arterioles, however, because there is a much larger number of capillaries in parallel with each other and this more than compensates for the higher resistance in a single capillary.

Blood viscosity

Different liquids offer different intrinsic resistances to flow, as measured in terms of their viscosity. The higher the viscosity, the greater the pressure gradient required to achieve a given flow rate. For this reason, any increase in the viscosity of blood will increase the peripheral resistance of the circulation. Such increases may occur following an increase in the erythrocyte count (polycythaemia). As the haematocrit rises (Section 2.1) blood viscosity may be more than doubled, and this can eventually lead to ventricular failure.

Regulation of arterial blood pressure

Arterial blood pressure (BP) is equal to the cardiac output multiplied by the peripheral resistance (Eq. 13). It follows from this that arterial blood pressure can be controlled by mechanisms which alter either the *cardiac output* or the *peripheral resistance* (Fig. 48). Peripheral resistance is regulated through changes in arteriolar constriction. Changes in cardiac output result from changes in cardiac function or changes in venous return. The latter often result from changes in blood volume, since these produce parallel changes in the rate of return of blood to the heart. Regulatory mechanisms may act quickly, to provide short term control of blood pressure from minute to minute, or over the longer term, setting the mean blood pressure from week to week.

Nervous control of blood pressure

The short term control of blood pressure relies on nerve-based reflexes which can detect and respond to changes in blood pressure within seconds. The efferent, or motor, limb of these reflexes involves autonomic nerves which are regulated from the *medulla oblongata*.

The vasomotor centre The vasomotor centre activates *sympathetic nerves*, which can raise blood pressure through four separate actions:

- Sympathetic nerves to the heart *stimulate heart rate* and *contractility*, increasing cardiac output.
- Sympathetic nerves to the arterioles release noradrenaline (norepinephrine), which causes contraction of the smooth muscle in their walls. This *vasoconstriction* increases peripheral resistance. A secondary effect of vasoconstriction is to reduce the hydrostatic pressure in

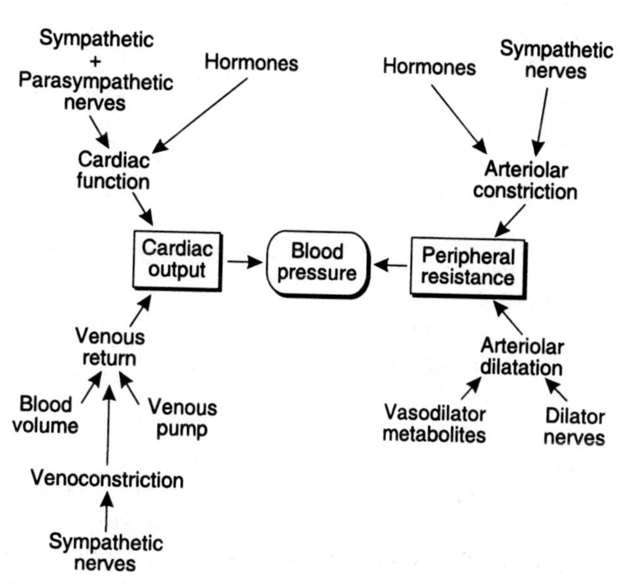

Fig. 48 A summary of the main controls that affect arterial blood pressure.

capillaries. This favours absorption of fluid from the interstitial space into the circulation (Section 3.8), increasing the circulating volume and thus promoting venous return.

- Sympathetic nerves to the venules and veins cause these to constrict. *Venoconstriction* has little effect on peripheral resistance but directly increases venous return, raising cardiac output.
- Preganglionic sympathetic nerves stimulate release of noradrenaline (norepinephrine) and adrenaline (epinephrine) from the adrenal medulla (Section 8.6). These hormones have similar cardiovascular effects to those of the postganglionic sympathetic transmitter noradrenaline (norepinephrine).

The cardiac parasympathetic nucleus The term cardiac parasympathetic nucleus will be used here to represent a variety of areas in the medulla oblongata which activate the cardiac parasympathetic fibres in the vagus, thus slowing the heart, decreasing cardiac output, and reducing blood pressure. There are very few parasympathetic nerves supplying blood vessels and so they have little effect on peripheral resistance or venous return.

Reflex regulation of autonomic activity

The activity in sympathetic and parasympathetic nerves is modulated by a number of important cardiovascular receptors and reflexes.

Baroreceptor reflexes Baroreceptors are stretch receptors located in the walls of the *carotid sinus*, where the common carotid artery divides, and in the *aortic arch*. Increases in arterial pressure lead to stretching of the aorta and the carotid sinus which stimulates the sensory output from the baroreceptors. This *inhibits sympathetic* outflow to the cardiovascular system, reducing cardiac output and peripheral resistance through the mechanisms outlined above, and *stimulates parasympathetic* nerves to the heart, again reducing cardiac output. All of these effects tend to reduce blood pressure. Conversely, reductions in arterial pressure lead to reduced baroreceptor activity, which increases sympathetic outflow and decreases parasympathetic outflow, thereby helping raise blood pressure back towards normal.

The sensitivity ranges of the two sets of baroreceptors are slightly different. In the carotid sinus, there is no baroreceptor activity at pressures below 60 mmHg, with output rising to a maximum at 180 mmHg. The equivalent range for the aortic arch receptors is higher, going from 90 to 200 mmHg. If arterial pressure remains high for a prolonged period, however, baroreceptor sensitivity *adapts* to a higher pressure range and so they provide very little long term control of mean blood pressure.

Baroreceptor responses are very rapid, acting within a few seconds to compensate for any sudden change in blood pressure, e.g., on getting up out of bed when arterial pressure tends to fall because of the effects of gravity. An inadequate baroreceptor reflex may lead to *postural (orthostatic) hypotension*, in which the blood pressure drops abnormally low when an individual stands up from a sitting or lying position, causing dizziness or fainting.

Low-pressure volume receptor reflexes There are also stretch receptors in the walls of the great veins, atria and the pulmonary trunk. These are sometimes called the *cardiopulmonary baroreceptors* and they are particularly sensitive to changes in circulating blood volume. An increase in blood volume (which increases venous return and thus raises cardiac output and arterial blood pressure) stretches the atria and the pulmonary vasculature, stimulating the volume receptors. This tends to reduce blood pressure by two main mechanisms.

- *Vasoconstrictor sympathetic nervous* activity is decreased, reducing peripheral resistance.
- *Release of antidiuretic hormone (ADH)*, which is also known as vasopressin, from the posterior pituitary is inhibited. This is controlled through the hypothalamus, the site of ADH production (Section 8.2). ADH causes direct *vasoconstriction* and, more importantly, also stimulates *water absorption* from the collecting ducts of the kidney, increasing blood volume (Section 5.4). Volume receptor-dependent inhibition of ADH release will, therefore, reduce both peripheral resistance and cardiac output. Abnormal decreases in blood volume produce the reverse effects, stimulating both sympathetic nervous activity and ADH secretion.

Chemoreceptor reflexes The peripheral chemoreceptors are found within the aortic and carotid bodies and are sensitive to changes in tissue O_2 levels. Their primary role is in the regulation of ventilation (Section 4.5). If arterial pressure is very low, however, tissue O_2 levels may drop, even though the arterial O_2 levels are normal, simply because tissue blood flow is inadequate to meet the metabolic needs of the chemoreceptor cells. This activates the receptors, which stimulate *vasoconstrictor sympathetic nerves* in an attempt to restore the blood pressure.

Response to cerebral ischaemia At extremely low blood pressures, the blood flow to the brain can no longer match metabolic demand. The resulting accumulation of CO_2 and H^+ in cerebral tissues acts as an extremely potent, direct stimulus to the vasomotor centre in the medulla oblongata and leads to greatly increased activity in sympathetic nerves supplying the cardiovascular system.

Cushing reaction This is also a response to cerebral ischaemia, but the cause is a *rise* in *intracranial pressure*, which compresses the cerebral arteries. The direct stimulation of sympathetic nervous activity through the medulla raises arterial pressure above intracranial pressure, restoring blood flow. The abnormally high blood pressure is detected by the baroreceptors, however, and these stimulate the cardiac parasympathetic nucleus, leading to a parasympathetically induced fall in heart rate. Elevated blood pressure with a reduced heart rate is known as the Cushing reaction and is a classical sign of raised intracranial pressure, e.g., caused by an intracranial tumour.

Hormonal control of blood pressure

A number of hormones are involved in the regulation of blood pressure. These may have relatively rapid actions (within minutes) or may exert their full effect over hours to days.

Catecholamines The release of adrenaline (epinephrine) and noradrenaline (norepinephrine) by the adrenal medulla has been discussed in the context of the sympathetic activity which stimulates this release. These hormones act within minutes, causing vasoconstriction and increasing cardiac rate and contractility.

Antidiuretic hormone (vasopressin) This has also been discussed as one limb of the reflex controlled by the low-pressure volume receptors. The vasoconstrictor effects of the hormone are experienced within minutes but the increased renal reabsorption of water, and the resultant expansion of the blood volume, may take longer to develop fully.

Renin–angiotensin–aldosterone system Reductions in arterial pressure lead to decreases in renal blood flow which are detected by the *juxtaglomerular apparatus* within each nephron (Section 5.4). This responds by secreting the proteolytic enzyme *renin*, which converts the inactive precursor peptide *angiotensinogen* to *angiotensin I*. Angiotensin-converting enzyme (produced by endothelial cells) catalyses formation of the more active *angiotensin II*, a potent *vasoconstrictor* which increases peripheral resistance. This relatively rapid response is augmented by the second action of angiotensin II, i.e., stimulation of secretion of the steroid hormone *aldosterone* from the adrenal cortex. Aldosterone promotes renal reabsorption of Na^+ and water in the distal convoluted tubule, thereby expanding blood volume and increasing blood pressure. This effect requires several hours to have significant impact.

Long term regulation of blood pressure

It is not clear how mean arterial pressure is kept constant over weeks to years in normal individuals. One element of control relies on long term regulation of *blood volume*, which helps keep venous return and cardiac output constant. A variety of renal mechanisms are involved, several of them hormonally controlled.

- Changes in blood pressure may produce small but appreciable changes in renal blood flow and glomerular filtration rate. If blood pressure falls, for example, the resulting decrease in excretion of Na^+ and water favours an increase in blood volume, raising blood pressure.
- Antidiuretic hormone promotes renal reabsorption of water when blood volume decreases.
- Aldosterone promotes Na^+ and water reabsorption when blood pressure falls.
- *Atrial natriuretic peptide* favours Na^+ and water excretion by the kidneys. This hormone is secreted by the heart in response to atrial stretch, as would occur if the blood volume were elevated.

Long term regulation of *peripheral resistance* is very poorly understood. It has already been pointed out that baroreceptors have little or no role here, even though they are very important in short term control of vascular resistance. Most people who suffer from raised blood pressure (hypertension) have an abnormally high peripheral resistance, but usually the underlying cause is unknown.

Blood pressure and age

From young adulthood onwards, there is a tendency for resting systolic, diastolic and pulse pressures to rise, and these effects can become quite marked in the elderly. This means that any definition of a normal blood pressure must take the age of the patient into account. The changes seen may be at least partly explained by the *decreasing elasticity* of the arterial wall with ageing, which reduces the degree of damping of the ventricular pressure wave.

3.7 Control of regional blood flow

Learning objectives

At the end of this section you should be able to:

- explain how perfusion pressure and vascular resistance regulate tissue blood flow
- use graphs to explain the terms hyperaemia and autoregulation. Outline how local control mechanisms may help explain these phenomena
- describe how nerves and hormones may affect vascular resistance
- describe the special features of blood flow regulation in different tissues.

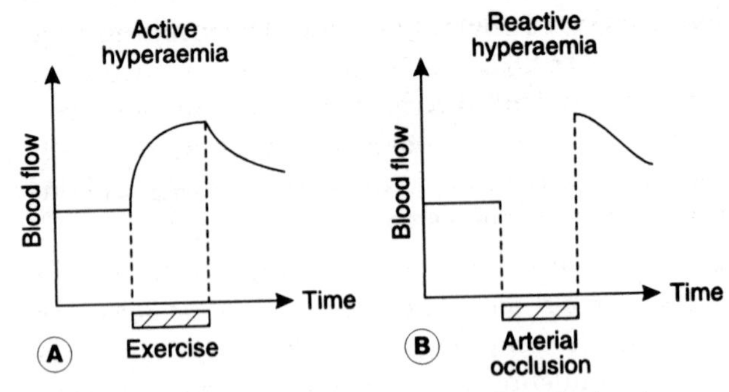

Fig. 49 Control of regional blood flow in response to changing conditions in a muscle. (A) Active hyperaemia. (B) Reactive hyperaemia.

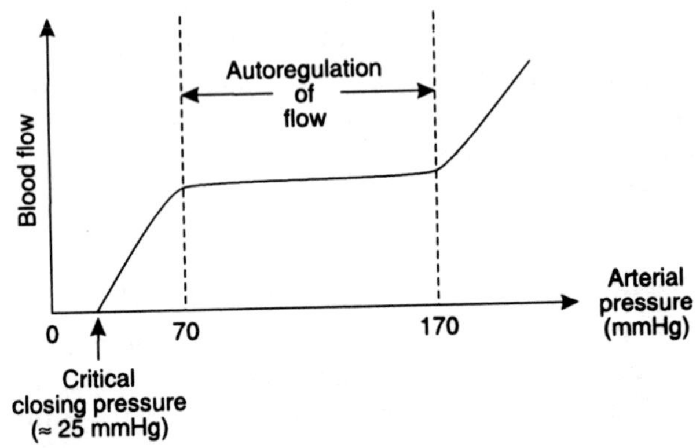

Fig. 50 Local blood flow control mechanisms help keep flow constant over a wide range of arterial pressures. At extremes of pressure this autoregulation breaks down.

The purpose of the circulation is to provide each organ with a blood supply adequate for its metabolic needs. This depends on two physical factors: the perfusion pressure and vascular resistance.

The perfusion pressure. A pressure gradient is required to perfuse the tissues with blood, and this perfusion pressure is equal to the difference between arterial and venous pressures in any tissue. Arterial pressure is particularly important since it is normally much higher than venous pressure, i.e., the maintenance of an adequate perfusion gradient depends on the maintenance of normal arterial pressure.

Vascular resistance. Local flow will depend not just on perfusion pressure but also on the local vascular resistance to flow through that organ. This relationship can be expressed as:

$$\text{Blood flow}\,(Q) = \frac{\text{Perfusion pressure}\,(\Delta P)}{\text{Resistance}\,(R)} \qquad \text{(Eq. 15)}$$

Vascular resistance is controlled by constriction or dilatation of local arterioles and is regulated both by *local* and *extrinsic* control mechanisms.

Local control of blood flow

Different tissues have different metabolic demands at different times and regional blood flow has to be varied to meet these changing needs. This depends on local factors within the relevant vascular bed. For example, blood flow to a muscle increases both during exercise (*active hyperaemia*) and following a period of arterial occlusion under resting conditions (*reactive hyperaemia*; Fig. 49). Local control of blood flow also helps to limit the effects of changes in arterial pressure. This is best demonstrated by plotting blood flow through an organ (e.g., the kidney) against arterial pressure (Fig. 50). There is relatively little change in flow over a wide pressure range (70–170 mmHg) despite a more than doubling of the perfusion pressure. This *autoregulation* of blood flow relies on local changes in vascular resistance, which largely compensate for the pressure changes. At very high or low pressures, autoregulation breaks down and blood flow becomes highly dependent on arterial pressure. It should be noted that flow drops to zero at a pressure of 20–30 mmHg. This *critical closing pressure* is thought to be the minimum pressure necessary to keep small blood vessels from collapsing.

Local control mechanisms

Several such mechanisms have been proposed. While all may be involved in controlling blood flow, the relative importance of each is unclear.

Myogenic responses Passive stretching of arterioles caused by a sudden rise in blood pressure can stimulate active contraction of their smooth muscle, leading to constriction and increased resistance to flow. Reduced stretch at lower pressures, by comparison, favours relaxation of the smooth muscle, arteriolar dilatation and decreased resistance to flow. Such responses may be involved in autoregulation of blood flow, since the changes in resistance would tend to compensate for any changes in arterial pressure.

Vasodilator metabolites Vasodilator metabolites are believed to be produced locally within body tissues whenever the supply of O_2 falls below O_2 consumption. These metabolites decrease the local vascular resistance and so increase blood flow to meet tissue needs. Such a mechanism could explain active hyperaemia (in which an increased workload leads to increased O_2 demand), reactive hyperaemia (in which arterial occlusion reduces O_2 supply) and autoregulation (in which any change in blood flow caused by a change in pressure produces an imbalance between O_2 supply and demand). No single metabolite has been identified which fully accounts for the effects seen, but candidate substances include CO_2, H^+, K^+, lactic acid, phosphate and adenosine, all of which may accumulate during periods of reduced perfusion. Different metabolites may be important in different tissues and several may be involved in any given tissue. It is also possible that a reduction in tissue O_2 may directly reduce contractile activity in arteriolar smooth muscle (by reducing the intracellular energy supply), favouring dilatation.

Other vasoactive substances Stimuli unrelated to the metabolic needs of a tissue may lead to local production of chemicals which alter vascular resistance. *Inflammation* produces a number of vasodilators which lead to increased blood flow at sites of trauma or infection, e.g., bradykinin (Section 2.5). *Endothelial cells* also produce substances which lead to contraction and relaxation of vascular smooth muscle. One example is endothelium-derived relaxing factor (EDRF), which causes dilatation. EDRF has now been identified as the gas nitric oxide, and is believed to play an important role in many vascular responses.

Extrinsic control of regional blood flow

A number of extrinsic nervous and hormonal mechanisms affect regional blood flow. These can override local tissue controls in order to maintain cardiovascular function during periods of physiological stress, e.g., when arterial pressure falls or during exercise.

Nervous control of blood flow

Nervous control of regional flow depends on autonomic nervous activity.

Sympathetic vasoconstrictor nerves release noradrenaline (norepinephrine) (*adrenergic nerves*) and regulate peripheral resistance. They are particularly dense in the vasculature of the gastrointestinal tract, skin and skeletal muscle but are present in most organs. Reflex activation of these nerves, in response to a fall in blood pressure for example, reduces local blood flow in affected tissues. The circulations to the heart and CNS are spared from such sympathetic constriction, as these vital organs must be supplied with blood. Nerve-mediated vasoconstriction effectively diverts flow away from temporarily dispensable regions in an attempt to maintain adequate perfusion of the coronary and cerebral circulations.

Sympathetic vasodilator nerves are believed to be present in skin. They cause arteriolar smooth muscle to relax, leading to an active vasodilatation. These nerves are normally activated in response to body heating, increasing cutaneous blood flow and heat loss. The neurotransmitter involved has not been identified but is thought likely to be a peptide transmitter released along with acetylcholine (a *co-transmitter*), since cholinergic sympathetic nerves are known to be responsible but active vasodilatation is not inhibited by cholinergic blocking drugs like atropine.

Parasympathetic nerves are not widely distributed to blood vessels and have no significant effect on peripheral resistance. Stimulation of *vasodilator parasympathetic nerves*, however, can produce an increase in local blood flow in exocrine glands, the blush region of the skin and the erectile tissue of the penis. These nerves are normally *cholinergic* but other vasodilating neurotransmitters may also be involved, e.g., vasoactive intestinal polypeptide.

Hormonal control of blood flow

Hormones which affect local vascular resistance, and thus alter local blood flow, have already been considered in the context of their role in blood pressure control. Vasopressin and angiotensin II have vasoconstrictor actions, as do the catecholamines in most tissues. Low concentrations of circulating adrenaline (epinephrine) may produce vasodilatation in skeletal muscle as the result of its action on receptors known as β-adrenoceptors. Adrenergic vasoconstriction, however, is mediated by α-adrenoceptors (Section 7.6).

Blood flow in specific circulations

A number of specializations in blood flow control are seen in different circulations and some of the more important of these will be considered here.

Skeletal muscle blood flow

At rest, about 15% of the cardiac output goes to skeletal muscle but this can rise to 70% in moderate exercise. Since there is also an associated increase in cardiac output this represents a 10–15 fold increase in absolute flow to muscle.

Local control, sympathetic nerves and circulating hormones all play a role in the control of blood flow through skeletal muscle.

Local metabolic control. This is very important, especially during exercise, producing vasodilatation and an increased blood flow to match the increased workload of the muscle cells (Fig. 49).

Neurohumoral control. This largely depends on sympathetic nerves and catecholamines released from the adrenal medulla. Changing levels of nerve activity and hormone concentration may lead to vasoconstrictor or vasodilator responses.

- *Vasoconstriction.* Increased activity in adrenergic sympathetic nerves is reflexly activated by the baroreceptors in response to a *fall in arterial blood pressure* (Section 3.6). The resulting α-adrenoceptor dependent vasoconstriction reduces skeletal muscle blood flow and helps raise peripheral resistance, favouring an increase in arterial pressure.
- *Vasodilatation.* Vascular resistance is reduced in skeletal muscle both during exercise and as part of the reflex response to fear or anger, the 'fight or flight' response. The sympathetic nerves contribute to this through reflex reduction in vasoconstrictor activity, leading to passive vasodilatation. Although there is also evidence for additional active vasodilatation in muscle under these circumstances, this is not mediated directly by nerves in humans but probably reflects the action of circulating adrenaline (epinephrine) on vascular β-adrenoceptors. During exercise, however, active vasodilatation mainly depends on local metabolite production.

Cutaneous blood flow

Blood flow to the skin represents about 10% of resting cardiac output. This greatly exceeds metabolic need, suggesting that local control is not very important in flow regulation. Nervous control is dominant and skin blood flow is adjusted to meet the *temperature regulation* requirements of the body through hypothalamically regulated changes in sympathetic activity. This acts on two different types of cutaneous resistance vessel: normal arterioles and *arteriovenous anastomoses*, which directly link small arteries to the cutaneous venous plexus, bypassing the capillaries (Fig. 51). This venous plexus has a large surface area and is well adapted for efficient heat loss from blood to the environment. Arteriovenous anastomoses are found in the palms of the hands, the soles of the feet, the ear lobes and around the lips.

If body temperature drops, the hypothalamus activates sympathetic vasoconstriction, which reduces cutaneous blood flow, decreasing heat loss. As core temperature rises, vasoconstrictor activity decreases, leading to passive dilatation of arterioles (i.e., reduced vasoconstriction) and a general increase in blood flow. Arteriovenous anastomoses also open, diverting blood into the venous plexus. Both these effects promote heat loss. Body heating is also associated with nervously mediated, active vasodilatation of cutaneous arterioles. This is probably caused by release of a co-transmitter (possibly a dilator peptide) from cholinergic sympathetic nerves.

It should also be noted that sympathetic vasoconstrictor nerves greatly decrease cutaneous blood flow during periods of low arterial pressure. This explains why individuals suffering from shock appear so pale.

Coronary blood flow

The coronary blood vessels supply the heart and so coronary blood flow is vital for survival. The heart receives about 4–5% of the resting cardiac output. Vascular resistance is under local metabolic control and flow increases with cardiac workload. Because the myocardium compresses the arteries running through it as it contracts during systole, this flow is restricted to diastole. This, along with reduced time for ventricular filling, limits maximal heart rate, since the time available for myocardial perfusion also decreases as the rate of contraction increases.

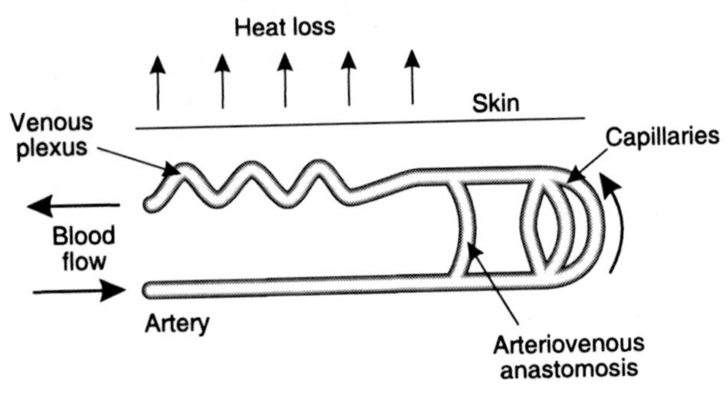

Fig. 51 Cutaneous blood vessels.

The coronary circulation is spared from sympathetic vasoconstriction, during exercise or periods of low blood pressure for example, since it is essential for cardiac function. Autonomic stimulation of the heart by the sympathetic nerves may lead to increased coronary blood flow but this is an indirect effect secondary to the sympathetically induced increase in cardiac workload.

Cerebral blood flow

The brain normally receives 15% of the resting cardiac output. Vascular control depends on local metabolic factors, with a fall in O_2 and rises in CO_2 or H^+ all leading to vasodilatation. This leads to a general fall in resistance within the cerebral circulation if perfusion pressure drops for any reason and can also raise blood flow to specific areas of the brain involved in increased neuronal activity. Cerebral blood vessels are minimally involved in sympathetic vasoconstrictor responses. This is hardly surprising since these reflexes generally operate so as to maintain a continuous blood supply to what is arguably the single most important organ in the body.

Pulmonary blood flow

Pulmonary blood flow is controlled locally by the alveolar gas tensions in a given region. Alveolar O_2 normally diffuses out of the alveoli into the blood (Section 4.3). If the partial pressure of O_2 in the alveoli falls (alveolar P_{O_2} decreases), or the partial pressure of CO_2 rises (alveolar P_{CO_2} increases), this stimulates pulmonary vasoconstriction. The resulting increase in resistance reduces flow to that region, diverting blood to areas of the lung with more normal gas pressures. This pattern of local control of blood flow is the reverse of that seen in the systemic circulation, where low tissue O_2 and high tissue CO_2 are associated with vasodilatation.

Box 11 Clinical note: Cor pulmonale

If pulmonary vasoconstriction is widespread, there may be an increase in pulmonary arterial pressure because of increased total pulmonary resistance, which is normally only about one-third of systemic resistance. This type of *pulmonary hypertension* is sometimes seen in patients with chronic respiratory disease. It can eventually lead to hypertrophy and then failure of the right ventricle. Right ventricular hypertrophy secondary to lung disease is called *cor pulmonale*.

Splanchnic blood flow

The splanchnic circulation comprises the vasculature of the entire gastrointestinal tract as well as that of the liver, pancreas and spleen, and receives about 25% of the resting cardiac output. Blood flow through this system is greatly reduced by reflex activation of adrenergic sympathetic vasoconstrictor nerves in response to hypotension or during exercise. This tends to raise peripheral resistance and blood pressure, ensuring that perfusion of other more essential areas is maintained. Under these conditions sympathetic nerves also constrict the splanchnic veins, increasing venous return and thus cardiac output. The considerable importance of this mechanism reflects the large splanchnic contribution to the total *capacitance* of the systemic circulation (Section 3.9). In contrast to these vasoconstrictor and venoconstrictor responses, vascular resistance falls and gastrointestinal blood flow is increased following ingestion of a meal. This probably depends on locally produced dilator hormones such as secretin, cholecystokinin and vasoactive intestinal polypeptide, as well as intestinal kinins, rather than the direct action of nerves.

3.8 Capillary function

Learning objectives

At the end of this section you should be able to:

- describe the mechanisms responsible for movement across the capillary wall
- outline the roles of hydrostatic and colloid osmotic forces in controlling fluid filtration; indicate approximate values for these forces in capillaries
- explain oedema in terms of these forces.

Fluid and solute exchange between blood and the interstitium occurs almost exclusively in capillaries, so it is at this level that the cardiovascular system fulfils its primary function, i.e., the support of cellular metabolism.

Organization of the microcirculation

Arterioles either open into capillaries directly, or via smaller metarterioles, each of which supplies a number of capillary vessels (Fig. 52). Spontaneous rhythmical contraction and relaxation of arterioles and metarterioles results in *vasomotion*, with continuously changing blood flow within individual

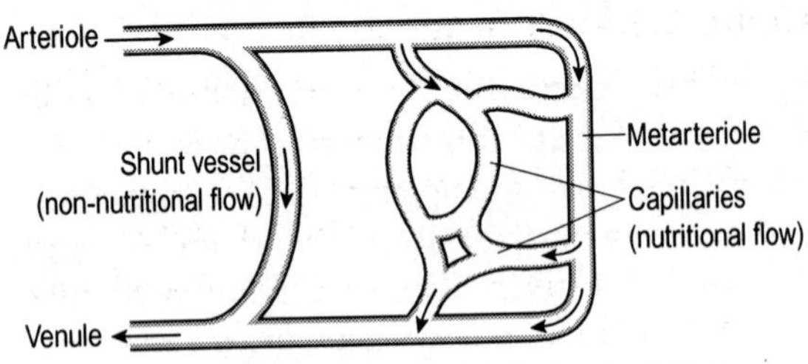

Fig. 52 Outline of the elements of the microcirculation.

capillaries. Mean flow and pressure within an entire capillary bed remain fairly constant, however, if there is no major change in conditions. Shunt vessels can also be opened, diverting blood away from adjacent capillaries.

Capillary structure

The capillary wall represents the main barrier to nutrient exchange in the microcirculation. It contains no smooth muscle and consists of a single layer of endothelial cells surrounded by a basement membrane. There are potential spaces, known as *intercellular clefts*, between adjacent cells. In some tissues these gaps are effectively closed, reducing the permeability of the capillary wall to plasma solutes. This is seen in the brain for example, and contributes to the functional blood–brain barrier. In some tissues there are holes, or *fenestrations*, through the endothelial cells themselves. Fenestrated capillaries offer much less resistance to fluid exchange across the capillary wall than normal and are found in a range of fluid secreting organs, e.g., the renal glomeruli, endocrine and exocrine glands, the lining of synovial joints, and choroid plexus, which secretes CSF into the ventricles of the brain.

Capillary exchange mechanisms

A number of different processes contribute to transport across the capillary wall allowing exchange to occur between plasma and interstitial fluid.

Diffusion

This mechanism accounts for the greatest total amount of transcapillary exchange, as solutes diffuse into or out of the capillaries. The direction and rate of diffusion for any molecule or ion depends on two main factors.

Transcapillary concentration gradients provide the driving force for diffusion. Glucose and O_2 are more concentrated in plasma than in the interstitium, and they diffuse out of the capillary, while waste products such as CO_2, which tend to accumulate around metabolizing cells, diffuse in.

Capillary permeability dictates the rate of diffusion under any given concentration conditions. Lipid-soluble substances can rapidly diffuse across the endothelial cells themselves, e.g., O_2 and CO_2. Polar substances, such as ions and glucose, are lipid insoluble but they also cross the capillary wall readily—through the intercellular clefts and, when present, capillary fenestrations. Such passage is limited by molecular size, however, and large polar molecules, like the plasma proteins, cannot easily escape from the capillary.

Pinocytosis

Pinocytotic activity (endocytosis and exocytosis; Section 1.3) in endothelial cells may account for some capillary transport of fluid and large molecules, but the total contribution is probably small.

Fluid filtration and absorption

Pressure gradients across the capillary wall lead to bulk flow of fluid, in which water and the small ions and molecules dissolved in it are driven across the capillary wall. It is important to note that while this process accounts for all net fluid movement, it accounts for only a small fraction of the total transcapillary solute transport, because solute diffusion is generally much faster than convective transport of solutes along with the filtered fluid ('wash-along'). The forces involved in fluid filtration are generally referred to as the Starling forces, in honour of the main proponent of this model of fluid exchange. The following factors are important in determining the resulting rates of capillary filtration or absorption.

The hydrostatic pressure gradient (ΔP) is the difference between capillary pressure (P_c) and interstitial fluid pressure (P_i), i.e., $\Delta P = P_c - P_i$. Since $P_c > P_i$, the hydrostatic gradient acts out of the capillary, favouring filtration (Fig. 53A).

The osmotic pressure gradient ($\Delta\pi$) is generated because the capillary wall acts as a semipermeable membrane with respect to *plasma proteins*, i.e., water can flow through the intercellular clefts and fenestrations freely but proteins penetrate these barriers much less easily. If unopposed, this would result in fluid flow into the capillaries. The osmotic pressure generated by the plasma proteins is often called the *colloid osmotic pressure*, or *oncotic pressure* of plasma. Protein accounts for a very small fraction of the total osmolality of plasma, which is largely caused by the dissolved electrolytes (Section 1.3), but proteins are

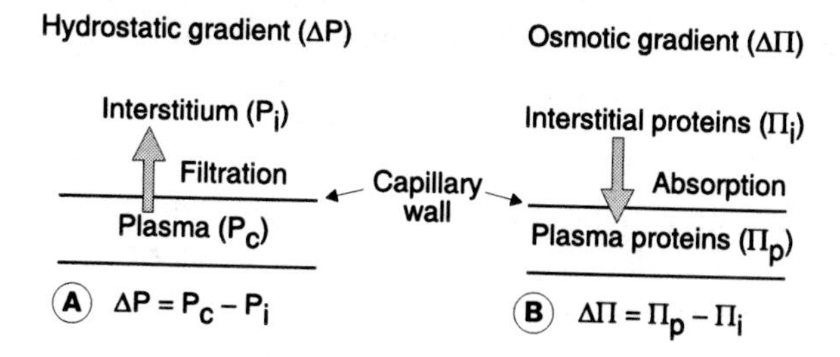

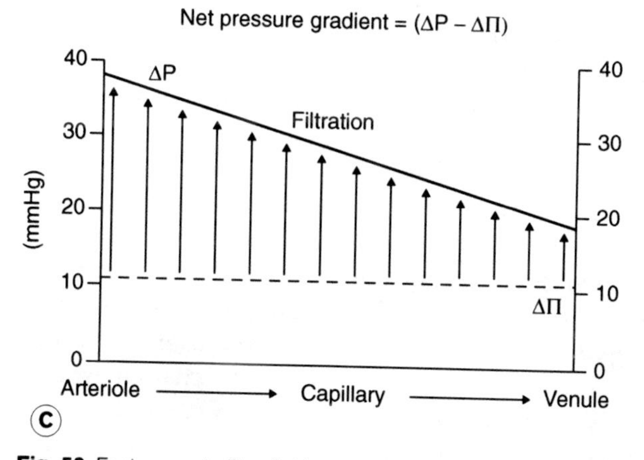

Fig. 53 Factors controlling fluid exchange across the capillary wall (see text). (A) The hydrostatic gradient. (B) The colloid osmotic gradient. (C) The net pressure gradient.

the only relevant osmotic influence at the capillary wall, which is readily permeable to smaller solutes. The colloid osmotic pressure is affected by two factors: the concentration of plasma protein and the reflection coefficient (σ) at the capillary wall. The reflection coefficient allows for the fact that the capillary wall is not completely impermeable to protein and so the osmotic effect of that protein is less than it would be at a perfect semipermeable membrane, i.e., one which 'reflects' all the protein, preventing any from escaping. The actual colloid osmotic pressure for a protein solution = $\sigma(\pi)$, where π is the osmotic pressure which would be generated across a perfect semipermeable membrane. The value for the reflection coefficient must lie between zero, where there is free diffusion of solute and no resulting osmotic pressure, and unity, for a perfect semipermeable membrane: i.e., $0 \leq \sigma \leq 1$. The value of σ for protein varies in different capillary beds, but a value of 0.9 is reasonable for most capillaries, i.e., the actual colloid osmotic pressure gradient across the capillary wall is 90% of the theoretical value predicted from the protein concentration gradient. Protein concentration, which determines the osmotic effect, is higher in plasma than in interstitial fluid, so osmotic pressure resulting from the plasma proteins (π_p) is greater than that due to the protein in the interstitium (π_i), i.e., the osmotic gradient ($\pi_p - \pi_i$) favours absorption of fluid into the capillary (Fig. 53B). The effective difference in colloid osmotic pressure across the membrane = $\sigma\Delta\pi = \sigma(\pi_p - \pi_i)$.

The net filtration pressure gradient ($\Delta P - \sigma\Delta\pi$) determines the direction of fluid transfer at any given point along the capillary. As we travel from the arteriolar to the venular end, capillary pressure falls from about 35 mmHg to about 15 mmHg. This decline represents the pressure gradient necessary to drive blood flow through the capillary resistance (Eq. 12; Fig. 53C). Interstitial fluid pressure is hard to measure but is probably subatmospheric in many tissues (about –2 mmHg), i.e., it actually increases the hydrostatic gradient across the capillary wall. The hydrostatic gradient acting out of the capillary, ΔP, falls, therefore, from 37 mmHg at the arteriolar end to 17 mmHg at the venular end, with a mean value of 27 mmHg. The colloid osmotic gradient reflects the difference between the colloid osmotic pressure of plasma (π_p = 25 mmHg) and that of the interstitial fluid (π_i = 12 mmHg). This latter value, which is based on the best current measurements, is larger than many texts allow for, and reduces the colloid osmotic pressure gradient to about 12 mmHg (i.e., $\sigma\Delta\pi = \sigma(\pi_p - \pi_i) = 0.9\,(25 - 12) = 12$ mmHg). Using these values for the hydrostatic and osmotic gradients to calculate the net pressure gradient, we see that this favours filtration along the whole length of the capillary. Although the size of the filtration gradient decreases as the capillary hydrostatic pressure falls, the outwardly acting hydrostatic gradient exceeds the inwardly directed osmotic gradient even at the venular end of the capillary (Fig. 53C). The filtered fluid is returned to the circulation by means of the *lymphatic system*, so the interstitial fluid volume remains constant. This situation applies in all but a small number of tissues specialized for absorption, e.g., in the capillaries of the intestinal mucosa and renal tubules. It should be noted that there are situations in which the balance of the Starling forces can also be reversed in other capillaries, resulting in fluid absorption. This occurs following major blood loss, for example, when reduced blood pressure and reflex arteriolar constriction dramatically lower the capillary hydrostatic pressure, resulting in a limited period of capillary fluid absorption, which has an important compensatory effect as it helps re-expand the blood volume, and thus raise arterial pressure.

The capillary filtration coefficient is a term which describes the efficiency of fluid exchange in a given tissue in terms of the rate of filtration per unit filtration pressure per unit mass of tissue. The filtration coefficient is high in regions containing capillaries with a *high permeability* to fluid, e.g., the fenestrated glomerular capillaries of the kidney, which are specially adapted for filtration (Section 5.3). If the *total surface area* of capillary wall available for exchange is large, i.e., if there is a high density of capillaries per unit of tissue volume, this will also increase the filtration coefficient.

Oedema

Oedema is swelling caused by an accumulation of interstitial fluid and is a common finding in a range of clinical conditions. It is best understood in terms of the factors which control transcapillary fluid exchange. The following list summarizes the main causes of oedema.

Increases in the *hydrostatic gradient* increase the rate of capillary filtration. This occurs whenever capillary pressure increases substantially over a long period, e.g., because of a rise in venous pressure. This may result from prolonged standing, heart failure or venous obstruction by a clot or tumour.

Decreases in the *osmotic gradient* reduce absorption, increasing net filtration. This commonly results from a low plasma protein concentration, e.g., because of liver failure, renal disease or malnutrition. If the *capillary permeability* to protein rises, this also reduces the effective osmotic gradient, both by decreasing σ and by increasing π_i. This helps account for the severe, rapid oedema seen in inflammatory responses to infection or trauma.

Increases in the *capillary filtration coefficient* also favour increased fluid filtration for any given net filtration gradient. This contributes further to the oedema of inflammation.

Lymphatic obstruction prevents both fluid and protein from being cleared from the tissues. The resulting swelling is referred to as *lymphoedema*.

3.9 Venous system

Learning objectives

At the end of this section you should be able to:

- explain the role of veins as high compliance vessels within the circulation
- describe and explain the jugular venous pressure waveform and list factors which affect central venous pressure
- outline the main factors which affect peripheral venous pressure
- list the main causes of fainting and outline the mechanisms responsible.

Veins have two important functions, acting as:

- conduit vessels
- reservoir vessels.

Veins act as *conduit vessels*, which complete the circulation of blood back to the heart. This is promoted by the presence of one-way valves which ensure that compression by contracting muscles forces blood centrally.

Veins also act as *reservoir* or *capacitance vessels*, storing blood which can be mobilized when needed. This reflects the fact that veins distend much more easily than arteries, allowing them to stretch as intravenous pressure rises. A plot of vascular volume against intravascular pressure for the systemic circulation (Fig. 54) shows that, as well as containing a greater total volume of blood (65% of the blood volume under normal pressure conditions as compared with only 15% in the systemic arterial system), the rate of change in volume with changing pressure, or the *compliance*, is also much higher in veins. If intravenous pressure falls, as occurs following blood loss for example, venous volume will drop, thus helping to restore the effective circulating blood volume. This effect can be greatly accentuated by sympathetic venoconstrictor nerves, which can mobilize an additional 500 ml of blood from the venous store, particularly from the liver, spleen and gastrointestinal tract.

Central venous pressure

This is the outflow pressure for the venous system and is measured as the pressure in the right atrium since this is where the venae cavae terminate. It

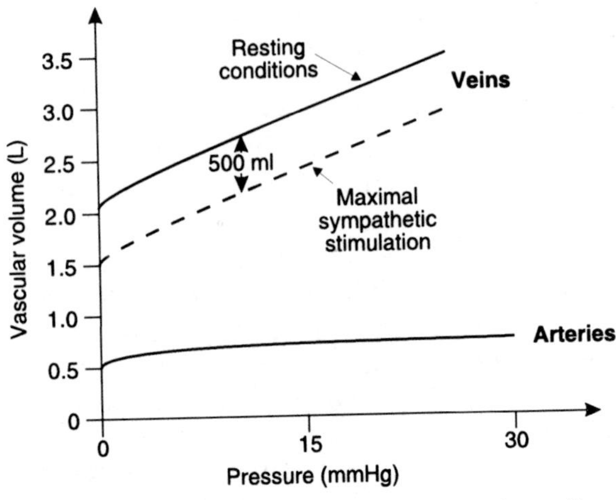

Fig. 54 Changes in systemic venous and arterial volume with changing intravascular pressure. The slope of each graph ($\Delta V/\Delta P$) equals the compliance. Stimulation of the sympathetic nerves can mobilize 500 ml of blood from the venous store.

demonstrates the 'a', 'c' and 'v' waves typical of atrial pressure changes (Fig. 41), and has a mean value in the range of 2–6 cmH_2O (1–5 mmHg). Central venous pressure depends on the balance between venous return, which provides the inflow to the right atrium, and the function of the right ventricle which pumps the blood away. Increases and decreases in blood volume produce parallel changes in venous return and central venous pressure, while right ventricular failure leads to a rise in central venous pressure. Direct measurement of central venous pressure is often used to monitor whether acutely ill patients are being under- or overinfused with intravenous fluids.

Jugular venous pulse

This can be observed clinically as a double pulsation in the jugular veins of the neck, with peaks corresponding to the 'a' and 'v' waves of right atrial pressure (the 'c' wave is rarely visible). The jugular veins act as physiological manometers and reflect the central venous pressure. The vertical height of the right jugular pulse relative to the *sternal angle* (a marker for the level of the right atrium) is normally recorded and is usually about 2–4 cm. Jugular venous pressure is raised in right ventricular failure.

Peripheral venous pressure

This refers to the pressure in the veins draining the various tissues and organs of the body. It is of particular clinical interest because venous pressure directly affects the pressure in capillaries, and elevated venous pressure is an important cause of oedema.

Three main factors affect peripheral venous pressure.

Central venous pressure. Normally central venous pressure contributes 2–3 mmHg to venous pressure, but if it rises or falls, so does peripheral venous pressure.

Venous resistance. There is a pressure gradient between the peripheral and central veins caused by venous resistance to blood flow. This resistance is low, so the pressure fall is only about 7–9 mmHg. This means that the peripheral venous pressure at heart level is about 9–12 mmHg (central venous pressure + pressure fall resulting from flow) (Fig. 55).

Posture. This has important effects on venous pressure. Gravity generates *hydrostatic* pressure gradients (=ρgh) in any vein above or below heart level (the right atrium is the zero reference level for hydrostatic pressures in the circulation). For example, the pressure in feet veins may increase from 10 mmHg

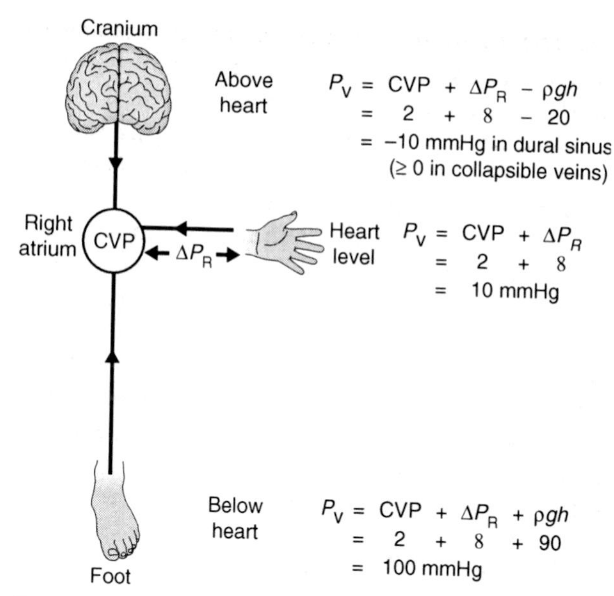

Fig. 55 Factors contributing to peripheral venous pressure for veins at different levels relative to the heart. Above the heart negative pressures may occur in semi-rigid intracranial veins.

when lying horizontally to 100 mmHg when standing upright because of the vertical column of blood between the heart and the foot (Fig. 55). The full hydrostatic effect is only seen if the subject stands perfectly still, however, since leg movement reduces the venous pressure through the venous pumping action of the skeletal muscles (Fig. 56 below). Hydrostatic effects reduce the pressure in veins above heart level, but this cannot fall below atmospheric pressure (0 mmHg) in the veins of the neck, since these will simply collapse (Fig. 55). This increases their resistance and may reduce blood flow. If a vein is effectively rigid, however, negative (subatmospheric) venous pressures may result. This occurs in the dural sinuses inside the skull in the upright position. Any accidental opening between the atmosphere and these sinuses, e.g., during intracranial surgery, may lead to air being sucked into the circulation. The resulting *air embolism* can be fatal.

Extrinsic venous pumps

Two mechanical pumping mechanisms assist venous return.

Ventilatory movements. These are important since inspiration generates negative intrathoracic pressures and positive intra-abdominal pressures (Section 4.1). This favours the transfer of blood from the abdominal veins to the inferior vena cava. Expiration reverses these pressure changes, but the *venous valves* prevent back flow of blood.

Skeletal muscle contractions. Rhythmical extrinsic compression by contracting skeletal muscle generates centrally directed blood flow because of the presence of venous valves (Fig. 56). This muscle pump increases venous return and reduces the peripheral venous pressure, e.g., venous pressure in the feet is reduced from 100 mmHg when standing still to 25 mmHg when walking. In patients with dilated varicose veins, the valves are no longer capable of preventing back flow (the valves are *incompetent*). The muscle pump is of little benefit in such veins, since each compression forces blood equally in both directions. This leads to chronically elevated venous pressures in the legs, resulting in oedema and ulceration of the overlying skin.

Posture and fainting (syncope)

The intravascular pressure increases greatly in the lower parts of the body on standing up. This can reduce arterial pressure (i.e., as measured at heart level) by two mechanisms, one rapid and one which develops more slowly. In either case, cerebral blood flow may be compromised leading to a brief loss of consciousness referred to as fainting or syncope.

Pooling of blood. Elevated venous pressure leads to rapid pooling of blood in the distensible veins of the lower body (Fig. 54). This effectively reduces the circulating blood volume, decreasing venous return and cardiac output. The resulting fall in arterial pressure is normally limited by baroreceptor activation of sympathetic cardiostimulatory and vasoconstrictor nerves, although cerebral blood flow may still drop by up to 20%. Baroreceptor reflexes may be impaired, for example in the elderly, or in patients suffering from neuropathies of the autonomic nerves, or as a side-effect of drug treatment, and the more pronounced fall in blood pressure on standing under these conditions (*orthostatic hypotension*) may reduce cerebral perfusion by as much as 40%, causing dizziness or fainting. On losing consciousness, the sufferer will normally fall to the ground. This is a protective mechanism since the hypotensive effects of gravity are reversed in the prone position, and both blood pressure and cerebral blood flow rapidly increase. For this reason it is unwise to try to sit a faint victim upright until recovery is complete.

Increased fluid filtration. Elevated venous pressure on standing increases capillary pressure and favours increased fluid filtration from the plasma to the interstitium (Section 3.8). This leads to a more slowly evolving depletion of blood volume, which tends to reduce arterial pressure. As a result, normal individuals forced to stand still for long periods of time, e.g., soldiers on public parade, may eventually faint. Walking or moving greatly diminishes venous (and therefore capillary) pressure through the action of the muscle pump, making fainting much less likely.

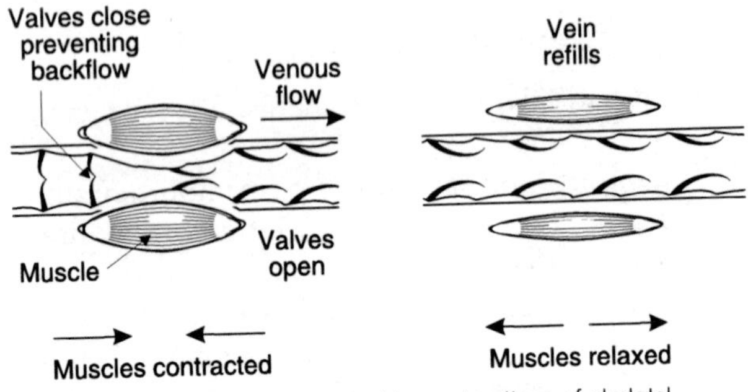

Fig. 56 Venous return is assisted by contractions of skeletal muscle provided that venous valves are competent.

Other causes of syncope

The common element is a reduction in blood pressure, which decreases cerebral blood flow.

- Intense emotions such as fear or anxiety may act through the hypothalamus to increase parasympathetic activity to the heart (reducing cardiac output) and inhibit vasoconstrictor sympathetic nerves to skeletal muscle (reducing peripheral resistance). This can lead to fainting associated with a low heart rate which is often called *vasovagal syncope.*
- *Micturition syncope* is caused by a combination of postural hypotension and a reflex reduction in heart rate associated with bladder emptying.
- *Cough syncope* is caused by the increased intrathoracic pressure generated during each cough, which tends to obstruct the venae cavae, reducing venous return.
- *Carotid sinus syncope* is caused by external pressure on the carotid sinus, e.g., from tight-fitting collars. This mechanically stimulates the baroreceptors and the resulting sympathetic inhibition and vagal stimulation reduce arterial pressure, causing fainting.
- *Exertion syncope* occurs when the metabolic demands of the body exceed the maximum cardiac output. Vasodilatation reduces peripheral resistance and blood pressure falls. Syncope becomes more likely if cardiac output is limited, e.g., in patients with aortic stenosis, who have pathological narrowing of the aortic valve which restricts the stroke volume.

3.10 Lymphatics

Learning objectives

At the end of this section you should be able to:

- outline the structure and main functions of the lymphatics.

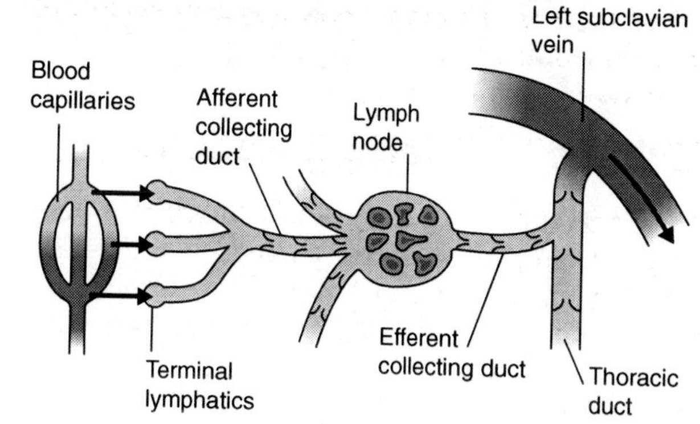

Fig. 57 The structure of the lymphatic system. Lymph is formed from interstitial fluid.

Relevant structure

The lymphatics begin as endothelial bulbs in the interstitium known as *lymphatic capillaries* or *terminal lymphatics*. These are much more permeable than blood capillaries, allowing proteins and even whole cells to enter. Tissue fluid which finds its way into these lymphatics is then called *lymph* (Fig. 57). It passes into larger, valved vessels, known as *collecting ducts*, which have smooth muscle in their walls. Afferent lymphatic ducts supply lymph nodes, which drain in turn by efferent ducts. Eventually the lymph empties into the central veins by means of major lymph vessels, e.g., the *thoracic duct* which drains lymph from the lower part of the body, the left arm and the left side of the head and neck into the left subclavian vein.

Lymphatic functions

- Every 24 hours, a load of filtered fluid and protein equivalent to that contained in the entire plasma volume is returned to the circulation by the lymphatics. Formation of lymph at the level of the terminal lymphatics can be greatly accelerated by compression of the tissues caused by movement, but subsequent propulsion seems to involve the *intrinsic contractile activity* of the lymph ducts. This produces unidirectional flow because of the regular lymphatic valves.
- The lymphatic system has important *immune functions*. Foreign antigens are carried from the tissues to regional lymph nodes by the lymphatic ducts and there they come into contact with antigen-specific lymphocytes (Section 2.5). There is also a continuous recirculation of lymphocytes through the tissues to the blood via the lymph.
- Lymphatics of the small intestine are important in the transport of *absorbed lipids* to the circulation (Section 6.4).

Self-assessment: questions

Multiple true/false questions

Each of the following statements consists of a stem followed a number of possible endings. State whether each statement is True or False. For each stem, all, several or none of the statements may be true.

1. The heart:
   a. is symmetrical about a plane along the ventricular septum
   b. converts chemical energy to mechanical energy
   c. contains cells which demonstrate spontaneous electrical activity
   d. contains cells which do not usually demonstrate spontaneous electrical activity
   e. receives most of its myocardial blood supply during diastole

2. Cardiac muscle cells:
   a. are highly dependent on glucose as a nutrient
   b. are highly dependent on aerobic metabolism
   c. have prolonged action potentials partly because they demonstrate a voltage-activated Ca^{2+} current
   d. are electrically insulated from one another
   e. are striated

3. The cardiac action potential:
   a. acts as the physiological stimulus for cell contraction
   b. is generated by voltage-sensitive ion channels all of which tend to open when the membrane is depolarized
   c. is conducted more slowly through the atrium than through the atrioventricular node
   d. is conducted throughout the atria by Purkinje fibres
   e. varies in length in different regions of the heart

4. The electrocardiogram:
   a. changes in shape when recorded from different sites on the body
   b. is generated by the conduction of the action potential across the heart
   c. has a QRS complex generated by the spread of repolarization across the atria
   d. should demonstrate a longer than normal gap between the P and QRS waves when conduction between the atria and ventricles is delayed
   e. has a peak amplitude of about 100 mV

5. In the cardiac cycle:
   a. the T wave of the ECG overlaps the decline in ventricular pressure
   b. the 'a' wave in atrial pressure coincides with the arterial pulse
   c. ventricular ejection reduces ventricular volume by more than 90% at rest
   d. the first heart sound immediately precedes the arterial pulse
   e. the aortic valve is closed whenever ventricular pressure exceeds aortic pressure

6. Cardiac output:
   a. can be determined from the heart rate
   b. decreases on standing upright from a sitting position
   c. can be accurately estimated using measurements of O_2 consumption, and the O_2 concentrations in the radial artery and vein
   d. increases when parasympathetic nerves to the heart are stimulated
   e. increases when sympathetic nerves to the veins are stimulated

7. Cardiac preload:
   a. increases during exercise
   b. is mainly dependent on venous return
   c. tends to reduce the force of cardiac contraction as preload increases under normal conditions
   d. is more important in determining cardiac output than is afterload
   e. would be reduced following removal of 500 ml of blood from a patient

8. Arterial blood pressure:
   a. is affected by the amount of elastin in the arterial walls
   b. is normally regulated solely through changes in peripheral resistance

c. is the sole determinant of the rate of blood flow through an organ
d. may be increased by increases in venous return
e. can be measured using an auscultatory method

9. Arterioles:
   a. have muscular walls
   b. are the most compliant type of blood vessel
   c. are important in the control of local blood flow
   d. control peripheral resistance
   e. can constrict leading to rises in arterial and capillary pressures

10. Baroreceptors:
    a. are a type of stretch receptor
    b. are located in the aortic and carotid bodies
    c. control the mean arterial pressure from week to week
    d. produce sensory signals which stimulate the vasomotor centre and cardiac parasympathetic nucleus in the medulla oblongata
    e. are stimulated by a sudden fall in blood pressure

11. The renin–angiotensin system:
    a. is stimulated following blood loss
    b. relies on angiotensinogen secretion from the juxtaglomerular apparatus
    c. promotes aldosterone secretion from the adrenal cortex
    d. can stimulate vasoconstriction
    e. is active, even under normal conditions

12. In the systemic circulation, tissue blood flow:
    a. increases linearly with arterial pressure over the range 20 to 200 mmHg
    b. may increase as a result of passive vasodilatation following inhibition of sympathetic nerves
    c. may increase as a result of active vasodilatation following stimulation of sympathetic nerves
    d. is increased following a period of reversible ischaemia
    e. may be increased by locally produced metabolites

13. Pulmonary:
    a. blood flow equals systemic blood flow
    b. vascular resistance is similar to systemic peripheral resistance
    c. blood flow decreases in areas of increased alveolar CO_2 tension
    d. hypertension can lead to left ventricular hypertrophy
    e. veins drain into the left atrium

14. During exercise:
    a. blood flow is reduced in the gastrointestinal tract
    b. parasympathetic nerves dilate muscle arterioles
    c. cardiac output increases mainly because of an increase in stroke volume
    d. cardiac β-adrenoceptors are stimulated
    e. systolic blood pressure tends to rise more than diastolic pressure

15. Resistance to blood flow:
    a. in skin is normally regulated from control centres in the hypothalamus
    b. in coronary blood vessels is increased during exercise
    c. in the brain may be increased by sympathetic nerves during hypotension (reduced arterial pressure)
    d. increases as blood haematocrit increases
    e. decreases dramatically as the diameter of a blood vessel decreases

16. Net capillary filtration:
    a. increases when the plasma protein concentration is increased
    b. increases as capillary pressure increases
    c. tends to increase following vasoconstriction
    d. increases as the capillary permeability to protein increases
    e. is more likely to occur at the arteriolar than the venular end of a capillary

17. On standing upright:
    a. peripheral resistance increases
    b. arterial pressure as measured in a vessel at heart level tends to fall
    c. neck veins collapse
    d. cerebral blood vessels dilate

e. the effective circulating blood volume decreases

Single best answer questions

For each of the following questions choose the single best answer.

1. The systemic and pulmonary circulations differ:
   a. in their mean total blood flow
   b. in their total rate of capillary gas transfer
   c. in their arterial pressure profile
   d. in their mean capillary pressure
   e. in their response to exercise

2. During the cardiac action potential:
   a. an increase in Ca^{2+} conductance produces rapid depolarization
   b. an increase in K^+ conductance slows the early depolarization
   c. an increase in Na^+ conductance produces repolarization
   d. the refractory period is restricted to the phase of rapid depolarization
   e. a decrease in K^+ conductance reduces the current required to maintain the plateau

3. Spontaneous electrical activity in the heart:
   a. is observed in sino-atrial node cells only
   b. is dependent on functioning nerves
   c. will be driven from the fastest pacemaker site
   d. could theoretically be generated by a decreasing inward current driving the pacemaker potential
   e. would be increased in rate if the threshold potential were increased

4. The electrocardiogram:
   a. is similar in amplitude to the cardiac action potential
   b. results from the simultaneous depolarization of a large number of myocytes
   c. would be expected to have a prolonged QRS wave if AV conduction were slowed
   d. would be expected to have a prolonged QRS wave if conduction in the branches of the bundle of His were slowed
   e. would be expected to show shortening of the PR interval if AV conduction were slowed

5. During the cardiac cycle:
   a. the 1st heart sound precedes the 'c' wave in the atrial pressure
   b. the 2nd heart sounds results from closure of the atrioventricular valve
   c. ventricular systole is normally longer than diastole
   d. isovolumetric relaxation immediately precedes ventricular ejection
   e. ventricular filling is doubled during atrial contraction

6. In the measurement of cardiac output by Fick's principle:
   a. an increase in the rate of O_2 absorption is indicative of an increase in cardiac output
   b. a decrease in systemic venous $[O_2]$ is indicative of an increase in cardiac output
   c. a decrease in pulmonary venous $[O_2]$ is indicative of an increase in cardiac output
   d. an increased systemic arteriovenous $[O_2]$ difference is indicative of an increase in cardiac output
   e. an increase in the rate of O_2 absorption with a parallel increase in arteriovenous $[O_2]$ difference is indicative of an increase in cardiac output

7. Increased cardiac preload:
   a. results from increased arterial pressure
   b. leads to an increase in end systolic volume
   c. increases cardiac output
   d. increases ventricular stroke volume
   e. is associated with a fall in atrial pressure

8. Arterial blood pressure may be increased by:
   a. decreased sympathetic tone
   b. venodilatation
   c. decreased parasympathetic tone
   d. decreased antidiuretic hormone release
   e. adrenoceptor blocking drugs

9. Vascular resistance is most markedly affected by:
   a. vessel lumen diameter
   b. vessel cross-sectional area
   c. blood viscosity
   d. vessel length
   e. vessel wall thickness

10. Which of the following is greater in veins than arteries:
    a. resistance
    b. wall thickness
    c. compliance
    d. density of innervation
    e. total blood flow

Matching item questions

Theme: Cardiovascular control

Options

A. Aldosterone
B. Antidiuretic hormone
C. Sympathetic nerves
D. Parasympathetic nerves
E. Angiotensin II
F. Local metabolites
G. Length–tension relationship
H. Atrial natriuretic peptide (hormone)

For each of the descriptions below choose the most appropriate option from the list above. Each option may be used once, more than once or not at all.

1. Explain(s) why cardiac output equals venous return.
2. Regulate(s) cutaneous blood flow in line with thermoregulatory needs.
3. Expand(s) plasma volume and raise(s) BP by increasing renal Na^+ absorption.
4. Adjust(s) vascular resistance to match blood flow to workload.
5. Stimulate(s) an increase in heart rate during exercise.
6. Pituitary hormone which causes vasoconstriction.

Theme: Vascular properties

Options

A. Aorta
B. Pulmonary trunk
C. Large systemic arteries
D. Large systemic veins
E. Capillaries
F. Arteriovenous shunts
G. Arterioles
H. Venules

For each of the descriptions below choose the most appropriate option from the list above. Each option may be used once, more than once or not at all.

1. Responsible for vasomotion.
2. Demonstrate(s) the largest pulse pressure.
3. Has/have the highest compliance.
4. Has/have the highest wall thickness to lumen ratio.
5. Decrease(s) nutritional flow in capillary beds.
6. Carry/carries an outer layer of pericytes.

Short notes

Write short notes on the following:

a. cardiac pacemaker potentials
b. atrioventricular valve function
c. pulmonary arterial pressure
d. Starling's law of the heart
e. the baroreceptor reflex
f. control of systemic peripheral resistance
g. central venous pressure

Modified essay

The cardiovascular system has to adapt to many different physiological conditions. One of the most important variables is the rate of energy expenditure by the body, since this changes the blood flow requirements of the tissues involved. Such changes are most frequently the result of changes in the level of exercise.

Questions

1. How is rate of fluid flow in a pipe related to pressure and resistance? Apply this to the cardiovascular system with regard to the factors which will determine blood flow in an organ. Which of these factors is normally varied to control regional blood flow?
2. Which type of blood vessel controls the resistance to flow through a vascular bed?

What makes these vessels suitable for this role? How do they change their resistance?

3. List the main types of control mechanism involved in regulating local vascular resistance.
4. What changes would you expect to see in blood flow to the leg muscles and the abdominal organs during a 5 mile run? How are these changes regulated? What is the likely direction of change of systemic peripheral resistance?
5. Exercise leads to an early increase in venous return. Outline two mechanisms which may contribute to this.
6. Nervous control of the heart also plays a role in exercise. What nerves are involved and what effects do they have?
7. How does coronary blood flow change during exercise and what brings this change about?
8. Systolic pressure is often said to be particularly affected by changes in cardiac output, while diastolic pressure is more sensitive to peripheral resistance. If this is so, can you explain why mean arterial pressure is often little changed even during periods of quite heavy exercise?
9. Measurements on a normal individual show that during moderate exercise there are increases in heart rate and cardiac output, but little change in stroke volume. In someone with a successful cardiac transplant, the same workload produces a similar rise in cardiac output, but in this case there is little or no change in heart rate, stroke volume increasing by the necessary amount instead. Given that a transplanted heart is detached from its nerve supply, how might these observations be explained?

Data interpretation

A patient attends her doctor complaining that her fingers feel extremely cold and often go very white, especially in cool weather when they can become painful or numb. She also notices that, on warming up, her fingers become intensely red for some time before returning to normal. Having excluded serious cardiac or arterial conditions, the doctor makes an initial diagnosis of Raynaud's syndrome. This is a condition in which cutaneous arterioles show an excessive tendency to vasoconstrict, especially in cool conditions.

Questions

1. Why might vasoconstriction lead to the symptoms described above?
2. What nerves might be contributing to the vasoconstriction? Does this suggest a possible therapy for the condition?

It was decided to carry out a special test in which the skin temperature of the middle finger was continuously measured. This gives some indication of blood flow to the finger. Basal measurements were carried out in a cool room (17°C) under conditions in which the patient's 'white hand' symptoms were seen. After a time, the rest of the patient's body was warmed using heated blankets (period marked by the shaded rectangle in Fig. 58). The resulting changes in finger temperature are plotted on graph A (Fig. 58). Recordings made under similar conditions in a female control subject of similar age are shown for comparison in graph B.

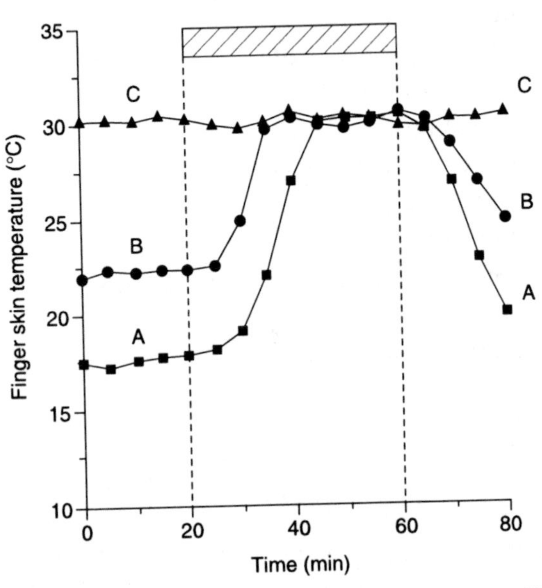

Fig. 58 Plots of finger skin temperature in two subjects sitting in a room at 17°C. Between 20 and 60 minutes the subjects were warmed by wrapping their bodies in blankets and placing their feet and legs in warm water. The results are shown for a patient suffering from Raynaud's syndrome, before (A) and several weeks after the sympathetic nerves to the arm had been cut surgically (i.e. after sympathectomy) (C) and for an age-matched control (B).

Questions

3. Why is skin temperature a useful measure of finger blood flow? Would temperature measurements from the skin of the forearm be a useful gauge of total forearm blood flow? Explain your answer.
4. Compare and contrast the experimental results for the patient and control subject.
5. Can you suggest a possible explanation for these findings on the basis of the nervous control of cutaneous blood flow? Why do you think this type of test is known as a sympathetic release test?
6. A sympathetic release test is sometimes used to gauge whether surgery to cut or block the sympathetic nerves may be of benefit in cases of Raynaud's syndrome. Based on these results, do you think surgery should be considered for this patient? Explain your answer.

It is decided to go ahead with surgery. Some months after the operation the sympathetic release test is repeated and the results obtained are shown in graph C (Fig. 58).

Questions

7. Describe the changes from before surgery. Why is there so little change in finger temperature during heating now?
8. Do you think it likely that this patient will have benefited from the operation? Why do you say this?

Clinical scenario

A 67-year-old man with a history of essential hypertension (high blood pressure) comes to hospital complaining of being out of breath. He has noticed increasing shortness of breath on light exercise (e.g., climbing the stairs at home) over a period of weeks, and is particularly breathless at night, wakening after a few hours sleep. Sitting up in bed relieves this a little. On examination he is breathless at rest and appears centrally cyanosed (blue colouration around the lips). He has a pulse rate of 76 min^{-1}, systemic BP of 158/97 mmHg, an elevated jugular venous pulse of 8 cm, and ankle oedema. There is evidence of fluid in his lungs both on physical examination and chest X-ray, and his heart appears enlarged.

Questions

1. Is the systolic or diastolic pressure of more interest in this case? Explain why.
2. Why do you think his heart was enlarged?
3. Which part of the heart would you have expected to be affected first in this situation?
4. Based on your understanding of capillary filtration, what has caused the oedema of the ankles?
5. What other observation supports this suggestion?
6. Which part of the heart is directly responsible for these changes in the systemic circulation?
7. Based on these observations in the systemic circulation, explain how a similar mechanism could explain the accumulation of fluid in the lungs.
8. This condition is often treated with diuretics as these increase urinary fluid excretion. Explain why this might help.
9. What is this condition known as?

Viva questions

1. What compensatory mechanisms would you expect to be activated following serious blood loss?
2. What factors control capillary fluid exchange?
3. Explain the terms 'resistance vessels' and 'capacitance vessels' as they apply in the circulation.

Self-assessment: answers

Multiple true/false answers

1. a. **False.** There are differences in wall thickness and atrioventricular valve structure.
   b. **True.** Mechanical contraction uses energy from ATP.
   c. **True.** Those in the sino-atrial node drive the rest of the heart since they demonstrate the highest frequency.
   d. **True.** This is the case for most cardiac cells.
   e. **True.** Contraction collapses myocardial blood vessels during systole.

2. a. **False.** Fatty acids are the major substrate, glucose and lactate can also be used.
   b. **True.** This makes ischaemia very damaging.
   c. **True.** Inward Ca^{2+} current maintains the action potential plateau.
   d. **False.** Electrical continuity occurs via gap junctions.
   e. **True.**

3. a. **True.** Depolarization triggers a rise in intracellular $[Ca^{2+}]$.
   b. **False.** Most of the cardiac ion channels open, or activate, as the membrane is depolarized but one population of K^+ channels actually close.
   c. **False.** Conduction is slower in the atrio-ventricular node than anywhere else in the heart.
   d. **False.** Purkinje fibres conduct in the ventricles.
   e. **True.** Atrial muscle cells have the shortest action potentials, Purkinje cells the longest.

4. a. **True.** Electrode pairs which are parallel to the main axis of electrical conduction show the largest deflections.
   b. **True.** This generates potential differences between polarized and depolarized areas.
   c. **False.** QRS reflects spread of ventricular depolarization.
   d. **True.** The PQ interval depends on the time between atrial and ventricular depolarization.
   e. **False.** ECG amplitude is normally 1–2 mV.

5. a. **True.** T wave marks ventricular repolarization which leads to relaxation.
   b. **False.** Atrial 'a' wave precedes the rise in arterial pressure.
   c. **False.** Stroke volume = 70 ml, end diastolic volume = 140 ml, ejection fraction = 70/140 = 50%.
   d. **True.** Closing of the atrioventricular valves at onset of ventricular systole.
   e. **False.** This would open aortic valve.

6. a. **False.** Also need to know the stroke volume.
   b. **True.** Because of blood pooling in leg veins.
   c. **False.** The Fick technique requires a mixed venous sample from the right ventricle or pulmonary trunk.
   d. **False.** This decreases heart rate and cardiac output.
   e. **True.** Sympathetic venoconstriction increases venous return, raising cardiac output.

7. a. **True.** Because of venous compression by muscles and sympathetic venoconstriction.
   b. **True.** Increasing venous return increases ventricular stretch.
   c. **False.** The reverse is true; Starling's law of the heart.
   d. **True.** Steady-state cardiac output is little affected by changes in afterload.
   e. **True.** Reducing blood volume reduces venous return.

8. a. **True.** Arterial elasticity damps the pressure changes, reducing pulse pressure.
   b. **False.** Cardiac output is also important.
   c. **False.** Local resistance to flow is also important.
   d. **True.** This increases cardiac output.
   e. **True.** Using the sphygmomanometer.

9. a. **True.** This smooth muscle can contract and relax, causing vasoconstriction and vasodilatation.
   b. **False.** Veins are much more compliant.
   c. **True.** They control local resistance.

d. **True.** By constricting or dilating.
e. **False.** Arterial pressure would rise but vasoconstriction reduces capillary pressure.

10. a. **True.** Increased arterial pressure stretches the walls of the carotid sinus and aortic arch.
b. **False.** These contain peripheral chemoreceptors.
c. **False.** Short term control of blood pressure from minute to minute.
d. **False.** Baroreceptor inputs stimulate the cardiac parasympathetic nucleus but inhibit sympathetic outflow from the vasomotor centre.
e. **False.** This would reduce baroreceptor activity.

11. a. **True.** Reduced renal perfusion pressure is detected by the juxtaglomerular apparatus.
b. **False.** The juxtaglomerular apparatus secretes renin; angiotensinogen is manufactured in the liver.
c. **True.** This stimulates renal Na^+/water reabsorption (Section 5.4).
d. **True.** Angiotensin II is a vasoconstrictor.
e. **True.** Changes in blood pressure can increase or decrease renin secretion.

12. a. **False.** Flow rises rapidly up to 70 mmHg and between 170 and 200 mmHg but is autoregulated between 70 and 170 mmHg.
b. **True.** This is seen in nearly all tissues except heart and brain.
c. **True.** This is seen in skin during body heating.
d. **True.** Reactive hyperaemia caused by metabolite accumulation.
e. **True.** This is the basis of hyperaemia in exercising tissues.

13. a. **True.** Cardiac output is the same for both ventricles.
b. **False.** Pulmonary resistance is much lower, so pulmonary arterial pressure is much lower than systemic pressure even though blood flow is the same.
c. **True.** This is 'reversed' local control.
d. **False.** Right ventricular hypertrophy.
e. **True.** Carrying oxygenated blood.

14. a. **True.** Reflex sympathetic vasoconstriction.
b. **False.** Parasympathetic nerves play no role in controlling vascular resistance other than in the penis and some exocrine glands.
c. **False.** Increases in heart rate normally predominate in increasing cardiac output. There is usually little change in stroke volume.
d. **True.** By noradrenaline (norepinephrine) from sympathetic nerves and by circulating catecholamines from adrenal medulla.
e. **True.** Cardiac output increases, raising systolic pressure, but the fall in peripheral resistance caused by muscle vasodilatation limits the rise in diastolic pressure and may cause it to fall.

15. a. **True.** Temperature regulating centres.
b. **False.** Increased cardiac workload leads to vasodilatation through local control mechanisms.
c. **False.** Cerebral and coronary circulations are exempt from sympathetic constrictor reflexes.
d. **True.** As a result of an increase in blood viscosity.
e. **False.** Resistance increases as diameter decreases, $R \propto 1/r^4$.

16. a. **False.** This increases the osmotic gradient favouring absorption.
b. **True.** Increases the hydrostatic gradient out of the capillary.
c. **False.** Vasoconstriction decreases the hydrostatic pressure in downstream capillaries.
d. **True.** Leakage of proteins decreases the absorptive osmotic gradient.
e. **True.** Because of the larger hydrostatic gradient.

17. a. **True.** Baroreceptor-initiated sympathetic vasoconstriction.
b. **True.** Venous pooling reduces venous return and cardiac output.
c. **True.** Venous pressure drops to atmospheric.
d. **True.** Local compensatory response to reduced cerebral flow.
e. **True.** As a result of rapid venous pooling in leg veins followed by slower losses through increased capillary pressure and filtration.

Single best answers

1. d. Total blood flow and gas exchange must be almost the same for both circulations. The arterial pressure profiles are similar in outline, although the absolute pressures are much lower in the pulmonary circulation. The reduced capillary pressure in the pulmonary circulation helps limit fluid around pulmonary capillaries, ensuring short diffusion distances for gas exchange.
2. e. There is reduced outward K^+ current during the plateau of the cardiac action potential, so the inward Ca^{2+} current can maintain the depolarization more easily.
3. c. Several regions in the heart can generate spontaneous activity but the sino-atrial node normally drives the heart because it has the fastest intrinsic rate of activity.
4. d. Since the branches of the bundle of His are part of the rapid conducting pathway responsible for spreading depolarization over the ventricles, slow conduction here will prolong the time taken for ventricular depolarization to be complete. This is known as 'bundle branch block' and shows up as a prolonged QRS complex, since this corresponds to ventricular depolarization. Slowed AV conduction will increase the delay between atrial and ventricular depolarization, i.e., increase the PR interval. The ECG is small in amplitude compared to the action potential and results from the spread of electrical activity over the heart. Simultaneous depolarization would not generate any surface potential differences.
5. a. Revise the main elements and timing of events during the cardiac cycle.
6. e. The cardiac output can only be assessed if both O_2 consumption and arteriovenous $[O_2]$ difference are measured simultaneously.
7. d. This is one of the more extreme examples of a 'single best answer question'. Increased preload increases stroke volume, in accordance with Starling's law of the heart. This increases cardiac output but the answer 'd.' is preferred to 'c.', since this highlights the mechanism responsible.
8. c. Decreased parasympathetic tone increases heart rate and cardiac output, increasing arterial blood pressure. All the other changes would decrease it.
9. a. Resistance is inversely proportional to r^4, so radius (or diameter) has the greatest effect. Resistance is inversely proportional to (cross-sectional area)^2 and blood viscosity, directly proportional to vessel length, and unaffected by wall thickness unless this limits lumen diameter.
10. c. Wall distensibility and vessel compliance are greater in veins than arteries. Total venous blood flow must equal arterial flow in any circulation. All the other parameters mentioned are greater in arteries than veins.

Matching item answers

Theme: Cardiovascular control

1. G. This is the mechanism which underlies Starling's law of the heart, i.e., stroke volume and, therefore, cardiac output increase as end diastolic volume increases. Since venous return is a crucial determinant of end diastolic volume, this relationship tends to match cardiac output to venous return.
2. C. Hypothalamically controlled sympathetic nerves control cutaneous vascular resistance. Vasoconstriction results from adrenergic nerve activity and reduces heat loss. Vasodilatation results from decreased vasoconstrictor activity as well as activation of sympathetic vasodilator nerves. The latter are cholinergic but the dilator is believed to be a peptide co-transmitter. Vasodilatation increases cutaneous heat loss.
3. A. Na^+ absorption leads to H_2O absorption, which expands plasma volume, increasing venous return, cardiac output and, thus, BP. Antidiuretic hormone has a similar action since it promotes H_2O absorption but it does not affect Na^+ absorption.
4. F. Production of vasodilator metabolites under anaerobic conditions is probably the main way in which tissue blood flow is increased to meet increased metabolic demands.
5. C. Increased sympathetic activity is accompanied by decreased parasympathetic activity. The increased heart rate accounts for most of the increase in cardiac output required to meet the needs of exercising muscle.

6. B. Antidiuretic hormone is also known as vasopressin. Angiotensin II is also a vasoconstrictor but it is not a pituitary hormone.

Theme: Vascular properties

1. G. The rhythmical contraction and relaxation of arterioles leads to oscillating flow in downstream capillaries. This is known as vasomotion.
2. C. The pulse pressure in the large systemic arteries is greater than that in the aorta, due to reflection of the pressure wave back from the smaller vessels downstream.
3. D. Compliance is defined as the change in volume per unit change in distending pressure.
4. G. As a lot of the wall thickness is contractile smooth muscle, this allows arterioles to generate marked changes in tone and vascular resistance.
5. F. Blood flow through arteriovenous shunts will bypass the normal, nutritional capillaries. This increases local blood flow, e.g., to promote heat loss from the skin in a warm environment or during exercise.
6. E. Pericytes are flat, branched, contractile cells found wrapped around the outside of most capillaries.

Short note answers

a. Pacemaker potential is a spontaneous depolarization from resting potential which leads to regular action potential production. The slope of the potential determines the rate of action potential production (draw diagram, e.g., Fig. 39) observed in different regions, e.g., sino-atrial node, atrioventricular node. In the intact heart, the most rapidly depolarizing potential produces highest frequency of action potentials (normally sino-atrial node) and drives the rest of the heart by action potential conduction.

b. Atrioventricular valves prevent backflow of blood from ventricle to atrium (tricuspid valve on right, mitral valve on left). They are open during ventricular diastole, allowing filling, but close at onset of ventricular systole when ventricular pressure rises above atrial pressure. Closure generates an audible vibration, the first heart sound. Leaking or incompetent atrioventricular valves lead to a regurgitation of blood into the atria during ventricular systole. This reduces stroke volume and may lead to ventricular failure.

c. Pulmonary arterial pressure is determined by the cardiac output and pulmonary vascular resistance. Normal values are 25/10 mmHg (systolic/diastolic) and the pressure profile has a brief upward deflection (dicrotic notch) caused by the dissipation of backflow momentum following closure of the pulmonary valve. Pulmonary pressure is much lower than systemic pressure, indicating that pulmonary vascular resistance is comparatively low (total blood flow is the same in both, i.e., equals cardiac output). This may rise because of inadequate pulmonary ventilation, since low P_{O_2} and high P_{CO_2} cause pulmonary vasoconstriction, leading to pulmonary hypertension.

d. Starling's law of the heart states that the force of myocardial contraction increases as a function of the resting fibre length, or preload. This leads to an increase in stroke volume and cardiac output as atrial filling pressure or end diastolic volume increase, e.g., in response to an increase in venous return. As a consequence, cardiac output equals venous return, and right and left ventricular outputs remain equal.

e. Baroreceptors in the carotid sinus and aortic arch are stretch receptors activated by increases in systemic arterial pressure. Their outputs inhibit sympathetic nerves from the medulla oblongata supplying the heart (reducing heart rate and decreasing myocardial contractility) and peripheral vasculature (reducing vasoconstriction and venoconstriction). Parasympathetic output to the heart is increased (slows heart rate). The overall result is a fall in cardiac output and peripheral resistance, reducing blood pressure. These effects are reversed if blood pressure falls below normal.

f. Systemic peripheral resistance is largely determined by arteriolar resistance and may be increased by vasoconstriction and decreased by vasodilatation. The main regulators of peripheral resistance are vasoconstrictor sympathetic nerves (increase resistance) and circulating vasoconstrictor hormones (catecholamines, angiotensin, antidiuretic hormone). These modulate resistance to maintain normal arterial pressure. Local vasodilator substances decrease local resistance, e.g., during exercise or in response to inflammation; if extreme, this may also decrease total peripheral resistance. Increased blood viscosity caused by an abnormally high haematocrit may also increase peripheral resistance in individuals with polycythaemia.

g. Central venous pressure (CVP) is the pressure in the right atrium and is normally about 2–6 cmH_2O. The CVP pressure wave demonstrates three waves, 'a' (atrial systole), 'c' (upwards pressure on the atrioventricular valve during isovolumetric ventricular contraction) and 'v' (venous filling with a closed atrioventricular valve), during each cardiac cycle. The mean CVP reflects the balance between the rate of venous return and right ventricular function. Elevated CVP suggests right ventricular failure or abnormal expansion of blood volume, e.g., as the result of overinfusion.

Modified essay answers

1. Flow = Pressure gradient/Resistance. In the body, the pressure gradient or perfusion pressure is largely dependent on the arterial pressure. In principle, blood flow may be altered by changes in pressure or vascular resistance. In practice, blood pressure is normally kept reasonably constant by the baroreceptor and associated reflexes, and local vascular resistance is varied to determine local blood flow.
2. Arterioles are the main resistance vessels in the body, providing over half of the total resistance in the systemic circulation (the peripheral resistance). They have very muscular walls which can contract or relax, decreasing and increasing their radius. This can increase and decrease their resistance dramatically, since resistance $(R) \propto 1/r^4$.
3. Local controls include vasodilator metabolites, whose concentrations depend on the balance between blood flow and tissue metabolism, and changes in arteriolar stretch caused by changing intravascular pressure leading to myogenic changes in arteriolar constriction. Extrinsic control is via nerves (sympathetic constrictor, sympathetic dilator in skin and parasympathetic dilator in the penis) and hormones (e.g., catecholamines, angiotensin II and vasopressin, also called antidiuretic hormone).
4. Blood flow goes up in the muscle because of vasodilatation. This depends on a combination of local metabolite formation and decreased sympathetic vasoconstriction. Blood flow to the gastrointestinal tract, kidney and other abdominal organs decreases as a result of activation of sympathetic vasoconstrictor nerves. (Sympathetic activity also inhibits gastrointestinal motility and secretion so eating immediately before or after heavy exercise may lead to nausea.) The massive decrease in muscle resistance and a thermoregulatory decrease in the cutaneous vascular resistance usually dominate, leading to a fall in peripheral resistance.
5. Repetitive external compression of the veins by the exercising muscles promotes venous return by the venous pump. Sympathetic venoconstriction in the abdominal organs has a similar effect.
6. Increased activity in sympathetic nerves increases heart rate (positive chronotropic effect) and stimulates ventricular contractility (positive inotropic effect). A parallel reduction in parasympathetic (vagal) activity contributes to the increased rate. The final outcome is an increase in cardiac output.
7. It increases in parallel with the increased cardiac workload. This is probably regulated by a decrease in myocardial arteriolar resistance caused by local control mechanisms.
8. We would expect exercise to raise systolic pressure (cardiac output rises), but reduce diastolic pressure (peripheral resistance falls). These effects tend to cancel out so that mean arterial pressure (i.e., diastolic pressure + pulse pressure/3) is little changed.
9. Under normal circumstances, the cardiac response results solely from a change in heart rate driven by the nervous reflexes activated during exercise. A denervated heart cannot produce such rate changes but increased workload still requires increased cardiac output. This is generated through increases in

stroke volume, since the increased venous return leads to increased stretch and stimulates ventricular contraction, as predicted by Starling's law.

Data interpretation answers

1. The increased vascular resistance reduces blood flow. This leads to blanching of the fingers (whiteness) and reduces finger temperature (coldness). If blood flow is inadequate to meet metabolic needs, ischaemic pain can develop. The intense reddening is probably caused by a reactive hyperaemia following the build-up of vasodilator metabolites.
2. Sympathetic vasoconstrictor nerves are important in the skin. Some people benefit from treatments which reduce the sympathetic nervous activity to the affected region, e.g., surgical interruption of the sympathetic nerves to that arm (sympathectomy).
3. Skin temperature will follow blood flow closely in regions where the surface area to volume ratio is large so that the majority of blood vessels are close to the surface and heat exchange with the environment is efficient. The finger fulfils these criteria. The forearm does not, since a change in blood flow to deep structures far from the skin will have much less effect on skin temperature.
4. In the absence of body heating, the patient's skin temperature is only a little above room temperature, while the control subject's finger is about 5°C warmer. In response to body heating, skin temperature increases in both control and patient, reaching a similar maximum plateau level in both. This increase is reversed when body heating is stopped.
5. The cool experimental environment has led to sympathetic vasoconstriction in both cases, but this is more pronounced in the patient with Raynaud's syndrome. Thermoregulatory centres in the hypothalamus respond to body warming by inhibiting sympathetic tone, causing vasodilatation and increased blood flow to the skin. This is recorded as an increase in skin temperature. Blood vessels are effectively released from the sympathetic vasoconstrictor tone, hence the term sympathetic release test. The effect of this is more obvious in the patient, suggesting that there was an exaggerated amount of sympathetic constriction prior to heating. The temperature plateau reflects maximal blood flow in the absence of any sympathetic constriction. The fact that the plateau is identical in control and patient suggests that the patient does not suffer from any other obstructive vascular disease, e.g., arterial obstruction by atheroma, which would be expected to reduce peak blood flow, and therefore, the peak skin temperature. When body heating stops, sympathetic vasoconstriction increases again and blood flow falls.
6. Since there is evidence for a reduction of blood flow under cool conditions which is reversed by heating, these results suggest that surgery may be of benefit since sympathetic inhibition seems to increase flow considerably.
7. Blood flow, as indicated by the finger temperature, is now higher under cool conditions in the treated patient than the control. This reflects the absence of sympathetic constriction following sympathectomy. With no sympathetic tone to be abolished, there is no increase in flow during heating; finger blood flow is maximal regardless of the environmental conditions.
8. The high skin temperature (caused by the increased blood flow) under conditions which previously produced 'white hand' symptoms suggests the operation has been of benefit. (Clinical studies have shown that this benefit is not always permanent, however, and symptoms may return over time.)

Clinical scenario answers

1. The diastolic pressure is of more interest as it is clearly elevated. Systolic pressure increases normally with age, and a systolic value = 100 + (Age in years) mmHg is acceptable. The increase in diastolic pressure shows relatively little change with age and a value >90 mmHg requires treatment.
2. Cardiac hypertrophy due to the sustained increase in afterload (i.e., the heart has been pumping against an elevated pressure). This man is now also in cardiac failure, and this is associated with stretch or dilatation of the heart, which causes further enlargement.

3. The left ventricle is affected first because it has to pump the blood against the elevated systemic pressure.
4. This is due to increased capillary hydrostatic pressure, secondary to the increased venous pressure in the systemic circulation.
5. The elevated jugular venous pressure is a direct indication that the central venous pressure is raised.
6. The right side of the heart, with failure of the right ventricle leading to an increase in right atrial pressure. This is communicated to the systemic veins as the central venous pressure is raised.
7. Failure of the left ventricle leads to elevated left atrial pressure, an increase in pulmonary venous pressure and an increase in pulmonary capillary pressure. This again increases capillary filtration, pushing excess fluid into the lungs. It is this pulmonary oedema which causes the breathlessness, mainly by increasing the diffusion distance between the alveolar air and pulmonary capillaries, thus reducing pulmonary gas exchange.
8. Reduction of the extracellular fluid volume in the body is paralleled by a reduction in blood volume, and reduced venous return. This helps to reduce venous pressure in both pulmonary and systemic circulations, reducing capillary filtration and decreasing the oedema.
9. This man is suffering from cardiac failure. He has both left ventricular failure (causing the pulmonary oedema and breathlessness) and right ventricular failure (causing the peripheral oedema). This combination is also known as congestive cardiac failure. The right ventricular failure is actually a consequence of the left ventricular failure, since this elevated the pressure in the pulmonary circulation. The underlying cause is the hypertension in the systemic circulation.

Viva answers

1. This is another way of asking about blood pressure control. The problem is that this is an enormous topic so you should have some way of organizing your answer. It is also very important to begin with the most important mechanisms, which should also be the ones you know most about. Points worth mentioning include the following:

   a. Compensation must involve increase in cardiac output or peripheral resistance since BP = CO × PR.

   b. Reduced blood volume reduces venous return and cardiac output by Starling mechanism, and so reduces arterial BP.

   c. Rapid reflexes (detectors = baroreceptors, low volume pressure receptors, direct effect of cerebral ischaemia) activate sympathetic nerves to heart and blood vessels and inhibit parasympathetic nerves to heart within seconds. This increases heart rate and contractility (increases cardiac output), increases vasoconstriction (increases peripheral resistance), increases venoconstriction (increases venous return and, therefore, cardiac output).

   d. Slower mechanisms include release of ADH and aldosterone, which promote renal fluid reabsorption, and low capillary pressure (accentuated by vasoconstriction) which favours reabsorption of tissue fluid into vasculature. These all help expand plasma volume over minutes to hours.

   e. Lost red cells will eventually be replaced because a low haematocrit raises erythropoietin levels. This may take many days after BP returns to normal.

2. The factors involved are the hydrostatic gradient (acting out of the capillary), the osmotic gradient (acting into the capillary) and the ease with which fluid can cross the capillary wall (the capillary coefficient). You should be able to state average values for hydrostatic and osmotic gradients. More detailed questioning may go on to consider the causes of oedema or the plasma and interstitial values of hydrostatic and osmotic pressure.
3. Both types of vessel make a contribution to the maintenance of normal blood pressure but by very different mechanisms. Resistance vessels make the largest contribution to the resistance of the circulation to blood flow, i.e., the arterioles. The examiner may point out that capillaries have a smaller diameter and so must surely have a higher resistance. This is true for a single arteriole versus a single capillary, but the resistance of all the arterioles is still greater than that of all the capillaries since there are many more capillaries arranged in parallel with each other. Arterioles are important in regulation of BP because

increasing or decreasing their radii (vasoconstriction and vasodilatation) has a large effect on total peripheral resistance. Their structure is adapted to this role since they have a relatively large amount of smooth muscle in their walls and are densely innervated by sympathetic nerves. Capacitance vessels (venules and veins) have a lower resistance but contain the largest percentage of the total blood volume. They have thin, distensible walls which increases their compliance (change in volume for a given change in pressure). They also have smooth muscle in their walls which is innervated by sympathetic nerves. Passive changes in venous volume with changes in intravenous pressure, accentuated by active venoconstriction, allow the venous system to minimize changes in venous return, and thus cardiac output, following small changes in blood volume.

Respiratory physiology

Overview

The main homeostatic responsibility of the respiratory system is maintenance of normal levels of O_2, CO_2 and pH within the systemic arterial blood. This blood is then distributed to actively metabolizing tissues by the circulation. Blood gas pressures, which determine the concentrations of O_2 and CO_2 in plasma and erythrocytes, are controlled by gas exchange between alveolar air and pulmonary blood. The pressure gradients which drive this exchange are maintained by the mixing of alveolar gas with fresh air transported into the lungs by the process of ventilation. Feedback control mechanisms regulate arterial gases through changes in the rate of pulmonary ventilation, which alter alveolar O_2 and CO_2 levels. Ultimately, cellular respiration uses O_2 to release energy through oxidation of food molecules, producing CO_2 as a byproduct, and the total rate of this energy release is referred to as the metabolic rate.

4.1 Mechanics of pulmonary ventilation

Learning objectives

At the end of this section you should be able to:

- name the main components of the lungs
- outline the functions of the airways
- write the equation relating pressure, resistance and airflow
- describe the mechanical forces acting on the lungs
- describe the functions of surfactant
- define lung compliance and identify factors which affect it
- name the muscles responsible for inspiration and expiration
- show how intrapleural and intra-alveolar pressures vary during quiet ventilation
- outline the factors which affect the work of breathing.

Pulmonary ventilation refers to the movement of air in and out of the lungs during breathing. Alveolar gas concentrations are kept at the required level by continuous alternation between expiration, which expels O_2-depleted and CO_2-loaded gas, and inspiration, which replaces this with normal air. The necessary flow of gas is driven by pressure gradients generated by the movements of the chest wall and diaphragm.

Relevant structures

Air enters the nose and is drawn through the nasopharynx into the *larynx*. It passes through the glottis before entering the trachea, which divides into the right and left main *bronchi* going to the two lungs. The airways continue to branch, decreasing in diameter (Fig. 59). Bronchi, which have cartilage in their

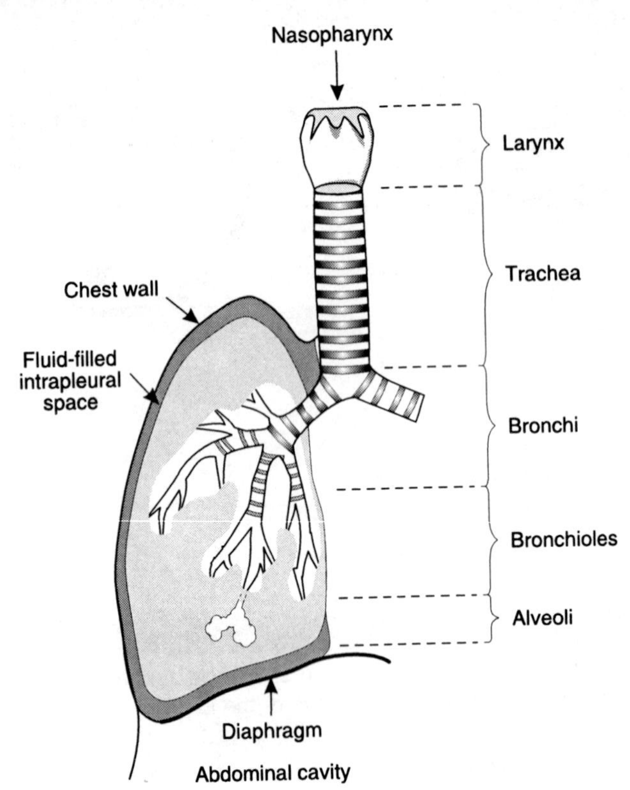

Fig. 59 The right lung, indicating the branching airways leading to the alveoli. Intrapleural fluid holds the pulmonary and parietal pleurae close together.

walls, lead into *bronchioles,* which do not, and are, therefore, collapsible. The airways finally terminate in grape-like collections of *alveoli*, the main site of gas exchange. Bronchi and bronchioles contain smooth muscle and the airways are lined by mucus-secreting cells within a *ciliated epithelium*. These have important protective functions. The alveoli have a thin epithelial wall and are covered on their inner surface by a narrow layer of alveolar fluid. Pulmonary connective tissue contains a large amount of elastic tissue which is stretched beyond its passive resting length at normal lung volumes and, therefore, generates tension.

The lungs lie within the thoracic cavity. The outer surface of the lung is covered by a membrane known as the visceral or *pulmonary pleura*, and this is separated from the *parietal pleura*, which lines the inside of the thoracic cavity, by the thin layer of pleural fluid filling the *intrapleural space* (Fig. 59). This is sometimes called a potential space, since it contains liquid rather than gas. Liquids cannot be easily expanded or compressed and so the two layers of pleura normally remain tightly adherent to one another. The outer surface of the lung is forced to follow the movements of the diaphragm and chest wall so that lung volume increases and decreases as thoracic volume changes.

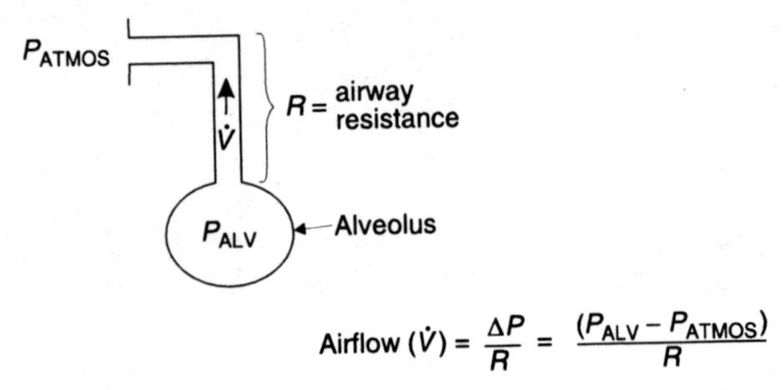

Fig. 60 The rate of airflow ($\dot{V}$) in or out of the lungs depends on the pressure gradient (ΔP) between the alveoli and the atmosphere, and the airways resistance (R).

Functions of the airways

The airways have three main functions

- conduits for gas passage
- protection of the lungs from entry of foreign matter
- warming and humidifying gases.

Conduits for gas passage

The airways act as conduits connecting the atmosphere to the alveoli. The flow of gas ($\dot{V}$) depends on the pressure gradient between the alveoli and the atmosphere (ΔP) and the resistance (R) of the airways (Fig. 60).

$$\text{Airflow} = \frac{\text{Pressure gradient}}{\text{Airways resistance}} \qquad \text{(Eq. 16)}$$

Normally the resistance to airflow is very low so only small pressure gradients (1–2 cmH_2O) are necessary to move gas in and out of the lungs during ventilation. Contraction and relaxation of smooth muscle in the bronchi and bronchioles can alter this resistance, however, with *bronchoconstriction* and *bronchodilatation* increasing and decreasing airways resistance, respectively. This is under autonomic nervous control, with adrenergic sympathetic nerves leading to bronchodilatation and cholinergic parasympathetic nerves stimulating bronchoconstriction. Increased airways resistance is the central feature of an obstructive lung disease such as asthma, in which the patient has difficulty in moving gas in and out of the lungs.

Protection of the lungs

Various mechanisms protect the lungs from the entry of foreign matter. The air is partially filtered by the nasal hairs, and bacteria and particles of atmospheric pollutants which escape this are usually trapped in the layer of *mucus* lining the airways. Continuous movement of the *motile cilia* on the surface of epithelial cells in the trachea, bronchi and bronchioles transports this mucus back up towards the larynx and out into the pharynx where it is swallowed. Excess mucus may contribute to the increased airways resistance in some obstructive lung diseases by partially blocking the lumen.

The *vocal folds* in the larynx, which are responsible for speech and sound production (phonation), also protect the lungs from inhalation of food by reflexly closing the glottis during swallowing. Should this fail, a cough reflex is triggered by any particles which make contact with the mucosa of the vocal folds or airways, producing an explosive expulsion of gas which expels the solid matter. This is vital since obstruction of even a small airway can lead to collapse of part of the lung and provide a focus for infection.

Warming and humidifying gases

As air passes through the airways it is warmed and humidified. Gases are completely saturated with water vapour before they reach the alveoli, producing a water vapour pressure of 6.3 kPa (47 mmHg) at 37°C (body temperature).

Forces acting on the lung

During quiet breathing, three forces act on the lungs. Two tend to collapse the lung but these are opposed by a third, distending force or pressure.

Elastic tissue. The elastic tissue of the lungs is stretched under normal conditions and the resulting tension acts as a collapsing force pulling inwards on the visceral pleura (Fig. 61A).

Surface tension. The surface tension of the fluid lining the alveoli also tends to collapse them, pulling inwards, away from the chest wall.

Negative intrapleural pressure. The elastic and surface tension effects in the lungs are normally opposed by a distending pressure caused by the negative (subatmospheric) pressure in the intrapleural space (the intrapleural pressure). This is developed as a consequence of the chest wall and diaphragm pulling outwards on the parietal pleura. Since the two layers of pleurae are being pulled in opposite directions, a negative pressure is developed in the intrapleural fluid. Consequently the pressure in the alveoli, which must equal atmospheric pressure on average since the alveoli are directly connected to the atmosphere through the airways, is greater than the pressure outside the pleural covering of the lung. The resulting transmural pressure gradient, or distending pressure, keeps the lungs inflated (Fig. 61B). If an airway ruptures or the chest wall is penetrated, however, air will be sucked into the pleural cavity, producing a *pneumothorax* (gas in the intrapleural space). Under these conditions, the intrapleural pressure rises to zero (atmospheric), or may even become positive, and the lung collapses.

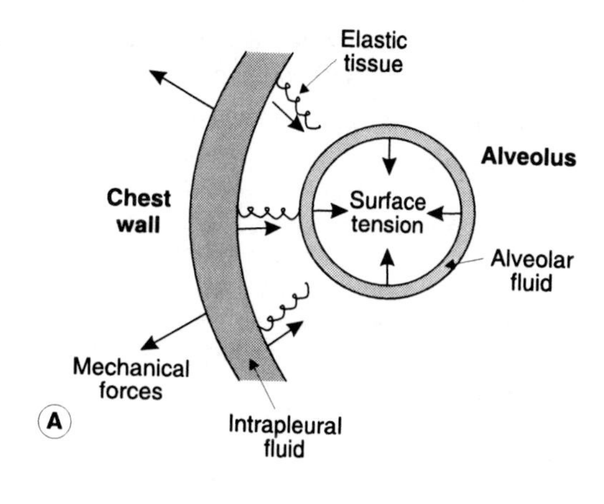

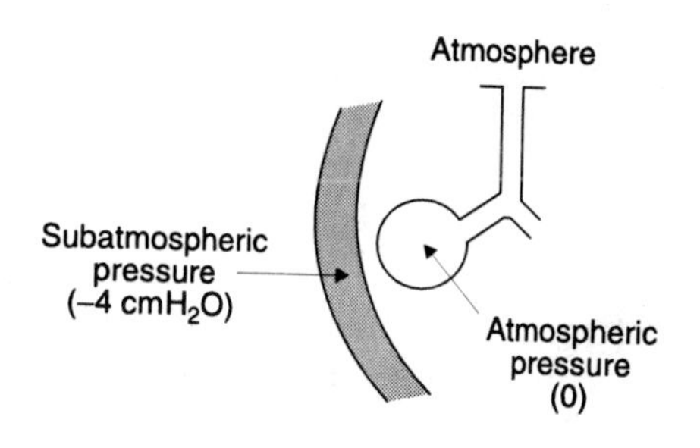

Distending pressure
= Alveolar pressure – Intrapleural pressure

(B) End of expiration = 0 – (–4) = +4 cmH_2O

Fig. 61 (A) The opposing forces acting on the lung generate a negative, or subatmospheric, intrapleural pressure. (B) The lung is kept inflated by the net distending pressure. An intrapleural pressure of –4 cmH_2O is typical at full expiration during quiet breathing.

Surfactant and alveolar surface tension

The magnitude of the surface tension resulting from the alveolar fluid is important in determining the forces necessary to keep the lungs inflated. Surfactant is a natural detergent-like substance, which is secreted into the alveoli by specialized epithelial cells in the alveolar wall known as *type II alveolar cells* (see Fig. 68). This reduces the surface tension relative to that of a simple electrolyte solution and allows the lungs to be kept expanded at a much less negative intrapleural pressure than would otherwise be possible. Infants born prematurely often produce inadequate amounts of surfactant and the increased effort of inflating the lungs leads to breathing difficulties, a condition known as *respiratory distress syndrome*.

Box 12 Clinical note: Pneumothorax

The term pneumothorax indicates the presence of gas in the intrapleural space. This gas may either have escaped from ruptured alveoli or entered from the atmosphere following traumatic damage to the chest wall. In each case the negative intrapleural pressure which normally keeps the lungs inflated is lost and the lung collapses. A simple pneumothorax may cause few problems and will resolve over time as the intrapleural gas is absorbed into adjacent capillaries. In a tension pneumothorax, however, a flap of damaged alveolar tissue acts as a valve allowing air to enter the pleural cavity during each expiration but preventing its escape during inspiration. The positive pressure which develops not only collapses the lungs but may prevent venous return to the heart leading to cardiovascular collapse. Tension pneumothorax is a medical emergency requiring immediate release of the pressure, e.g. by puncturing the chest wall with a needle.

Surfactant and alveolar stability The surface tension in the alveoli is not just important in determining the overall effort necessary to inflate the lungs but also helps reduce the tendency of individual alveoli to collapse as they change diameter. This phenomenon, which is called alveolar instability, is best understood by considering the relationship between the distending pressure (ΔP) necessary to keep an alveolus (or any distensible sphere) inflated, its radius (r) and the tension in the walls of the alveolus (T), as described by the *law of Laplace* (Eq. 17 in Fig. 62). The necessary distending pressure is proportional to the tension in the wall of the sphere but inversely proportional to its radius (Fig. 62A).

As alveoli become smaller, i.e., as r decreases, pressure will tend to increase, assuming that the surface tension, which contributes significantly to the wall tension (T), remains constant. This means that the pressure should be higher in small alveoli, causing them to collapse by driving air into larger alveoli with lower internal pressures (Fig. 62B). Surfactant helps prevent this since, as alveolar diameter decreases, the concentration of surfactant in the alveolar fluid increases, reducing surface tension. Thus, alveolar wall tension (T) and radius (r) decrease (or increase) in parallel with each other and alveolar pressure (ΔP) is little affected.

Lung compliance

Compliance is a measure of the ease with which the lungs can be inflated and is determined from the gradient of a plot of lung volume against distending pressure (Fig. 63, containing Equation 18).

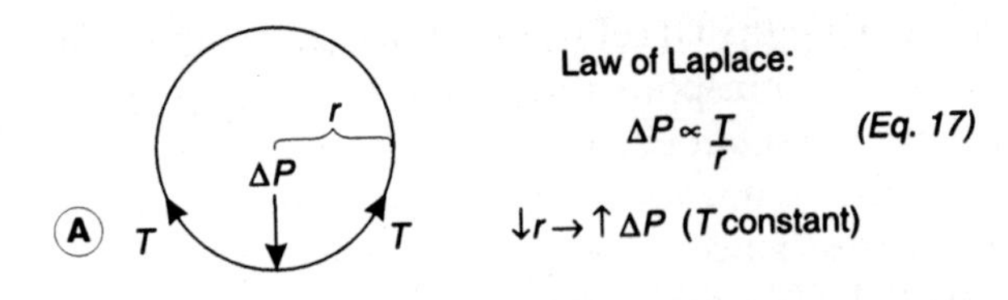

ΔP = Alveolar pressure – Intrapleural pressure

Intrapleural pressure does not change with r:

$\therefore\ \downarrow r \rightarrow \uparrow \Delta P \equiv \uparrow$ Alveolar pressure

Fig. 62 Alveolar pressure may be affected by alveolar size. (A) Laplace's law tells us that if surface tension (T) remains constant, alveolar distending pressure (ΔP) must increase as radius (r) decreases. (B) These changes in ΔP must reflect changes in intra-alveolar pressure (P), since intrapleural pressure is independent of alveolar size. The resulting alveolar instability is limited by the effects of surfactant (see text).

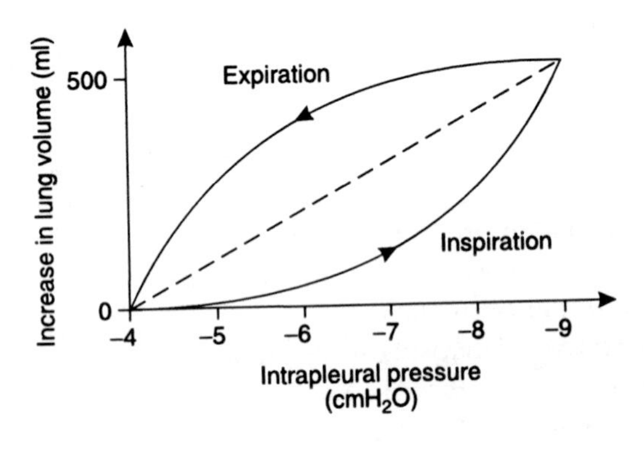

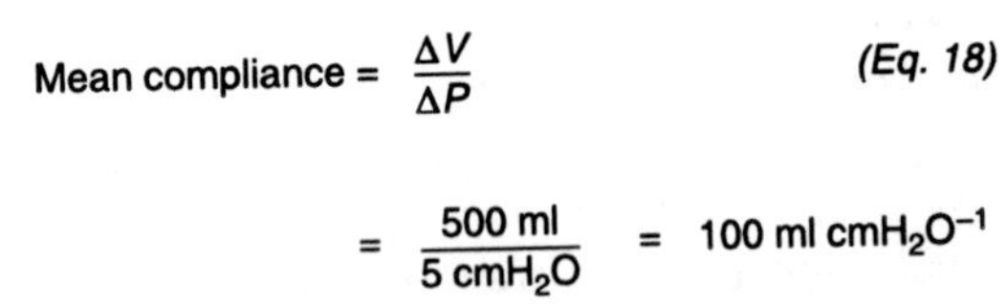

Fig. 63 The relationship between lung volume and intrapleural pressure during quiet respiration. Intrapleural pressure = –distending pressure (Fig. 60B), so the average slope of this plot (dotted line) gives the compliance under these conditions.

The relationship between lung volume and pressure demonstrates hysteresis, i.e., the shape is different during inspiration and expiration. An average compliance can be determined over the relevant range of pressures, however, using a linear interpolation (the dashed line in Fig. 63). Physiologically, the most important measure is the compliance of the lung and chest wall together, i.e., the compliance of the intact respiratory system, which is normally about 1 L kPa^{-1} (100 ml cmH$_2$O^{-1}). This may be reduced by diseases of the lung itself, e.g., those causing lung fibrosis, or by abnormalities of the chest wall.

Muscles of respiration

Inspiration is an *active* process in which the thoracic volume is increased by the action of the relevant muscles. The dome of the *diaphragm* is pulled down during diaphragmatic contraction, thereby increasing the vertical height of the thoracic cavity. This is augmented by contraction of the *external intercostal* muscles between the ribs, which raises them into a more horizontal position, increasing the width of the thorax from front to back. Accessory muscles in the neck, including sternocleidomastoid and scalenus, may also be used during maximal inspiration to elevate the sternum and first two ribs.

The intercostal muscles are innervated by *intercostal nerves* from the thoracic spinal cord but the diaphragm is innervated by the *phrenic nerves*, which arise from cervical spinal nerve roots 3, 4 and 5 before descending through the thorax to their destination. Because of this, high cervical cord damage is likely to be fatal, since respiration ceases. Lower level spinal injuries may leave a patient severely handicapped without being immediately life-threatening, since the phrenic output to the diaphragm is adequate for resting inspiration even in the absence of intercostal muscle contraction.

Quiet expiration is *passive* and relies on elastic recoil of the stretched lungs as the inspiratory muscles relax. The outward flow of gas may be actively accelerated during *forced expiration* by contraction of the abdominal muscles, which increases the intra-abdominal pressure forcing the diaphragm upwards, and of the *internal intercostals*, which actively pull the ribs downwards.

Pressure changes during ventilation

Intrapleural pressure is about −4 cmH$_2$O at the end of quiet expiration (Fig. 64). Contraction of the inspiratory muscles increases the outward force exerted by the diaphragm and chest wall on the parietal pleura and causes the intrapleural pressure to fall to about −9 cmH$_2$O. This raises the distending pressure acting across the walls of the alveoli (i.e., the difference between the intra-alveolar and intrapleural pressures, Fig. 61B) and so causes the lung to expand. A small, negative *intra-alveolar pressure* is generated by the expansion of the alveolar walls and this draws air in from the atmosphere. (This is analogous to drawing fluid into a syringe by pulling outwards on the plunger.) In forced inspiration, maximal contraction of the inspiratory muscles is reinforced by use of the accessory inspiratory muscles in the neck, which elevate the sternum and upper ribs as well. This leads to the generation of much more negative intrapleural pressures (e.g., −30 cmH$_2$O) and more rapid airflow.

During passive expiration, the elastic recoil of the stretched lungs produces a positive intra-alveolar pressure which drives air out (Fig. 64). Active contraction of abdominal and internal intercostal muscles during forced expiration generates strongly positive intrapleural pressures (e.g. +20 cmH$_2$O) which deflate the lungs more rapidly.

Work of breathing

Three elements contribute to this work.

- *Compliance work* is necessary to expand the lungs against elastic and surface tension forces.
- *Airways resistance* to airflow has to be overcome. This is usually very low but may be markedly increased in certain pulmonary diseases.
- *Tissue resistance* results from frictional forces which oppose the movement of one layer of pulmonary and pleural tissue past another during chest expansion.

Under normal conditions, the relative magnitudes of these components are:

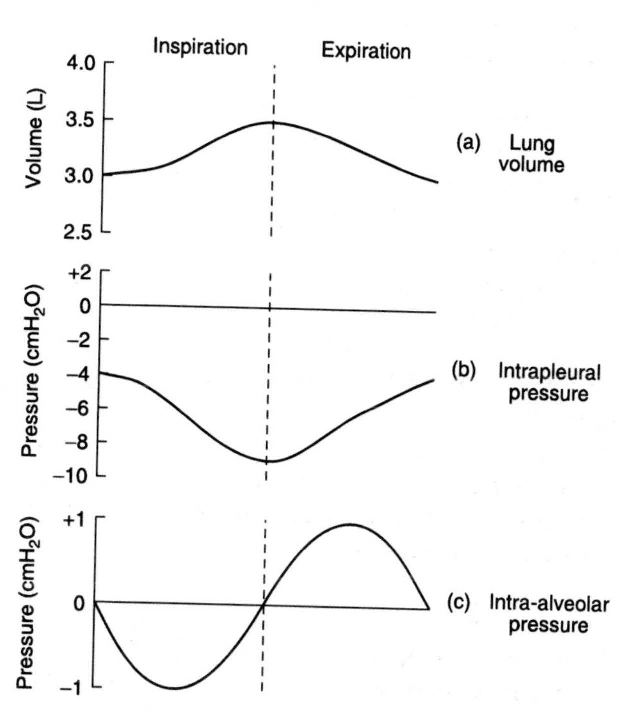

Fig. 64 The relationship between (a) lung volume, (b) intrapleural pressure and (c) intra-alveolar pressure during quiet breathing.

Four

Compliance work >> Airways resistance > Tissue resistance.

Work is usually only performed during inspiration. During rapid ventilation, however, increased airways resistance and tissue resistance effects necessitate energy expenditure during expiration as well and this raises the work of breathing in heavy exercise. Respiratory disease may have similar effects at rest, e.g., if the airways resistance is increased.

4.2 Pulmonary function tests

Learning objectives

At the end of this section you should be able to:

- define and give values for the named lung volumes and capacities
- explain the term dead space and how it affects alveolar ventilation
- outline how FVC, FEV_1 and PEFR are measured and how they change in lung disease.

Measurements of lung function are of great clinical significance because they allow respiratory disease to be classified pathophysiologically, its severity to be assessed and the benefits of therapy to be monitored objectively. Appropriate interpretation of these tests requires a knowledge of the normal pulmonary function test values which provide a benchmark for comparison.

Lung volumes

The volume of gas which can be moved in and out of the lungs during breathing is highly dependent on age, sex, body build and level of fitness, making it difficult to quote a single normal value for most of these measures.

Ventilatory volumes

The gas movement during ventilation can be recorded using a device called a spirometer (Fig. 65). This measures functionally important changes in lung volume.

- *Tidal volume* is the amplitude of the oscillation in lung volume during quiet respiration. It is usually about 400–500 ml.
- *Inspiratory reserve volume* (IRV) is the maximum volume of air which can be inspired in excess of normal inspiration. It is about 30% lower in females than males.
- *Expiratory reserve volume* (ERV) is the maximum volume of air which can be expired in excess of normal expiration. It is lower in females than males.

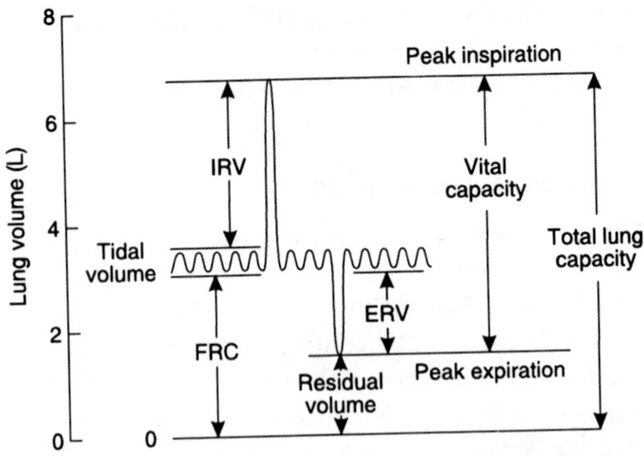

Fig. 65 Spirometer trace showing changes in lung volume during ventilation. Tidal volume is typical for quiet respiration. IRV, inspiratory reserve volume; ERV, expiratory reserve volume; FRC, functional residual capacity. As FRC and residual volume cannot be determined directly by spirometry, an indicator gas dilution method has to be used to determine the FRC to convert these measurements to total lung volumes.

Residual volume

Not all the gas can be expired from the lungs and the volume remaining after maximal expiration is called the residual volume. This cannot be measured directly using a spirometer but can be estimated from separate measures of the functional residual capacity (FRC, see below) and the expiratory reserve volume.

Residual volume = FRC – ERV

This gives a value in the region of 1–2 L, but this tends to increase with age.

Lung capacities

- *Total lung capacity* represents the sum of all the ventilatory volumes plus the residual volume (Fig. 65).
- *Vital capacity* is the sum of the ventilatory volumes, i.e., it is the volume of gas that is expelled from the lungs from peak inspiration to peak expiration. Values of 5 L in men and 3.5 L in women would be fairly typical, but vital capacity depends on body build, being larger in tall thin individuals as opposed to more obese subjects, decreases with age and is increased in athletically fit individuals. Body position is also important since, when lying down, the abdominal contents are forced against the diaphragm, restricting its movement (splinting

the diaphragm) and thus reducing the vital capacity.

- *Functional residual capacity* is the volume of gas left in the lungs at the end of quiet expiration; this can be estimated using a variation of the indicator dilution technique (Section 1.1). A known amount of the inert gas helium is equilibrated with the gas in the lungs. Since helium is insoluble in water and, therefore, does not diffuse out of the alveoli, the final concentration achieved can be used to calculate the functional residual capacity, which is typically about 3 L. This means that the normal tidal volume of incoming air (about 450 ml) is mixed with a 6–7 times larger volume of residual gas in the lungs, greatly reducing the breath-to-breath changes in alveolar O_2 and CO_2 concentrations during quiet respiration.
- *Inspiratory capacity* equals tidal volume plus inspiratory reserve volume.

Ventilation rates and minute volumes

The respiratory minute volume is the amount of new air being brought into the lungs each minute and can be calculated from the respiratory rate (about 12–15 min^{-1}) and the tidal volume (450 ml).

$$\text{Respiratory minute volume} = \text{Tidal volume} \times \text{Respiratory rate} \quad \text{(Eq. 19)}$$
$$= 450 \times 13 = 5850\,\text{ml min}^{-1} = 5.85\,\text{L min}^{-1}.$$

Dead space and alveolar ventilation rate

Not all the inspired air will actually reach the areas where gas exchange with the pulmonary circulation can take place. The volume which has to be ventilated but which does not participate in gas exchange is called the dead space. The *anatomical dead space* includes all the airways down to bronchiolar level. The air which enters these during inspiration is immediately expelled again at the beginning of the next expiration without contributing to pulmonary oxygenation. In some areas of the lung, there may also be alveoli which are ventilated but receive very little pulmonary perfusion. These regions cannot contribute to gas exchange either, and when their volume is included, we refer to the resulting total as the *physiological dead space*. It is this measurement which is of functional significance. Normally the dead space is about 150 ml.

The purpose of ventilation is to regulate alveolar gas concentrations. The rate at which the gas in the alveoli is renewed is determined by the alveolar ventilation rate where:

$$\text{Alveolar ventilation rate} = (\text{Tidal volume} - \text{Dead space}) \times \text{Respiratory rate}$$
$$= (450 - 150) \times 13 = 3.9\,\text{L min}^{-1}. \quad \text{(Eq. 20)}$$

It can be seen from this that any increase in the dead space automatically increases the respiratory minute volume necessary to achieve a given rate of alveolar ventilation. This is important in disease states which increase the physiological dead space. Also, in patients requiring artificial ventilation, a considerable volume of tubing is used to connect the airways to the ventilator itself. This adds to the dead space of the system and the tidal volume delivered by the machine has to be increased appropriately to ensure adequate alveolar ventilation.

Forced vital capacity and forced expiratory volume

In clinical practice, these measurements are often used to classify respiratory disease as either *restrictive* or *obstructive* in type. The patient is asked to take a maximal inspiration and then to breathe out as rapidly and fully as possible into a specialized spirometer which plots the expired volume against time in seconds (Fig. 66). Two measurements are commonly made from this.

Forced vital capacity (FVC) is the total volume of expired gas. This is similar to the vital capacity but is measured during forced expiration. The FVC is reduced in restrictive lung diseases. These may arise from conditions which limit the compliance of the lungs, e.g., lung fibrosis, or from a reduction in available lung volume caused by removal or collapse of part of the lung.

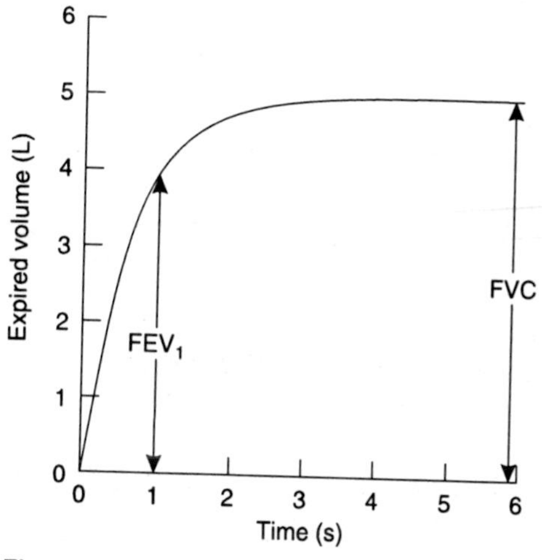

Fig. 66 Measurement of forced vital capacity (FVC) and forced expiratory volume in 1 second (FEV_1) from a forced expiration following maximal inspiration.

Four

Forced expiratory volume (FEV) is the volume of gas expelled in a given time; if this is measured for the first second it is called the forced expiratory volume in 1 second (FEV_1). This is limited by the speed with which gas can be forced through the airways and is decreased in obstructive lung disease. Since the actual magnitude of FEV_1 is always reduced in parallel with any reduction in FVC, even in the absence of obstruction, it is the ratio of FEV_1/FVC which is most useful diagnostically. This ratio should normally exceed 0.75 (75%) in healthy individuals but often falls below 0.5 (50%) when there is increased airways resistance, e.g., in asthma.

Peak expiratory flow rate (PEFR)

This is a simple test of ventilatory function which is widely used in clinical practice. The patient is simply asked to blow air out of their fully inflated lungs as rapidly as they can and the peak flow rate achieved is recorded with a flow meter. Normal values are again very dependent on age, sex and build but are of the order of $400\,L\,min^{-1}$. This may fall dramatically in cases of obstructive airways disease.

Box 13 Clinical note: Obstructive airways disease

Obstructive airways disease is a pathophysiological diagnosis characterized by reductions in FEV_1/FVC and PEFR. It is seen in two main groups of patients, asthmatics and those with chronic obstructive pulmonary disease (COPD). Asthma is characterized by wheezing episodes caused by an acute increase in airways resistance. There is an association with allergic conditions such as hay fever and atopic eczema and attacks may be triggered by a variety of stimuli including inhaled dust and antigens. Between episodes lung function is often normal, especially in young patients. In COPD, however, airways resistance is persistently elevated but function deteriorates further during chest infections. COPD is often caused by smoking and atmospheric pollution and is associated with chronic bronchitis (cough with spit) and emphysema (destruction of alveolar tissue). Despite these differences, both asthma and COPD may be treated with adrenergic agonists which mimic sympathetic nervous activity and cause relaxation of bronchial smooth muscle.

4.3 Gas exchange in the lungs and tissues

Learning objectives

At the end of this section you should be able to:

- outline the physical factors controlling gas diffusion
- quantify the pressure gradients driving O_2 and CO_2 diffusion in the lungs and tissues
- list components of the pulmonary diffusion barrier
- define ventilation:perfusion ratio ($\dot{V}_A/\dot{Q}$)
- outline how abnormal $\dot{V}_A/\dot{Q}$ affects pulmonary gas exchange.

The movement of O_2 and CO_2 in and out of the capillaries both in the lungs and in the peripheral tissues depends on gas diffusion. This is affected by three main factors.

Pressure gradients drive gas movements. The relevant SI unit of pressure is the kPa ($1000\,N\,m^{-2}$), although mmHg is also often used in respiratory physiology (1 mmHg = 0.133 kPa). Differences in total pressure lead to bulk flow of a gas mixture, e.g., air flows from regions of high pressure to regions of low pressure. At the interface between two gas mixtures, however, the tendency for any individual gas to diffuse from one region to another is determined by differences in the partial pressure for that gas, i.e., *diffusion is driven by partial pressure gradients*.

Partial pressure is defined as the pressure a gas would exert if it alone occupied the total volume available to the mixture of gases. It is determined both by the *concentration* of a gas, expressed as a fraction of the total mixture, and the *total pressure* of the mixture. For example, if we consider a mixture of gases at a total pressure of 1 kPa, and if 20% of the mixture is O_2, then the partial pressure of oxygen (P_{O_2}) can be calculated as follows:

$$\text{Partial pressure} = \text{Total pressure} \times \text{Fractional concentration} \qquad \text{(Eq. 21)}$$

$$P_{O_2} = 1 \times 0.2 = 0.2\,\text{kPa}$$

We could double the partial pressure of O_2 to 0.4 kPa either by increasing its concentration to 40%, or by doubling the total pressure of the mixture to 2 kPa. It is also important to note that the partial pressures of all the gases in a gas mixture must always add up to the total pressure of the mixture.

The diffusion coefficient for each gas is a measure of the ease with which it can diffuse through the aqueous body fluids, and is determined by its solubility in water and its molecular weight.

$$\text{Diffusion coefficient} \propto \frac{\text{Solubility in } H_2O}{\sqrt{\text{Molecular weight}}}$$

(Eq. 22)

Even though CO_2 has a higher molecular weight, it diffuses about 20 times more easily than O_2 since it is much more soluble in water. Both gases are highly lipid soluble and so pass easily through cell membranes.

Tissue properties. The physical properties of the tissues at the site of exchange, particularly the *surface area* over which diffusion can occur and the *diffusion distances* between one compartment and the other, also influence the total rate of gas exchange. The lungs are well adapted for gas diffusion, with a very large alveolar surface area and a very thin layer of fluid and tissue separating alveolar gas from pulmonary blood.

Gas exchange in the lungs

Gas exchange depends on the partial pressure gradients between alveolar air and pulmonary arterial blood for O_2 and CO_2. Effective gas exchange also requires easy diffusion between alveoli and blood and an appropriate balance between ventilation and perfusion within the lungs.

Alveolar gases

Room air is chiefly a mixture of N_2 and O_2, with variable amounts of water vapour and a tiny percentage of CO_2 (Table 5). Although the average total pressure in the alveoli is equal to that in the atmosphere, alveolar air differs from the air we breathe (the *inspirate*) in a number of ways. Firstly, it has a higher water vapour pressure, since the inspired gases become fully saturated as they pass through the airways. The partial pressure of water vapour (P_{H_2O}) in the alveoli always equals the saturated water vapour pressure at 37°C (6.3 kPa; 47 mmHg), irrespective of the total alveolar pressure. Since water molecules have been added to the inspirate, the concentrations of all the other gases in the mix are reduced by dilution (Table 5). This decreases their partial pressures, as illustrated by the changes in gas pressure which result from simply humidifying previously dry room air (Table 6).

The other major differences between alveolar gas and room air reflect constant removal of O_2 by diffusion into the pulmonary blood and constant addition of CO_2 from the same source. This reduces the P_{O_2} and elevates the P_{CO_2} as compared with humidified air (Table 6). The actual values achieved depend on the balance between the rate of ventilation and the rates of O_2 consumption and CO_2 production by the body. If ventilation were to increase without any change in gas use or production, alveolar air would become more like humidified air, with a rise in P_{O_2} and a fall in P_{CO_2}. This limits the maximum possible alveolar P_{O_2} to about 20 kPa (150 mmHg), i.e., the value in saturated room air (Table 6), at least while breathing air at normal atmospheric pressure. Higher partial pressures can only be achieved by breathing an oxygen-enriched gas mixture or by increasing the total pressures of the inspirate.

Normally, of course, ventilation is homeostatically regulated so that alveolar P_{O_2} and P_{CO_2} remain

Table 5 Concentrations of the gases in dry room air, air completely humidified at 37°C and alveolar air

Gas	Gas concentration (% of total)		
	Room air	Humidified	Alveolar air
N_2	79.00	74.09	74.9
O_2	20.96	19.67	13.6
CO_2	0.04	0.04	5.3
H_2O	0.00	6.20	6.2

Table 6 Partial pressures of the gases in dry room air, air completely humidified at 37°C and alveolar air

Gas	Room air kPa (mmHg)	Humidified air kPa (mmHg)	Alveolar air kPa (mmHg)
N_2	79.79 (600.4)	74.83 (563.1)	75.6 (569.2)
O_2	21.17 (159.3)	19.87 (149.5)	13.7 (103.4)
CO_2	0.04 (0.3)	0.04 (0.3)	5.3 (40.3)
H_2O	0 (0)	6.3 (47.1)	6.3 (47.1)
Total	101 (760)	101 (760)	101 (760)

Atmospheric pressure is 101 kPa (760 mmHg).

relatively constant (Section 5). As previously explained, the large functional residual capacity during quiet respiration also acts as a buffer against large changes in alveolar gas pressures during each respiratory cycle, since the incoming tidal volume is diluted in a much larger volume of residual pulmonary gas (Section 4.2).

Pulmonary blood gases

Gas diffusion gradients in the lung are determined by the differences between the partial pressures within the alveoli and those within the pulmonary capillaries. The partial pressures of O_2 and CO_2 in pulmonary arterial blood entering these capillaries are determined by the levels in systemic venous blood from peripheral tissues. Venous blood is mixed in the right ventricle, giving a P_{O_2} in the pulmonary arteries of 5.3 kPa (40 mmHg) and a P_{CO_2} of 6 kPa (45 mmHg). Since the P_{O_2} in the alveoli is 13.7 kPa (103 mmHg), O_2 diffuses into the pulmonary blood. Carbon dioxide, also driven by a partial pressure gradient (Fig. 67), diffuses in the opposite direction, from the capillaries (P_{CO_2} = 6 kPa, or 45 mmHg) into the alveoli (P_{CO_2} = 5.3 kPa, or 40 mmHg). Gas diffusion across the combined alveolar and capillary wall system is rapid, so that pulmonary blood normally equilibrates with alveolar gases before leaving the pulmonary capillary. Thus, the gas pressures in pulmonary venous blood equal those in the alveoli under physiological conditions (Fig. 67).

The P_{O_2} in systemic arterial blood is somewhat lower than that in the pulmonary veins because of mixing with deoxygenated blood from the bronchial veins. Bronchial blood vessels provide the nutritive needs of the bronchial system, losing O_2 and gaining CO_2 as they do so, and do not take part in alveolar gas exchange. Bronchial venous blood drains into the left atrium and reduces systemic arterial P_{O_2} to about 13 kPa (98 mmHg). Thus, the bronchial circulation acts as a *shunt*, bypassing pulmonary oxygenation.

The diffusion processes described above lead to a change in the gas content of pulmonary blood, as determined by the dissociation curves for O_2 and CO_2 transport in blood (Section 4.4). Oxygen content normally rises from 15 to 20 ml O_2 100 ml^{-1} blood. At the same time CO_2 levels drop from about 52 to 48 ml CO_2 100 ml^{-1} blood. The ratio of CO_2 release (4 ml CO_2 100 ml^{-1} blood) to O_2 uptake (5 ml O_2 100 ml^{-1} blood) is called the *respiratory exchange ratio*, which has a value of 0.8 in this case. This ratio is also, though not strictly correctly, referred to as the *respiratory quotient*. Under steady state conditions the respiratory exchange ratio is determined by the substrates being metabolized by the body and 0.8 is fairly typical for someone on a mixed, Western-style diet (Section 4.7).

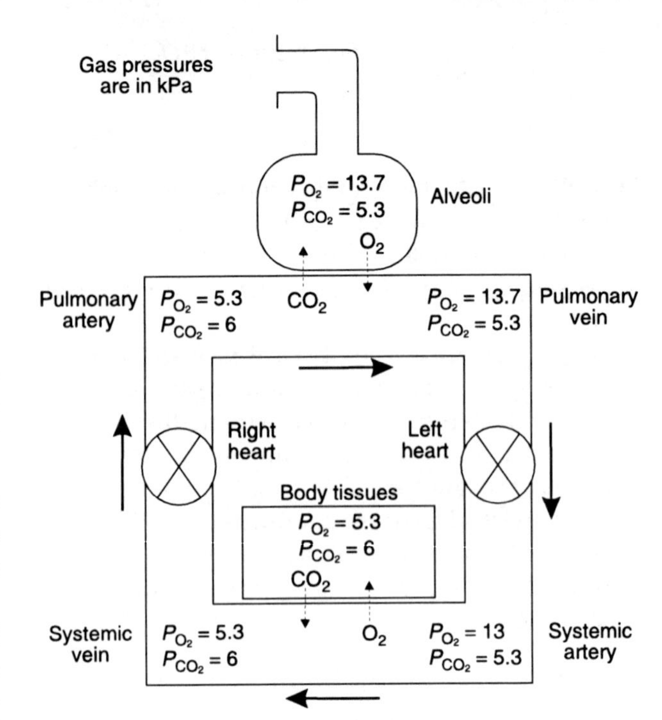

Fig. 67 Gas exchange in the lungs and tissues. Diffusion of O_2 and CO_2 from high to low pressure occurs until capillary blood equilibrates with the gases in the alveoli or in the interstitial fluid of peripheral tissues.

Pulmonary diffusion

Pulmonary gases must diffuse through several structures between the alveoli and the capillary blood (Fig. 68). A layer of alveolar fluid lies over the epithelium, which sits on a basement membrane. This may actually fuse with the basement membrane of the capillary endothelium so that the total thickness of the diffusion barrier can be as little as 0.2 μm. Coupled with the large alveolar surface area (about 70 m^2), this short diffusion distance makes the lungs very efficient gas exchange units, i.e., they have a *high diffusion capacity*. If the thickness of the diffusion barrier is increased, e.g., because of an increase in alveolar fluid during pulmonary oedema, or the available alveolar area is decreased, e.g., in emphysema, which involves alveolar destruction, gas exchange will be impaired as a result of the reduced diffusion capacity. This may cause abnormalities in the systemic arterial blood gas levels, even when alveolar ventilation is perfectly adequate.

Ventilation : perfusion ratio

Normal gas exchange requires both that alveoli are adequately ventilated and that they are perfused with pulmonary blood at an appropriate rate. One way of quantifying this relationship is to measure the alveolar ventilation:perfusion ratio, $\dot{V}_A/Q$.

$$\frac{\dot{V}_A}{Q} = \frac{\text{Alveolar ventilation rate}}{\text{Pulmonary blood flow}} \qquad \text{(Eq. 23)}$$

This may deviate from normal, a situation which is referred to as *ventilation–perfusion mismatch*. If an area of the lung is perfused but inadequately ventilated, $\dot{V}_A/Q$ will obviously be reduced. As a result, the alveolar P_{O_2} falls and the P_{CO_2} rises since less 'fresh' air is brought into the alveoli than normal. In the most extreme case, where $\dot{V}_A/Q = 0$ (perfusion with no ventilation), alveolar gas pressures will become equal to those in pulmonary arterial blood (Fig. 69). Blood passing through such a region will be inadequately oxygenated and this will reduce the final P_{O_2} in the systemic arterial blood. Thus, areas with a decreased $\dot{V}_A/Q$ increase the *physiological shunting* of blood. This is a major factor contributing to the abnormal blood gases seen in many respiratory diseases.

Where alveoli are ventilated but not properly perfused, $\dot{V}_A/Q$ will be increased. Such regions contribute less than they should to pulmonary gas exchange so they add to the physiological dead space. In the absence of gas diffusion in and out of the pulmonary blood, alveolar P_{O_2} rises and P_{CO_2} falls. In the most extreme case, where pulmonary flow equals zero (i.e., $\dot{V}_A/Q = \infty$), the alveolar gas pressures become equal to those in humidified room air (Table 6; Fig. 69). This can occur in regions of the lung affected by a pulmonary embolus, a travelling blood clot which completely obstructs one of the pulmonary arteries.

For the lungs as a whole under normal conditions, the mean value for $\dot{V}_A/Q = 4\,\text{L min}^{-1}$ (alveolar ventilation rate) $\div$ $5\,\text{L min}^{-1}$ (pulmonary blood flow) = 0.8. The local value of $\dot{V}_A/Q$ varies with posture. The upper regions of the lung are more poorly ventilated and more poorly perfused than the bases. When standing, however, the increase in ventilation between the apex and the base of the lung is much smaller than the parallel increase in perfusion, so that $\dot{V}_A/Q$ decreases as one moves towards the base. At the apices of the lungs $\dot{V}_A/Q$ is about 3.0 (contributing to the physiological dead space), but this falls to about 0.6 at the bases (which contribute to the physiological shunt).

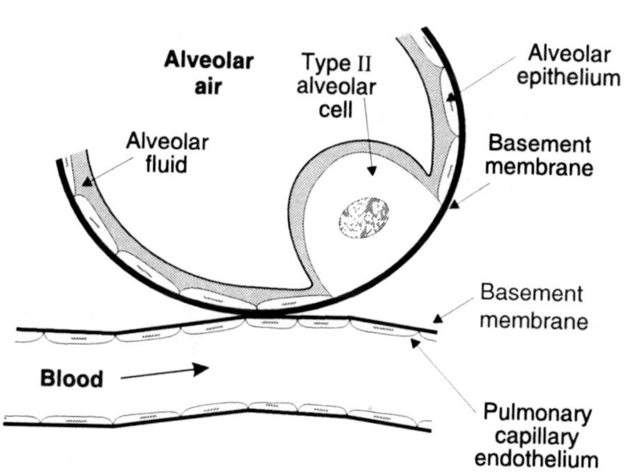

Fig. 68 The alveolar wall through which pulmonary diffusion must occur. The basement membrane of the respiratory epithelium and the capillary endothelium have fused. A surfactant-secreting type II pneumocyte is also shown.

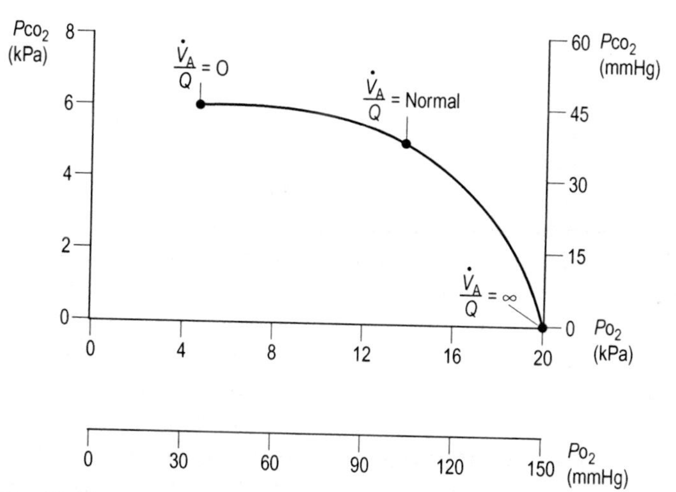

Fig. 69 Changes in the alveolar gas concentrations with changing $\dot{V}_A/Q$ ratio. As $\dot{V}_A/Q$ increases, P_{O_2} increases and P_{CO_2} decreases. Three specific points on the curve are indicated. When $\dot{V}_A/Q = 0$ (i.e., perfusion with no ventilation), alveolar P_{O_2} (5.3 kPa) and P_{CO_2} (6 kPa) equilibrate to the values in pulmonary arterial blood. When $\dot{V}_A/Q = \infty$ (i.e., ventilation with no perfusion), alveolar P_{O_2} (20 kPa) and P_{CO_2} (0 kPa) become equal to the values in humidified air. The intermediate point plots the alveolar P_{O_2} (13.7 kPa) and P_{CO_2} (5.3 kPa) when $\dot{V}_A/Q$ is normal.

Measurement of pulmonary gas exchange

The efficiency of gas exchange in the lungs is often measured in terms of the *transfer factor* for carbon monoxide (T_{CO}). This is added to the inspired air at very low concentrations (CO is poisonous; Section 4.4). The increase in arterial CO content over a short period of time is then measured to provide a value for the rate of CO absorption. Since haemoglobin has an extremely high CO affinity, essentially all the absorbed CO becomes attached to the haemoglobin leaving none in the plasma. This means that the pulmonary arterial *P*co can be assumed to be zero, so that the diffusion gradient driving CO absorption equals the alveolar *P*co, which is measured. From this the transfer factor for CO can be calculated.

$$T_{CO}\left(\text{ml CO min}^{-1}\ \text{kPa}^{-1}\right) = \frac{\text{Rate of CO absorption}\ \left(\text{ml CO min}^{-1}\right)}{\text{Alveolar}\ P_{CO}\ (\text{kPa})} \quad \text{(Eq. 24)}$$

The transfer factors for O_2 and CO_2, though different from that for CO, will be proportional to T_{CO}, which can, therefore, be used clinically to assess whether pulmonary gas exchange is impaired. Defects may arise because of impaired diffusion across the alveolar walls themselves but are more commonly caused by a ventilation–perfusion mismatch.

Gas exchange in peripheral tissues

The pressure gradients driving peripheral gas exchange are normally dictated by P_{O_2} and P_{CO_2} levels within a given tissue, since systemic arterial blood gas pressures are kept constant, with a P_{O_2} of about 13 kPa (98 mmHg) and a P_{CO_2} of 5.3 kPa (40 mmHg). Tissue gas pressures may vary widely from organ to organ and time to time, depending on the balance between blood flow and local metabolic activity, which consumes O_2 and produces CO_2. Average values are about 5.3 kPa (40 mmHg) for P_{O_2} and 6 kPa (45 mmHg) for P_{CO_2}, so the pressure gradients favour O_2 diffusion from the blood into the interstitium, while CO_2 diffuses in the opposite direction. Systemic capillary blood equilibrates with the interstitial fluid so that systemic venous gas tensions equal those in the interstitium (Fig. 67). On average, resting body tissues take up O_2 at a rate of about 5 ml 100 ml^{-1} blood flow, and release CO_2 at about 4 ml 100 ml^{-1} blood.

4.4 Gas transport in blood

Learning objectives

At the end of this section you should be able to:

- describe how haemoglobin carries O_2
- differentiate between O_2 content and saturation
- draw an O_2 dissociation curve and use it to calculate O_2 uptake and release
- describe the major factors which affect O_2 dissociation
- explain how CO_2 is transported in the blood
- draw a CO_2 dissociation curve and use it to calculate CO_2 uptake and release
- describe the major factors which affect CO_2 transport
- explain the relationship between CO_2 content and blood pH.

Both O_2 and CO_2 are transported between the lungs and the tissues in the blood. The mechanisms involved will be considered separately for each gas.

Oxygen transport

Haemoglobin

The vast majority of the O_2 in blood is transported within red cells and is bound to haemoglobin, with only a negligible additional amount dissolved in the plasma. One haemoglobin molecule consists of four polypeptide chains (the *globin* elements), each of which is attached to a pigmented *haem* group made up of a protoporphyrin ring surrounding a *ferrous* ion (Fe^{2+}). Oxygen can bind reversibly with these haem elements, forming *oxyhaemoglobin*, so a single molecule of haemoglobin can carry up to four O_2 molecules.

Various factors may reduce the O_2-carrying capacity of blood. Simplest, and most common, is a reduction in the concentration of haemoglobin, a condition known as *anaemia*. This may arise in various ways, although nutritional deficiencies and chronic blood loss are probably the commonest causes in Western society (Section 2.3). The O_2-carrying capacity of the blood is proportional to the haemoglobin concentration, and this explains the main symptoms of anaemia, which include weakness, tiredness and reduced exercise tolerance. The normal haemoglobin concentration is about 15 g dl^{-1} in males and 13 g dl^{-1} in females (1 dl = 0.1 L = 100 ml). Since each gram of haemoglobin can carry 1.34 ml of O_2, this equates to O_2 capacities of 20 and 17.5 ml 100 ml^{-1} blood, respectively.

Some conditions reduce the ability of the haemoglobin molecules to carry O_2 (reduce their O_2 *affinity*). For example, if the Fe^{2+} is converted to Fe^{3+}, then the O_2-binding capability is greatly reduced. Certain drugs can oxidize haemoglobin in this way, forming methaemoglobin. Carbon monoxide can also interfere with O_2 transport since it binds very tightly to haemoglobin, leading to a build-up of carboxyhaemoglobin. Little unbound haemoglobin is left for O_2 transport.

Oxygen dissociation curve

This sigmoid, or S-shaped curve describes the relationship between the partial pressure of O_2 and the concentration of O_2 in blood (Fig. 70A). As P_{O_2} is raised from zero to 2 kPa the O_2 content of the blood also rises, slowly at first and then more rapidly. The shape of this initial part of the curve reflects the fact that there are four O_2 binding sites on each haemoglobin molecule. When one site attaches to O_2, the O_2-binding ability of the other sites is increased and so the rate of rise of O_2 concentration also increases. This phenomenon is known as cooperativity, since the separate haemoglobin subunits cooperate in their function. As the P_{O_2} is increased further, more and more O_2 is carried in the blood until all the available sites on haemoglobin are occupied (saturation); the curve reaches a plateau at a P_{O_2} of around 16 kPa (120 mmHg). Further increases in P_{O_2} have little effect on O_2 transport.

The quantity of O_2 carried depends on the haemoglobin concentration, so that a 50% reduction in haemoglobin (from 15 to 7.5 $g\,dl^{-1}$) also reduces the oxygen content at the plateau (the *O_2-carrying capacity*) by half (Fig. 70A). It is important to appreciate, however, that this only alters the height of the dissociation curve; its shape is not affected by the haemoglobin concentration. Plotting the O_2 concentrations in the normal and anaemic samples as percentages of their own maximum values, i.e., in terms of the *O_2 saturation* of the blood, produces identical curves (Fig. 70B). Any change in this saturation curve indicates a change in the O_2-carrying properties of the haemoglobin itself, rather than a change in haemoglobin concentration.

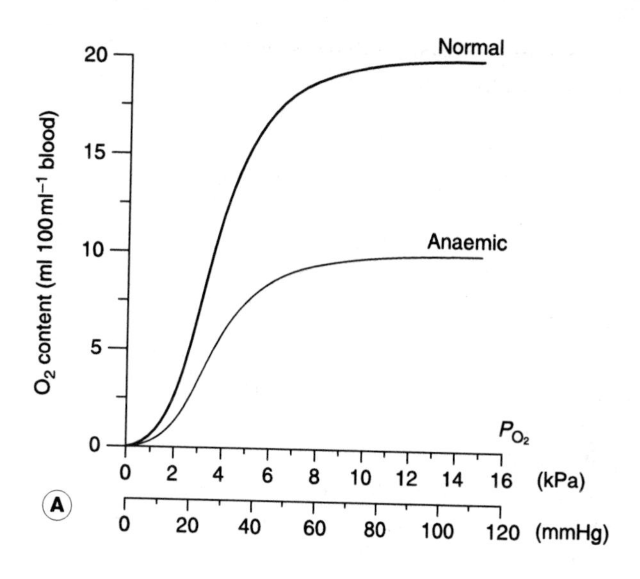

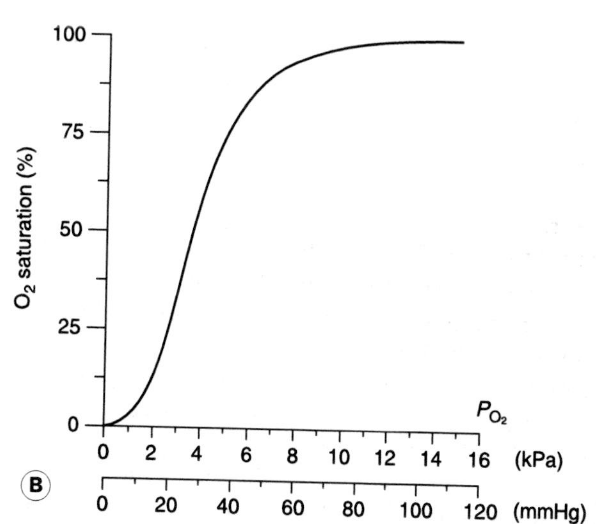

Fig. 70 Oxygen dissociation curves. (A) Oxygen content for a normal blood sample ([Hb] = 15 $g\,dl^{-1}$), and for an anaemic sample ([Hb] = 7.5 $g\,dl^{-1}$). (B) Oxygen saturation curve showing O_2 content as a percentage of the maximum in each sample. This is independent of [Hb].

Using the O_2 dissociation curve to determine O_2 uptake The O_2 dissociation curve can be used to determine how much O_2 will be released into the body tissues for a given fall in O_2 pressure. Under normal conditions, blood is 97% saturated at the P_{O_2} of systemic arterial blood (13 kPa, 98 mmHg). Saturation only falls to 75% on reducing the P_{O_2} to 5.3 kPa (40 mmHg), the value found in systemic venous blood (Fig. 70B). With a haemoglobin concentration of 15 $g\,dl^{-1}$, this is equivalent to a tissue O_2 uptake of 5 ml O_2 100 ml^{-1} blood (Fig. 70A). These processes are reversed as deoxygenated blood passes through the pulmonary capillaries. The P_{O_2} is initially raised to 13.7 kPa (105 mmHg), although this is then reduced to 13 kPa (98 mmHg) by mixing with deoxygenated blood shunted through the bronchial circulation (Section 4.3). The net result is the absorption of the same amount of O_2 from the alveoli as was given up to the peripheral tissues.

Much higher than average rates of O_2 uptake may be achieved in rapidly metabolizing organs, in which case the tissue P_{O_2} falls below normal. If the P_{O_2} were to fall to 2.7 kPa (20 mmHg), for example, O_2 saturation would fall to 25% (Fig. 70B), yielding an additional 10 ml O_2 100 ml^{-1} blood with a normal haemoglobin concentration. Quantitatively, this is the most important mechanism whereby tissue extraction of O_2 from blood may be increased.

Factors which alter the O_2 dissociation curve (Bohr effect) Changes in the conditions under which the O_2 dissociation curve is measured can change its shape. For example, if the CO_2 level in the blood is increased then the curve is shifted to the right, so that O_2 saturation is lower for a given P_{O_2} than would otherwise be the case (Fig. 71). This is the Bohr effect and, as a result, the O_2 dissociation curve for systemic venous blood, which has an elevated P_{CO_2}, lies slightly to the right of that for arterial blood. This increases O_2 extraction as blood passes through actively respiring tissues, although the effect is relatively small. As blood passes through the lungs the P_{CO_2} falls again, shifting the curve back to the left.

A number of factors other than CO_2 produce similar shifts in the dissociation curve. Increasing the temperature and reducing the pH both produce rightward shifts. These may contribute to increased O_2 extraction in metabolically active tissue, where the local temperature rises, and increased CO_2 and lactic acid production leads to a fall in pH. Intracellular 2,3-diphosphoglycerate (2,3-DPG), which accumulates in red cells during prolonged periods of hypoxia, has a similar effect. This provides a compensatory mechanism which favours O_2 release to the tissues, e.g., in those living at altitude. Unlike the other factors, however, the shift in the dissociation curve resulting from raised 2,3-DPG levels is not reversed in the lungs, so oxygen uptake may also be limited as a result.

Fetal haemoglobin and adult myoglobin

The ability of molecules like haemoglobin to bind O_2 is not just affected by external factors, such as the level of CO_2, but also by the structure of their polypeptide chains. For example, these chains differ in the fetal and adult forms of haemoglobin, and this results in a higher O_2 affinity in fetal blood. This is seen when the O_2 dissociation curves for the two kinds of haemoglobin are compared, with that in the fetus lying to the left of that in adults (Fig. 72). This increases oxygen transfer across the placenta, maintaining high levels of saturation in fetal blood even at relatively low values of P_{O_2}. Similarly, the dissociation curve for myoglobin, an O_2 storage molecule found in skeletal muscle fibres, is also considerably to the left of that for haemoglobin, favouring O_2 transfer from haemoglobin to myoglobin at any given P_{O_2} (Fig. 72). This makes myoglobin a useful O_2 source, particularly in muscles which produce sustained (tonic) contractions, thus compressing blood vessels and restricting blood flow. It is the relatively high concentration of myoglobin in the muscles associated with the maintenance of posture which gives them their distinct dark red colour (Section 1.6).

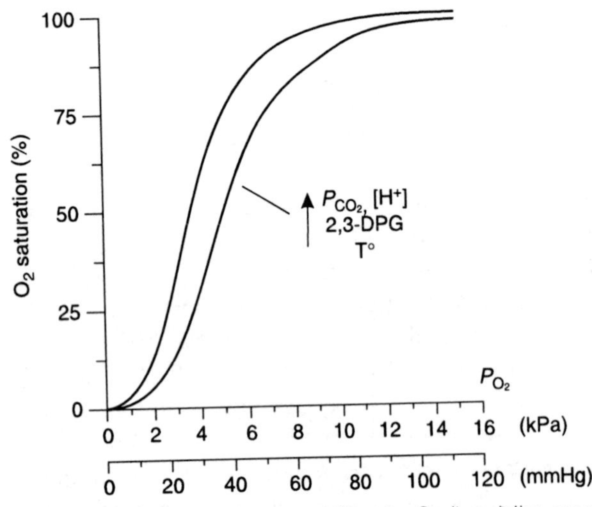

Fig. 71 The Bohr effect. Factors shifting the O_2 dissociation curve to the right: elevation of CO_2, H^+, 2,3-diphosphoglycerate (2,3-DPG) or a rise in temperature.

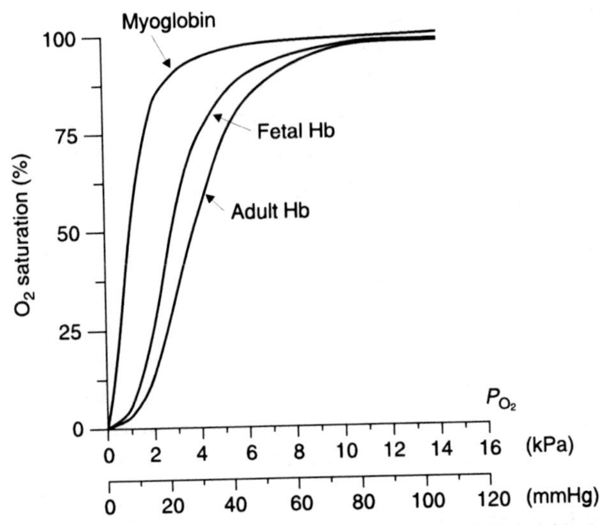

Fig. 72 Changes in O_2 affinity caused by changes in the globin peptide sequence of O_2-binding molecules. At any given P_{O_2}, O_2 will be transferred from the lower- to the higher-affinity carrier molecule.

Carbon dioxide transport

Chemistry of CO_2 transport

Carbon dioxide is carried in three main forms in the blood; as bicarbonate ions (HCO_3^-), in carbamino groups and dissolved as free gas.

Transport as HCO_3^-. This accounts for about 60–70% of the total CO_2 carried in the blood. The gas diffuses into the red blood cells where it reacts with water to form carbonic acid. This reaction is catalyzed by the presence of the enzyme *carbonic anhydrase*.

$$CO_2 + H_2O \rightleftharpoons H_2CO_3 \rightleftharpoons H^+ + HCO_3^- \quad \text{(Eq. 25)}$$

Carbonic acid rapidly dissociates into H^+ and HCO_3^-. The H^+ mainly binds to haemoglobin molecules inside the cell, limiting the change in pH, i.e., haemoglobin acts as an important buffer in this system (Section 5.9). Bicarbonate ions diffuse out of the cell into the plasma (Fig. 73), but this is balanced by a reverse movement of Cl^- into the red cells, which maintains electrical neutrality. This *chloride shift* raises the intracellular concentration of Cl^- in venous samples of blood, since these have a higher CO_2 content than systemic arterial blood. The total number of intracellular particles is also increased so there is an osmotically driven movement of water into the erythrocytes under these conditions. As a result, both red cell volume and haematocrit are raised in venous samples.

Transport as carbamino groups. These are formed by a reaction between CO_2 and amino acid residues in peptides and proteins. About 20–30% of circulating CO_2 is in this form. The reaction is readily reversible, releasing CO_2 if the P_{CO_2} falls. Although some carbamino groups are found on plasma proteins, most are formed in association with the peptide chains of haemoglobin. Deoxygenated haemoglobin has a greater capacity for forming carbamino groups than oxyhaemoglobin.

Transport as dissolved CO_2. This only accounts for about 10% of the total CO_2 transport. As for any gas, the concentration of dissolved CO_2 increases linearly with the P_{CO_2} of the solution (Henry's law).

Carbon dioxide dissociation curve

This is a plot of the total concentration of CO_2 in blood at different values of P_{CO_2}, regardless of what form it is transported in (Fig. 74). When this is compared with the O_2 dissociation curve (Fig. 70A) several differences are obvious.

- The solubility of CO_2 in blood is considerably higher than that of O_2, as indicated by higher concentrations at a given partial pressure, e.g., at a P_{CO_2} of 5.3 kPa (40 mmHg), blood contains about 50 ml CO_2 100 ml^{-1}, as compared with 15 ml O_2 100 ml L^{-1} at a similar P_{O_2}.
- The physiological range of P_{CO_2} is much narrower than that for P_{O_2}: P_{CO_2} is normally 5.3 kPa (40 mmHg) in systemic arterial blood, rising to 6 kPa (45 mmHg) in systemic venous blood. This compares with P_{O_2} ranging from 13 kPa (98 mmHg) in arteries to 5.3 kPa (40 mmHg) in veins.
- There is no upper plateau on the CO_2 dissociation curve. Blood cannot be saturated

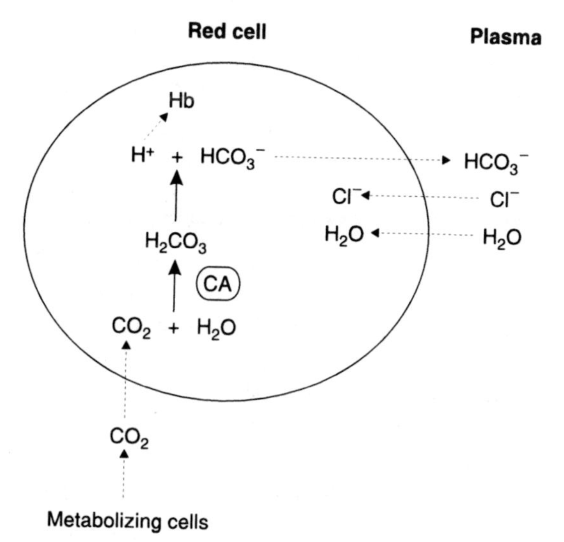

Fig. 73 Schematic representation of the conversion of CO_2 to HCO_3^- by the enzyme carbonic anhydrase (CA) in red blood cells.

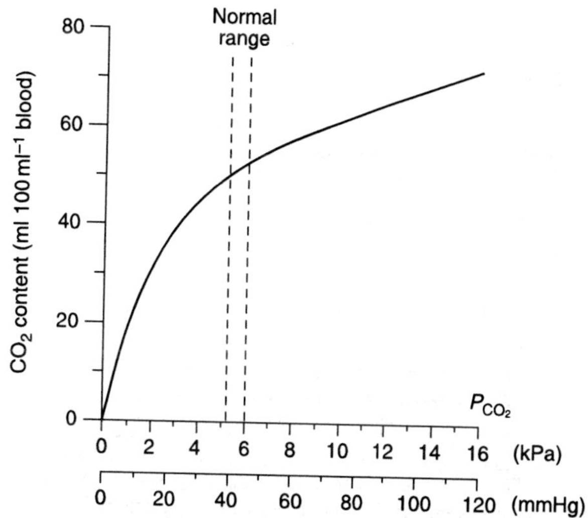

Fig. 74 The CO_2 dissociation curve.

Four

with CO_2 as it can with O_2; rather, as the P_{CO_2} continues to rise, so too does the CO_2 content. As we shall see, this means that P_{CO_2} normally has to be more tightly controlled than P_{O_2}.

Factors which alter the CO_2 dissociation curve The position of the CO_2 dissociation curve is mainly influenced by the O_2 tension in the blood, shifting downwards and to the right as P_{O_2} increases. This is called the *Haldane effect* and, as a result, the amount of CO_2 carried in the blood at a given P_{CO_2} falls as oxygen levels rise (Fig. 75). One consequence is that the CO_2 dissociation curve for systemic venous blood (low P_{O_2}) lies above that for systemic arterial blood (high P_{O_2}), increasing the amount of CO_2 that can be carried from the tissues back to the heart. The significance of this is best appreciated by considering some physiologically relevant values. Arterial blood has a P_{CO_2} of 5.3 kPa (40 mmHg), which is equivalent to a CO_2 content of about 48 ml 100 ml^{-1} (point X in Fig. 75). As blood passes through the systemic capillaries the P_{CO_2} rises to 6 kPa (45 mmHg), which would raise the CO_2 concentration to 50 ml 100 ml^{-1} if there were no change in the dissociation curve. At the same time, however, the P_{O_2} falls, shifting the CO_2 dissociation curve upwards, so that the actual CO_2 content in venous blood is 52 ml 100 ml^{-1} (point Y in Fig. 75). In other words, the Haldane shift increases the uptake of CO_2 from the tissues from 2 to 4 ml CO_2 100 ml^{-1} of blood. This is of much greater physiological significance than the relatively small changes in O_2 transport caused by the Bohr shift. Obviously, as blood passes through the lungs both the P_{CO_2} and P_{O_2} levels are reversed, taking us back to point X again.

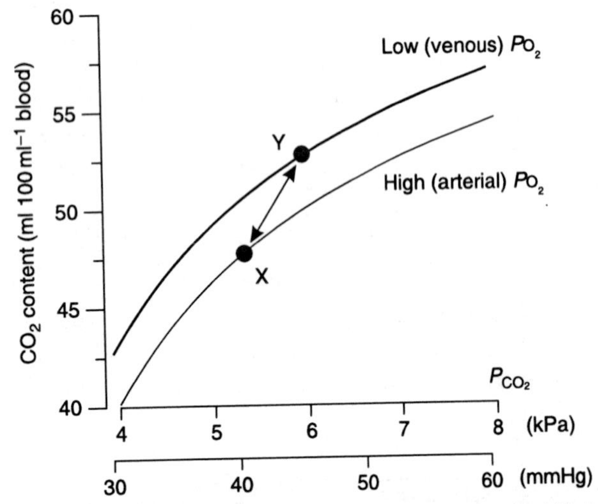

Fig. 75 The Haldane effect: the CO_2 dissociation curve is shifted down and to the left as O_2 increases. The CO_2 content in systemic arterial blood may be represented by point X, while point Y indicates that in venous blood.

The mechanisms underlying the Haldane shift depend on differences in the chemistry of oxyhaemoglobin and deoxyhaemoglobin.

- Deoxyhaemoglobin is a more effective pH buffer than oxyhaemoglobin. Since the conversion of CO_2 to HCO_3^- also leads to the production of H^+ (Eq. 25 and Fig. 73), the total quantity of CO_2 which can be carried in this form is partly limited by the tendency for pH to fall as a result. With more effective buffering in deoxygenated (i.e., venous) blood, more CO_2 can be converted to HCO_3^- than would otherwise be possible. Nevertheless, venous pH (7.36) is lower than that in arterial blood (7.40).
- Deoxyhaemoglobin is also able to carry more CO_2 in the carbamino form.

Carbon dioxide and acid–base balance

One important aspect of the control of CO_2 levels in the extracellular fluid is the influence this has on the pH of the body fluids. This subject is very important, given the effects H^+ concentration has on the function of enzymes and other proteins, and control of acid–base balance will be considered in more detail elsewhere (Section 5.9). The importance of respiratory control of CO_2 levels in pH regulation will only be outlined here.

As discussed earlier, CO_2 and water react to form carbonic acid, which then dissociates producing H^+ and HCO_3^- (Eq. 25 and Fig. 73). Increasing the CO_2 content drives this reaction towards the right, liberating H^+. As CO_2 levels rise, therefore, the pH falls. This may occur locally in the interstitium if the rate of CO_2 removal in the blood does not match the metabolic production of CO_2. The pH of systemic arterial blood can also fall if arterial P_{CO_2} rises above normal, a condition known as *respiratory acidosis*. This is a common feature of respiratory disease, especially when airway obstruction impairs ventilation. Similarly, reducing CO_2 levels drives the equilibrium (Eq. 25) towards the left, removing protons and causing a *respiratory alkalosis*. This occurs whenever ventilation is increased relative to CO_2 production, so that alveolar and arterial P_{CO_2} levels fall and the pH rises.

Box 14 Clinical note: Ventilation–perfusion mismatch and arterial blood gases

Many clinical conditions give rise to ventilation–perfusion mismatch, often with regions of high and low ventilation:perfusion ratio in different parts of the lung. Blood from the overventilated regions has a high P_{O_2} and a low P_{CO_2}, while this pattern is reversed for underventilated areas. This might suggest that the two effects should cancel out, giving approximately normal partial pressures in systemic arterial blood. In fact this is often the case for CO_2, but the systemic P_{O_2} is nearly always reduced. This can be explained by the different shapes of dissociation curve for the two gases. Increased ventilation lowers the CO_2 content by reducing the P_{CO_2}, but the parallel increase in alveolar P_{O_2} cannot appreciably increase the O_2 content of the blood above normal because of the plateau on the O_2 dissociation curve. When blood from the under- and overventilated regions of the lung is mixed in the left ventricle, the average CO_2 content is brought back towards normal, but the low O_2 content of blood from underventilated regions remains uncompensated, i.e. hypoxia results.

4.5 Regulation of respiration

Learning objectives

At the end of this section you should be able to:

- outline the role of the brain in controlling ventilation
- explain how arterial blood gases/pH can affect ventilation
- outline the role of central and peripheral chemoreceptors
- briefly identify other factors which influence ventilatory control.

Since the rates of O_2 uptake and CO_2 production by body cells vary widely with changing metabolic demand, respiration has to be controlled so as to maintain appropriate levels of O_2 and CO_2 (and H^+) in the tissues. This depends on the regulation of ventilation through the interaction of *neurological* and *chemical control* mechanisms so that mean alveolar gas pressures remain constant. Since pulmonary blood normally equilibrates with alveolar gases before entering the systemic circulation, systemic arterial blood gas concentrations can also be controlled by regulating ventilation appropriately.

Neurological control from respiratory centres in the brain

Although ventilation may be consciously controlled, it is normally regulated via involuntary nervous mechanisms. Breathing is regular and cyclical, inspiration alternating with expiration, and this rhythmical activity is dependent on the activity of neurones within the brainstem. The neurones which activate the inspiratory and expiratory muscles are located in the medulla oblongata but are influenced by centres in the pons, which modify their electrical activity, and thus alter the pattern of ventilation.

Inspiratory respiratory neurones in the medulla oblongata demonstrate spontaneous, rhythmical activity, firing regular bursts of action potentials separated by periods of inactivity. These neurones stimulate the motoneurones which pass out from the spinal cord to the diaphragm and external intercostal muscles and so activate contraction in these inspiratory muscles. During the pauses in inspiratory neurone activity, the inspiratory muscles relax and expiration occurs passively. In quiet respiration, therefore, it is the *inspiratory centre* which plays the major role in stimulating ventilation.

Expiratory respiratory neurones in the medulla are normally quiescent and only become active during episodes of increased ventilation involving active, or forced, expiration. Under these conditions, bursts of expiratory neurone activity coinciding with the lulls in inspiratory neurone activity may be recorded. These stimulate motoneurones causing the internal intercostal and abdominal muscles to contract (Section 4.1).

Respiratory cells in the pons are not essential for respiration but can modify the pattern of breathing. Stimulating the *pneumotaxic centre* tends to inhibit the inspiratory neurones and so shortens inspiration. Damage to this pneumotaxic area, by comparison, can lead to *apneusis*, in which inspiration is prolonged and is only interrupted by short, expiratory gasps.

Chemical factors modifying respiratory centre activity

The spontaneous firing pattern in the medullary respiratory neurones is regulated by a number of chemical factors which control the depth and rate of ventilation.

Arterial blood gases and pH

Chemical control by the gases and pH in arterial blood, as monitored by *respiratory chemoreceptors*, is the dominant factor in the regulation of ventilation.

Arterial P_{CO_2} is most important, with pH playing a secondary role allowing respiratory compensation for metabolic acid–base disturbances. Surprisingly, arterial P_{O_2} plays little or no role in the normal regulation of ventilation, although abnormally low levels can powerfully stimulate ventilation.

Elevated arterial P_{CO_2}. Any elevation of arterial P_{CO_2} stimulates ventilation, leading to a compensatory reduction in alveolar P_{CO_2} as excess CO_2 is blown off (it has already been noted that increasing the ventilation : perfusion ratio, $\dot{V}_A/Q$, decreases alveolar P_{CO_2}; Fig. 69). This is the primary mechanism responsible for the regulation of breathing, and ventilation is adjusted to keep the arterial P_{CO_2} close to 5.3 kPa (40 mmHg). Close regulation of CO_2 levels is important since any rise in P_{CO_2} will increase the CO_2 content of the blood appreciably (see the CO_2 dissociation curve; Fig. 74), promoting acidosis (Eq. 25).

Depressed arterial pH. Any fall in arterial pH (increase in $[H^+]$) leads to an increase in ventilation. This reduces alveolar and arterial P_{CO_2} and so elevates the systemic pH by driving the carbonic acid dissociation reactions towards the left, removing H^+ (protons) from the extracellular fluid (Eq. 25). Respiratory changes can, therefore, compensate for an acidosis (low pH) caused by a nonrespiratory problem (a metabolic acidosis; Section 5.9). Similarly, ventilation may decrease in response to a metabolic alkalosis, favouring CO_2 accumulation and a reduction of the pH back towards normal.

Depressed arterial P_{O_2}. Ventilation is also stimulated by low levels of arterial P_{O_2}. This only happens, however, if the P_{O_2} drops well below the normal value (13 kPa; 98 mmHg): increased respiratory drive is only significant when the P_{O_2} is less than about 8 kPa (60 mmHg). The shape of the O_2 dissociation curve removes any need to regulate P_{O_2} more tightly than this since haemoglobin remains 85% saturated with O_2 even at this relatively low O_2 pressure (Fig. 70B). Oxygen delivery to the tissues is not greatly compromised unless arterial P_{O_2} falls below these levels and only then is ventilation stimulated. Oxygen is not directly involved, therefore, in the normal regulation of ventilation. Nevertheless, the arterial P_{O_2} remains relatively constant under physiological conditions, as a secondary consequence of the close control of P_{CO_2} levels.

Chemoreceptors

The changes in arterial CO_2, O_2 and H^+ levels which lead to the respiratory changes described above are detected by respiratory chemoreceptors which regulate ventilation through their connections with the respiratory centres. These receptors are divided into two main groups based on their anatomical location, and each has its own pattern of sensitivity to changes in arterial blood gases and pH levels.

Central chemoreceptors. These are located within the CNS itself, close to the respiratory centre in the medulla. These receptors are particularly sensitive to changes in the *arterial* P_{CO_2} and are less affected by changes in arterial pH or P_{O_2}. Experiments on the mechanism of the CO_2-stimulatory effect suggest that the cells of the central chemoreceptors are actually sensitive to H^+. Carbon dioxide rapidly diffuses from the blood into the brain where it reacts with water to produce H^+ (Eq. 25), and it is the resulting drop in pH that directly stimulates the central chemoreceptors. An acidosis of the arterial blood itself, however, has little immediate effect on central chemoreceptors, because H^+ cannot easily cross the blood–brain barrier.

Peripheral chemoreceptors. These are located within the carotid bodies, close to the bifurcation of the common carotid arteries, and in the aortic bodies along the aortic arch. These receptors are less important than central chemoreceptors in the responses to an increase in P_{CO_2} but they are sensitive to changes in *arterial pH*, stimulating or inhibiting ventilation in response to arterial acidosis and alkalosis, respectively. Peripheral chemoreceptors play a further role as part of the fail-safe response to *very low* O_2 levels. They are only activated when the P_{O_2} falls well below the physiological range (i.e., at or below 8 kPa; 60 mmHg), however, and this mechanism is not important in normal ventilatory control.

Other factors in respiratory control

Although arterial blood gases are the primary controlling factor in respiratory control, a number of other influences should be mentioned.

The Hering–Breuer reflex depends on stretch receptors within the lungs which send inhibitory impulses to the inspiratory centre in the medulla via the vagus nerve. This prevents excessive inflation of the lung but is only significant at high tidal volumes (more than 1.5 L) and probably plays no role in the regulation of quiet breathing.

Voluntary control of the respiratory muscles by the motor cortex can override the medullary respiratory centre. This allows us either to speed up ventilation (hyperventilate) or to hold our breath at will. Voluntary control is only partial, however, since if we stop breathing for more than a minute or so, arterial P_{CO_2} rises and P_{O_2} falls providing a chemoreceptor-driven stimulus to ventilation that can no longer be resisted.

Vasomotor centre. Increased activity in the vasomotor centre is primarily responsible for cardiovascular adjustments to low blood pressure but also stimulates ventilation. This explains the increased breathing, known as air hunger, which is seen in surgically shocked patients.

4.6 Respiratory effects of exercise

Learning objectives

At the end of this section you should be able to:

- identify factors which affect gas exchange during exercise
- outline the control of ventilation during exercise
- explain what is meant by 'excess post-exercise O_2 consumption'.

The rates of gas exchange in both the peripheral tissues and the lungs are greatly increased during exercise.

Tissue O_2 supply. Increased tissue O_2 supply in exercising muscles depends on:

- increases in cardiac output and local blood flow (Sections 3.5 and 3.7)
- increased O_2 extraction from the capillaries. The increase in O_2 extraction reflects the lower P_{O_2} that occurs in rapidly metabolizing tissues, e.g., at a P_{O_2} of 2.7kPa (20mmHg), 15ml of oxygen is extracted from every 100ml of arterial blood, as compared with 5ml under normal conditions (Fig. 70A). Increased levels of CO_2, H^+ and increased tissue temperature also shift the O_2 dissociation curve to the right in exercising muscle (the Bohr effect, Fig. 71), further augmenting O_2 release.

Pulmonary O_2 uptake. Increased pulmonary O_2 uptake depends on:

- increased pulmonary blood flow (which always equals the cardiac output)
- an increased O_2 diffusion gradient in the lungs. The pulmonary arterial P_{O_2} equals systemic venous P_{O_2} (Fig. 67) and, as we have seen, this falls during exercise. The mean alveolar P_{O_2} remains constant at about 13.7kPa (103mmHg) as a result of the increase in ventilation, which almost exactly matches the increased O_2 needs of the body. This change in ventilation is the most important feature of the respiratory response to exercise.

Control of ventilation during exercise

The resting minute volume is about 5 L min^{-1}, but increases in both tidal volume and respiratory rate may raise this to over 100 L min^{-1} during maximal exercise. Under these conditions, O_2 absorption can rise from the resting rate of about 250 ml min^{-1} to over 4000 ml min^{-1}, while CO_2 release may increase from 200 ml min^{-1}, to as much as 8000 ml min^{-1}. This makes it tempting to propose that the increase in ventilation is essentially a chemoreceptor response to exercise-induced changes in blood gases and pH. In fact, the increase in ventilation is so closely matched to the changes in metabolic demand that there is actually very little alteration in the average levels of the arterial P_{O_2}, P_{CO_2} or pH. This suggests that a simple model in which chemoreceptors alone control ventilation during exercise is not appropriate, since deviations from normal arterial conditions which might produce an adequate chemoreceptor response do not occur.

Although the control of ventilation during exercise is not well understood it seems that several mechanisms may contribute.

- *Nervous controls:* these involve collateral connections from the motor cortex of the brain to the respiratory centres, which stimulate ventilation at the onset of exercise (Fig. 76). Reflexes from receptors within both contracting muscles and moving joints may also contribute to nervous stimulation of ventilation. These cortical and peripheral stimuli are lost as soon as exercise stops, which may help explain the initial, rapid fall in ventilation which is seen at the end of a period of exercise.
- *Chemical control:* this depends on chemoreceptors and probably plays some role in the ventilatory

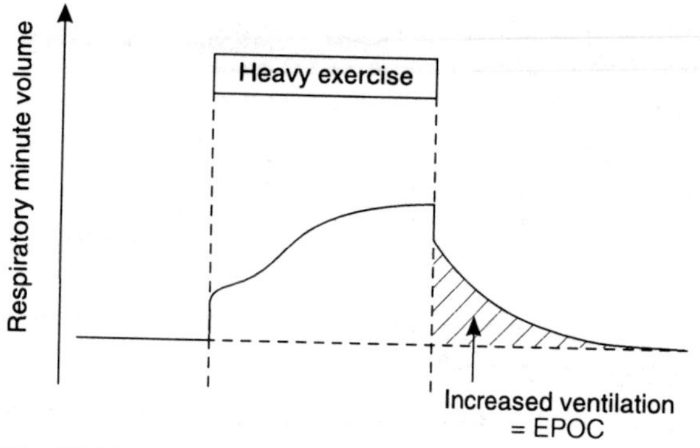

Fig. 76 Ventilatory responses to exercise, including the period of excess post-exercise O_2 consumption (EPOC).

Four

Box 15 Clinical note: Respiration under abnormal conditions

Altitude and respiration

The alveolar P_{O_2} depends on the P_{O_2} in the gas which is being breathed (the inspirate; Section 4.3). The percentage of O_2 in the atmosphere is essentially constant but atmospheric P_{O_2} may still vary because of changes in total atmospheric pressure (Eq. 21), since:

$$P_{O_2} = P_{atmos} \times \frac{\%O_2}{100} \qquad \text{(Eq. 26)}$$

This is relevant at high altitude, where the fall in total pressure leads to a fall in P_{O_2} in the inspired air and in the alveoli. If the resulting reduction in arterial P_{O_2} (*hypoxia*) is large enough, it stimulates ventilation via the peripheral chemoreceptors, thus blowing off CO_2 at a more rapid rate than normal so that arterial P_{CO_2} falls (*hypocapnia*) and pH rises (*respiratory alkalosis*) resulting in an unwanted reduction in the overall drive to breathe. Over a period of days to weeks adaptations take place. Erythrocyte production is accelerated as a result of erythropoietin release from the kidneys (Section 2.3), increasing the haemoglobin concentration and thus the O_2-carrying capacity of the blood. Anoxia also raises the level of 2,3-diphosphoglycerate in red cells, producing a Bohr shift to the right in the O_2 dissociation curve (Section 4.4). This promotes O_2 release to the peripheral tissues but may limit O_2 uptake in the lungs.

Respiratory failure

Respiratory failure may be defined as an abnormality of the arterial blood gases when breathing a normal inspirate; in other words there is failure of ventilation or gas exchange in the lungs. This may involve a reduction in arterial P_{O_2} alone, but P_{CO_2} often also deviates from normal and this produces an abnormal arterial pH (a respiratory acid–base disorder). These abnormalities often stimulate respiratory drive, so that ventilation is usually increased at rest. The resulting breathlessness is known as *dyspnoea*. Abnormal O_2, CO_2 and H^+ levels may also produce additional symptoms of respiratory failure.

Oxygen

Reduction in arterial P_{O_2} (*hypoxia*) reduces oxygen delivery to the tissues and limits aerobic metabolism. Hypoxia of adequate degree (P_{O_2} < 8 kPa; 60 mmHg) will stimulate ventilation through the peripheral chemoreceptors. The amount of circulating deoxyhaemoglobin increases and this produces a purple–blue discoloration known as *cyanosis*. Respiratory failure is one cause of central cyanosis, which is best judged from the colour of the lips and tongue and is caused by reduced arterial O_2 levels. Clinically, this is distinguished from peripheral cyanosis of the fingers and toes caused by a reduced blood supply. Cyanosis depends on the absolute concentration of deoxyhaemoglobin and so is less pronounced in anaemic patients. Hypoxia may be treated symptomatically by allowing the patient to breathe a gas mixture containing an increased percentage of O_2. This increases the alveolar P_{O_2}, promoting O_2 exchange in the lungs. If prolonged, hypoxia leads to compensatory increases in haemoglobin concentration and red cell 2,3-diphosphoglycerate levels, as already described for the effects of altitude.

Carbon dioxide

CO_2 levels may be increased or decreased in respiratory failure. Increases in arterial P_{CO_2} (***hypercapnia***) are most commonly seen in patients with reduced levels of alveolar ventilation, e.g., as a result of obstructive lung disease. Increased CO_2 levels stimulate ventilation and cause peripheral vasodilatation. If severe, cerebral activity may be depressed. Oxygen therapy has no effect on hypercapnia, which can only be relieved by treating the underlying cause, e.g., by decreasing airways resistance and improving alveolar ventilation. Artificial ventilation may be necessary in extreme cases. Some patients with chronic respiratory disease may actually become adapted to a high level of P_{CO_2} and they then rely on a low arterial P_{O_2} to provide the necessary ventilatory drive. It is dangerous to correct the arterial P_{O_2} completely in such patients since this may depress the medullary respiratory centre by removing the drive from peripheral chemoreceptors. This explains why 24–28% O_2 is often used in patients with chronic respiratory failure.

Respiratory failure may also be associated with a reduced arterial P_{CO_2} (*hypocapnia*). This can occur in pneumonia or pulmonary oedema, for example, and seems to be caused by reflex stimulation of ventilation via sensory fibres from the lung rather than any ventilatory response to hypoxia.

pH

The arterial pH abnormality most often encountered in respiratory failure is a ***respiratory acidosis***. This is a direct consequence of elevated CO_2 levels and may lead to compensation by renal mechanisms which increase the plasma concentration of HCO_3^- (Section 5.9). An elevated HCO_3^- level suggests chronic rather than acute respiratory failure, since renal compensation takes time to develop. More rarely, ***respiratory alkalosis*** associated with hypocapnia is encountered response to exercise. During sustained exercise, the initial increase in ventilation is often followed by a slower rise up to a plateau level (Fig. 76). Chemoreceptors may be involved in this.

- *Body temperature:* this tends to rise during exercise and may also stimulate ventilation.

Excess post-exercise O_2 consumption (EPOC)

It is common experience that ventilation rates do not return to resting levels immediately after a period of prolonged heavy exercise; it may take several minutes to 'get one's breath back'. Measurements show that, following an initial fall in ventilation, the minute volume may remain elevated for some time after the energy expenditure by the muscles has returned to normal (Fig. 76). This phenomenon, which used to be referred to as the O_2 debt, is now generally referred to by the descriptive title of 'excess post-exercise O_2 consumption', or EPOC. It is at least partly explained by the fact that the rate of adenosine triphosphate (ATP) consumption often exceeds production through oxidative metabolism during moderate to heavy exercise. Additional ATP is formed from the muscle stores of creatine phosphate and through anaerobic glycolysis. Anaerobic metabolism leads to the formation of lactic acid, however, and this has subsequently to be converted back to pyruvate in the liver and further metabolized via reaction pathways which use O_2 (Section 4.7). Oxygen continues to be consumed at an increased rate after exercise has stopped, therefore, until oxidative metabolism restores the creatine phosphate levels and reduces lactic acid levels to normal.

4.7 Cellular respiration, energy exchange and metabolic rate

Learning objectives

At the end of this section you should be able to:

- identify the major energy sources in food and the body
- describe the role of ATP in energy exchange
- explain the term metabolic rate and how it may be measured
- explain the term respiratory exchange ratio and its meaning
- list the main factors affecting metabolic rate
- explain what is meant by basal metabolic rate and list factors which alter it
- outline how weight is regulated.

Table 7 Caloric values of different food types as determined by complete oxidation in a bomb calorimeter

Food type	Caloric value ($kJ\,g^{-1}$)	($kcal\,g^{-1}$)
Carbohydrate	17.2	4.1
Protein	22.3	5.3
Fat	39.0	9.3

The primary function of the cardiovascular and respiratory systems is to provide the cells of the body with O_2 and food molecules. Cells use these to generate adenosine triphosphate (ATP), the immediate energy source for most active cell processes, through oxidative reactions which produce CO_2 and water as byproducts. This is respiration at the cellular level. The function of the intact respiratory system is to absorb O_2 and remove CO_2 at appropriate rates so as to maintain arterial levels appropriate for gas exchange in the tissues. This requirement is affected by the rate of energy use in the tissues, i.e., the metabolic rate.

Metabolism is a general term used to cover all the biochemical reactions going on within the body. These may be subdivided into anabolic processes, which use energy to generate complex molecules from simpler subunits (*anabolism*), and catabolic reactions, which break molecules into simpler products, releasing energy in the process (*catabolism*).

Food as an energy source

One major function of food is to provide the body with an external source of energy. The chemical structures of food molecules allows them to be classified into three main groups, namely, carbohydrates, fats and proteins. The total energy available from complete oxidation of any type of food can be determined by burning (completely oxidizing) a known quantity of that food and measuring the heat released in a bomb calorimeter. Such measurements indicate that fat is by far the richest energy source, with a *caloric value* more than twice that of carbohydrate (Table 7). The SI unit for energy is the joule (J) but the calorie (cal) is still in widespread dietetic use. In fact, the scale of human energy consumption means that kilojoules and kilocalories are more useful, and the 'big calorie' (C) is often quoted, where 1 C = 1 kcal. It is 'big calories' which are in common use so that an '800 calorie diet' actually means restricting intake to 800 kcal.

Efficiency of energy transfer

The energy actually made available for cellular work by oxidation of food in the body is considerably less than the relevant caloric value. In the case of protein, for example, metabolic oxidation is incomplete, reducing the theoretically available energy to 4.1 kcal g^{-1}. More importantly, energy exchange processes in cells are not 100% efficient, so that, at best, less than 50% of the available energy from glucose oxidation is converted into a chemically usable form (ATP) inside a cell. Further losses occur in the transfer of energy between ATP and other molecules such as contractile proteins in muscle cells. Therefore, the final *mechanical efficiency* (percentage of the total chemical energy used for mechanical work) is only 25–30% under ideal conditions. The residual, or 'wasted', energy appears as *heat* which is important in maintaining body temperature.

Energy stores in the body

Once absorbed from the gastrointestinal tract, food molecules may either be metabolized for energy or stored for future use (Section 6.5). The main carbohydrate energy store is in the form of *glycogen*, a polymer of glucose, which is particularly concentrated in skeletal muscle and liver. Glycogen can be rapidly broken down to maintain blood glucose levels but the store becomes depleted after about 12 hours of starvation. Fats act as a long term energy store, with *triglycerides* being broken down into glycerol, which can be used to synthesize glucose, and fatty acids, which are oxidized for energy. Circulating amino acids and some proteins can also be used as an energy source but most of the body's protein has important structural and functional roles and is spared from catabolic breakdown until fat stores are almost exhausted.

Cellular energy exchange

During aerobic metabolism glucose is completely oxidized as it would be in a calorimeter. Within the body, however, this reaction is divided into a series of steps allowing energy to be released in small packets rather than in an explosive burst. Energy release is not coupled directly to energy use in the cell but drives the formation of high energy phosphate bonds, producing ATP by coupling inorganic phosphate groups (P_i) with adenosine diphosphate (ADP).

$$ADP + P_i + Energy \rightarrow ATP$$

This reaction is readily reversible and, as ATP is degraded back to ADP and P_i by ATPase enzyme systems in the cell, the chemical energy in the phosphate bond is released for use in energy-consuming processes, e.g., ion transport against concentration gradients, cell contraction or synthesis of complex compounds.

Aerobic and anaerobic metabolism

Approximately 95% of carbohydrate metabolism is aerobic, i.e., the glucose is oxidized to produce CO_2 and water. This ultimately relies on the reactions of the Krebs cycle, which are coupled to the generation of ATP within *mitochondria* in a process known as *oxidative phosphorylation*. Glucose cannot enter the Krebs cycle directly but must first be converted to pyruvate in a series of reactions known as *glycolysis*. Glycolysis does not require O_2 and, since it also produces ATP, it can be used to meet some of a cell's energy needs under anaerobic conditions. Pyruvate is converted to *lactic acid* as a result, however, and this has later to be converted back to glucose in the liver in a series of O_2-consuming reactions. Because of this, it is often said that anaerobic metabolism generates an O_2 debt which has to be paid back later.

The total amount of ATP which can be produced from anaerobic glycolysis (two ATPs per glucose molecule) is only a small fraction of that available from aerobic metabolism (38 ATPs per molecule), but this may still be important when the supply of O_2 is inadequate. Different cells, however, have very different capacities to use anaerobic metabolism. Red blood cells rely totally on glycolysis, thus sparing the O_2 they carry, while anaerobic ATP production often boosts the work output from skeletal muscle during heavy exercise (Section 4.6). Cardiac and cerebral cells, in contrast, are totally dependent on aerobic metabolism so that O_2 deprivation leads to loss of function within seconds, and irreversible damage within minutes.

Metabolic rate

This refers to the total rate of energy consumption by an individual and is measured using *calorimetry*.

Direct calorimetry

This relies on the direct measurement of *heat production* and is based on the principle that all the energy used by the body will ultimately be converted to heat provided an individual sits still and carries out no external work. Internal mechanical work is always being carried out, e.g., as blood is pumped through the circulation and air is moved in and out of the lungs, but this work is eventually converted to heat through viscous and other frictional forces. It is also assumed that the energy consumed by the synthesis of complex molecules, such as proteins, is exactly matched by that released from their breakdown in steady-state conditions. Although the method is accurate, the heat produced by a human is technically difficult to measure and so indirect calorimetry is more commonly used.

Indirect calorimetry

This relies on measurement of the rate of O_2 *consumption*, which can then be converted to a rate of energy use if the food substrate being oxidized is known. In the case of carbohydrate, for example, complete oxidative metabolism releases 21 kJ (5 kcal) of energy for every litre of O_2 consumed. This relationship is unaffected by the rate of external work, providing all the metabolism is aerobic. Different energy values apply for fat and protein, however, so that a figure of 20.2 $kJ\,L^{-1}$ O_2 consumed (4.8 $kcal\,L^{-1}$ O_2) is probably more appropriate for someone on a mixed diet.

Respiratory exchange ratio and respiratory quotient

Some indication of the actual proportions of each food type being metabolized, and therefore a better estimate of the energy released by the consumption of a given volume of O_2, can be gained from measurement of the respiratory exchange ratio (RER).

$$RER = \frac{\text{Rate of } CO_2 \text{ production}}{\text{Rate of } CO_2 \text{ consumption}} \qquad \text{(Eq. 27)}$$

If we consider the general reaction for the oxidation of carbohydrate, for example, we can see that each O_2 molecule consumed produces a single CO_2 molecule, giving a respiratory exchange ratio of 1.00.

$n(CH_2O) + nO_2 \rightarrow nCO_2 + nH_2O$
(e.g. $n = 6$ for glucose)

The values for fat and protein are lower than this so that, although the respiratory exchange ratio may be close to 1.00 immediately after a meal when a large amount of carbohydrate is being metabolized, it tends to fall as time passes and more fat and protein are used. An average value of about 0.8 is typical on a normal mixed diet.

The respiratory exchange ratio does not always provide a good indication of the substrate being metabolized in the body tissues since other factors also affect it. For example, following anaerobic metabolism, O_2 is consumed in the conversion of lactate to glucose, without any parallel CO_2 production. This may reduce the respiratory exchange ratio to as low as 0.5. Alternatively, periods of hyperventilation may blow off additional CO_2 without changing O_2 absorption, raising the respiratory exchange ratio above 1.00. It is for these reasons that the term respiratory exchange ratio is preferred over respiratory quotient, which more strictly refers to the ratio of CO_2 production to O_2 consumption during catabolism of substrates within the tissues. Under steady state conditions, however, the respiratory exchange ratio must equal the respiratory quotient.

Factors affecting metabolic rate

The metabolic rate is affected by several factors.

- *Exercise*: this has very dramatic effects on the metabolic rate, raising it up to 50 fold over short periods. This explains the enormous differences in nutritional energy requirements for a manual labourer as compared with someone working at a desk job.
- *Food consumption*: the metabolic rate rises immediately after a meal, especially one containing a large quantity of protein. This is referred to as the *specific dynamic action* of food or protein and it helps explain the warming effect of eating.
- *Environmental temperature*: the metabolic rate rises when the environmental temperature falls below about 20°C, because of shivering, and when it rises above 27°C, when cooling mechanisms such as sweating are activated. Temperatures between these extremes are said to lie in the *thermoneutral* range.
- *Nervous stress*: anxiety, fear and other forms of nervous stress elevate the metabolic rate through increased sympathetic nervous activity.
- *Sleep*: metabolism is slowed during sleep.
- *Climate*: the metabolic rate falls in individuals who have become acclimatized to a warm climate after moving from a colder one.

Basal metabolic rate

This refers to a prescribed set of conditions which minimize the metabolic rate. Subjects should have just woken and be at complete mental and physical rest, having fasted for 12 hours. The room temperature should be in the middle of the comfortable (thermoneutral) range. Body size obviously affects total energy use so the basal metabolic rate is expressed in kJ (kcal) m^{-2} of body surface area, allowing comparisons to be made between the underlying metabolism in different individuals. Surface area is more closely correlated with energy use than either height or weight.

The basal metabolic rate (BMR) is altered by several factors:

- it is higher in men than women
- it declines with age, being especially high in early childhood
- it is altered by the hormonal state of an individual. Thyroid hormone is particularly important in the stimulation of metabolism, although growth hormone and male sex steroids have similar, but less pronounced effects. The increased metabolic rate in response to sympathetic nervous activity is mainly the result of catecholamine secretion from the adrenal medulla, although there may also be stimulation of brown fat metabolism, particularly in babies.

Self-assessment: questions

Multiple true/false questions

Each of the following statements consists of a stem followed by a number of possible endings. State whether each statement is True or False. For each stem, all, several or none of the statements may be true.

1. Pulmonary airways:
   a. normally offer a high resistance to airflow
   b. are all collapsible
   c. may constrict in response to cold air
   d. dilate in response to adrenergic antagonist drugs
   e. humidify inspired air

2. The chest wall:
   a. is normally partially compressed as a result of the negative pressure in the intrapleural space
   b. actively decreases thoracic volume through contraction of the internal intercostal muscles during quiet expiration
   c. is supplied by spinal nerves from the cervical region
   d. generates a force acting outwards on the parietal pleura during inspiration
   e. tends to protrude on the side of a pneumothorax

3. The surface tension of the alveolar fluid:
   a. tends to decrease lung compliance
   b. tends to decrease the work of breathing
   c. is decreased by the presence of surfactant
   d. tends to decrease as alveoli shrink
   e. is the only collapsing force in the lungs

4. Intra-alveolar pressure:
   a. is positive with respect to the atmosphere during expiration
   b. is more positive than the intrapleural pressure throughout quiet respiration
   c. is the sole factor determining gas flow rates during ventilation
   d. is zero (atmospheric) at peak inspiration
   e. is zero (atmospheric) at peak expiration

5. During quiet breathing:
   a. a volume of gas approximately equal to the tidal volume enters the alveoli during each respiratory cycle
   b. a volume of fresh inspirate approximately equal to the tidal volume enters the alveoli during each respiratory cycle
   c. the functional residual capacity can be measured directly using a spirometer
   d. the residual volume exceeds the tidal volume during quiet breathing
   e. the respiratory minute volume can be determined from the tidal volume and the respiratory rate

6. The forced expiratory volume in 1 second (FEV_1):
   a. is lower, on average, in men than women
   b. is likely to be reduced in restrictive lung disease
   c. is likely to be reduced in obstructive lung disease
   d. is, on its own, a useful measure of the severity of obstructive lung disease
   e. can be used clinically as an alternative measure to the peak expiratory flow rate (PEFR)

7. Carbon dioxide:
   a. has a higher diffusion coefficient than O_2
   b. is present at a higher partial pressure in mixed systemic venous blood than pulmonary arterial blood
   c. is mainly transported in blood in the form of HCO_3^-
   d. is important in the regulation of arterial pH
   e. is released at the same rate as O_2 is consumed when pure carbohydrate is metabolized aerobically

8. With regard to alveolar gases:
   a. increased ventilation tends to increase P_{O_2} and reduce P_{CO_2}
   b. increased altitude affects all gas pressures similarly
   c. a reduced ventilation:perfusion ratio increases both P_{O_2} and P_{CO_2}

d. an increased ventilation:perfusion ratio adds to the physiological dead space
e. moderate exercise tends to cause a reduction in the mean P_{O_2} and an increase in the mean P_{CO_2}

9. The oxygen dissociation curve:
 a. is sigmoidal, indicating the presence of both cooperativity and saturation in the O_2-binding system
 b. has a characteristic shape which is independent of the haemoglobin concentration when plotted in terms of % saturation
 c. is shifted to the right under conditions in which the O_2 affinity of haemoglobin is increased
 d. can be determined by measuring the O_2 content of samples of blood equilibrated with gas mixtures of different P_{O_2} values
 e. may be altered in shape by changes in the structure of the haemoglobin protein chains

10. The Haldane effect:
 a. promotes the transport of O_2 in systemic arterial blood
 b. approximately doubles the amount of CO_2 which is taken up by blood as it passes through systemic capillaries
 c. promotes CO_2 release in the pulmonary capillaries as the blood P_{O_2} rises
 d. depends on increased CO_2 binding by oxygenated haemoglobin
 e. shifts the CO_2 dissociation curve to the right as blood pH falls

11. An increased systemic P_{CO_2}:
 a. can cause a respiratory acidosis
 b. is likely in individuals living at altitude
 c. stimulates ventilation
 d. is a more potent ventilatory stimulus than a decreased arterial P_{O_2}
 e. tends to increase peripheral resistance

12. Respiratory chemoreceptors:
 a. in the carotid and aortic bodies are most important in the ventilatory response to an elevated P_{CO_2}
 b. in the carotid and aortic bodies are strongly stimulated by the low arterial O_2 content in anaemic patients
 c. in the medulla are responsive to changes in arterial P_{CO_2}
 d. transduce chemical changes into electrical signals
 e. may be sensitive to H^+

13. Ventilation may be increased:
 a. in response to a metabolic (nonrespiratory) acidosis
 b. in response to a marked fall in arterial pressure
 c. in respiratory failure
 d. following a period of anaerobic metabolism
 e. following an emotional upset

14. During exercise:
 a. a reduction in muscle P_{O_2} could promote O_2 delivery to the tissue
 b. the respiratory exchange ratio (RER) initially increases
 c. signals from the motor cortex stimulate ventilation
 d. ventilation is closely matched to metabolic needs
 e. the Hering–Breuer reflex helps stimulate inspiration

15. Metabolic rate:
 a. can be accurately determined by indirect calorimetry during anaerobic exercise
 b. is increased during sleep
 c. tends to fall as individuals become acclimatized to a warm climate
 d. is correlated with body surface area
 e. is depressed by thyroid hormone

Single best answer questions

For each of the following questions choose the single best answer.

1. The total resistance to pulmonary airflow:
 a. is greater for laminar airflow than for turbulent airflow
 b. of the large airways is greater than that of the small airways
 c. of an airway increases in proportion to its diameter
 d. decreases during expiration
 e. is increased by sympathetic nerve activity

2. With regard to intrathoracic pressures:
   - **a.** negative intra-alveolar pressures are largely generated by alveolar surface tension
   - **b.** the increased intrathoracic pressure during expiration promotes venous return
   - **c.** intra-alveolar pressure remains negative throughout the ventilatory cycle during quiet breathing
   - **d.** the pressure tends to be larger in large alveoli than small ones
   - **e.** the change in intrapleural pressure during ventilation is greater than the change in intra-alveolar pressure

3. Lung compliance:
   - **a.** is increased by lack of alveolar surfactant
   - **b.** decreases as the lung inflates
   - **c.** is decreased in airways obstruction
   - **d.** is decreased in emphysema due to loss of alveolar tissue
   - **e.** is decreased following a spinal injury affecting the intercostal muscles

4. Spirometry reveals that a patient's FEV_1 is 48% of the predicted value while their FVC is 91% of predicted. From this we may conclude that:
   - **a.** they will be cyanosed
   - **b.** they are suffering from restrictive airways disease
   - **c.** they are suffering from emphysema
   - **d.** they are suffering from obstructive airways disease
   - **e.** they are suffering from chronic bronchitis

5. If the arteriovenous O_2 difference (arterial P_{O_2} – venous P_{O_2}) increases for a given organ we may conclude that:
   - **a.** O_2 consumption by that organ has increased
   - **b.** blood flow through that organ has decreased
   - **c.** O_2 extraction from blood perfusing that organ has increased
   - **d.** blood flow through the organ has increased
   - **e.** pulmonary arterial P_{O_2} will be decreased

6. Ventilation:perfusion:
   - **a.** is lower at the apices of the lungs than at the bases in the upright position
   - **b.** would be expected to increase, on average, in obstructive airways disease
   - **c.** may be determined from the respiratory minute volume and the pulmonary blood flow
   - **d.** would be expected to increase following blockage of a major pulmonary artery by a pulmonary embolus
   - **e.** mismatch tends to affect P_{CO_2} more than P_{O_2}

7. Anaemia:
   - **a.** reduces the P_{O_2} in systemic arterial blood
   - **b.** reduces the O_2 saturation in systemic arterial blood
   - **c.** reduces the O_2 delivery to the tissues
   - **d.** reduces the O_2 content of systemic arterial blood
   - **e.** reduces the O_2 affinity of the blood

8. The O_2 dissociation curve is shifted to the left:
   - **a.** by an increase in P_{CO_2}
   - **b.** by an increase in the affinity for O_2
   - **c.** by a decrease in pH
   - **d.** by an increase in temperature
   - **e.** by an increase in haemoglobin concentration

9. In an arterial blood sample with an elevated P_{CO_2}:
   - **a.** P_{O_2} is likely to be low
   - **b.** bicarbonate concentration is likely to be high
   - **c.** the pH is likely to be reduced
   - **d.** the haematocrit is likely to be reduced
   - **e.** plasma Cl^- concentration is likely to be elevated

10. Respiratory exchange ratio:
    - **a.** is increased when the rate of O_2 consumption is increased
    - **b.** is increased when the rate of CO_2 production is increased
    - **c.** is increased when metabolism shifts from carbohydrate to fatty acid utilization
    - **d.** is increased following anaerobic metabolism
    - **e.** is increased during hyperventilation

Matching item questions

Theme: Gas transport

Options

A. Deoxygenation
B. Haldane effect

C. O_2 saturation
D. O_2 content
E. Carbonic anhydrase
F. Cl^- shift
G. Carbamino group
H. Bicarbonate group
I. pH fall

For each of the descriptions below choose the most appropriate option from the list above. Each option may be used once, more than once or not at all.

1. Increases the H^+ buffering power of haemoglobin.
2. Is decreased by a fall in P_{O_2} but not by anaemia.
3. Decreases the O_2 affinity of haemoglobin.
4. Results from increased HCO_3^- production in erythrocytes.
5. Accounts for about 25% of CO_2 transport in blood.
6. Is increased at any given P_{O_2} in fetal, as compared with adult haemoglobin.

Theme: Control of respiration

Options

A. Inspiratory respiratory neurones
B. Expiratory respiratory neurones
C. Pneumotaxic centre
D. Central chemoreceptors
E. Peripheral chemoreceptors
F. Reduced systemic arterial P_{O_2}
G. Elevated systemic arterial P_{CO_2}
H. Reduced systemic arterial pH
I. Hering–Breuer reflex

For each of the descriptions below choose the most appropriate option from the list above. Each option may be used once, more than once or not at all.

1. Stimulate(s) ventilation in response to inceases in P_{CO_2}.
2. Inhibit(s) inspiratory activity at high lung volumes.
3. Receive(s) a blood flow which greatly exceeds their metabolic demands.
4. Promote(s) hyperventilation.
5. Increase(s) alveolar ventilation by reducing pH within the medulla.
6. Stimulate(s) efferent activity in the phrenic nerve.

Short notes

Write short notes on the following:

a. lung compliance
b. residual volume
c. haemoglobin function
d. peripheral respiratory chemoreceptors
e. respiratory exchange ratio
f. energy stores in the body

Modified essay

Ventilation involves the movement of gas in and out of the alveoli against a variety of resisting forces. This effort expended in this process may be quantified in terms of the work of breathing.

Questions

1. Name the three components which contribute to the work of breathing, briefly explaining what each involves.
2. Which of these components would be increased in surfactant deficiency and why?
3. What is the functional purpose of ventilation?
4. What are the normal partial pressures of blood gases in the systemic arteries?
5. Explain how these are maintained at their normal values.
6. What physical mechanism is responsible for gas exchange in systemic capillaries?
7. What information would you need to calculate the rate of O_2 consumption in a given organ using blood gas measurements? How, in principle, would you make this calculation?
8. Comment on the main differences between gas exchange in the pulmonary and systemic circulations.
9. Explain how the principles outlined in your response to Question 7 can be applied to the pulmonary circulation to estimate cardiac output.

Data interpretation

The trace shows spirometry recordings from a normal individual (Fig. 77). Changes in lung volume are plotted against time under a number of different conditions. Initially the subject is breathing normally at rest. At the point marked A he is asked to inhale fully, exhale fully and then breathe normally again.

Question

1. What is the value of the (a) minute volume, (b) inspiratory reserve volume, (c) expiratory reserve volume and (d) vital capacity, in this subject? If the airways dead space is 150 ml, (e) what is the alveolar ventilation rate at rest?

Samples of room air, and alveolar gas from the subject above are compared for their O_2 content. The room air has an $[O_2]$ of 19.67% when corrected for saturation with water vapour at 37°C, while that in a sample of alveolar air, taken at the end of expiration, has an $[O_2]$ of 13.8%.

Question

2. (a) Why does the $[O_2]$ in room air have to be corrected to allow for water vapour saturation for comparisons with alveolar air? (b) Why is the alveolar sample taken at the end, rather than the beginning of expiration? (c) Using the O_2 values given, and the relevant result from Question 1 above, calculate the rate of O_2 consumption at rest in this subject. (d) What is the resting rate of energy consumption in this individual? (Assume 1 L of O_2 is equivalent to 20.2 kJ.)

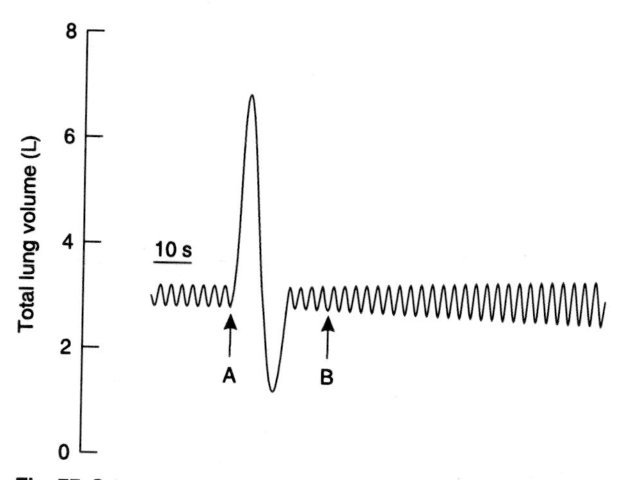

Fig. 77 Spirometer recording from a normal individual. Initially the subject was breathing at his usual resting rate. At point A he was required to inhale fully and then exhale maximally. At point B he was given a 50 cm length of hollow tubing to breathe through.

At the point marked B on the spirometer trace, the subject was asked to breathe in and out through a wide bore, cylindrical tube 50 cm in length. The effects this had on respiration are shown.

Question

3. (a) What effect does breathing through additional tubing have on the effective dead space? (b) Why does this lead to an increase in the tidal volume? (c) Assuming the alveolar ventilation rate remains constant when breathing through the tube, calculate the volume of the tube; what is its radius? (d) Do you think that alveolar ventilation could have been maintained by increasing respiratory rate alone in this case? Explain your answer.
(e) Do increases or decreases in the ventilation : perfusion ratio add to the physiological dead space?

Clinical scenario

An elderly man is admitted to the accident and emergency unit complaining of breathlessness and a cough which produces green sputum. He has a long history of chronic obstructive airways disease, which tends to become particularly severe whenever he gets a pulmonary infection. His tongue and lips appear blue. His fingers and toes are warm and his peripheral pulses are full, or bounding, in character.

Questions

1. What effect does airways obstruction have on the partial pressures of O_2 and CO_2 in the alveoli? How does this contribute to breathlessness (dyspnoea)?
2. Is there any clinical evidence suggesting that the changes in P_{O_2} and P_{CO_2} described above have occurred in this patient?
3. How would the changes in intra-alveolar pressure during the respiratory cycle in a patient with obstructive airways disease compare with those in a normal individual?

Explain your answer in terms of the physics of gas flow through the airways.

4. What changes, if any, would you expect to find in the following measurements in a patient with pure obstructive airways disease? (a) Peak expiratory flow rate, (b) forced expiratory volume in 1 second (FEV_1), (c) plasma HCO_3^- concentration, (d) CO transfer factor (T_{CO}).
5. Which of the following agents might be helpful in treating this patient and why? (a) Oxygen, (b) β-adrenoceptor agonists (these mimic sympathetic nerve activity in the bronchi), (c) β-adrenoceptor antagonists, (d) muscarinic cholinergic agonists (these mimic parasympathetic nerve activity in the bronchi), (e) muscarinic cholinergic antagonists, (f) an antibiotic.

Viva questions

1. A patient has obstructive airways disease and looks very cyanosed. What abnormalities would you expect to find in the arterial blood gases?
2. Draw me an O_2 dissociation curve. How do you think it would be changed by changes in P_{CO_2}/pH/temperature/changes in the globin chains/anaemia (you may choose any of these)?
3. Explain how respiration is controlled.

Self-assessment: answers

Multiple true/false answers

1. a. **False.** Resistance is normally very low.
   b. **False.** Bronchi contain cartilage and do not collapse.
   c. **True.** Cold may trigger airways obstruction in asthmatic patients.
   d. **False.** Adrenergic agonists dilate airways; this is used clinically in the treatment of airways obstruction.
   e. **True.** Inspirate becomes saturated with water vapour.

2. a. **True.** Compression is caused by the gradient between atmospheric pressure outside and the negative, or subatmospheric, intrapleural pressure inside.
   b. **False.** Quiet expiration is driven by passive elastic recoil of lungs.
   c. **False.** Thoracic spinal nerves supply the chest wall; cervical roots supply the diaphragm via the phrenic nerves.
   d. **True.** This makes the intrapleural pressure more negative, expanding the lung.
   e. **True.** Caused by the loss of the negative intrapleural pressure; see (a).

3. a. **True.** Surface tension increases the pressure required for lung expansion.
   b. **False.** Increases work by decreasing compliance.
   c. **True.** Secreted by type II alveolar cells.
   d. **True.** Caused by increased surfactant concentration with decreased fluid surface area.
   e. **False.** Stretching of pulmonary elastic tissue also contributes to the collapsing force.

4. a. **True.** Drives alveolar air out of the lungs.
   b. **True.** This may not be true during forced expiration when intrapleural pressures become positive.
   c. **False.** Also depends on airways resistance.
   d. **True.** Airflow is momentarily zero as direction of airflow reverses. With zero airflow, pressure must be equal in alveoli and atmosphere.
   e. **True.** Airflow is again momentarily zero as direction of flow reverses.

5. a. **True.** This must be the case since the airway volume is almost constant during quiet respiration. Some of the gas entering the alveoli, however, is dead space gas.
   b. **False.** The volume of fresh inspirate which reaches the alveoli = Tidal volume – Dead space.
   c. **False.** This requires an indirect indicator dilution technique.
   d. **True.** Typical values are 2 L for residual volume and 0.5 L for tidal volume.
   e. **True.** Respiratory minute volume = Tidal volume × Respiratory rate. (Remember that this is not the same as alveolar minute volume.)

6. a. **False.** It is higher.
   b. **True.** As the forced vital capacity (FVC) falls in restrictive disease, so does FEV_1.
   c. **True.** Air is expired more slowly.
   d. **False.** FEV_1 is also reduced in restrictive disease; FEV_1/FVC is a more useful measure.
   e. **False.** Again the relevant comparison is with FEV_1/FVC; low values of PEFR and FEV_1/FVC both indicate airways obstruction.

7. a. **True.** As a result of its higher solubility in water.
   b. **False.** Equal; mixed systemic venous blood is pumped into the pulmonary arteries by the right ventricle.
   c. **True.** Relies on carbonic anhydrase in red cells.
   d. **True.** Reacts with water to form carbonic acid, which releases protons.
   e. **True.** I.e. respiratory exchange ratio equals 1.00.

8. a. **True.** Tends to make alveolar air more like room air.

b. **False.** P_{O_2} falls as a result of reduced pressure in the inspirate and P_{CO_2} falls because of a hypoxically driven increase in ventilation, but alveolar P_{H_2O} is unaffected since it depends on temperature rather than total pressure.
c. **False.** P_{O_2} falls.
d. **True.** Consider the most extreme example, i.e., perfusion = 0. Ventilation occurs without gas exchange, so this is, by definition, dead space.
e. **False.** There is little or no change in mean alveolar or arterial gases since increased ventilation closely matches increased metabolic need.

9. a. **True.** The initially increasing slope with increasing P_{O_2} indicates cooperativity, the plateau indicates saturability.
b. **True.** When absolute O_2 concentrations are plotted, values increase with haemoglobin concentration but overall shape remains the same.
c. **False.** Shift to the right indicates decrease in O_2 affinity.
d. **True.** Dissociation curve is a plot of O_2 content against P_{O_2}.
e. **True.** For example, this shifts the curve to the left in fetal haemoglobin and adult myoglobin.

10. a. **False.** Haldane effect relates to CO_2 transport; Bohr effect relates to O_2 transport.
b. **True.** From 2 to 4 ml 100 ml^{-1}.
c. **True.** CO_2 dissociation curve shifts down and to the right.
d. **False.** Oxyhaemoglobin shows decreased CO_2 binding.
e. **False.** Haldane shift is dependent on changes in P_{O_2} rather than pH.

11. a. **True.** Caused by increased carbonic acid formation.
b. **False.** Increased ventilation caused by low atmospheric P_{O_2} leads to reduced P_{CO_2}.
c. **True.** Arterial CO_2 is the primary chemical regulator of normal ventilation.
d. **True.** P_{O_2} has to be very low, e.g., < 8.0 kPa (60 mmHg), to stimulate ventilation.
e. **False.** Directly stimulates vasodilatation, reducing peripheral resistance.

12. a. **False.** This is a function of the central chemoreceptors.
b. **False.** Chemoreceptors are not directly sensitive to arterial gas content or concentration but, rather, to the partial pressure of gas, i.e., gas tension.
c. **True.** Main regulatory mechanism during normal breathing.
d. **True.** Sensory receptors convert, or transduce, the relevant stimulus into an electrical signal within the nervous system.
e. **True.** This explains the ventilatory response to acidosis.

13. a. **True.** This decreases the P_{CO_2} helping raise the pH towards normal, i.e., provides respiratory compensation for metabolic acidosis.
b. **True.** This effect is mediated via chemoreceptor stimulation in conditions of very low blood flow when the tissue P_{O_2} falls.
c. **True.** In an attempt to correct abnormal arterial gases.
d. **True.** To pay back the O_2 debt.
e. **True.** Hyperventilation of this kind may lead to a low P_{CO_2} (hypocapnia), alkalosis and faintness. The simplest solution is to get the subject to breathe in and out of a paper bag so that CO_2 is rebreathed and alveolar P_{CO_2} rises.

14. a. **True.** As P_{O_2} falls, more O_2 is released from haemoglobin.
b. **True.** Increased ventilation blows off CO_2 more quickly than it is produced and this can more than quadruple RER.
c. **True.** This increases ventilation before arterial blood gases can change.
d. **True.**
e. **False.** This inhibitory reflex normally prevents overinflation of the lungs.

15. a. **False.** This technique assumes that all metabolism is aerobic.
b. **False.** Decreased.
c. **True.** Mechanism is unclear.
d. **True.** Metabolic rates are normalized by dividing energy use by surface area to allow comparisons between different individuals.

e. **False.** One major action of thyroid hormone is to stimulate metabolism.

Single best answers

1. b. Although the resistance of individual small airways is obviously high, the total resistance is reduced because there are a very large number of them in parallel, reducing their total resistance. Turbulence greatly increases resistance, as do sympathetic nerves and expiration, both of which reduce the diameter of medium to small size airways. Resistance decreases with diameter, and the relationship is $R \propto 1/r^4$.
2. e. A change of about $\pm 5\,cmH_2O$ in intrapleural pressure produces a change in intra-alveolar pressure of $\pm 1\,cmH_2O$. Surface tension effects contribute to the generation of a negative intrapleural pressure, rather than the intra-alveolar pressure, and alveolar pressure must always be positive during expiration. Large alveoli would be expected to have a lower pressure than small ones, but relative concentration differences in surfactant help minimize this effect.
3. b. There is less stretch on the elastic tissue of the lungs at low volume and so the compliance falls as you approach maximum inspiration. Lack of surfactant decreases compliance, by increasing opposing surface tension, and emphysema increases compliance, due to loss of elastic tissue in the lungs. Airways obstruction and paralysis of the ventilatory muscles do not affect lung compliance.
4. d. Based on their normalized (% predicted) values, the ratio of FEV_1:FVC is 0.53 in this case, indicating obstruction. This may be associated with emphysema or chronic bronchitis, but one cannot tell this from the pulmonary function tests. Again, the patient may or may not be cyanosed, depending on the P_{O_2} and the haemoglobin concentration. The reduction in FVC probably results from the obstruction, and cannot be used as an indicator of restrictive disease if there is appreciable obstruction.
5. c. An increased arteriovenous O_2 difference must indicate greater O_2 extraction from the blood, as the venous P_{O_2} must be reduced. This could result from an increased O_2 consumption or a decreased blood flow, or even an increased O_2 consumption which outstripped a simultaneous increase in blood supply. In the absence of any additional information, we cannot safely decide between these. Although pulmonary arterial P_{O_2} may be decreased, we cannot know this for certain either, as this represents the average P_{O_2} for mixed systemic venous blood, and opposite changes in the O_2 content of venous blood from other organs will affect this as well.
6. d. A pulmonary embolus will reduce pulmonary perfusion without affecting ventilation. The $\dot{V}_A/Q$ ratio is determined by the alveolar ventilation volume, rather than the respiratory minute volume (which includes anatomical dead space ventilation); is highest at the apices in the upright position, and will tend to decrease in obstructive airways disease (reduced alveolar ventilation). $\dot{V}_A/Q$ mismatch affects P_{O_2} more than P_{CO_2}, as areas of high $\dot{V}_A/Q$ can compensate for areas of low $\dot{V}_A/Q$ to a much greater extent in the case of CO_2 than O_2 transport. This reflects the different shapes of the dissociation curves for the two gases.
7. d. Anaemia reduces the O_2 content of blood as there is less haemoglobin to carry the O_2. This has no effect on systemic arterial P_{O_2}, which depends on pulmonary ventilation and gas exchange, or on the O_2 affinity or saturation of the blood, both of which reflect the properties, rather than the amount, of haemoglobin. Anaemia may decrease O_2 delivery to the tissues, but this will also be a function of blood flow, which will usually be increased to ensure that metabolic needs are supplied. Thus, resting heart rate and cardiac output are often elevated.
8. b. A leftward shift in the O_2 dissociation curve is the hallmark of increased O_2 affinity. The curve is shifted to the right (affinity is decreased) by increased P_{CO_2}, reduced pH and increased temperature (the Bohr effect). Increasing haemoglobin concentration does not affect the shape of the curve, it simply increases the O_2 content at any given P_{O_2}.

9. c. Increased CO_2 leads to a respiratory acidosis. Although bicarbonate is slightly increased by the dissociation of carbonic acid, the effect is too small to be clinically detectable. Haematocrit tends to increase when CO_2 is elevated due to the Cl^- shift into erythrocytes, which increases the intracellular osmotic load, leading to fluid uptake and cell swelling. This will tend to reduce plasma Cl^-. In an arterial blood sample with an elevated P_{CO_2}, P_{O_2} may or may not be reduced, depending on the underlying pathophysiology.
10. e. Hyperventilation blows off more CO_2 than is being produced by O_2 consumption in the steady state, increasing the RER. Fatty acid utilization and anaerobic metabolism decrease the RER. Increased O_2 consumption may or may not be associated with a decreased RER. Similarly, increased CO_2 production does not, on its own, lead to increased RER, e.g., CO_2 production increases during aerobic exercise without any change in RER, since O_2 consumption goes up by the same proportion. It is the ratio of CO_2 production to O_2 consumption which matters.

Matching item answers

Theme: Gas transport

1. A. This helps explain why deoxygenated venous blood can carry more CO_2 than arterial blood.
2. C. O_2 content is reduced in anaemia but O_2 saturation is not.
3. I. This is one example of the Bohr shift.
4. F. The HCO_3^- is exchanged for Cl^- at the red cell membrane.
5. G.
6. C. O_2 saturation is higher because the O_2 affinity is higher; C is a better answer than D because fetal O_2 content will only be higher if the haemoglobin levels in fetus and adult samples are similar.

Theme: Control of respiration

1. D. This is the primary mechanism whereby ventilation is regulated under normal circumstances.
2. I. Stretch receptors in the lung send inhibitory signals via the vagus nerve to the inspiratory neurones at high lung volume.
3. E. This ensures that the tissue P_{O_2} and pH reflects that in the arterial blood and is not affected by the metabolic activity of the chemoreceptor cells themselves.
4. H. This is how a metabolic acidosis stimulates respiratory compensation, as hyperventilation reduces the P_{CO_2}, and so helps elevate pH. Note that use of the term 'hyperventilation' implies that the rate of ventilation is greater than that required to normalize CO_2 levels. The increased ventilation in response to elevated P_{CO_2} itself is not, therefore, regarded as hyperventilation.
5. G. This is thought to be the main mechanism whereby CO_2 stimulates ventilation. Note that reduced arterial pH does not have this effect as the H^+ ions do not cross the blood–brain barrier readily, whereas CO_2 molecules do. These then react with water and dissociate to release H^+ within the medulla oblongata, stimulating the central chemoreceptors.
6. A. This leads to diaphragmatic contraction during inspiration.

Short note answers

a. Lung compliance is a measure of the ease with which the lungs can be inflated and is defined as: Change in lung volume ÷ Change in distending pressure. Compliance work makes the largest single contribution to the total work of breathing, which increases as compliance decreases. Compliance is largely determined by the elastic and surface tension forces which resist lung expansion and is increased by surfactant (reduces surface tension) and decreased by lung fibrosis. Compliance decreases with ageing.

b. Residual volume equals the volume of gas remaining in the lungs at the end of maximal expiration. It is measured using dilution of an indicator gas like helium and is normally of the order of 1.5 L. This helps reduce the size of breath to breath variations in pulmonary gas pressures as the fresh inspirate is mixed with a larger volume of alveolar air (= residual volume + expiratory reserve volume, i.e., the functional residual capacity). Residual volume increases with age.

c. Haemoglobin's primary function is to bind O_2 via the Fe^{2+} within the porphyrin ring of the haem structure. The total O_2-transport capacity of blood is determined by the haemoglobin concentration. Four O_2 molecules are bound per haemoglobin molecule, one per subunit, and binding is cooperative and saturable. This accounts for the sigmoid shape and plateau of the O_2 dissociation curve for blood. Haemoglobin also transports CO_2 in the form of carbamino groups on the globin chains and acts as an important pH buffer in blood through reversible reactions between basic amino acid residues in its peptide sequence and H^+ (Section 5.9).

d. Peripheral respiratory chemoreceptors are found in two sites, the carotid bodies adjacent to the carotid bifurcation and the aortic bodies along the arch of the aorta. They are sensitive to increased arterial [H^+] and so stimulate ventilation in response to arterial acidosis. They may also be activated by very low P_{O_2} levels. Their main sensory output is directed to the respiratory centres in the medulla, where they modulate the activity in respiratory neurones controlling ventilation, increasing both respiratory rate and tidal volume.

e. Respiratory exchange ratio (RER) = Rate of CO_2 production ÷ Rate of O_2 consumption. The actual value of RER largely depends upon the substrate metabolized and this is of some value when converting O_2 consumption to energy consumption in indirect calorimetry, since different food types have different caloric values. RER is 1.0 for carbohydrate but is less for proteins and fats. The average value on a mixed diet is 0.8. RER values may also be altered by changes in ventilation (RER rises when ventilation first rises during exercise) or by changes in metabolism (RER falls when the O_2 debt is being paid back following anaerobic metabolism).

f. Energy is stored in a variety of forms in the body. Very short term storage occurs in the form of high-energy phosphate bonds, e.g., in adenosine triphosphate (ATP), within metabolizing cells. These are generated by the stepwise breakdown and oxidation of nutrient molecules such as glucose and fatty acids. A secondary cellular reserve in the form of creatine phosphate may be used to prevent ATP depletion during periods of increased cell activity, e.g., in a contracting skeletal muscle. Within the body as a whole, there are also stores which can help maintain the circulating level of nutrients during fast periods. Glycogen is a polymer of glucose and is stored in liver and muscle. Muscle glycogen can only be used within the muscle cells themselves, but liver glycogen can be broken down to maintain plasma glucose levels over the short to medium term (up to about 12 hours). Fat deposits act as longer term stores, releasing fatty acids and glycerol for energy use over a period of days to weeks. In a very prolonged fast, proteins and amino acids will also be used for energy, but this involves catabolizing structural and functional cell elements.

Modified essay answers

1. Compliance work (work expended against elastic and surface tension forces during inspiration), airways resistance work (to overcome the resistance to airflow through the airways), and tissue resistance work (to overcome the frictional resistance to relative tissue movement within the lungs and between the pleura).
2. Surfactant deficiency would increase compliance work as it increases alveolar surface tension.
3. Ventilation maintains normal alveolar gas pressures.
4. P_{O_2} = 13 kPa (98 mmHg), P_{CO_2} = 5.3 kPa (40 mmHg).
5. Blood in the pulmonary capillaries equilibrates with alveolar gases, which are maintained at the appropriate levels through regulation of ventilation. P_{O_2} is slightly reduced between the pulmonary veins and the systemic arteries by shunting of deoxygenated blood into the left atrium from the bronchial veins.
6. Exchange occurs by diffusion driven by the gradients in gas pressure between the arterial blood and the tissue fluids. This leads to diffusion of O_2 into the tissues and diffusion of CO_2 into the blood.
7. The rate of O_2 consumption can be calculated from the change in O_2 concentration between

the arterial and venous blood and the blood flow to that organ. The calculation would be given by the equation: O_2 consumption = (Arterial [O_2] – Venous [O_2]) × Regional blood flow.

8. The diffusion gradients for O_2 and CO_2 between the pulmonary capillaries and the alveoli are reversed when compared with the systemic circulation. This leads to uptake of O_2 into the blood and loss of CO_2 into the alveoli. Thus, pulmonary venous O_2 is higher than pulmonary arterial O_2, while pulmonary venous CO_2 is lower than pulmonary arterial CO_2.
9. In the pulmonary circulation the total blood flow equals the cardiac output. Since it is possible to measure the pulmonary O_2 consumption, and the [O_2] in pulmonary arterial and venous blood, the relationship in Answer 7 above can be rearranged to give: Pulmonary blood flow = Cardiac output = Pulmonary O_2 consumption/(Pulmonary venous [O_2] – Pulmonary arterial [O_2]).

   This is the application of Fick's principle to the measurement of cardiac output.

Data interpretation answers

1. (a) Minute volume = Tidal volume × Respiratory rate = 450 ml × 14 min^{-1} = 6.30 L min^{-1}. (b) IRV = 3.25 L. (c) ERV = 1.55 L. (d) VC = 5.25 L. (e) Alveolar ventilation = (Tidal volume – Dead space) × Respiratory rate = (450 – 150) × 14 = 4.20 L min^{-1}.
2. (a) This corrects for the dilution of O_2 (and other gases) which occurs as the inspired room air is saturated with water vapour in the airways. (b) This avoids contamination by gas from the dead space which is expelled first during expiration. (c) The O_2 consumption rate = Alveolar ventilation rate × (%O_2 entering alveoli – %O_2 leaving alveoli) ÷ 100 = 4.20 × (19.67 – 13.8) ÷ 100 = 0.247 ml O_2 min^{-1}. (d) Energy consumption = 20.2 × 0.247 = 4.99 kJ min^{-1}.
3. (a) Dead space is increased by an amount equal to the volume of the tube. (b) The increase in tidal volume is necessary to maintain a constant rate of alveolar ventilation in the face of a greatly increased dead space. (c) Tidal volume increased by 550 ml when breathing through the tubing but respiratory rate did not change. If there is no change in alveolar ventilation, this must mean that the additional dead space (i.e., the volume of the tube) equals 550 ml.
   Cylinder volume = $\pi r^2 l$
   (r, radius; l, length = 50 cm)
   550 ml = 550 cm^3 = 50 πr^2
   $r = (550/50\pi)^{1/2} = 1.87$ cm.
   (d) No, because the total dead space (550 + 150 = 700 ml) exceeds the resting tidal volume (450 ml). Without an increase in tidal volume, there can be no alveolar ventilation under these conditions, regardless of the respiratory rate. (e) Abnormal increases in the ventilation : perfusion ratio add to the physiological dead space (ventilation with reduced perfusion leads to reduced gas exchange).

Clinical scenario answers

1. Decreased ventilation causes alveolar P_{O_2} to fall and alveolar P_{CO_2} to rise. This produces abnormal gas pressures and a fall in pH in the arterial blood, all of which stimulate ventilation through the chemoreceptors, increasing respiratory rate and tidal volume and leading to a sensation of breathlessness. (Clinical note: increases in P_{CO_2} are usually the important stimulus, since P_{CO_2} is normally much more closely regulated than P_{O_2}. If [CO_2] is chronically elevated, however, adaptation may occur and respiration may become dependent on a hypoxic drive owing to low P_{O_2}.)
2. Reduced arterial P_{O_2} (hypoxia) is suggested by central cyanosis (blue discoloration) caused by increased deoxyhaemoglobin. Increased P_{CO_2} (hypercapnia) is suggested by the warm peripheries (indicating increased blood flow) and bounding pulses caused by CO_2-induced vasodilatation.
3. The increased airways resistance means that a larger pressure gradient has to be developed between the alveoli and the atmosphere for any given flow rate. The alveolar pressure fluctuations would, therefore, be larger in airways obstruction.
4. (a) Reduced. (b) Reduced. (FEV_1/FVC is a better guide to the degree of obstruction in mixed obstructive/restrictive disease.) (c) This may be normal if the respiratory problem is acute, but is likely to be elevated if it is chronic, since renal compensation for the respiratory acidosis provoked by the elevated

CO_2 levels leads to increased HCO_3^- levels.
(d) T_{CO} is likely to be reduced by the reduced ventilation:perfusion ratio. This contributes to the physiological shunting of blood and so reduces the arterial CO concentration. Ventilation:perfusion mismatches are a common cause of T_{CO} abnormalities.

5. (a) Augmented O_2 increases the P_{O_2} in the inspirate and this helps increase alveolar and systemic arterial P_{O_2}. (b) Sympathetic nerves cause bronchodilatation through the action of noradrenaline (norepinephrine) on β-adrenoceptors. β-adrenoceptor agonists are therapeutically useful because they mimic this effect, reducing airways resistance and improving alveolar ventilation. (c) β-adrenoceptor antagonists are likely to increase airways obstruction by inhibiting sympathetically induced bronchodilatation, and so might make things worse.
(d) Parasympathetic nerves cause bronchoconstriction by the action of acetylcholine on smooth muscle muscarinic receptors. Muscarinic agonists are bronchoconstrictors and would increase airways obstruction, making things worse.
(e) Muscarinic antagonists may be useful since they reduce parasympathetic bronchoconstriction. (f) An antibiotic would help clear any underlying bacterial infection (suggested by green sputum; the green colour comes from dead polymorphonuclear leucocytes).

Viva answers

1. Possible effects on O_2 and CO_2 should be considered in turn:
   - The first point to make is that the cyanosis (blue discoloration of lips/tongue, etc.) is caused by deoxygenated haemoglobin and indicates that the P_{O_2} must be low. This is likely to lead the examiner to ask what a normal value might be and if things are going very well you may be asked to hazard a guess as to how low P_{O_2} might have fallen. Normal values for systemic arterial P_{O_2} are about 13 kPa (95–100 mmHg). Cyanosis is only readily visible with about 5 g/dl of deoxygenated haemoglobin; at normal haemoglobin concentrations (a little under 15 g/dl) this represents about two-thirds saturation which occurs around 5 kPa (35 mmHg; Fig. 70B). (As an aside, this indicates that you can have considerable hypoxia without cyanosis.)
   - Obstructive disease will reduce alveolar ventilation causing P_{CO_2} to rise, and again you would be expected to know the normal value (5.3 kPa; 40 mmHg). Discussion may well go on to consider what form the CO_2 is likely to be transported in and what pH changes may result (respiratory acidosis).
2. You should at least know the shape of the dissociation curve and be able to label the axes. Ideally, you will be able to use a few important points to position the curve, e.g., the plateau region is reached at about 10 kPa (75 mmHg) and 90% saturation, with a slow rise towards 100% above this. 75% saturation equates to the venous P_{O_2} of 5.3 kPa (40 mmHg) and then falls rapidly to about 25% just below 3 kPa (20 mmHg). If you place these points on the graph first and then draw an S-shaped curve through them you will have a fairly accurate plot. The curve is shifted to the left (increased O_2 affinity) by a fall in P_{CO_2} / a rise in pH/ a fall in temperature/ in myoglobin and fetal haemoglobin. O_2 saturation is unaffected by anaemia, but the O_2 content is reduced in parallel with the fall in haemoglobin concentration (Fig. 70A).
3. The main focus here is likely to be the role of chemoreceptors in regulating ventilation through their inputs to the respiratory centres in the medulla oblongata. Emphasize the role of CO_2, as detected by the central chemoreceptors in the CNS, as the main stimulus to ventilation. Then explain how a fall in pH, as detected by peripheral chemoreceptors in the carotid and aortic bodies, can also increase ventilation, blowing off CO_2 and helping to compensate for arterial acidosis. Last of all you should mention that detection of hypoxia by peripheral chemoreceptors can also lead to increased ventilation, but only at levels below about 8 kPa or 60 mmHg. Further discussion may consider why P_{CO_2} is more tightly regulated than P_{O_2} (related to the different shapes of the dissociation curves for the two gases; i.e., no plateau versus plateau) and may go on to consider exercise, where factors other than a change in average P_{CO_2} seem to be important.

Renal physiology

Overview

Maintenance of steady-state conditions for fluid volume and composition depends on a balance between the relevant inputs and outputs. The renal system plays a major role in this by removing excess fluid and electrolyte from the body along with a variety of metabolic waste products. This depends on the kidney's ability to form a plasma filtrate; this filtrate is then modified by absorption and secretion within the nephron to produce urine. These processes can be regulated so that the volume, osmolality and pH of urine vary widely to meet the homeostatic needs of the body. Other, non-renal mechanisms are also involved in the maintenance of normal fluid and acid–base balance.

5.1 Body fluids and their distribution

Learning objectives

At the end of this section you should be able to:
- outline how body fluids are distributed
- summarize the ionic composition of intra- and extracellular fluids.

Water accounts for 60% of body mass in males, i.e., over 40 L in a 70 kg adult. This fraction is somewhat lower in females, at about 50%, because of the increased average body content of fatty tissue, which contains less water than other tissue types. Total body water is divided between *intracellular* (two-thirds of total body fluid) and *extracellular* compartments (one-third of total body fluid), with plasma forming a major subcompartment (one-quarter of extracellular fluid) within the extracellular space (Section 1.1, Fig. 2). Water can move freely between these fluid spaces and this exchange is controlled by pressure gradients across the cell membrane (which separates intracellular and extracellular spaces) and the capillary wall (which separates plasma from interstitial fluid). In each case, the distribution of fluid depends on the net effect of hydrostatic and osmotic forces. The solutes controlling osmosis at the two sites, however, are quite different. Only those species which cannot diffuse from one compartment to another will exert an osmotic effect (Section 1.3) and the relevant barriers to diffusion demonstrate very different solute permeabilities.

Charged ions. These cannot easily cross the lipid cell membrane and so the concentration or osmolality of *extracellular electrolytes* (about 300 mosmol kg^{-1} in total) is the dominant factor determining the distribution of fluid between extracellular and intracellular spaces. Since Na^+ normally accounts for nearly 95% of the total cations in the extracellular fluid and Cl^- is passively distributed with Na^+ to maintain electrical neutrality, regulation of body Na^+ content is central to control of extracellular fluid volume.

Proteins. Ions diffuse rapidly across the capillary wall but proteins do not, so the osmotic pressure gradient affecting distribution of fluid between plasma and other, extravascular spaces within the extracellular compartment depends on the transcapillary gradient in protein concentration. It should be remembered, though, that protein only accounts for a tiny fraction (<1%) of total plasma osmolality (Section 3.8).

Ionic composition of extracellular fluid

The extracellular concentrations of a number of important ions are commonly determined using plasma from venous blood samples. Normal values are summarized in Table 8. Although the total charge carried by all the cations and anions must be equal, a significant fraction of the anion load is not accounted for in routine measurements. It can be seen that the sum of the measured cations exceeds the sum of the measured anions, i.e.,

$[Na^+] + [K^+] + [Ca^{2+}] = 147\,mmol\,L^{-1}$

$[Cl^-] + [HCO_3^-] = 130\,mmol\,L^{-1}$

The difference between these totals is called the *anion gap* (17 mmol L^{-1} in this case). This represents the contribution made by anions not normally measured in the clinical laboratory, e.g., sulphate, phosphate, proteins and other organic anions. Any condition which generates additional anions of this sort, e.g., the anions released by dissociation of ketoacids in diabetes mellitus (Section 8.7), will increase the anion gap.

Ionic composition of intracellular fluid

This is extremely difficult to determine accurately and may differ between cell types. The major anions and cations are different from those found outside the cell.

- Intracellular $[Na^+]$ is low (about 10 mmol L^{-1}) and $[K^+]$ is high (about 150 mmol L^{-1}).
- Intracellular $[Cl^-]$ is very low while proteins, phosphate and sulphate account for the majority of intracellular cations.

Table 8 Typical plasma electrolyte values in a sample of venous blood from a normal individual. (Note: the contribution of ions labelled below as 'others' is not routinely measured.)

Cations	(mmol L^{-1})	Anions	(mmol L^{-1})
Na^+	140	Cl^-	105
K^+	4.5	HCO_3^-	25
Ca^{2+}	2.5		
Others	3	Others	20
Total	150	Total	150

5.2 Relevant renal structure

Learning objectives

At the end of this section you should be able to:
- identify the main regions of the kidney
- draw a labelled diagram of a nephron
- draw a labelled diagram of the blood supply of the nephron
- summarize the ultrastructural features of different parts of the nephron.

There are two kidneys, each with an outer *cortex* and inner *medulla* (Fig. 78A). Urine is formed within functional subunits known as *nephrons* (Fig. 78B). Each nephron contains a *glomerulus*, consisting of a tuft of capillaries, along with the afferent and efferent arterioles through which blood enters and leaves (Fig. 78C). The glomerulus is surrounded by the epithelium of the *Bowman's capsule* and these structures combine to form the renal corpuscle. This is the site of initial plasma filtration. The resulting filtrate is then modified by a variety of secretory and reabsorptive processes as it passes in turn through the *proximal convoluted tubule*, the *loop of Henle*, the *distal convoluted tubule* and the *collecting duct* (Fig. 78B). The glomeruli and convoluted tubules lie within the renal cortex while the loops of Henle and collecting ducts extend down into the medullary region. The end product, urine, is eventually delivered via the *renal pelvis* to the *ureter, which empties into the bladder.*

Ultrastructural features of the nephron

The walls of each nephron are made up of a single layer of epithelial cells, although the appearance of these cells is not uniform throughout. Specialized epithelial cells in Bowman's capsule called *podocytes* surround the glomerular capillaries and contribute to the filtration barrier (Section 5.3, Fig. 79). The cells lining the convoluted tubules carry a brush border of *microvilli* on their luminal surface and contain large numbers of intracellular mitochondria. These features are particularly pronounced in the proximal convoluted tubule and provide a large surface area and rich supplies of ATP, appropriate features for a region specializing in active transport. In the descending limb of the loop of Henle, by contrast, the epithelium is flat, with few organelles. This is consistent with a low level of active transport and high permeability to fluid and electrolytes. Cells in the more distal, thick portion of the

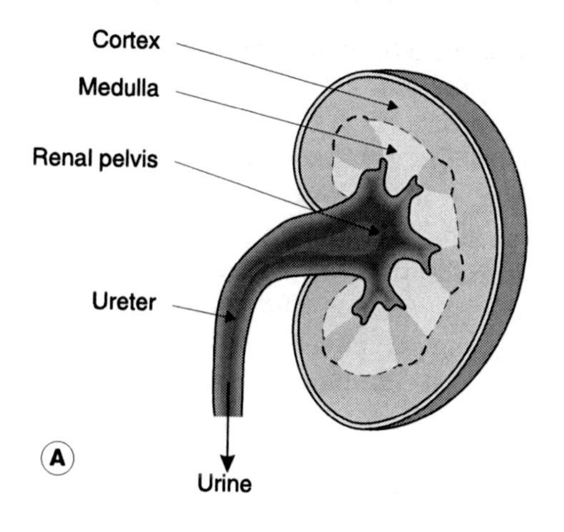

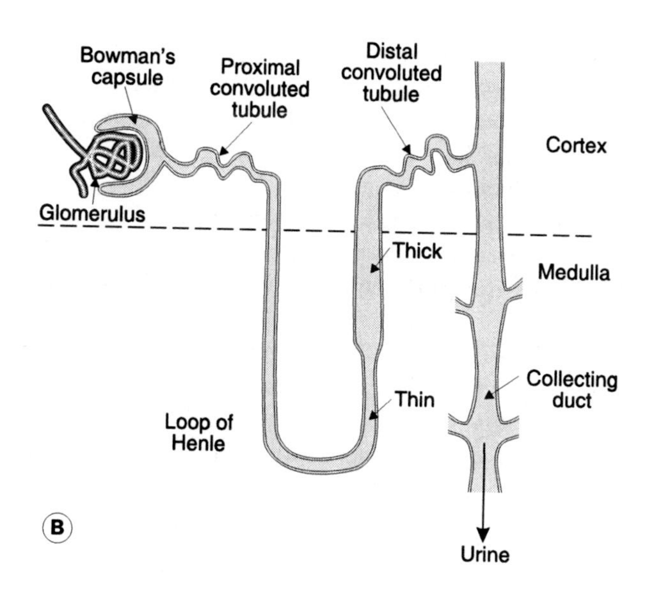

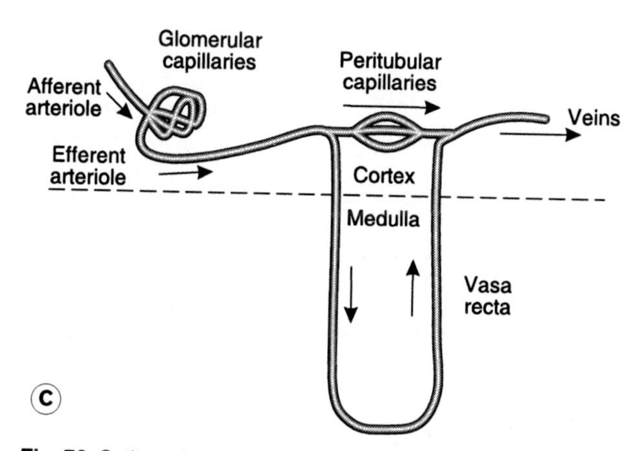

Fig. 78 Outline of renal structure. (A) The kidney. (B) The nephron. (C) The vascular supply of the nephron. Arrows indicate the direction of blood flow.

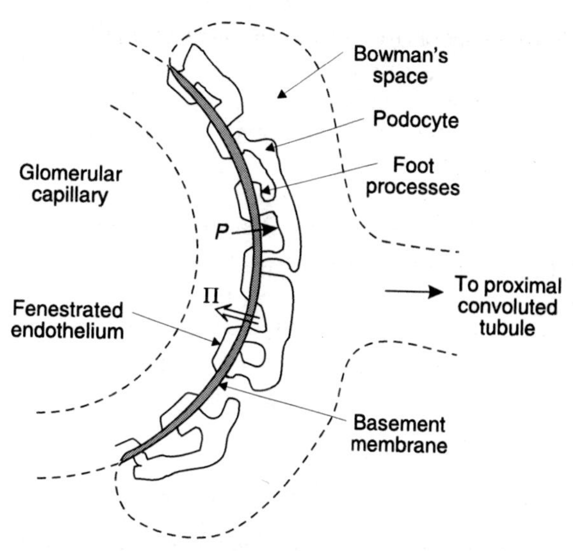

Fig. 79 The main structures separating plasma in the glomerular capillaries from the Bowman's space in the glomerulus. The colloid osmotic pressure (Π) opposes the hydrostatic gradient (*P*) acting into the Bowman's space.

ascending limb, however, are more like those in the tubules. These cells are involved in active ion transport and show a very low level of permeability to water, probably because of well-developed tight junctions between adjacent epithelial cells in this area.

Blood supply to the nephron

The blood supply to the kidney is delivered via the renal artery. The glomerular capillaries are supplied through an *afferent arteriole* but drain through an *efferent arteriole* rather than a venule (Fig. 78C). Blood from the efferent arteriole enters the peritubular capillary beds and the *vasa recta*. The latter vessels descend into the medulla to supply the loops of Henle and collecting ducts. Eventually all these vessels drain through the cortex into the renal veins.

5.3 Glomerular filtration

Learning objectives

At the end of this section you should be able to:

- identify factors which contribute to the high glomerular filtration rate
- summarize with approximate values the forces responsible for filtration
- describe the main mechanisms which control glomerular blood flow and filtration.

The initial step in the formation of urine is production of a plasma filtrate within the glomerulus. As is the case for capillary exchange elsewhere in the body (Section 3.8), the driving forces involved are the hydrostatic and osmotic pressure gradients between the glomerular capillaries and Bowman's space (Fig. 79). Glomerular filtration differs from filtration in general systemic capillary beds, however, in that these gradients result in the efflux of much larger volumes of fluid, with a total filtration rate of approximately 120 ml min^{-1} for both kidneys. This reflects structural specializations in the glomerulus which favour filtration.

High glomerular filtration coefficient

The kidneys are adapted to permit a high rate of filtration, i.e., they demonstrate a high filtration coefficient. Two factors contribute to this:

- a large glomerular surface area
- a low resistance to fluid movement across the glomerular walls (*high hydraulic conductivity*). This reflects the discontinuous, or *fenestrated* structure of the glomerular endothelium (Fig. 79).

The outer aspect of the endothelial basement membrane is in contact with the foot processes of the specialized epithelial cells known as podocytes which line Bowman's capsule. Water, electrolytes and low-molecular-weight solutes pass easily through the filtration barrier formed by these structures, but the plasma proteins and cellular elements of blood are excluded. As a result, the fluid entering Bowman's space, which is sometimes referred to as an *ultrafiltrate*, has the same composition as plasma but without the protein.

Forces controlling filtration

The balance between filtration (hydrostatic) and absorption (colloid osmotic) forces in the glomerulus differs from that in other capillaries (Section 3.8).

Hydrostatic pressures. The hydrostatic pressure within glomerular capillaries is higher than elsewhere (about 50 mmHg). This is largely caused by the high outflow resistance offered by the efferent arteriole, which raises the upstream pressure. At the same time, there is very little decline in pressure between the afferent and efferent arterioles as a result of the comparatively low resistance of the glomerular capillaries (Pressure fall = Flow × Resistance; Section 3.6). Since the pressure within Bowman's space is about 10 mmHg, there is a net *hydrostatic pressure gradient* of 40 mmHg favouring glomerular filtration along the entire length of the glomerular capillary.

Osmotic pressure. Because proteins cannot cross the glomerular wall they exert a colloid osmotic pressure opposing filtration, just as they do in other capillaries (Section 3.8). At the afferent arteriolar end of the glomerulus this equals the normal plasma colloid osmotic pressure of about 25 mmHg, so there is a *net filtration gradient* (hydrostatic gradient – osmotic gradient) of about 15 mmHg at that point. This is not dissimilar to the gradient at the arteriolar end of a normal capillary, but in the glomerulus this results in a much faster rate of filtration because of the very high hydraulic conductivity of the capillary wall. As a result of this fluid removal, the concentration of the plasma proteins rises and this increases the osmotic gradient opposing filtration. Eventually the osmotic gradient exactly matches the hydrostatic gradient so no further filtration occurs.

Interaction of forces in the glomerulus. The values of the relevant forces at different points along the length of the glomerular capillary can be plotted graphically to indicate the region over which filtration will occur (Fig. 80). It should be noted that the osmotic gradient rises to balance the hydrostatic gradient exactly but never exceeds it, so there is no absorption of fluid at any point along the glomerular capillaries.

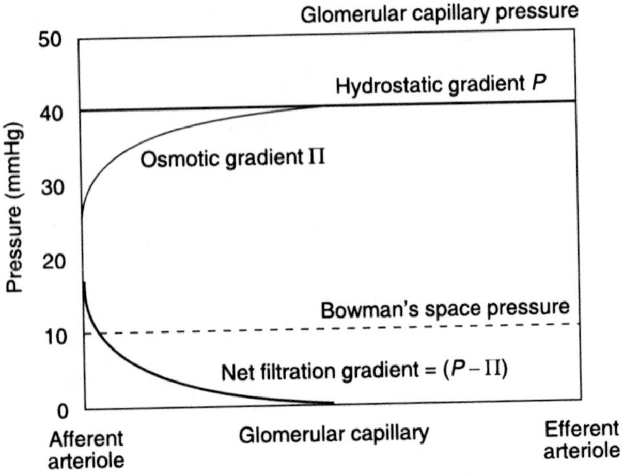

Fig. 80 Graphical representation of the forces acting on plasma along the length of the glomerular capillary. The hydrostatic gradient (P) = (Glomerular capillary pressure – Bowman's space pressure). This remains constant but as the opposing osmotic gradient increases, the net filtration gradient falls.

Regulation of glomerular filtration rate and renal blood flow

The glomerular filtration rate is the total rate of renal filtration in both kidneys and normally equals about $120\,ml\,min^{-1}$. This is remarkably constant over a wide range of conditions. Physiological control of filtration is a complex and poorly understood subject, but glomerular filtration rate will be increased by factors which increase any of the following variables:

- glomerular capillary hydrostatic pressure
- glomerular capillary flow rate
- glomerular capillary surface area.

The maintenance of *renal blood flow* is particularly important for normal filtration, since this is the main determinant of glomerular capillary pressure and flow. Renal blood flow is high (about $1.2\,L\,min^{-1}$) and shows remarkably little variation over a wide range of arterial pressures, providing us with one of the best examples of *autoregulation* of local blood flow (Section 3.7, Fig. 50). Therefore, if arterial pressure falls, dilatation of the afferent arterioles reduces renal vascular resistance and so helps to limit the decrease in blood flow. At very low pressures (e.g., mean arterial pressure < 80 mmHg), autoregulation breaks down and renal blood flow declines more rapidly. Reflex stimulation of sympathetic nerves to the kidney under these conditions may further reduce both glomerular blood flow and hydrostatic pressure by constricting the afferent arterioles. Therefore, both glomerular filtration and urine production may be dramatically reduced in surgical shock with marked arterial hypotension.

5.4 Modification of glomerular filtrate

Learning objectives

At the end of this section you should be able to:

- explain how tubular mechanisms modify glomerular filtrate
- describe how Na^+ and H_2O are reabsorbed in different parts of the nephron
- explain the countercurrent multiplier mechanism and its effect on the medullary osmotic gradient
- summarize how Na^+ and H_2O reabsorption are hormonally regulated
- describe the main features of renal handling of K^+, Ca^{2+} and phosphate, glucose and urea
- describe how H^+ is secreted
- list the urinary buffers and explain how they work.

The formation of urine can be thought of as a two-stage process. Filtration produces an essentially protein-free, but otherwise plasma-like fluid with an osmolality of about $300\,mosmol\,kg^{-1}$. This is then modified by reabsorption and secretion as it passes through the rest of the nephron. Some of these processes are hormonally controlled and this allows the volume, ionic composition and total osmolality of urine to be varied in accordance with the homeostatic needs of the body.

Mechanisms leading to reabsorption of Na^+ and water

Each of the main structural components of the nephron contributes to control of the extracellular fluid volume and composition through mechanisms which lead to the reabsorption of Na^+ and water.

Proximal convoluted tubule

Renal transport of Na^+ and water are closely linked. The mechanisms involved are particularly important since more than 170 L of fluid is filtered out of the plasma each day in the glomeruli, and the vast bulk of this has to be rapidly reabsorbed to avoid fatal depletion of the extracellular fluid. Most of this reabsorption depends on tubular cells, which are specialized for active transport. Energy released by the breakdown of ATP drives Na^+ across the tubular epithelium into the peritubular fluid. Chloride ions follow passively through electrostatic attraction and the resulting osmotic gradient draws water out of the tubule (Fig. 81). The net effect is reabsorption of a NaCl solution essentially isosmotic with plasma. As a result, the volume of the filtrate is decreased but its osmolality does not change.

About 70% of the filtered Na^+ is automatically reabsorbed from the proximal convoluted tubule in this way. This fraction remains relatively fixed across a wide range of Na^+ filtration rates, i.e., a fixed fraction of the total quantity of Na^+ presented to the transport system is reabsorbed (Fig. 82). Renal transport processes of this kind are referred to as gradient–time limited systems.

Loop of Henle

In the loop of Henle active reabsorption of Na^+ leads to passive Cl^- transport, but this is confined to the thick portion of the ascending limb (Fig. 81). Unlike the convoluted tubules, however, this region is impermeable to water and so ion transport raises the concentration of NaCl and, therefore, the osmolality in the medullary fluid around the whole of the loop of Henle. The thin descending limb is permeable

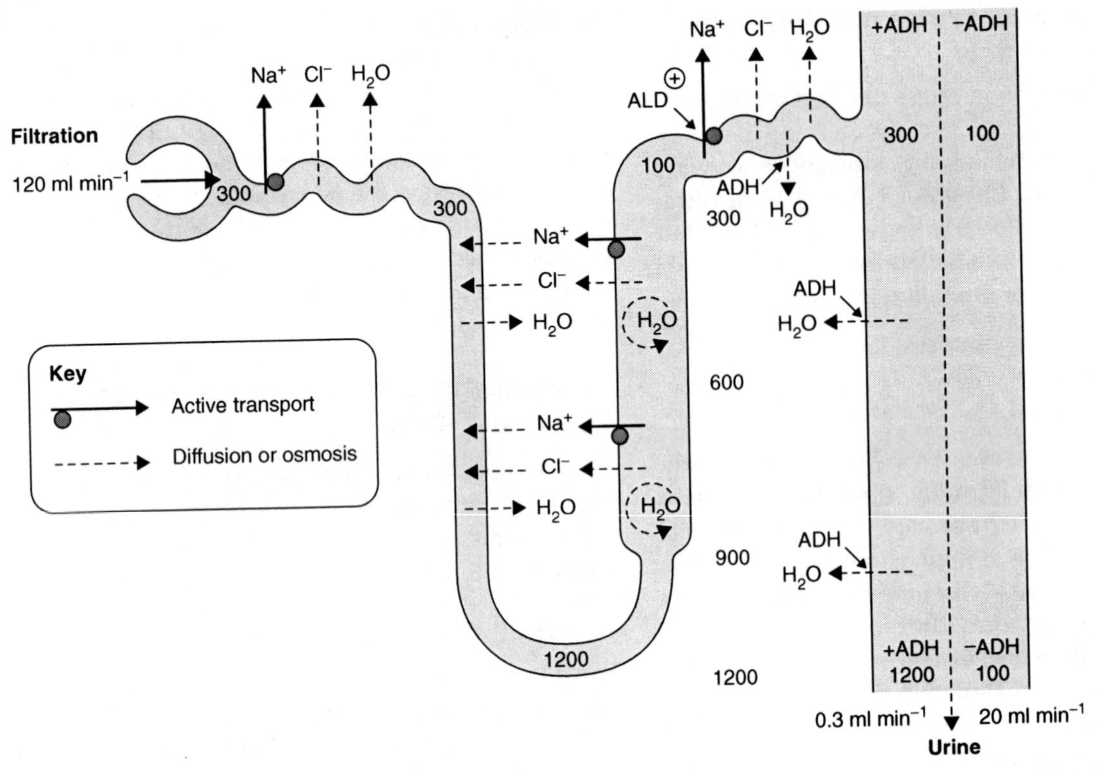

Fig. 81 Sodium and water reabsorption in the nephron. Numbers refer to the osmolality of the tubular fluid and medullary interstitium and are in mosmol kg^{-1}. Significant H_2O reabsorption from the collecting ducts only occurs in the presence of antidiuretic hormone (ADH). Osmolalities and urinary flow rates are indicated for maximal (+ADH) and minimal (–ADH) levels of ADH. A fraction of the Na^+ reabsorption in the distal convoluted tubule is stimulated by aldosterone (ALD).

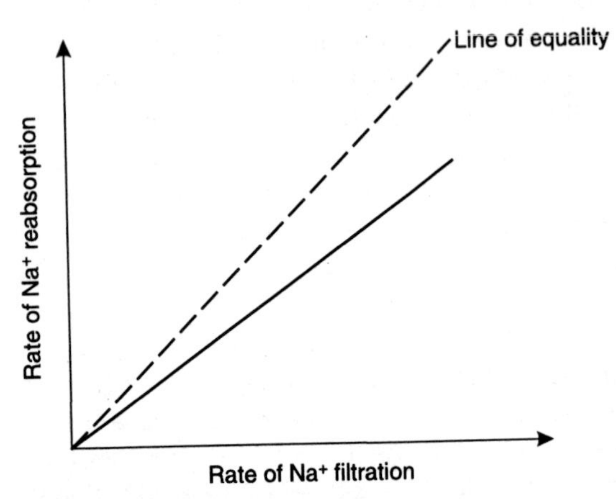

Fig. 82 Reabsorption of Na^+ as a function of its filtration in the proximal convoluted tubule. Approximately 70% of the filtered load is reabsorbed (solid line). The dashed line represents equal rates of filtration and absorption.

both to ions and water, so as the fluid from the proximal convoluted tubule descends through the medulla it is both concentrated and reduced in volume by the passive influx of Na^+ and Cl^- and the osmotic removal of water. This concentrated solution becomes more dilute as it ascends the opposite limb of the loop of Henle because of the active removal of solute, so that the solution entering the distal convoluted tubule actually has a lower osmolality (about 100 mosmol kg^{-1}) than that leaving the proximal tubule (300 mosmol kg^{-1}). This mechanism allows for the production of hypo- as well as hypertonic urine.

The most important effect of ion transport by the loop of Henle is to increase the osmolality of the interstitial fluid in the medulla of the kidney, which rises to a maximum of about 1200 mosmol kg^{-1} at the bottom of the loop (Fig. 81). This provides the driving force for subsequent reabsorption of water from the collecting ducts.

Distal convoluted tubule and collecting ducts

There are two important mechanisms which play a role in Na^+ and water reabsorption in these regions.

Sodium reabsorption. Most of the remaining Na^+ load is actively reabsorbed from the distal convoluted tubule and collecting ducts so that less than 1% of the filtered Na^+ is excreted in the urine. These transport processes are analogous with those in the proximal convoluted tubule, with isosmotic reabsorption of NaCl and water. A small but important fraction of this resorption depends on stimulation by the hormone *aldosterone*, allowing Na^+ excretion to be regulated so as to maintain a normal extracellular fluid volume.

Water reabsorption. The low osmolality of the fluid entering the distal convoluted tubule allows for osmotic reabsorption of water into the extracellular fluid of the cortex. Further osmotically driven water absorption occurs from the collecting ducts as they descend through the medulla towards the renal pelvis (Fig. 81). This depends on the high osmolality generated in the surrounding medullary interstitium by the ionic pumps of the loop of Henle. Since this water reabsorption occurs without any parallel ion transport, the volume of the urine is decreased and its osmolality is increased. The distal convoluted tubule and collecting duct remain relatively impermeable to water, however, unless stimulated by circulating *antidiuretic hormone* (ADH), and a large volume of urine (a *diuresis*) with a low osmolality is produced in the absence of this hormone. With maximal ADH secretion, a small volume of highly concentrated urine is excreted (Fig. 81). This provides an endocrine control mechanism which can rapidly alter renal water absorption so as to correct abnormalities of extracellular fluid volume or osmolality.

The countercurrent multiplier mechanism in the loop of Henle Ion transport in the loop of Henle is responsible for the generation of the very hypertonic conditions within the renal medulla, with osmolalities reaching a maximum of about 1200 mosmol kg^{-1}. Although the principles underlying this have already been discussed, i.e., active reabsorption of Na^+ in the thick ascending limb, with little or no parallel movement of water, and free diffusion across the walls of the descending limb, we will now briefly consider the countercurrent multiplier mechanism, which helps explain the very high osmolalities achieved. This reflects a design which effectively allows filtered fluid to be concentrated not once but several times as it moves round the loop.

Let us assume that the system is initially filled with isotonic solution (300 mosmol kg^{-1}) from the proximal convoluted tubules. The ion pumps will then transfer solute from the ascending limb into the interstitium until the concentration gradient generated across the epithelium exactly opposes the energy available from the active transport system (Section 1.3). In a simple, single conduit system, this might limit the gradient to about 200 mosmol kg^{-1}, with a maximal interstitial osmolality of about 400 mosmol kg^{-1} (Fig. 83A). In the countercurrent system, however, the solution in the permeable descending limb equilibrates with the interstitium and so becomes concentrated. As more (isotonic) fluid enters at the top of the descending limb, concentrated fluid is forced round into the ascending

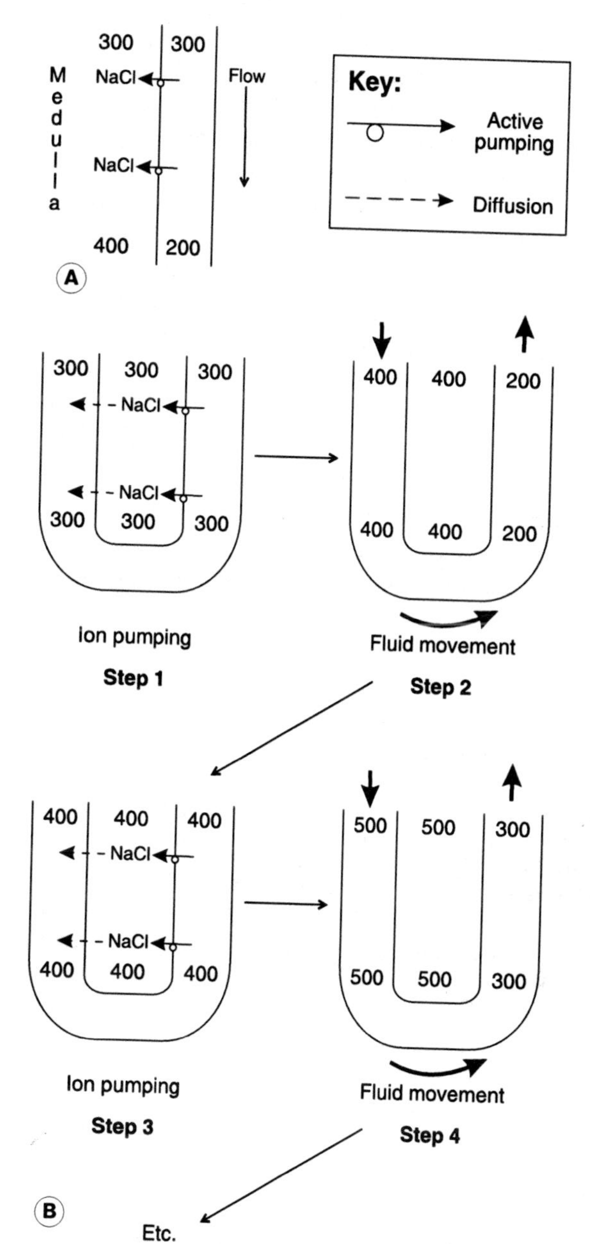

Fig. 83 Principles of the countercurrent multiplier mechanism. The numbers indicate the osmolality in mosmol kg^{-1}. (A) In a single tubule the ionic gradient between a tubule and the surrounding interstitium (200 mosmol kg^{-1}) is fixed by the available ion pump energy. (B) In the loop of Henle, the same osmotic gradient can be generated by the ion pumps in the ascending limb (Step 1). Fluid in the permeable descending limb equilibrates with the interstitium and concentrated solution is then transferred to the ascending limb (Step 2). The pumping process is repeated (Step 3). The absolute osmolality in the medulla (500 mosmol kg^{-1}) is now higher than before, i.e., the countercurrent arrangement multiplies the overall effect of ion pumping on osmolality.

limb, reducing the transepithelial osmotic gradient while maintaining a high total interstitial osmolality (Fig. 83B). Ion pumping can proceed once more, until a 200 mosmol kg^{-1} gradient is again achieved. In this example, the osmolality in the medullary interstitium and lower part of the descending limb is raised to 500 mosmol kg^{-1} as a result. Repeated cycles of ion pumping followed by fluid transfer can thus generate a higher medullary osmolality than would otherwise be possible. At the same time, a gradient in osmolality develops along the length of the loop of Henle with low values in the upper medulla and high values adjacent to its tip.

Countercurrent exchange in the vasa recta The blood vessels supplying the medullary regions of the kidney (the vasa recta) are arranged in a U configuration so that blood descends one limb before ascending back to the cortex (Fig. 78C). As blood descends through areas of increasing osmolality, NaCl diffuses across the capillary walls, increasing plasma osmolality (Fig. 84). If these vessels drained into the venous system directly from the medulla, they would leach away a great deal of solute, increasing the energy needed to maintain the high medullary osmolality. As the hypertonic plasma ascends through the medulla again, however, ions diffuse back into the surrounding interstitium. The net effect is the continuous recycling of fluid and solutes as blood flows in opposite directions through the parallel limbs of the vasa recta, a process known as countercurrent exchange.

Regulation of renal Na^+ and water reabsorption

Three hormones play an important role in the regulation of extracellular fluid volume and osmolality through their actions on renal reabsorption of Na^+ and water (Fig. 85).

Antidiuretic hormone

This peptide hormone is manufactured by cells in the supraoptic and paraventricular nuclei of the *hypothalamus* and is released from the *posterior pituitary* (Section 8.2). Antidiuretic hormone (ADH) secretion is stimulated by:

- increased osmolality in the extracellular fluid, which is detected by osmoreceptors within the hypothalamus; these also stimulate drinking
- decreased circulating blood volume, as detected by cardiovascular volume receptors (Section 3.6)
- decreased arterial pressure as detected by cardiovascular baroreceptors (Section 3.6).

Antidiuretic hormone increases the water permeability in the collecting ducts and the distal convoluted tubule by stimulating the insertion of water-selective channels in epithelial cell membranes. These channels, known as aquaporins, promote reabsorption of water due to the osmotic gradients across the walls of the distal convoluted tubule (low intratubular osmolality) and the collecting ducts

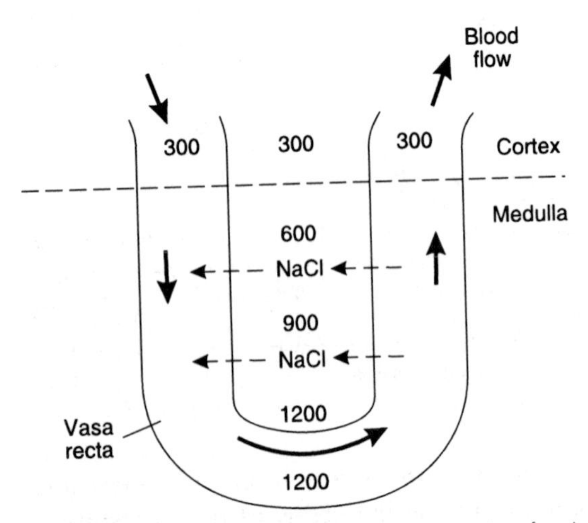

Fig. 84 Countercurrent exchange in the vasa recta refers to the passive diffusion of Na^+ and Cl^- from the ascending to the descending blood capillary loop. This avoids excess washout of medullary solute. (The numbers refer to osmolality in mosmol kg^{-1}.)

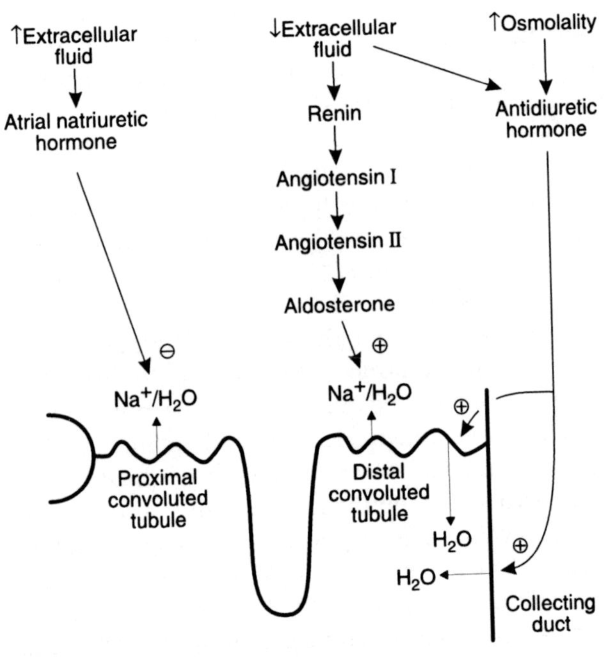

Fig. 85 Endocrine control of renal Na^+ and water reabsorption.

(high medullary interstitial osmolality). As a result, the extracellular fluid volume is expanded while its osmolality is reduced. Antidiuretic hormone also helps to elevate arterial blood pressure by direct vasoconstriction, leading to its alternative name of *vasopressin* (Section 3.6).

Under normal conditions there is continuous release of antidiuretic hormone but this is suppressed if plasma osmolality falls or extracellular fluid volume expands. This occurs within minutes after absorption of a large volume of liquid and the resulting decrease in renal water reabsorption leads to increased production of low concentration urine, i.e., decreased secretion of antidiuretic hormone leads to a diuresis, as its name implies.

The renin–angiotensin–aldosterone system

This depends on release of the enzyme renin from the juxtaglomerular apparatus. Release is activated by:

- decreases in afferent arteriolar pressure
- reductions in the filtered Na^+ load, e.g., caused by reduced plasma [Na^+] or reduced glomerular filtration
- stimulation of renal sympathetic nerves.

All of these stimuli are particularly relevant with regard to regulation of extracellular fluid volume, since this directly controls plasma volume and so affects blood pressure (Section 3.6). In each case changes are detected by the *juxtaglomerular apparatus*, a structure formed by the close association of specialized juxtaglomerular cells in the wall of the afferent arteriole and the macula densa in the distal convoluted tubule. Renin is released from the juxtaglomerular cells and catalyzes the conversion of the plasma precursor protein angiotensinogen into angiotensin I. This is further converted to angiotensin II by *angiotensin converting enzyme* on the surface of vascular endothelium. Angiotensin II promotes release of the steroid hormone aldosterone from the adrenal cortex (Section 8.5) and this in turn stimulates reabsorption of Na^+ and water from the distal convoluted tubule and collecting ducts, leading to expansion of the extracellular space and increasing blood pressure. These effects take several hours to develop fully. Although aldosterone only influences the reabsorption of the final 2–3% of the total load of filtered Na^+, this represents a volume of up to 5 L of isotonic fluid in 24 hours and so provides an important mechanism for the control of extracellular volume. It should also be noted that high levels of aldosterone favour renal conservation of fluid by reducing urine volume but have limited effects on plasma or urine osmolality.

Angiotensin II has a variety of other actions with complementary effects, for example:

- it stimulates systemic arteriolar vasoconstriction (raises peripheral resistance and blood pressure)
- it promotes antidiuretic hormone release (expands extracellular volume)
- it promotes drinking (expands extracellular volume).

Atrial natriuretic hormone

This peptide hormone is released from the atria of the heart when the extracellular fluid volume is greater than normal, probably as a result of increased atrial stretch. It promotes glomerular filtration and inhibits tubular reabsorption of Na^+, leading to increased urinary loss of Na^+ (*natriuresis*) and water (*diuresis*). This reduces both extracellular and plasma volumes towards normal.

Box 16 Clinical note: Diuretics

A variety of clinical conditions are characterized by the accumulation of excess fluid within the body, e.g., in left ventricular failure this leads to pulmonary oedema which impairs gas exchange and causes hypoxia. Diuretics are pharmacological agents which promote Na^+ and H_2O loss from the kidney, increasing the volume of urine. Although different drugs act at different sites along the nephron, in each case the diuresis results from inhibition of ionic transport mechanisms normally responsible for reabsorption of NaCl. As a result H_2O is retained osmotically within the tubules and excreted along with the electrolytes.

Transport of K^+

Potassium ions are reabsorbed and secreted in different parts of the nephron. Active reabsorption from both the proximal convoluted tubule and the ascending limb of the loop of Henle reduces the K^+ load to less than 10% of the filtered K^+. It is the rate of K^+ secretion in the distal convoluted tubule, however, which largely determines the rate of K^+ excretion in the urine. *Aldosterone* stimulates this secretion and is the most important regulator of plasma [K^+] since adrenocortical secretion of aldosterone is directly stimulated by K^+ (Section 8.5).

Reabsorption of Ca^{2+} and phosphate ions

Both these ions are actively reabsorbed from the tubular fluid, particularly in the proximal convoluted tubule, so that normally only about one-third of the filtered load reaches the urine. Reabsorption is hormonally controlled by *parathormone*, which promotes Ca^{2+} reabsorption while inhibiting that of phosphate. This is an important mechanism in the regulation of plasma Ca^{2+} concentrations (Section 8.4).

Transport of H^+ and HCO_3^-

The kidney plays an important role in the regulation of extracellular pH through several different, if related, tubular mechanisms:

- active secretion of H^+
- H^+ buffering within the tubular lumen
  —buffering with HCO_3^-, which results in absorption of filtered HCO_3^-
  —buffering with HPO_4^{2-} or NH_3, which results in H^+ excretion and generation of additional HCO_3^-.

These processes lead to the excretion of excess H^+ or HCO_3^- in the urine and the resulting urinary pH can vary between 4.5 and 8.2, depending on the conditions. Under normal circumstances the end result is reabsorption of all filtered HCO_3^- within the proximal convoluted tubule, and urinary excretion of an acid load equivalent to 50–80 mmol of H^+ each day, mostly as $H_2PO_4^-$ and NH_4^+ formed within the distal convoluted tubule. This metabolic acid is generated during catabolism of the proteins in an average mixed diet. Net excretion of acid is coupled with tubular generation of additional HCO_3^-, thus tending to raise plasma $[HCO_3^-]$.

H^+ secretion and HCO_3^- absorption

Bicarbonate is one of the most important pH buffers in the body (Section 5.9). Since it is freely filtered in the glomerulus it is vital that the renal tubules should have an efficient mechanism for HCO_3^- absorption, otherwise its loss in the urine would quickly lead to fatal acidosis. Protons and HCO_3^- are generated within the tubular epithelial cells by dissociation of carbonic acid formed by the reaction of CO_2 with water (Fig. 86A). This reaction is catalyzed by the enzyme carbonic anhydrase. The H^+ is actively transported into the tubular lumen while the HCO_3^- diffuses into the peritubular capillaries, maintaining electrical neutrality inside the cell. Secreted H^+ has to be buffered in the urine to limit the $[H^+]$ gradient between the tubular cells and the lumen, otherwise it would quickly become too large for the active

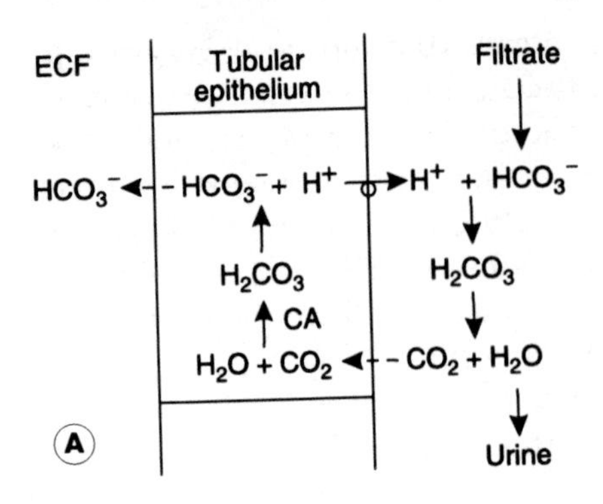

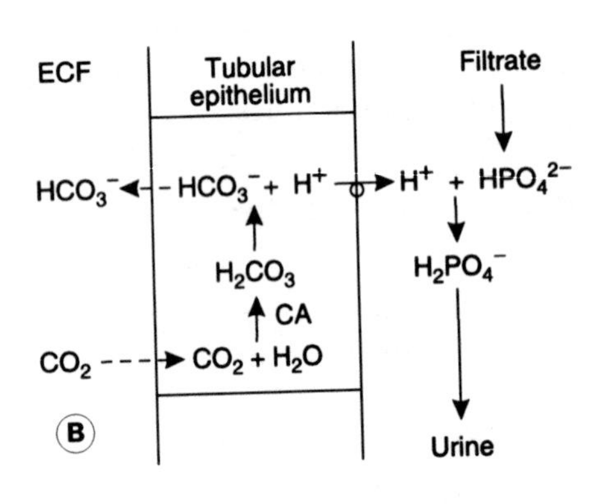

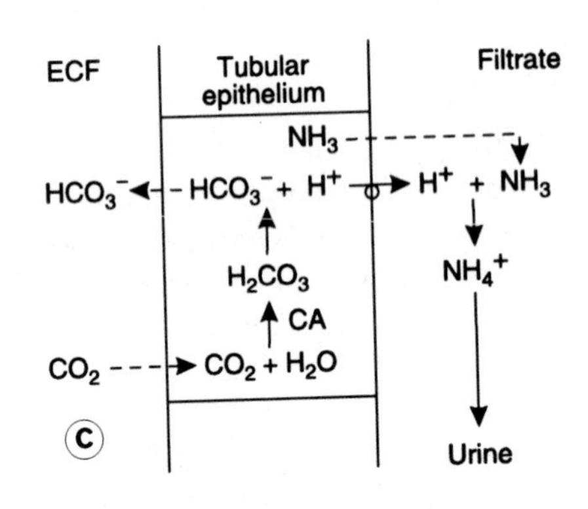

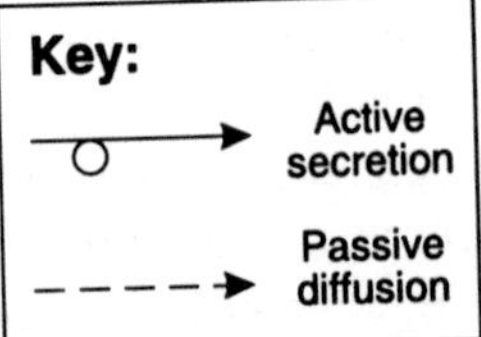

Fig. 86 Regulation of extracellular pH through renal secretion of H^+ and absorption of HCO_3^-. This relies on carbonic anhydrase (CA). Tubular secretion of H^+ leads to reabsorption of filtered HCO_3^- (A). Secreted H^+ can also protonate filtered phosphates (B) or ammonia (C) and is thus excreted. The remaining HCO_3^- from the dissociation of carbonic acid in the tubular epithelium is absorbed into the extracellular fluid (ECF), thus increasing its HCO_3^- content.

transport system, preventing further secretion. Filtered HCO_3^- reacts with H^+ to form CO_2 and H_2O in the tubular fluid. Carbon dioxide can diffuse freely back across the tubular cell membrane, completing the cycle. The net effect is that one HCO_3^- ion is absorbed into the peritubular plasma from the filtrate for each H^+ which is secreted.

This mechanism conserves HCO_3^- and, under normal conditions, all the filtered HCO_3^- is absorbed (mainly in the proximal convoluted tubule) and none is excreted in urine. If the rate at which HCO_3^- is filtered exceeds the rate of H^+ secretion, however, then the excess cannot be absorbed and will be excreted, making the urine alkaline. This mechanism tends to reduce plasma HCO_3^- during alkalosis, and thus tends to lower the pH (see Eq. 41 in Section 5.9).

H^+ excretion and HCO_3^- generation

Although the mechanisms described above couple H^+ secretion to HCO_3^- absorption, they do not lead to any net excretion of H^+ in the urine since the CO_2 produced diffuses back out of the tubular lumen (Fig. 86A). This underlines the point that renal secretion (a tubular cellular mechanism) and excretion (permanent removal in the urine) are not the same thing. Renal excretion of acid still depends on tubular secretion but in this case the H^+ is buffered by phosphate species such as HPO_4^{2-}, or ammonia (NH_3). The $H_2PO_4^-$ and ammonium (NH_4^+) ions formed cannot cross the lipid membrane of the tubular epithelium and are excreted in the urine (Fig. 86B, C). Phosphate enters the lumen by filtration but the NH_3 is actually secreted by the tubular cells, much of it having been synthesized from glutamine within the cell (ammoniagenesis).

One important feature of H^+ buffering by phosphate and ammonia is that it leaves the HCO_3^- produced by dissociation of carbonic acid within the tubular cells to diffuse into the extracellular fluid. Since this amounts to generation of additional HCO_3^- over and above that which was filtered in the glomerulus, it can actually raise plasma [HCO_3^-], rather than simply preventing it from decreasing as absorption of filtered HCO_3^- does. Bicarbonate is an important base within the body, and this mechanism provides for renal compensation of a respiratory acidosis. For example, during a chronic respiratory acidosis both the rate of H^+ secretion and the rate of NH_3 production are upregulated within the first week, leading to an increase in plasma [HCO_3^-]. This tends to reverse the fall in blood pH, at least partially compensating for the original defect (Sections 4.4 and 5.9).

Reabsorption of glucose and other nutrients

Glucose is freely filtered in the glomerulus so the concentration in the filtrate equals that in plasma. This means that the glucose load presented to the tubules is proportional to the plasma glucose concentration (Fig. 87). Normally, the rate of active glucose reabsorption in the proximal convoluted tubule equals the rate of filtration, so no glucose appears in the urine. This reuptake can eventually be saturated, however, and the maximum achievable rate of glucose reabsorption is known as the *transport maximum* (T_{max}). If the rate of glucose filtration rises above T_{max}, the excess load will all be excreted in the urine (Fig. 87). This occurs when the plasma glucose concentration rises above about 11 mmol L^{-1}, so this is regarded as the *renal threshold* concentration, above which glucose will be excreted in the urine (*glycosuria*). Glycosuria is an important feature of diabetes mellitus, in which a deficiency of the hormone insulin leads to abnormally high glucose concentrations (Section 8.7).

The relationship between plasma glucose concentration and urinary excretion rate is curved over a range of concentrations immediately above threshold. This region of *splay* reflects small variations in the maximum transport rate for different tubules within the kidneys. Eventually all glucose transport systems are saturated, and urinary excretion then

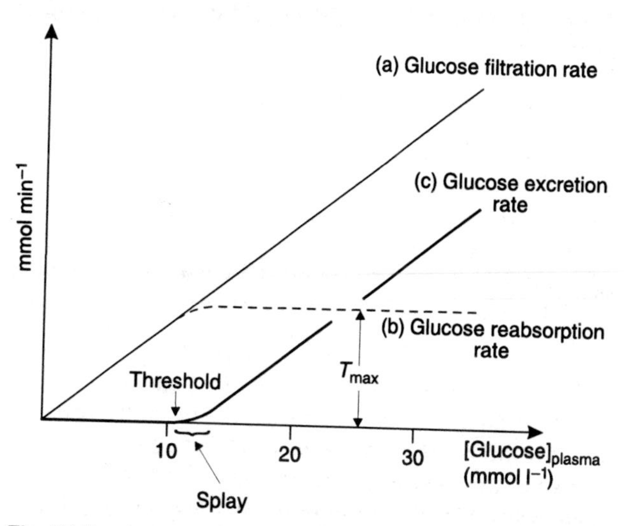

Fig. 87 Renal glucose handling varies with plasma glucose concentration. (a) Glucose filtration rate equals glomerular filtration rate × [Glucose]$_{plasma}$. (b) Glucose reabsorption rate. (c) Urinary glucose excretion rate is zero over the normal range but rises linearly at high plasma concentrations when filtration exceeds T_{max}.

Five

rises linearly with plasma glucose since any increase in the filtered glucose load now results in an equal increase in glucose excretion.

Tubular maximum transport mechanisms of this sort are also involved in the reabsorption of other filtered nutrients from the proximal convoluted tubule, especially amino acids. Systems of this kind should be contrasted with the gradient–time dependent resorption of Na^+, in which a fixed fraction of the filtered load is reabsorbed (Fig. 82).

Passive reabsorption of urea

Urea is produced in the liver as the end product of protein metabolism, and one of the functions of the kidney is to clear it from the body in the urine. Indeed, plasma urea concentration is often used clinically in the assessment of renal function, since urea levels rise in renal failure. As a small solute, urea is freely filtered in the glomerulus and it is not actively reabsorbed. This does not mean that all the filtered urea is excreted, however, since the reabsorption of water from the tubular filtrate increases the intratubular urea concentration. The resulting concentration gradient leads to passive reabsorption since the tubular epithelium has a low but finite permeability to urea. Other filtered solutes which become concentrated by tubular removal of water will also be passively reabsorbed, providing they can cross the tubule wall (i.e., other lipid-soluble substances).

Box 17 Clinical note: Chronic renal failure

The features of chronic renal failure reflect the wide range of normal renal functions. Failure of body fluid regulation may lead to fluid depletion or fluid accumulation, depending on the nature of the renal defect. Toxic protein metabolites which are normally excreted by the kidney accumulate, as indicated by a rise in urea levels (a harmless marker of impaired renal function). Serum K^+ also rises. There may be increased renin production due to reduced glomerular filtration and this can lead to renal hypertension through increased angiotensin activity. Failure of acid secretion favours metabolic acidosis while inadequate erythropoietin production leads to anaemia. The major fluid and electrolyte imbalances can be managed using control of dietary protein, K^+ and fluid intake along with renal dialysis techniques, but only a successful transplant can restore all normal functions.

5.5 Renal clearance

Learning objectives

At the end of this section you should be able to:

- define the term 'renal clearance'
- explain how renal clearance may be used to estimate glomerular filtration rate
- explain how renal clearance may be used to estimate renal plasma/blood flow rate.

Different substances are removed from the circulation at different rates by the kidneys. One measurement which allows the renal handling of materials to be compared quantitatively is renal clearance. This may be defined as the volume of plasma which would contain the same amount of any substance as is excreted in the urine in a given time. In fact, the formal definition refers to the volume of plasma which would be completely cleared of any given substance by the kidney in a given time; hence the term renal clearance. These two definitions are equivalent, since the volume of plasma which would be cleared must be the volume of plasma containing the relevant amount of solute. The first definition may avoid unnecessary confusion, however, since very few substances are completely removed from the plasma as it flows through the kidney, as the term 'cleared' may seem to imply. If one calculates how much plasma would have contained the same amount of any substance as is excreted in the urine over a certain period then that is its clearance value, regardless of how it is handled by the kidney.

Suppose we wish to determine the clearance for some substance X (C_x). We first need to collect urine over a known period of time and measure both its volume (V) and the urinary concentration of the relevant solute (U_x).

The total amount of solute in the urine = $U_x \cdot V$

$$\therefore\ C_x = \text{Plasma volume containing } (U_x \cdot V) \text{ mmol of solute}$$

$$\text{Volume} = \frac{\text{Amount}}{\text{Concentration}}$$

$$\therefore\ C_x = \frac{U_x \cdot V}{P} \qquad \text{(Eq. 28)}$$

where P is the plasma concentration of the solute.

Clearance has the same units as V since the units of the urinary and plasma concentrations cancel each other out. Since V is the volume of urine collected in a given period of time, clearance is measured in ml min^{-1}.

Measurement of glomerular filtration rate

Glomerular filtration rate (GFR) can be estimated by measuring the clearance of any substance whose renal handling fulfils the following criteria:

- It must be freely filtered in the glomerulus.
- It must not be reabsorbed from the nephron.
- It must not be secreted into the nephron.
- It must not be metabolized in the nephron.

Under these circumstances:

$$\text{Rate of filtration} = \text{Rate of excretion} \qquad \text{(Eq. 29)}$$

$$\text{Rate of filtration} = \text{Concentration in filtrate} \times \text{GFR}$$

But, if a substance is freely filtered:

$$\text{Concentration in filtrate} = \text{Plasma concentration } (P)$$

$$\therefore \text{Rate of filtration} = P \times \text{GFR}$$

$$\text{Rate of excretion} = \text{Urine concentration } (U) \times \text{Urine volume } (V)$$

Using Equation 29 it follows that:

$$\text{GFR} \times P = U \times V$$

$$\therefore \quad \text{GFR} = \frac{U \times V}{P} \qquad \text{(Eq. 30)}$$

It can be seen by comparison with Equation 28 that this is actually the clearance value for the substance concerned.

The main difficulty in making a determination of the GFR using this method is to find a substance which obeys the required criteria for renal handling. The polysaccharide *inulin* is one such material which may be injected into the subject, allowing plasma concentration and urinary excretion to be measured. In clinical practice, however, *creatinine clearance* is usually measured. Creatinine is a metabolite of muscle creatine and, as a naturally occurring substance, does not have to be introduced into the circulation artificially. Urine is collected over 24 hours and a plasma sample taken during this period is used to determine the circulating creatinine concentration. The resulting value for creatinine clearance, which is normally about $120\,\text{ml min}^{-1}$, is a reasonable estimate of GFR. This represents total filtration in both kidneys.

Measurement of renal plasma flow

Renal plasma flow may be estimated by measuring the clearance of a substance which is completely removed from the plasma in a single circuit through the kidneys, i.e., its concentration is reduced to zero before the plasma reaches the renal venous system. This requires active secretion. Under these circumstances:

$$\text{Rate of delivery in arterial blood} = \text{Rate of urinary excretion} \qquad \text{(Eq. 31)}$$

$$\text{Rate of delivery} = \text{Renal plasma flow} \times \text{Plasma concentration } (P)$$

$$\text{Rate of excretion} = \text{Urine concentration } (U) \times \text{Urine volume } (V)$$

Using Equation 31 it follows that:

$$\text{Renal plasma flow} \times P = U \times V$$

$$\therefore \quad \text{Renal plasma flow} = \frac{U \times V}{P} \qquad \text{(Eq. 32)}$$

Once again, this is a clearance measurement (Eq. 28). Although very few substances are completely cleared in a single passage through the kidney, filtration and secretion combine to remove about 90% of a chemical called *para-aminohippuric* acid (PAH), at least at low arterial plasma concentrations. This agent may be injected into the bloodstream and its clearance measured in the normal way. The resulting value is about $650\,\text{ml min}^{-1}$, which provides an estimate of the total plasma flow to both kidneys. Assuming a haematocrit of 45%, this equates to a blood flow of $1200\,\text{ml min}^{-1}$, i.e., over 20% of the total cardiac output.

5.6 Micturition

Learning objectives

At the end of this section you should be able to:

- describe the reflex arcs involved in control of micturition
- explain how micturition is initiated and controlled.

Urine is formed at an average rate of about $1\,\text{ml min}^{-1}$ with normal hydration. It is transported from the pelvis of the kidney to the bladder by peristaltic waves of contraction in the ureter. The bladder can accommodate 200–300 ml of urine with little increase in pressure but eventually sensory inputs from stretch receptors in the bladder wall stimulate reflex contraction of the bladder smooth muscle (*detrusor muscle*) through *parasympathetic nerves* from the *sacral spinal cord* (Fig. 88). These contractions are associated with an awareness of the urge to urinate and may force urine into the urethra. Subsequent relaxation of the striated muscle of the external urethral sphincter, coupled with continued contraction of the bladder muscle, allows urine to flow through the urethra. Voluntary contraction of the abdominal muscles can accelerate the process by increasing the intra-abdominal pressure. Once emptied to a small residual volume (about 10 ml), the bladder muscle relaxes and the external sphincter constricts allowing urine collection to recommence.

As with defecation, micturition is an automatic reflex in the first year of childhood but can later be regulated by inhibitory inputs from higher centres, which allow us to retain both our urine and our dignity. Voluntary control of micturition is referred to as urinary continence and depends on descending spinal pathways which inhibit the parasympathetic output to the bladder while stimulating the somatic nerves supplying the external urethral sphincter. Such control may be lost following damage to the spinal cord, leading to *incontinence* of urine. Also, if the pelvic nerves carrying the parasympathetic motor supply to the detrusor muscle itself are damaged, the ability to empty the bladder efficiently is lost. Retention of urine results and may often lead to recurrent urinary tract infections.

5.7 Other renal functions

Learning objectives

At the end of this section you should be able to:

- summarize the endocrine functions of the kidney.

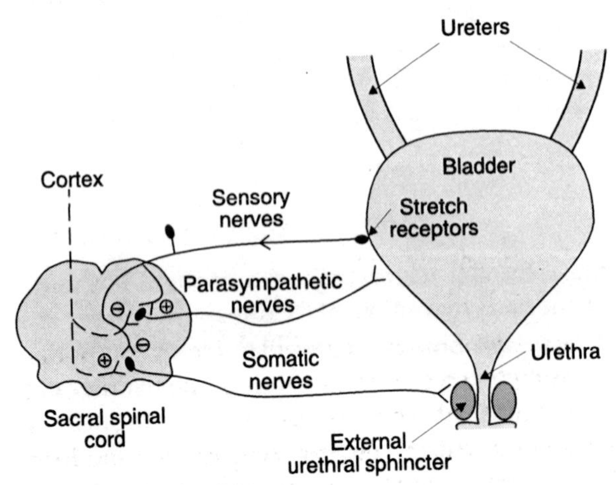

Fig. 88 Neural circuits involved in control of micturition. The reflex parasympathetic control (solid lines) may be overridden by cortical controls (broken lines).

As well as playing a central role in the excretion of waste metabolites and the maintenance of fluid and electrolyte balance, the kidneys also fulfil a number of other important endocrine roles. These are all discussed elsewhere and will only be listed here for completeness.

- The juxtaglomerular apparatus acts as the source of *renin*, which regulates aldosterone activity by activating angiotensinogen. This is important in regulation of total body Na^+ levels and extracellular fluid volume (Sections 5.4 and 8.5).
- The kidneys produce *erythropoietin*, which stimulates erythrocyte production in the bone marrow and thus maintains normal haemoglobin concentrations in the blood (Section 2.3).
- The kidneys act as the major site for activation of vitamin D by hydroxylation to form *1,25-dihydroxycholecalciferol*, or calcitriol, which is responsible for most of the vitamin's activity in the body. This is an important mechanism in body Ca^{2+} regulation (Section 8.4).

5.8 Fluid and electrolyte balance

Learning objectives

At the end of this section you should be able to:

- list and quantify fluid outputs from the body
- list and quantify fluid inputs to the body
- outline how output and intake are regulated to achieve fluid balance.

The excretion of water and dissolved electrolytes in urine is one of the mechanisms regulating the volume and ionic constitution of the extracellular fluid. Such control is important since it determines both the circulating plasma volume and the concentrations of a variety of functionally important ions, e.g., Na^+, K^+, H^+ and HCO_3^-. Also, since the interstitial osmolality (largely determined by the [Na^+]) regulates the osmotic movement of water across the plasma membrane, any abnormality in this variable will lead to a redistribution of fluid between the intracellular and extracellular spaces. For example, loss of water from the extracellular compartment will tend to increase extracellular osmolality, favouring osmotic reduction in cell volume. This will, in turn, alter intracellular conditions and may interfere with normal cell function. Cerebral neurones, for example, are particularly sensitive to osmotic changes. It is desirable, therefore, to maintain a steady state in which the intake of fluid and electrolytes each day exactly balances the losses. The term *fluid balance* refers to the relationship between fluid intake and output from the body. Excess intake represents a *positive* fluid balance while excess loss is a *negative* fluid balance.

Fluid output

Total fluid output can vary widely from a minimum of about 1 L $24h^{-1}$ up to values over 7 L $24h^{-1}$. Fluid loss occurs via the following routes.

- *Urine excretion* normally amounts to 1–1.5 L $24h^{-1}$ but this figure may rise to clear a heavy fluid load or fall if water supplies are limited.
- *Insensible fluid loss* refers to the continuous evaporation from the surface of the skin and airways. Daily losses of this sort total at least 400 ml, although increased ventilation during heavy exercise may double the respiratory component.
- *Faecal excretion* accounts for approximately 100 ml of fluid each day.
- *Sweating* leads to additional fluid and electrolyte losses, separate from insensible skin losses. Sweating is varied as required for core temperature regulation (Section 1.1). With strenuous exercise, hot environmental conditions or during a fever, it can be the dominant fluid output, with daily volumes over 5 L.

Of these elements, only urinary volume can be regulated to help achieve fluid balance through fluid loss or conservation. It should also be noted that total daily fluid output cannot drop much below approximately 1 L. This is made up of insensible losses and faecal excretion (about 500 ml) plus the minimum urinary loss compatible with adequate clearance of toxic waste materials from the body (about 500 ml).

Fluid intake

Water is added to the body fluid compartments by two main mechanisms.

Drinking allows fluid to be ingested and absorbed to replace fluid losses. This is controlled by a thirst centre in the hypothalamus. Not all the absorbed water comes directly from drinking; as much as 50% may be ingested as solids, e.g., fruit and vegetables.

Metabolic production of water is a by-product of substrate oxidation in body cells (Section 4.7), e.g., carbohydrate metabolism. This normally generates about 300 ml of metabolic water each day.

Regulation of fluid balance

Any mismatch between fluid intake and output will lead to fluid accumulation or depletion within the body. This alters the extracellular fluid volume and may change its osmolality, and it is these effects which are detected by receptors, leading to compensatory changes in fluid intake and loss.

Fluid intake

Fluid intake is mainly controlled by altering drinking behaviour. This is regulated through the *thirst centre* in the hypothalamus. Thirst may be stimulated by:

- an increase in osmolality, as detected by hypothalamic osmoreceptors
- a decrease in extracellular fluid volume. This leads to parallel reductions in blood volume and venous return, which reduce cardiac output in accordance with Starling's law of the heart (Section 3.5). The resultant fall in renal perfusion increases renin release from the juxtaglomerular apparatus (Section 5.4). Plasma levels of angiotensin II rise as a result and this stimulates thirst.

Fluid output

Fluid output is regulated through modulation of renal Na^+ and water resorption, as discussed in an earlier section (Section 5.4). In summary, this depends on:

- stimulation of antidiuretic hormone release from the posterior pituitary in response to elevated plasma osmolality or decreased plasma volume, thus increasing water reabsorption
- stimulation of aldosterone secretion through the renin–angiotensin system, leading to increased Na^+/water reabsorption in response to reduced extracellular fluid volume
- stimulation of atrial natriuretic hormone release from the heart, leading to decreased Na^+/water reabsorption in response to abnormal expansion of the plasma volume

Antidiuretic hormone responses are much more rapid than those caused by aldosterone, so isotonic changes in fluid volume (i.e., in which there is no change in plasma osmolality) may not produce any change in renal function for several hours, whereas changes in osmolality will affect urine output within minutes.

5.9 Acid–base balance

Learning objectives

At the end of this section you should be able to:

- define the terms pH, acid, base and buffer
- identify the main pH buffers in the body
- write down the Henderson–Hasselbalch equation and explain its implications for pH control
- explain how respiratory mechanisms affect pH
- explain how renal mechanisms affect pH
- summarize how different acid–base disorders affect arterial blood biochemistry.

The maintenance of a normal pH within the body is very important since enzyme function is highly sensitive to changes in the concentration of H^+. A number of features contribute to the control of pH, particularly acid–base buffer systems, respiratory regulation of P_{CO_2} and renal regulation of $[HCO_3^-]$. These elements will be considered in turn, but first we will revise a few definitions which are relevant to the ideas discussed below.

- $pH = -\log_{10}[H^+]$. Neutral pH is 7.0. The normal pH in the extracellular fluid is 7.4.
- An acid is a substance which can donate protons (H^+) while a base is one which can accept them. This is the Brönsted acid–base model and may be represented by the general reaction:

$$HB\ (\text{Acid}) \rightleftharpoons H^+\ (\text{Proton}) + B^-\ (\text{Base}) \qquad (\text{Eq. 33})$$

Thus, acids and bases occur in pairs, so that for every acid there is a conjugate base and vice versa.

- A strong acid is one which completely dissociates into protons and base in solution, e.g., HCl and H_2SO_4.
- A weak acid is one which only partially dissociates in solution, e.g., carbonic acid. Weak acids in the body exist in chemical equilibrium with their conjugate bases.
- pH buffering depends on chemical reactions which reduce the effect of the addition or removal of H^+ on the final $[H^+]$. Any weak acid and its conjugate base can act as a pH buffer system. If H^+ is added it can combine with the base to form more acid, as predicted by the law of mass action. This removes H^+ from solution and so there is less of a reduction in pH than would otherwise have occurred. Alternatively, if H^+ is removed from the system this will favour dissociation of the acid, releasing protons which limit the increase in pH.

Buffers in blood

There are a number of different acid–base systems which act as important pH buffers in the blood.

Bicarbonate

Carbon dioxide and water react reversibly to form carbonic acid, which then dissociates, releasing H^+ and HCO_3^-.

$$CO_2 + H_2O \rightleftharpoons H_2CO_3 \rightleftharpoons H^+ + HCO_3^- \qquad (\text{Eq. 34})$$

This reaction is catalyzed by carbonic anhydrase in erythrocytes and is very important both in the transport of CO_2 as HCO_3^- (Section 4.4) and for H^+ buffering. If H^+ is added to blood, for example, it reacts with HCO_3^- to form carbonic acid (and hence CO_2), thus buffering the pH. Under physiological conditions the buffering power of HCO_3^- is greatly increased over that predicted for a closed chemical system because the additional CO_2 formed does not accumulate in the blood but is cleared from the body by ventilatory control mechanisms which closely regulate arterial P_{CO_2} (Section 4.5). This allows the chemical reaction to continue towards the left, removing more H^+ from solution. Respiratory and renal control mechanisms can adjust CO_2 and HCO_3^- concentrations to help compensate for abnormalities of acid–base balance. This will be considered in more detail below.

Proteins

The acidic and basic amino acid residues in plasma proteins and red cell proteins act as buffers. In the case of a free carboxyl group, for example, the relevant reaction is:

$$R-COOH \rightleftharpoons H^+ + R-COO^- \qquad (\text{Eq. 35})$$

The haemoglobin in erythrocytes is the most significant proteinaceous buffer in blood and provides more buffering capacity than any other single buffer. Unlike the bicarbonate buffer, however, this system cannot be physiologically regulated to compensate for abnormal acid–base conditions.

Phosphates

Both extracellular and intracellular phosphates act as buffers. Under physiological conditions phosphate makes a much smaller contribution to buffering than either haemoglobin or HCO_3^-. The relevant reaction for inorganic phosphate in plasma is:

$$H_2PO_4^- \rightleftharpoons H^+ + HPO_4^{2-} \qquad (\text{Eq. 36})$$

Effectiveness of pH buffering in blood

The significance of the buffers outlined above is underlined by the observation that addition of 1 mmol of H^+ to a litre of solution at normal arterial pH (7.4) would be expected to reduce the pH to 3.0 in the absence of any buffers. Needless to say, this would be immediately fatal. Acid–base buffering is so effective in the body, however, that adding this amount of acid to each litre of blood would only reduce the pH to 7.38, providing ventilation maintained a normal P_{CO_2}.

Intracellular buffers

Only buffering reactions in blood and extracellular fluid have been considered in the preceding discussion. Within the body there is also the possibility of H^+ uptake into cells and intracellular buffering. With the notable exception of erythrocytes, protons do not readily cross the plasma membrane, so intracellular buffering by other body cells is not immediately effective following addition or removal of acid or base. After a few hours, however, cellular transport and buffering of H^+ or HCO_3^- further reduces any change in extracellular pH. The main intracellular buffers are proteins and organic phosphates and these represent the greatest chemical buffering reserve in the body. The intracellular pH itself varies between cell types but is generally more acid than the extracellular pH, at about 7.2.

The Henderson–Hasselbalch equation

Consider the general acid–base buffering reaction (Eq. 33) restated in the following form:

$$H^+ + Base^- \rightleftharpoons Acid \qquad \text{(Eq. 37)}$$

At equilibrium, the concentrations of the weak acid and conjugate base in this reaction will be related as follows:

$$[H^+]\,[Base^-] = K_a\,[Acid] \qquad \text{(Eq. 38)}$$

where K_a is an equilibrium constant known as the dissociation constant for the acid.

This expression can be rearranged as follows:

$$[H^+] = K_a \frac{[Acid]}{[Base^-]}$$

Taking $\log_{10}$ of both sides and then multiplying by −1 we get:

$$-\log_{10}[H^+] = -\log_{10} K_a - \log_{10}\frac{[Acid]}{[Base^-]}$$

$$\text{i.e., } pH = pK_a + \log_{10}\frac{[Base^-]}{[Acid]} \qquad \text{(Eq. 39)}$$

where $pKa = -\log_{10} K_a$

This equation is known as the Henderson–Hasselbalch equation and if we apply it to the bicarbonate buffer system (Eq. 34) we get:

$$pH = pK_a + \log_{10}\frac{[HCO_3^-]}{[H_2CO_3]} \qquad \text{(Eq. 40)}$$

where K_a is the dissociation constant for carbonic acid.

The concentration of carbonic acid is not easily measured but since it is the product of the reaction between CO_2 and H_2O (Eq. 34), it can be expressed in terms of $[CO_2]$.

$$[H_2CO_3] = K[CO_2]$$

where K is another equilibrium constant.

In turn, $[CO_2]$ is dictated by P_{CO_2}, since the concentration of gas in aqueous solution is proportional to the partial pressure of that gas (Henry's law). So the Henderson–Hasselbalch equation for the HCO_3^- system (Eq. 40) can finally be expressed in the form:

$$pH = pK' + \log_{10}\frac{[HCO_3^-]}{P_{CO_2}} \qquad \text{(Eq. 41)}$$

where K' is a new constant.

One important insight from this relationship is that plasma pH can be controlled by appropriate regulation of $[HCO_3^-]$ and P_{CO_2}. These two variables are affected by different systems, with *respiration* controlling CO_2 levels while *renal* function helps determine $[HCO_3^-]$. Each of these plays an important role in the physiological control of acid–base balance.

Respiratory contribution to pH control

We have already seen that the CO_2 in blood influences pH by forming carbonic acid which dissociates to liberate H^+ (Eq. 34 and Section 4.4). Consideration of the Henderson–Hasselbalch equation as applied to the HCO_3^- buffer system (Eq. 41) emphasizes this point by demonstrating that, providing $[HCO_3^-]$ remains constant, any rise in P_{CO_2} will lead to a fall in pH, while a fall in P_{CO_2} will elevate it. Since the arterial levels of CO_2 are normally dictated by the alveolar P_{CO_2}, respiratory function plays an important role in pH regulation. There are two aspects to this.

Respiratory maintenance of normal pH. The respiratory system maintains a constant arterial P_{CO_2} in the face of varying rates of CO_2 production. This depends both on normal pulmonary gas exchange and an appropriate rate of alveolar ventilation, as controlled by respiratory chemoreceptors sensitive to CO_2 (Section 4.5). Pulmonary removal of CO_2

equates to an acid load of 15000 mmoles of H^+ every 24 hours. This is sometimes referred to as volatile acid because it is produced and removed as gaseous CO_2. Failure of CO_2 removal mechanisms may lead to an abnormal arterial P_{CO_2}, causing an acid–base disturbance of respiratory origin.

Respiratory compensation for abnormal pH. Arterial pH can directly affect ventilation through chemoreceptors sensitive to $[H^+]$ (Section 4.5). This alters P_{CO_2} in such a way as to compensate for pH changes which are not of a respiratory origin, i.e., it provides *respiratory compensation* for metabolic acid–base disturbances. Thus, ventilation is stimulated by a metabolic acidosis, reducing P_{CO_2}, and inhibited by an alkalosis, elevating P_{CO_2}. These compensatory effects take several hours to develop fully.

Renal contribution to pH control

The kidney helps control pH in several ways.

Renal maintenance of normal pH. Tubular secretion of H^+ helps maintain a normal pH despite the continuous production of acid metabolites (other than CO_2, i.e., nonvolatile acid) in the body. This normally results in an acid urine which excretes about 50–80 mmol of H^+ every 24 hours. Although small when compared with the volatile acid released as CO_2, this load is still vital to acid–base homeostasis. Kidney absorption of HCO_3^- occurs in exchange for secreted acid (Section 5.4). This maintains the physiological $[HCO_3^-]$ (25 mmol L^{-1}) which is crucial for pH control since any reduction in $[HCO_3^-]$ favours dissociation of carbonic acid (Eq. 34) and a fall in pH (Eq. 41).

Renal compensation for abnormal pH. Any change in plasma $[H^+]$ leads to parallel changes in the rates of both tubular acid secretion and HCO_3^- absorption. When the $[H^+]$ increases, for example, there is increased proton secretion. The H^+ is effectively exchanged for HCO_3^-, but not all of this is derived from the tubular filtrate. In fact, HCO_3^- exchange can continue through mechanisms involving the generation of additional HCO_3^- within the tubular epithelial cells (Fig. 86B, C), even after all the filtered HCO_3^- has been reabsorbed. In this way, the kidney can raise the $[HCO_3^-]$ during a prolonged (chronic) respiratory acidosis, and this tends to elevate the pH back towards normal. Similar but opposite effects are evoked by a sustained respiratory alkalosis, i.e., H^+ secretion decreases so there is a parallel decline in HCO_3^- absorption, causing plasma $[HCO_3^-]$ to fall. Overall, these mechanisms provide *metabolic compensation* for respiratory acid–base disturbances through appropriate alterations of plasma $[HCO_3^-]$. This compensation may take several days to develop fully.

Biochemical analysis of acid–base status

Acid–base balance may be investigated using biochemical analysis of an arterial blood sample. Several important variables are measured.

pH. The central feature of an acid–base disturbance is an abnormal pH. This is normally 7.4 in an arterial blood sample. A reduced pH indicates an acidosis, while an elevated pH is an alkalosis. Further analysis of the cause of such defects is based on measurement of arterial P_{CO_2}, $[HCO_3^-]$ and base excess.

P_{CO_2}. Any deviation of P_{CO_2} from its normal value (5.3 kPa; 40 mmHg) indicates a respiratory contribution to the acid–base disturbance. Reductions in P_{CO_2} lead to a rise in pH, while increases cause pH to fall (Eq. 41).

$[HCO_3^-]$. Deviations of $[HCO_3^-]$ from normal (25 mmol L^{-1}) are often used as a measure of the metabolic contribution to an acid–base upset. In a metabolic acidosis, $[HCO_3^-]$ falls since any increase in $[H^+]$ from a nonrespiratory (i.e., a metabolic) source will drive the HCO_3^- buffering reaction towards carbonic acid, removing HCO_3^- from solution (Eq. 34). Similarly, a metabolic alkalosis (reduced $[H^+]$) favours release of HCO_3^-, raising its concentration. These effects can be large since any P_{CO_2} generated or removed from the blood will be replaced during alveolar gas exchange, allowing more HCO_3^- to be consumed or generated. Abnormalities in P_{CO_2} itself also have a direct effect on $[HCO_3^-]$; an increase in P_{CO_2}, for example, would be expected to increase the formation of carbonic acid and HCO_3^- (Eq. 34). Although this effect is actually very small in comparison with metabolic and renal changes in $[HCO_3^-]$, it can be avoided altogether by determining the *standard* $[HCO_3^-]$, i.e., the $[HCO_3^-]$ expected if the arterial blood sample had been equilibrated to a normal P_{CO_2} (i.e., 5.3 kPa, 40 mmHg).

The base excess. This is measured by titrating arterial blood samples with a strong base or acid until a normal pH is achieved. The blood is first equilibrated with a physiological P_{CO_2}, so any respiratory contribution to the pH abnormality is removed. Since the resulting value reflects alterations in the total buffer base, only about half of which is actually HCO_3^-, base excess provides a more accurate measure of the metabolic contribution to an acid–base abnormality than does standard $[HCO_3^-]$ alone. Base excess is normally zero, so that a positive base excess indicates alkalosis, while a negative base excess, i.e., a base deficit, indicates a metabolic acidosis.

Box 18 Clinical note: Acid–base disorders

Abnormalities of arterial pH may be divided into acidoses, in which pH is lower than normal, and alkaloses, in which it is elevated (Table 9).

Acidosis

A reduction in arterial pH may be caused by respiratory or metabolic defects.

Respiratory acidosis

An accumulation of CO_2 (arterial P_{CO_2} is elevated) leads to a fall in pH. If respiratory acidosis is prolonged, there may be a secondary rise in $[HCO_3^-]$ and base excess because of increased renal absorption of HCO_3^- in exchange for secreted H^+ (Section 5.4). This increases buffering and helps raise the pH back towards normal (Eq. 41). Metabolic compensation of this sort takes days to develop fully, however, and is a feature of chronic rather than acute respiratory acidosis.

Metabolic acidosis

This is suggested by reductions in plasma $[HCO_3^-]$ and base excess. It results from two main types of problem. Acid may be added to the system, e.g., lactic acid produced by anoxic metabolism (Section 4.7) or ketoacids in diabetes mellitus (Section 8.7). Alternatively, base may be lost from the body, e.g., excessive loss of alkaline gastrointestinal secretions in diarrhoea. Renal failure can produce metabolic acidosis by both these mechanisms, since there is inadequate secretion of acid (equivalent to acid addition) and failure of HCO_3^- reabsorption (loss of base). Prolonged metabolic acidosis may stimulate ventilation via pH-sensitive chemoreceptors (Section 4.5), leading to a reduction in P_{CO_2}. This respiratory compensation tends to elevate the pH, limiting the effect of the metabolic defect.

Alkalosis

An abnormally high arterial pH may also be caused by either respiratory or metabolic defects.

Respiratory alkalosis

This results from increased ventilation leading to a reduction in P_{CO_2}. This may be caused by hysterical overbreathing but is also a feature of a variety of lung diseases, such as pneumonia. Metabolic compensation may occur through decreased renal HCO_3^- resorption, leading to a reduction in plasma $[HCO_3^-]$.

Metabolic alkalosis

This is suggested by an increase in $[HCO_3^-]$ and base excess. It may be caused by loss of acid, e.g., gastric acid loss during persistent vomiting, or addition of base, e.g., excess therapeutic administration of $NaHCO_3$. Over a period of hours, a metabolic alkalosis will depress ventilation, allowing P_{CO_2} to rise and providing respiratory compensation for the metabolic pH abnormality.

Table 9 Summary of abnormalities of arterial biochemistry associated with acid–base disturbances

a. Acidoses (low pH)

Abnormalities					
P_{CO_2}	⇑	⇑	⇑	⇔	⇓
$[HCO_3^-]$	⇔	⇑	⇓	⇓	⇓
Type	Resp	Resp	Mixed	Metab	Metab
Compens.	None	Metab	–	None	Resp

b. Alkaloses (high pH)

Abnormalities					
P_{CO_2}	⇓	⇓	⇓	⇔	⇑
$[HCO_3^-]$	⇔	⇓	⇑	⇑	⇑
Type	Resp	Resp	Mixed	Metab	Metab
Compens.	None	Metab	–	None	Resp

Arrows indicate whether a variable is high (⇑), low (⇓) or normal (⇔). 'Type' indicates whether the primary cause is respiratory (Resp) or metabolic (Metab) and 'Compens.' indicates what kind of compensation, if any, is suggested by the biochemistry.

Self-assessment: questions

Multiple true/false questions

Each of the following statements consists of a stem followed by a number of possible endings. State whether each statement is True or False. For each stem, all, several or none of the statements may be true.

1. In extracellular fluid (ECF):
   - **a.** the total concentration of cations exceeds that of anions
   - **b.** the concentration of Na^+ is an important determinant of osmotic activity at the cell membrane
   - **c.** the ionic compositions of plasma and interstitial fluid are identical
   - **d.** the concentration of K^+ is much lower than it is in intracellular fluid
   - **e.** the greatest single fraction of the pH buffer capacity is provided by protein

2. The nephron:
   - **a.** has endocrine functions
   - **b.** is lined by a single layer of epithelial cells throughout its length
   - **c.** is under endocrine control
   - **d.** lies exclusively within the renal cortex
   - **e.** absorbs more ions and molecules than it secretes

3. The glomerulus:
   - **a.** has both afferent and efferent arterioles
   - **b.** contains capillaries which are at a higher hydrostatic pressure than the peritubular capillaries
   - **c.** filters 20% rather than 10% of the renal plasma flow
   - **d.** contains renin-secreting cells
   - **e.** is lined by fenestrated endothelium

4. Glomerular filtration:
   - **a.** normally occurs at an overall rate of about $120\,ml\,min^{-1}$
   - **b.** pressure gradient equals the difference between the hydrostatic and plasma oncotic pressures in the glomerular capillary
   - **c.** drives fluid into Bowman's space
   - **d.** increases the haematocrit in glomerular capillary blood
   - **e.** normally occurs along the entire length of the glomerular capillary

5. The proximal convoluted tubule:
   - **a.** reabsorbs water using active Na^+ secretion
   - **b.** is lined by epithelial cells rich in mitochondria and bearing luminal microvilli
   - **c.** absorbs glucose and amino acids using carrier systems of the transport maximum (T_{max}) type
   - **d.** absorbs phosphate
   - **e.** only absorbs substantial amounts of fluid when stimulated by aldosterone

6. The loop of Henle:
   - **a.** receives hypertonic fluid from the proximal convoluted tubule
   - **b.** absorbs NaCl and water isosmotically from its ascending limb
   - **c.** generates high osmolalities within the renal medulla
   - **d.** is very permeable to water and electrolytes along its descending limb
   - **e.** receives most of its blood supply from the vasa recta

7. Renin release:
   - **a.** is promoted by reduced arteriolar stretch in the glomeruli
   - **b.** increases when systemic arterial pressure rises
   - **c.** directly activates angiotensin converting enzyme (ACE)
   - **d.** tends to expand the plasma volume
   - **e.** tends to increase renal perfusion

8. Renal clearance values:
   - **a.** always rise as the plasma concentration of the relevant solute rises
   - **b.** for glucose are normally zero
   - **c.** greater than the glomerular filtration rate (GFR) are always indicative of secretion
   - **d.** less than the glomerular filtration rate are always indicative of reabsorption

e. represent the volume of blood which would be cleared of a substance by the kidneys per unit time

9. When considering fluid balance:
 a. the minimum daily fluid output may normally be estimated by measuring the urine output and adding 500 ml
 b. urine output is always the single largest contributor to fluid output
 c. fluid intake and output may be altered in response to a change in extracellular osmolality
 d. fluid intake and output may be altered in response to a change in extracellular volume
 e. the hypothalamus is an important regulator of fluid intake

10. The acid buffering power:
 a. of HCO_3^- in blood is increased by respiratory control of P_{CO_2}
 b. of haemoglobin exceeds that of HCO_3^- in blood
 c. of HCO_3^- is increased by the action of carbonic anhydrase
 d. of venous blood is normally greater than that of arterial blood
 e. of plasma is greater than that of whole blood

11. In arterial blood:
 a. a raised pH with a low P_{CO_2} suggests a respiratory alkalosis
 b. a negative base excess (a base deficit) and a reduced pH is consistent with metabolic compensation for a respiratory acidosis
 c. a metabolic alkalosis may be caused by tissue anoxia
 d. metabolic compensation is a feature of chronic rather than acute respiratory abnormalities
 e. the combination of a reduced pH, elevated P_{CO_2} and elevated [HCO_3^-] is consistent with metabolic compensation for a respiratory acidosis

Single best answer questions

For each of the following questions choose the single best answer.

1. The hydrostatic pressure within the glomerular capillaries:
 a. is greater than the colloid osmotic pressure along the entire length of the capillary
 b. increases when the efferent arteriole constricts
 c. increases when the afferent arteriole constricts
 d. is the sole determinant of the glomerular filtration rate
 e. is greater than the pressure within the afferent arterioles

2. The renal clearance of a substance:
 a. increases as its plasma concentration increases
 b. increases as its urinary concentration increases
 c. increases as the urinary volume increases
 d. increases as its plasma concentration decreases
 e. increases in proportion to total urinary excretion if the plasma concentration remains fixed

3. If a substance is freely filtered and actively absorbed in the nephron:
 a. its concentration in the urine will be less than its concentration in the plasma
 b. its concentration in the urine will exceed its concentration in the plasma
 c. its clearance will exceed the GFR
 d. its clearance will increase in the presence of metabolic poisons
 e. its clearance will decrease in the presence of metabolic poisons

4. Fluid entering the loop of Henle from the proximal convoluted tubule:
 a. has a higher osmolality than plasma
 b. has a higher [Na^+] than plasma
 c. has a higher [glucose] than plasma
 d. has a higher [urea] than plasma
 e. has a higher [HCO_3^-] than plasma

5. Within the distal convoluted tubule:
 a. the osmolality of the tubular fluid is regulated by antidiuretic hormone (vasopressin)
 b. any filtered glucose is absorbed
 c. the osmolality of the tubular fluid is regulated by aldosterone
 d. any filtered HCO_3^- is absorbed
 e. the final composition of the urine is determined

6. Micturition:
   a. is a learned response
   b. requires relaxation of the external urethral sphincter via somatic nerves
   c. requires contraction of the detrusor muscle by sympathetic nerves
   d. depends on spinal reflex arcs coordinated at the sacral level
   e. depends on sympathetically mediated relaxation of the urethral sphincter

7. Aldosterone:
   a. mainly acts on the collecting ducts
   b. promotes K^+ excretion
   c. promotes Na^+ absorption from the proximal convoluted tubule
   d. favours metabolic acidosis
   e. increases urinary osmolality

8. Antidiuretic hormone (vasopressin):
   a. decreases urinary osmolality through the osmotically driven absorption of H_2O
   b. is secreted from the adrenal cortex
   c. is released via the posterior pituitary when plasma osmolality rises
   d. is a steroid hormone
   e. acts at intracellular receptors within the epithelium of the distal convoluted tubule

9. If systemic blood pressure falls by 50%:
   a. glomerular filtration rate decreases by 50% because of the reduced renal blood flow
   b. glomerular filtration rate is maintained within 10% of normal because there is sympathetically driven constriction of the afferent arteriole
   c. glomerular filtration rate is maintained within 10% of normal because there is sympathetically driven constriction of the efferent arteriole
   d. glomerular filtration rate drops to zero because there is sympathetically driven constriction of the afferent arteriole
   e. glomerular filtration rate drops to zero because there is sympathetically driven constriction of the efferent arteriole

10. Urinary excretion of an acid load:
   a. relies on intratubular buffering by HCO_3^-
   b. relies on intratubular buffering by phosphate and ammonia
   c. relies on intratubular buffering by carbonic anhydrase
   d. relies on intratubular buffering by phosphate and glutamine
   e. relies on active secretion of protons in the proximal convoluted tubule

Matching item questions

Theme: Transport mechanisms

Options

A. Glomerulus
B. Proximal convoluted tubule
C. Descending limb of the loop of Henle
D. Ascending limb of the loop of Henle
E. Distal convoluted tubule
F. Collecting duct

For each of the descriptions below choose the most appropriate option from the list above. Each option may be used once, more than once or not at all.

1. Can concentrate urine due to the high osmolality in the renal medulla.
2. Reabsorbs Na^+ by an aldosterone-sensitive mechanism.
3. Pumps NaCl but prevents water movement.
4. Is responsible for 70% of Na^+ reabsorption.
5. Contains fluid with an osmolality similar to plasma throughout its length.
6. Has little active transport but is freely permeable to water and electrolytes.

Theme: Endocrine control

Options

A. Aldosterone
B. Atrial natriuretic peptide
C. Renin
D. Angiotensinogen
E. Vasopressin
F. Erythropoietin
G. Angiotensin II
H. Parathormone

For each of the descriptions below choose the most appropriate option from the list above. Each option may be used once, more than once or not at all.

1. Inhibits Na^+/H_2O absorption in the proximal convoluted tubule.
2. Is released from the juxtaglomerular apparatus when blood volume decreases.
3. Promotes aquaporin insertion in the membranes of the distal convoluted tubule and the collecting duct.
4. Is released by the kidney in response to low tissue O_2 levels.
5. Upregulates Na^+/K^+ and Na^+/H^+ exchange in the distal convoluted tubule.
6. Promotes renal K^+ excretion in the urine.
7. Promotes thirst.

Short notes

Write short notes on the following:

a. the glomerular membrane
b. control of antidiuretic hormone release
c. renal handling of K^+
d. micturition
e. urinary buffers

Modified essay

The formation of urine involves filtration of plasma in the glomeruli with subsequent modification of the filtrate within the renal tubules.

Questions

1. Use a labelled diagram to illustrate the main features of the glomerular filtration membrane. Explain how this is adapted for high filtration.
2. List the forces which affect filtration at the proximal end of the glomerulus, indicating whether each force favours fluid movement into, or out of the capillary. How do these forces differ from their functional equivalents in less specialized capillary systems, e.g., within the skin?
3. Draw a graph showing how the forces listed in Answer 2 vary as you move along the length of a glomerular capillary. Explain any differences between these features in the glomerulus and in general body capillaries?
4. Explain how glomerular filtration is affected by (a) small, and (b) large decreases in arterial pressure?
5. What is the normal glomerular filtration rate in an adult? How is this measured normally?
6. What change would you expect to see in the composition of blood immediately after it has passed through a glomerulus? Would you expect these changes to be reflected in the composition of blood in the renal vein? Explain your answers.

Data interpretation

Question

1. Define renal clearance. What measurements must be made in order to calculate the clearance of any given substance? What formula would you use to make the calculation and what units would the result have?

In a measurement of creatinine clearance, the following results were obtained: Total amount of creatinine excreted in the urine in 24 hours = 15 mmol; plasma creatinine concentration = 100 μmol L^{-1}.

Questions

2. What is the creatinine clearance value in this individual? What aspect of renal function is this a measure of? Why can creatinine clearance be interpreted in this way?
3. If the plasma threshold concentration for glycosuria in this individual is 11 mmol L^{-1}, use the result from Question 2 to calculate the maximum rate of renal glucose reabsorption in mmol min^{-1}. State any assumptions made in your calculation.
4. Calculate the likely glucose clearance in this individual when (a) plasma glucose concentration equals 5 mmol L^{-1}, and (b) plasma glucose concentration equals 20 mmol L^{-1}. (Assume that changing plasma glucose concentration has no effect on GFR.)

Clinical scenarios

Four patients have been admitted to a medical ward during the evening. The house officer is asked to take a history of their complaints and to obtain an arterial blood sample for biochemical analysis. Each is found to have an acid–base disorder.

Patient A is a 57-year-old man who has been depressed for some time. In a suicide attempt early that morning he had swallowed the entire contents of a bottle of aspirin (salicylate). His respiration is appreciably increased and an arterial blood sample gives the following results:

pH = 7.30, P_{CO_2} = 4 kPa (30 mmHg), standard $[HCO_3^-]$ = 16 mmol L^{-1}

Patient B is a 76-year-old woman who is suffering from an acute exacerbation of chronic obstructive airways disease precipitated by a lower respiratory tract infection. She appears centrally cyanosed (Section 4.7). Arterial blood analysis gives the following results:

pH = 7.30, P_{CO_2} = 11 kPa (83 mmHg), standard $[HCO_3^-]$ = 34 mmol L^{-1}

Patient C is a 40-year-old man who is known to be an insulin-dependent diabetic. He was unconscious at admission and is found to be extremely dehydrated. Arterial blood analysis is as follows:

pH = 7.10, P_{CO_2} = 3.3 kPa (25 mmHg), standard $[HCO_3^-]$ = 10 mmol L^{-1}

Patient D is a 16-year-old girl who is complaining of breathlessness of sudden onset. She appears agitated and is breathing very rapidly but there is no other obvious abnormality on examination. Arterial blood analysis gives the following results:

pH = 7.55, P_{CO_2} = 3.3 kPa (25 mmHg), standard $[HCO_3^-]$ = 24 mmol L^{-1}

Questions

Use the information given in each case to identify:

1. the type of the acid–base upset
2. the nature of any compensatory response
3. the likely cause underlying the acid–base upset.

Viva questions

1. Explain how the kidney achieves such high filtration rates. What effect might you expect a decrease in plasma protein concentration to have on filtration?
2. What effects would drinking 500 ml of water have on fluid handling in the kidney? What difference would there be if you drank 500 ml of an isotonic sports drink instead?
3. There is normally no glucose in urine but in the disease diabetes mellitus there is high blood glucose concentration and glycosuria develops. Could you explain why this is?

Self-assessment: answers

Multiple true/false answers

1. a. **False.** These must be equal; the apparent 'anion gap' (Section 5.1) represents unmeasured anions and is not a true anion deficit.
   b. **True.** It is the main determinant since it is the major impermeant cation in extracellular fluid.
   c. **True.** Ions and small solutes readily cross capillary walls.
   d. **True.** This gradient is important in determining the resting membrane potential (Section 1.4).
   e. **False.** HCO_3^- has the greatest buffer capacity in ECF which mainly consists of plasma plus interstitial fluid. Protein provides the greatest single buffer capacity in whole blood because of the contribution here of haemoglobin.

2. a. **True.** For example, renin production.
   b. **True.**
   c. **True.** For example by aldosterone, ADH and atrial natriuretic factor.
   d. **False.** Loop of Henle and collecting duct extend into medulla.
   e. **True.** Particularly NaCl and water.

3. a. **True.**
   b. **True.** Because of the high-resistance efferent arteriole.
   c. **True.** GFR W 120 ml min^{-1}; renal plasma flow W 650 ml min^{-1}; ratio of GFR : RPF = 18%.
   d. **True.** The juxtaglomerular cells of the glomerular arterioles, which form part of the juxtaglomerular apparatus.
   e. **True.** This makes glomerular capillaries very permeable to water and small solutes.

4. a. **True.**
   b. **False.** Also have to subtract pressure in Bowman's space.
   c. **True.**
   d. **True.** Plasma volume decreases so haematocrit increases.
   e. **False.** Increasing oncotic pressure balances hydrostatic gradient before this (Fig. 80).

5. a. **False.** Active Na^+ reabsorption.
   b. **True.** Specialized for active transport.
   c. **True.**
   d. **True.** Also Ca^{2+}; both controlled by parathormone.
   e. **False.** Aldosterone acts on the distal convoluted tubule. There may be some hormonal control of proximal tubular fluid reabsorption (atrial natriuretic hormone inhibits it) but this only affects a tiny fraction of total reabsorption, most of which is obligatory.

6. a. **False.** Absorption in the proximal tubule is isotonic.
   b. **False.** NaCl absorption occurs with little water absorption because of low water permeability.
   c. **True.** By pumping ions but retaining water.
   d. **True.** Allows fluid to equilibrate with medullary interstitium.
   e. **True.**

7. a. **True.** This occurs when renal arterial pressure falls.
   b. **False.** Reduces renin release.
   c. **False.** Renin catalyses angiotensin I formation. This increases ACE substrate levels but the enzyme is already active.
   d. **True.** Through aldosterone-dependent Na^+/water absorption.
   e. **True.** Increased plasma volume leads to increased arterial pressure.

8. a. **False.** They may rise, fall or stay the same.
   b. **True.** Urinary concentration is 0.
   c. **True.** This must be true since the amount appearing in the urine must exceed the amount filtered if the clearance is greater than the GFR.
   d. **False.** This need not be true, e.g., clearance may also be less than GFR because the substance is not freely filtered.
   e. **False.** Clearance is defined in terms of plasma volume, not blood volume.

9. a. **True.** 500 ml is the likely minimum for unmeasured losses, e.g., insensible evaporation and faecal water. Sweating may raise this figure dramatically.
   b. **False.** Often true but sweat volume may be greater.
   c. **True.** Through changes in thirst and ADH secretion.
   d. **True.** Through changes in thirst, and ADH and aldosterone secretion.
   e. **True.** The site of the thirst centre.

10. a. **True.** This prevents P_{CO_2} from rising and so allows further reaction between H^+ and HCO_3^-.
    b. **True.**
    c. **False.** Carbonic anhydrase accelerates the reaction between CO_2 and H_2O to form H_2CO_3 but it does not affect the concentrations at equilibrium and so does not affect the buffering power. It is important, nevertheless, because it ensures rapid buffering.
    d. **True.** Levels of HCO_3^- are slightly higher in venous blood (CO_2 transport; Section 4.4) and deoxyhaemoglobin is a better buffer than oxyhaemoglobin.
    e. **False.** No buffering by haemoglobin.

11. a. **True.**
    b. **False.** A base deficit indicates a metabolic contribution to the acidosis, not compensation for it.
    c. **False.** Anoxia may cause a metabolic acidosis through anaerobic metabolism and lactate production.
    d. **True.**
    e. **True.**

Single best answers

1. b. The increases in outflow resistance raises the upstream pressure in the glomerulus; constricting the afferent arteriole would have the reverse effect. The filtration rate is determined by hydrostatic and colloid osmotic pressures, and by the filtration coefficient of the glomerulus. The high filtration rate through the protein impermeable filtration barrier raises the protein concentration and colloid osmotic pressure within the capillaries until the pressure gradients balance and no further fluid exchange occurs. Glomerular pressure must be lower than afferent arteriolar pressure or no blood flow would occur into the glomerulus.

2. e. Changes in urinary concentration (*U*), urinary volume (*V*), or plasma concentration (*P*) cannot be safely interpreted in terms of clearance changes unless all three variables are known. Increases in urinary excretion ($U \cdot V$) must indicate an increase in clearance if *P* is constant.

3. d. Metabolic poisons, which inhibit cellular ATP production, will inhibit active transport mechanisms in the tubular epithelium. In the proposed case, this would reduce absorption, increasing the rate of excretion and thus the clearance. The concentration in urine will normally depend more on the relative rate of H_2O absorption than the absolute rate of absorption of the substance in question. The substance's clearance would be likely to be less than the GFR, although this assumes there is no active parallel secretory process in the tubules, unless there is also active secretion at another site.

4. d. The [urea] is increased as a consequence of the large amount of Na^+/H_2O absorption in the proximal tubule (about 70% of the filtered load). The resorption of Na^+ causes osmotic absorption of H_2O, this leaves the osmolality and [Na^+] in the proximal tubule similar to that in the plasma and interstital fluid in the renal cortex. All the glucose and HCO_3^- are normally reabsorbed within the proximal tubule.

5. a. Antidiuretic hormone acts on the distal convoluted tubules and the collecting ducts. All filtered glucose and HCO_3^- are normally reabsorbed in the proximal tubule. Aldosterone regulates a small fraction of the total Na^+/H_2O reabsorption but this has little effect on the osmolality of the tubular fluid. The final composition of urine will also depend on the rate of H_2O absorption in the collecting ducts, which are distal to the distal convoluted tubules.

6. d. Micturition is a sacral spinal reflex coordinated present from birth (and in

utero before birth). It depends on parasympathetically mediated contraction of the detrusor smooth muscle. Somatic and sympathetic neurons can stimulate sphincter constriction, not relaxation.

7. b. This is the main mechanism controlling [K^+] in the body. Aldosterone acts on the distal portion of the distal convoluted tubule where it stimulates Na^+/H_2O absorption and K^+ and H^+ secretion. Deficiency favours metabolic acidosis, whereas excess promotes metabolic alkalosis. The effects on urinary osmolality are very small compared to those of antidiuretic hormone (vasopressin).

8. c. Antidiuretic hormone (vasopressin) is secreted in response to high plasma osmolality and promotes absorption of H_2O from the distal convoluted tubule and the collecting ducts down an osmotic gradient by increasing their H_2O permeability. It is a peptide hormone and acts on receptors on the outer cell membrane.

9. d. This is part of the generalized sympathetic vasoconstrictor response to hypotension. Although renal blood flow and glomerular filtration are highly autoregulated, they cannot withstand such a large fall in blood pressure. Although both afferent and efferent arterioles are constricted, it is constriction of the afferent arteriole which reduces glomerular pressure, and thus filtration. Filtration must fall to zero when the mean glomerular pressure falls below the sum of the intracapsular pressure and the colloid oncotic pressure, i.e., approximately 35 mmHg. This is almost certain to be the case when the upstream mean arterial pressure is less than 50 mmHg.

10. b. Intratubular buffering of secreted H^+ by HCO_3^- is crucial for the effective reabsorption of HCO_3^- in the proximal convoluted tubule, but has no effect on total body acid, as CO_2 is reabsorbed into the cell, where it reacts with H_2O, releasing protons. This reaction is catalysed by the enzyme carbonic anhydrase. Excess acid is secreted into the distal convoluted tubule where it is buffered by phosphates and ammonia, the latter being formed from glutamine in the tubular epithelium.

Matching item answers

Theme: Transport mechanisms

1. F. Absorption requires ADH and is driven by the osmotic gradient between the lumen of the ducts and the surrounding tissue fluid. ADH-sensitive, osmotically driven water reabsorption also occurs in the distal convoluted tubules but these are in the renal cortex.
2. E. This accounts for only 2–3% of the filtered Na^+ but is important in the regulation of body Na^+/H_2O.
3. D. This increases medullary osmolality while decreasing the osmolality of the fluid which enters the distal convoluted tubule.
4. B. This is the dominant site for total ion/water/nutrient absorption in the nephron.
5. B. Because reabsorption of NaCl drives parallel osmotic H_2O reabsorption, there is little change in osmolality in the proximal tubule. Osmolality is increased in C, decreased in D and E, and variable in F.
6. C. The descending limb equilibrates with the adjacent medullary fluid.

Theme: Endocrine control

1. B. This may limit the effects of Na^+/fluid overload. Not a major part of normal fluid homeostasis.
2. C. Renin then converts angiotensinogen to angiotensin I, which is converted to angiotensin II by angiotensin converting enzyme. Angiotenisin II promotes aldosterone release, leading to increased Na^+/H_2O reabsorption.
3. E. Also known as antidiuretic hormone. Aquaporin forms water channels, increasing the permeability to H_2O, and thus promoting osmotically driven H_2O reabsorption.
4. F. Stimulates red cell production and so increases [haemoglobin].
5. A. Provides a negative feedback response to reduced blood volume; see Question 2 above.
6. A. Aldosterone secretion is directly stimulated by increases in [K^+]. This provides the main control mechanism for regulation of K^+ levels.

7. G. This also helps to expand the extracellular fluid volume as part of the renin–angiotensin–aldosterone response to fluid loss or depletion.

Short note answers

a. The glomerular membrane separates plasma from Bowman's space within the renal corpuscle. It consists of the glomerular endothelium and an overlying layer of specialized epithelial cells called podocytes. Membrane permeability to water and small solutes is extremely high because of the discontinuities, or fenestrations, in the endothelium and the open spaces between the foot processes of the podocytes where they abut onto the outer aspect of the endothelial basement membrane. This provides for a high filtration coefficient with selective retention of plasma proteins in the glomerular capillaries. (A labelled diagram similar to Fig. 79 would be useful.)

b. Antidiuretic hormone is secreted from the posterior pituitary in response to stimuli affecting the hypothalamus, which is responsible for its manufacture and subsequent axonal transport and release. Release is stimulated by hypothalamic osmoreceptors sensitive to increases in plasma osmolality and inhibited by the sensory output from low-volume stretch receptors in the cardiovascular system sensitive to increases in blood volume. Thus, ADH concentration is increased when plasma osmolality is high or extracellular volume is low and is reduced when osmolality falls or volume rises.

c. Potassium ions are freely filtered in the glomerulus, actively reabsorbed in the proximal convoluted tubule and loop of Henle and actively secreted in the distal convoluted tubule. This last mechanism is controlled by aldosterone, which favours increased K^+ secretion in exchange for absorbed Na^+, and is responsible for regulation of plasma $[K^+]$ (Section 8.5). High plasma $[K^+]$ stimulates aldosterone secretion from the adrenal cortex, and this lowers $[K^+]$ in a classical example of feedback control.

d. Micturition is a sacral spinal cord reflex involving parasympathetically induced contraction of the bladder (detrusor) smooth muscle in response to stretch of the bladder wall as urine volume increases >300 ml. Urine is driven through the urethra by the increased intravesical pressure, and the bladder empties, leaving a residual volume <10 ml. Conscious control is achieved through somatic nerves, which contract the external urethral sphincter, preventing urination.

e. Urinary buffers react with secreted H^+ in the renal tubules, thus limiting changes in urinary pH. The main buffers are:
   - filtered HCO_3^-, which is effectively reabsorbed in exchange for the H^+
   - filtered phosphates such as HPO_4^{2-} (a titratable acid) which are protonated and excreted in the urine
   - NH_3 manufactured in the tubular epithelial cells from glutamine and other amino acids and secreted into the tubular lumen where it reacts with protons to form NH_4^+. This is excreted in the urine.

Modified essay answers

1. A diagram similar to Fig. 79 is required, with the glomerular capillary, basement membrane, podocytes and Bowman's space clearly labelled. The glomerulus is adapted for high filtration rates by having a fenestrated endothelium (high glomerular permeability) and a large surface area.

2. The forces are:
   - the hydrostatic pressure within the glomerular capillary acting outwards
   - the colloid osmotic pressure due to plasma protein acting inwards
   - the hydrostatic pressure within Bowman's space or capsule, opposing filtration from the glomerulus, i.e., acting inwards.

   The capillary hydrostatic pressure is higher than that in general body capillaries, while the colloid osmotic pressure is the same, at the proximal end of the glomerulus at least (see Answer 3 below). The intracapsular pressure in Bowman's space is positive, whereas the extracapillary pressure (interstitial fluid pressure) in most tissues is slightly negative, i.e., subatmospheric.

3. Should be similar to Fig. 80, with appropriate values and units marked. The glomerular capillary pressure is high and remains high mainly because the capillary resistance is low

compared with the resistance of the efferent arteriole at the outflow. The venules which drain general body capillaries have a much lower resistance. You should also comment on the increase in osmotic gradient as you move along the capillary, which is caused by a high rate of fluid filtration combined with a very low permeability to protein. This increases the protein concentration and so reduces the net filtration gradient until no more filtration can occur.

4. (a) Small decreases in arterial pressure have little effect due to well developed autoregulation of renal blood flow. Increases in efferent arteriolar constriction can also help maintain filtration pressures when arterial pressure changes are small. (b) Large decreases in arterial pressure cause a fall in renal blood flow and are associated with sympathetically stimulated vasoconstriction of the afferent arterioles, at the inflow end of the glomerular capillaries. This causes glomerular capillary pressure to drop dramatically, and glomerular filtration decreases, or stops altogether. This is an important feature of hypovolaemic shock.
5. The normal glomerular filtration rate is $120\,\text{ml}\,\text{min}^{-1}$. This is usually measured using the creatinine clearance rate (see Data interpretation question below).
6. Because of the high rate of fluid filtration, there will be an increase in haematocrit and plasma protein concentration. Since GFR is approximately 10% of the renal blood flow, and nearly 20% of the renal plasma flow, haematocrit will be increased by approximately 11% while plasma albumin increases by 22%. The concentrations of ions and other small molecules (e.g., glucose), which are freely filtered, will be unchanged. The glomerular changes in haematocrit and plasma protein are not reflected in the renal venous blood because tubular reabsorption returns the vast majority of filtered fluid to the circulation.

Data interpretation

1. Clearance is the volume of plasma which would be cleared of a given substance by the kidney in a given time. In order to calculate a clearance value, measurements of the total volume of urine voided in a given time (V), and the concentrations of the relevant substance in urine (U) and plasma (P), have to be made.

$$C_x = \frac{U_x \cdot V}{P} \qquad \text{(Eq. 28)}$$

The normal units are those of volume per unit time, e.g., $\text{ml}\,\text{min}^{-1}$.

2. In Equation 28 (above):

$$U \cdot V = 15\,\text{mmol}\,24\,\text{h}^{-1} = 10.4\,\mu\text{mol}\,\text{min}^{-1}$$

$$P = 100\,\mu\text{mol}\,\text{L}^{-1}$$

$$\therefore \text{ Clearance} = \frac{\text{UV}}{\text{P}} = \frac{10.4}{100}(\text{L}\,\text{min}^{-1})$$
$$= 0.104\,\text{L}\,\text{min}^{-1} = 104\,\text{ml}\,\text{min}^{-1}$$

This may be used as a measure of glomerular filtration rate (GFR) because creatinine obeys the necessary criteria for renal handling, i.e., it is freely filtered but is neither secreted nor reabsorbed.

3. Maximum reabsorption (T_{max}) = Rate of filtration at threshold = GFR × Threshold concentration. Using the creatinine clearance as a measure of GFR: $T_{max} = 104\ (\text{ml}\,\text{min}^{-1}) \times 11\ (\text{mmol}\,\text{L}^{-1}) = 0.104\ (\text{L}\,\text{min}^{-1}) \times 11\ (\text{mmol}\,\text{L}^{-1}) = 1.144\,\text{mmol}\,\text{min}^{-1}$. The main assumptions are: that glucose is freely filtered (so concentration in filtrate equals that in plasma); that there is no secretion of glucose adding to the tubular reabsorptive load; and that creatinine clearance equals GFR.
4. a. If rate of filtration < T_{max}, then reabsorption will equal filtration, and excretion ($U \cdot V$) will equal 0. This is the case for all concentrations less than the threshold concentration, $11\,\text{mmol}\,\text{L}^{-1}$ in this case. So, for a plasma concentration of $5\,\text{mmol}\,\text{L}^{-1}$, $\text{Clearance}_{glucose} = 0\,\text{ml}\,\text{min}^{-1}$.

   b. If rate of filtration > T_{max}, then $U \cdot V$ is the amount of glucose excreted in urine, i.e., the amount of glucose filtered less the amount of glucose reabsorbed.

   With a plasma concentration of $20\,\text{mmol}\,\text{L}^{-1}$:

$$\text{Amount filtered} = \text{GFR} \times 20 = 0.104 \times 20 = 2.08\,\text{mmol}\,\text{min}^{-1}$$

$$\text{Amount reabsorbed} = T_{max} = 1.144\,\text{mmol}\,\text{min}^{-1} \text{ (from Question 3.)}$$

$$\therefore 1.144\ U \cdot V = 2.08 - 1.44 = 0.936\,\text{mmol}\,\text{min}^{-1}$$

$$\text{Clearance glucose} = \frac{\text{U} \cdot \text{V}}{\text{P}} = \frac{0.939}{20} = 0.0468\,\text{L}\,\text{min}^{-1} = 47\,\text{ml}\,\text{min}^{-1}$$

Clinical scenario answers

Patient A

1. There is a metabolic acidosis (reduced pH and low standard [HCO_3^-]).
2. There has been respiratory compensation (increased respiration, reduced P_{CO_2}).
3. The cause is gastrointestinal absorption of salicylic acid.

Patient B

1. There is a respiratory acidosis (reduced pH and raised P_{CO_2}).
2. There is metabolic compensation (raised standard [HCO_3^-]).
3. The cause is reduced ventilation caused by chronic obstructive airways disease.

Patient C

1. There is a metabolic acidosis (reduced pH and reduced standard [HCO_3^-].
2. There is respiratory compensation (reduced P_{CO_2}).
3. The cause is likely to be diabetic ketoacidosis caused by insulin deficiency (Section 8.7).

Patient D

1. There is a respiratory alkalosis (elevated pH and reduced P_{CO_2}).
2. There is no metabolic compensation (normal standard [HCO_3^-]).
3. The cause is likely to be hysterical overbreathing.

Viva answers

1. The main items to be covered include the forces responsible for filtration, i.e., glomerular capillary pressure – (capsular pressure + plasma colloid osmotic pressure), and the high filtration coefficient of the capillaries due to their fenestrated structure and large surface area. Capillary pressure is high because of the downstream resistance in the efferent arteriole, so the net pressure gradient favours filtration, with no absorption under normal circumstances. Decreasing plasma protein concentration would decrease the colloid osmotic pressure, exaggerating the net filtration gradient further and increasing filtration rate.

2. Drinking 500 ml of water will lead to a prompt diuresis with increased production of a low osmolality urine. Absorption of the water increases body water and dilutes the body fluids leading to a fall in plasma osmolality which is detected by hypothalamic osmoreceptors. This causes a reduction in the secretion of ADH from the hypothalamus/posterior pituitary and this reduces the H_2O permeability in the distal convoluted tubules and collecting ducts, producing a large volume of dilute urine. The response is activated within 10–15 min.

   Drinking 500 ml of isotonic solution expands the extracellular fluid volume and increases total body solute without affecting osmolality, so there is less affect on ADH secretion. Plasma expansion does reduce activity of the renin–angiotensin–aldosterone system, so reabsorption of salt and water in the distal convoluted tubule is decreased. This results in an increased volume of urine of normal concentration, thus clearing both the water and electrolyte load. The response is much slower, however, and the diuresis may be delayed for 2 hours or more.

3. This is a slightly elaborate way of asking about glucose transport mechanisms. Glucose is freely filtered in the glomerulus and then actively reabsorbed from the proximal tubule. Reabsorption uses a high affinity saturable carrier so all the filtered glucose is reabsorbed up to a maximal load, beyond which any remaining glucose will be lost in the urine, i.e., this is a T_{max} transport system. Since the amount of glucose filtered depends on the glucose concentration in plasma (assuming GFR is relatively constant) glycosuria will occur if the renal threshold concentration is exceeded, i.e., 11 mmol L^{-1}.

   If things are going very well, you may be asked to contrast glucose and Na^+ reabsorption in the proximal tubule. Sodium reabsorption is also active but the load required to saturate the carriers is never reached. Also, because Na^+ can diffuse back into the tubule after it has been transported out, a fixed fraction (about 70%) and not all of the filtered Na^+ is reabsorbed. This is characteristic of gradient–time limited transport, although gradient–time limited mechanisms may also be passive.

Gastrointestinal and digestive physiology

Overview

The main function of the gastrointestinal tract is the provision of the body's nutritional requirements. This involves the mechanical propulsion of food along the gastrointestinal tract, its digestion (the breakdown of long-chain or complex food molecules into simpler units) and absorption of the products of digestion into the blood. Various exocrine secretions are involved in digestion and absorption, while the liver is important for the metabolic processing of the absorbed materials. These features will be considered in turn after a discussion of the nature and regulation of our daily diet.

6.1 Nutrition and appetite

Learning objectives

At the end of this section you should be able to:

- list the main requirements of a healthy diet
- describe the mechanisms responsible for weight control.

Nutritional requirements

These are difficult to define since different individuals have different requirements. A daily requirement figure appropriate for most of the population (e.g., 97.5%) is normally used. Our diet has to provide adequate energy for body metabolism and a supply of amino acids for the maintenance of functional and structural proteins. Energy may be derived from carbohydrate, fat or protein (Section 4.8), but some amino acids cannot be synthesized in the body and these *essential amino acids* must be supplied from digested protein. Thus, basic dietary requirements are usually specified in terms of *total energy* and *protein needs*. This equates to 12.6 MJ (3000 kcal) of energy and 46 g of protein per day for a young adult male with a moderate workload, but these values may be affected by many factors.

- Increased physical activity requires a higher energy intake.
- Pregnancy and lactation increase energy and protein needs in women.
- Age is an important factor. Total dietary requirements increase from birth to young adulthood. When the effect of increasing size is eliminated by expressing energy needs in $MJ\,kg^{-1}$ of body mass, however, values fall throughout life, from birth onwards.
- Disease states can also have dramatic effects, e.g., hyperthyroidism increases metabolic rate (Section 8.3) and, therefore, increases the required energy intake.

Vitamin and *mineral* requirements may also be specified but with even less certainty than energy needs. The dietary requirements of the vitamins folate and B_{12} and of the mineral iron are of particular clinical interest since deficiencies of these are relatively common and can lead to haemoglobin deficiency, i.e., anaemia (Section 2.3). The recommended daily requirements of folate (200 µg) and B_{12} (2 µg) are similar in adult men and women. Women have a higher iron requirement, at 15 mg day^{-1} compared with 10 mg for men. This reflects increased iron loss resulting from menstruation each month.

It may be that other aspects of the diet, e.g., fibre and lipid content, are also important for health. Studies to test such ideas unambiguously are, however, extremely difficult to design and carry out.

Control of eating

Following sustained growth during childhood and adolescence, body weight remains remarkably stable over many years of adulthood. This stability can only be achieved by matching energy intake to energy use. If the food we eat contains more energy than we need then the excess has to be stored, mainly in the form of fat, and weight increases. To maintain a stable weight, therefore, food intake has to be controlled. This is coordinated by the *hunger centre* and the *satiety centre* in the hypothalamus (Fig. 89). Following a meal, distension of the stomach and increased levels of circulating nutrients stimulate the satiety centre which inhibits feeding, while hunger develops as these inputs wane. Fat stores also release a protein hormone called leptin which acts on the hypothalamus to inhibit eating. This provides a feedback system tending to regulate the total lipid content (and thus the weight) of the body. Emotional factors also affect the hypothalamic centres, e.g., anxiety can increase or decrease appetite, and eating disorders, such as anorexia nervosa or bulimia, may reflect abnormalities in central nervous control of their activity.

Box 19 Clinical note: Obesity

Obesity is generally defined as a body mass index (BMI) of more than 30 kg m^{-2}, where:

BMI = (Weight in kg)/(Height in m)^2

Obesity is a serious health problem since it is associated with an increased risk of diabetes mellitus, hypertension, cardiac and joint disease. Although hypothyroidism causes weight gain by decreasing metabolism, most obese people have a normal or slightly increased basal metabolic rate. Abnormal feedback control of lipid store mass is another possible explanation but leptin levels are increased in obesity in proportion with fat content, so if there is reduced negative feedback it must reflect a diminished response to leptin in the hypothalamus. This remains under study but the fact that the number of seriously overweight people in Western society has increased dramatically over the last few decades suggests that an environmental factor is important, with our increasingly sedentary lifestyle being a likely candidate.

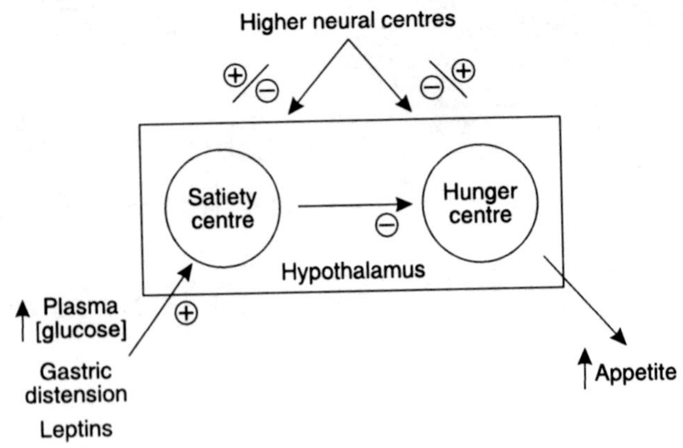

Fig. 89 Summary of factors controlling appetite.

6.2 Gastrointestinal motility

Learning objectives

At the end of this section you should be able to:

- draw a diagram showing the main components of the gut wall
- describe the mechanical activity in each gut region
- outline how gut motility is regulated
- describe the swallowing, vomiting and defecation reflexes.

Food has to be propelled through the gastrointestinal tract from mouth to anus. The time taken for this to occur is referred to as the *transit time*. This can vary from less than 20 hours to over 140 hours and much of this variation appears to depend on the fibre content of the diet (i.e., the amount of indigestible carbohydrate present). A high-fibre diet is usually associated with a short transit time as well as a large fecal mass.

General features of gut motility

The propulsion of gastrointestinal contents depends on the contractile activity of the gastrointestinal smooth muscle, which is generally organized as an *inner circular* and an *outer longitudinal* layer in the intestinal wall (Fig. 90). The electrical activity in this muscle consists of spontaneous waves of depolarization lasting several seconds, which are known as *slow waves* (Fig. 91). These may have action potential spikes superimposed on their peaks. The electrical activity in smooth muscle is dependent on Ca^{2+}, rather than Na^+ as is the case in nerves and striated muscle. The frequency of the slow waves varies in different parts of the gastrointestinal tract, with average values of about $3\,\text{min}^{-1}$ in the stomach, $12\,\text{min}^{-1}$ in the upper small intestine, falling to $9\,\text{min}^{-1}$ in the terminal ileum. Slow waves excite spontaneous contractility, which is modified by the action of nerves, hormones and local factors such as chemical stimulation and mechanical stretch. Nervous control comes both from *extrinsic autonomic nerves* and *intrinsic nerves*, with the latter forming the myenteric and submucosal nerve plexuses within the wall of the gut itself.

Mouth and oesophagus

Chewing is essentially a voluntary activity involving the skeletal muscles of the mouth and jaw. It has the effect of:

- reducing the risk of choking by breaking up large lumps of food
- mixing food with saliva and mucus, thus lubricating it prior to swallowing
- decreasing the particle size of the food, allowing it to be more easily mixed and digested in the stomach and intestine.

Swallowing is divided into three phases. During the *oral phase*, a bolus of food is forced backwards voluntarily with the tongue until pressure on the pharyngeal wall initiates the *swallowing reflex* (Fig. 92). This reflex cannot subsequently be interrupted and is coordinated by the swallowing centre in the medulla oblongata. During the *pharyngeal phase* the soft palate is deflected upwards, sealing off the nasal passages from the pharynx. Muscle contraction pulls the larynx upwards, the glottis closes and the epiglottis deflects the food posteriorly, away from the laryngeal opening. These events protect against aspiration of food into the airways. A travelling wave of constriction (*a peristaltic wave*) drives the food through the relaxed oesophageal sphincters and along the oesophagus itself in the *oesophageal phase* of swallowing. The reflex is controlled via both *somatic nerves*, supplying the striated muscle in the pharynx, larynx and upper oesophagus, and *parasympathetic nerves* innervating smooth muscle in the mid and lower oesophagus. These nerves stimulate contraction of most of the muscles involved, but nerves also stimulate relaxation of the oesophageal sphincters.

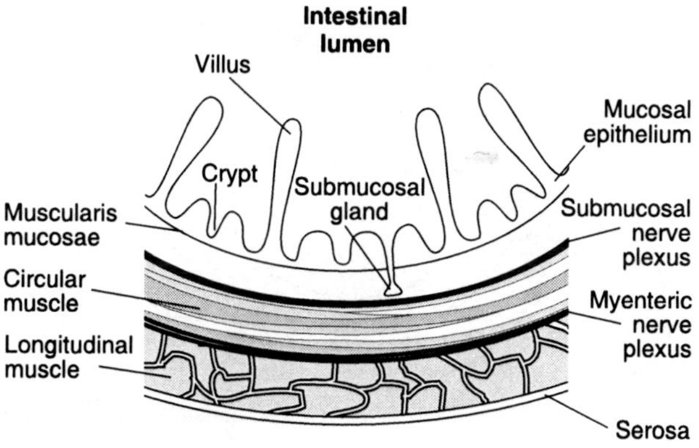

Fig. 90 Transverse section of intestinal wall showing the main structural features.

Box 20 Clinical note: Hiatus hernia

Between meals, a functional sphincter at the lower end of the oesophagus protects it from damage caused by entry of gastric acid. This is assisted by the acute angle of entry between the oesophagus and stomach, which produces a functional flap valve (Fig. 93). Protection is diminished in a condition known as hiatus hernia, in which the gastro-oesophageal junction lies above the diaphragm. One of the main symptoms of this condition is pain (heartburn), which is made worse on bending or lifting. This is caused by acid reflux into the lower oesophagus.

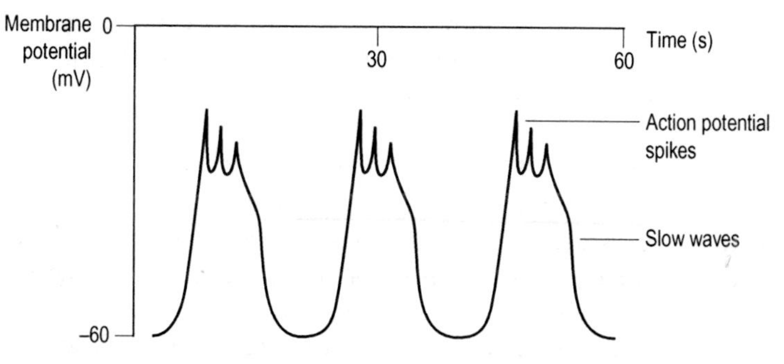

Fig. 91 Spontaneous electrical slow waves and action potential spikes. The frequency is typical of stomach.

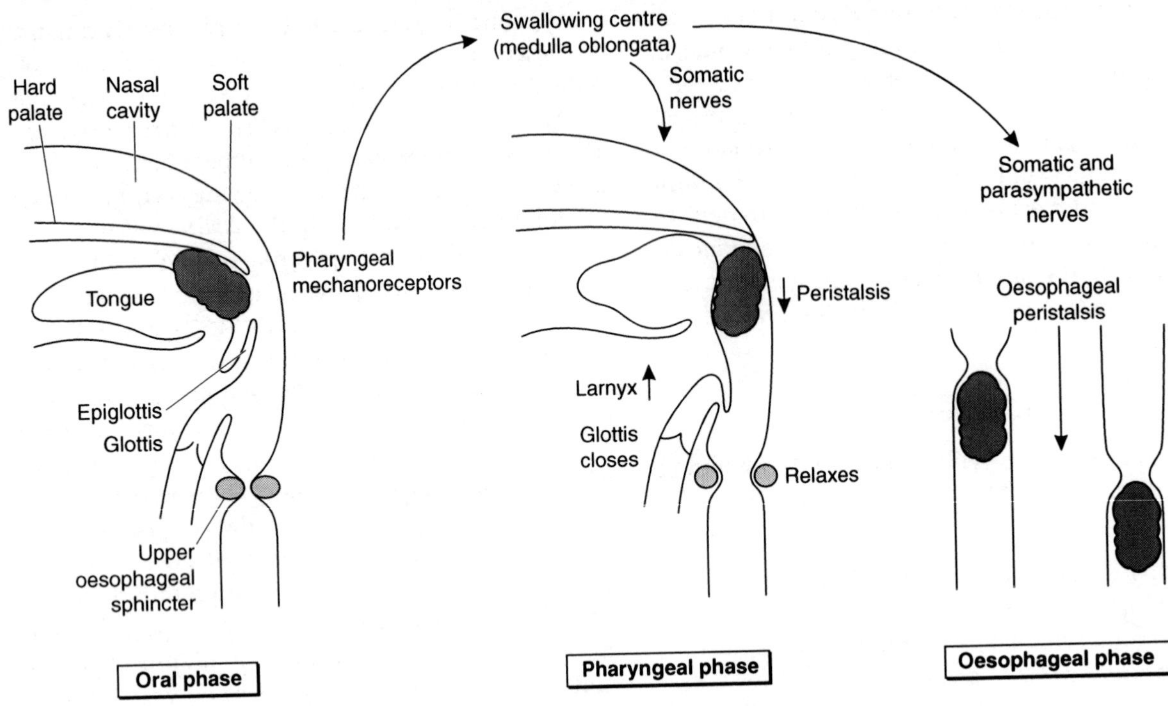

Fig. 92 Main structural and functional elements involved in the three phases of swallowing.

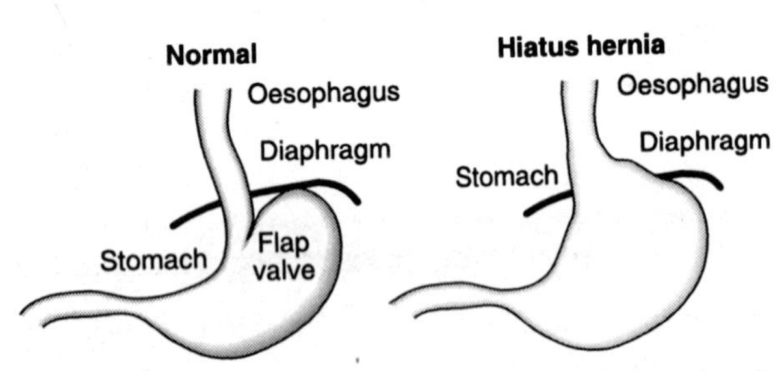

Fig. 93 The gastro-oesophageal junction in a normal stomach and in hiatus hernia.

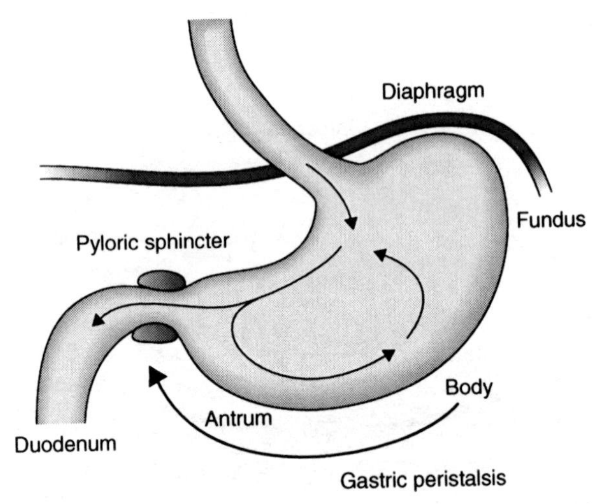

Fig. 94 Mechanical mixing in the stomach is promoted by gastric peristaltic waves and pyloric contraction.

Stomach

From the mechanical point of view, the stomach has two functions.

- It is a *reservoir for food* during meals, a function assisted by parasympathetically mediated relaxation of gastric muscle during swallowing (receptive relaxation).
- It is a *mixing chamber*, which churns food into semiliquid *chyme* and then dispenses it into the duodenum in relatively small volumes in the period following a meal.

Mechanical activity in the stomach consists of regular, peristaltic waves generated within the stomach muscle at a rate of about $3\,min^{-1}$. These spread from the body to the antrum, where the strongest contractions occur (Fig. 94). As the contractions reach the pylorus, the pyloric sphincter closes and prevents excessive emptying into the small intestine. So, although each wave forces some chyme into the duodenum, the bulk is mixed back into the body of the stomach.

Regulation of gastric motility

Regulatory factors influence the strength of gastric contraction more than its frequency. Factors which increase contractility tend to accelerate gastric emptying.

During a meal gastric contractility is stimulated by mechanical distension and increased parasympathetic nervous activity as well as by the hormone gastrin, which is secreted by the G cells in the gastric mucosa.

The presence of acid or fat in the duodenum slows gastric emptying, thus allowing time for pH neutralization and intestinal lipid absorption. This effect may be mediated by release of the hormones cholecystokinin and secretin from the small intestine in response to chyme.

Increased sympathetic nervous activity inhibits gastric motility, e.g., during heavy exercise or following blood loss.

Small intestine

Three patterns of contractile activity are commonly seen in the small intestine.

- The spontaneous mechanical activity in intestinal smooth muscle mainly consists of *segmentation* (Fig. 95). In segmentation there is no travelling wave of contraction. Instead, the muscle contracts simultaneously at regular intervals along the gut wall, thus dividing the lumen of the gut into a series of discontinuous segments and displacing the gut contents. The muscle then relaxes again before contracting at adjacent sites, once more pushing the chyme sideways in both directions. This mixes the intestinal contents and promotes efficient digestion and absorption. The frequency of contraction decreases as one moves distally from the duodenum ($12\,min^{-1}$) to the ileum ($9\,min^{-1}$). The force of contraction is increased in response to parasympathetic stimulation and decreased by sympathetic nerves and circulating catecholamines.
- The main propulsive force in the small intestine comes from localized waves of *peristalsis*. Unlike those in the oesophagus or stomach, these only travel a few centimetres along the small intestine before dying out and so the progress of intestinal contents is slow, particularly immediately after the ingestion of a meal. This helps to ensure that there is adequate time for digestion and absorption.

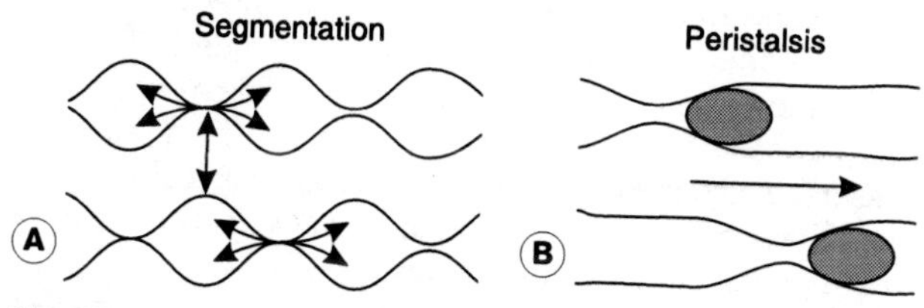

Fig. 95 Mechanical activity in the small intestine. (A) Segmentation. (B) Localized peristaltic waves.

- In the fasted state, i.e., several hours after a meal when absorption is essentially complete, more strongly propulsive *migratory motor complexes* develop. These help sweep the intestinal residue into the large intestine via the ileocaecal valve. Eating suppresses this form of activity, however, with a return to segmentation.

Large intestine

The colon normally maintains a slow rate of segmentation (2–4 contractions h^{-1}). Three to four times a day, however, the large bowel undergoes a contraction known as a *mass movement*. This is a synchronized and sustained contraction of the circular muscle which does not travel as a peristaltic wave would. It often occurs immediately after a meal, possibly triggered by the gastrin-dependent *gastrocolic reflex*.

Mass movements force the colonic contents into the rectum, distending it and thereby initiating the *defecation reflex* (Fig. 96). The sensory output from stretch receptors in the rectal walls stimulates parasympathetic nerves in the *sacral spinal cord* which, in turn, increase the contraction of the colon while relaxing the smooth muscle of the internal anal sphincter. Somatic nerves to the striated muscle of the external anal sphincter are inhibited, allowing it to relax. Rectal distension also gives rise to a conscious awareness of the urge to defecate and in the early years of life we learn to respond to this by voluntarily contracting the external anal sphincter. This allows us to control defecation (*faecal conti-*

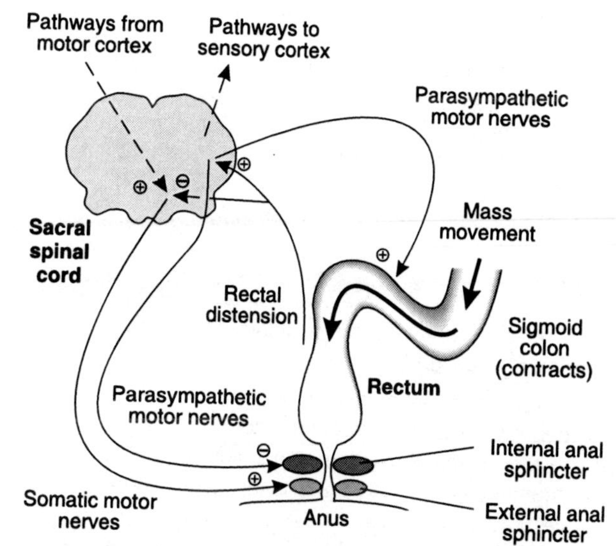

Fig. 96 Summary of the defecation reflex. Both excitatory (+) and inhibitory (–) connections are involved.

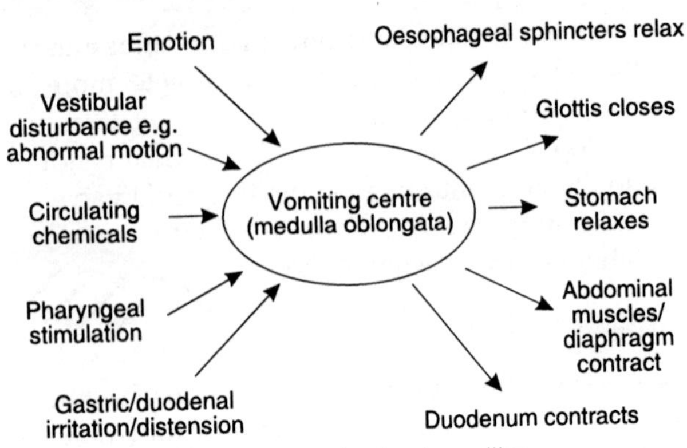

Fig. 97 The inputs and outputs leading to vomiting.

nence). When appropriate, however, the external sphincter is allowed to relax and defecation proceeds. Expulsion is often assisted by using the abdominal muscles and diaphragm to increase intra-abdominal pressure. This is referred to as abdominal straining.

6.3 Digestion and secretion

Learning objectives

At the end of this section you should be able to:

- list the components in exocrine secretions associated with the gut
- outline how secretion is controlled in different regions of the gut
- describe the role played by each secretion in digestion
- explain how bilirubin is excreted.

Digestion is the term given to the processes whereby the complex molecules in food are broken down into simpler subunits which can be absorbed from the gut and metabolized in the body. This largely depends on a variety of secretions produced by the gastrointestinal tract and associated organs. The

Box 21 Clinical note: Abnormal gastrointestinal motility

Vomiting

Vomiting is a complex set of motor functions coordinated by the vomiting centre in the medulla oblongata. The area postrema in the medulla oblongata is particularly implicated as a chemoreceptor zone which can trigger vomiting. The inputs stimulating this centre, and the resulting motor effects, are summarized in Fig. 97. Vomiting is usually associated with the sensation of nausea and is often preceded by sweating, pallor and an elevated heart rate. These are typical signs of sympathetic nervous activity. Although vomiting is preceded by reverse or antiperistalsis which can drive intestinal contents back into the stomach, the vomiting act is generated by an increase in abdominal pressure, which compresses the relatively relaxed stomach. There is little or no contribution from gastric contraction. In vomiting, NaCl, water and H^+ are all lost, so that dehydration and alkalosis may result.

Diarrhoea

Diarrhoea is an increased frequency of defecation. It is commonly caused by:

- increased bowel motility in response to inflammation, e.g., because of a bowel infection
- failure to absorb nutrient molecules from the intestinal lumen, leading to osmotic retention of excess fluid in the intestine
- excess secretion by the intestinal mucosa, e.g., in response to bacterial toxins.

In contrast to vomiting, diarrhoea does not involve coordination by regulatory centres in the brain. Diarrhoea causes loss of K^+ and HCO_3^-, both of which are secreted by the colon. Hypokalaemia (K^+ deficiency) and acidosis may result, although dehydration (loss of NaCl and water) is the main consequence. Diarrhoea and vomiting caused by infected food and water are among the most common causes of infant death in the developing nations of the world.

Constipation

Infrequent or difficult defecation is a common complaint, particularly in the elderly. The underlying cause is often inadequate fibre (roughage) in the diet so that only a small volume of dietary residue enters the large intestine. Rectal stretch is reduced under these conditions so the major sensory stimulus for the defecation reflex is lost. Slow passage through the large intestine also favours increased water absorption leading to further compaction of the faeces. Treatment usually involves the prescription of high-fibre foods and supplements. Constipation may also arise due to repeated voluntary inhibition of the defecation reflex, e.g., in those suffering from a condition such as an anal fissure, which makes defecation painful. This can eventually cause the reflex itself to wane, leading to a more prolonged constipated state.

whole of the gut is lined by exocrine glands and cells, secreting a total volume of about 7–8 L into the lumen each day. Each secretion is a mixture of aqueous and organic components. The *aqueous component* is derived from the extracellular fluid surrounding the secretory cells but differs from plasma because of modifications caused by the actions of a variety of ion pumps and carriers. Since different secretory cells contain different pumps, the composition of the aqueous component varies from gland to gland. Also, in glands with ducts, the secretion may be further modified by the duct epithelium (e.g., salivary glands and pancreas; Fig. 98). Digestive secretions usually also have an *organic component*, often in the form of one or more digestive enzymes manufactured by the secretory cells.

The overall activity of each gland is controlled by a combination of local and distant factors acting through nerves (*nervous control*), blood-borne hormones (*endocrine control*) and locally secreted chemicals (*paracrine control*).

Saliva

Saliva is secreted by the parotid, submandibular, sublingual and buccal salivary glands. The aqueous component is formed by primary secretion of a solution similar to extracellular fluid. This is modified as it passes along the gland ducts, with removal of Na^+ and Cl^-, and addition of K^+ and HCO_3^-. *Bicarbonate* makes saliva alkaline and helps buffer the acid in food, protecting against dental caries. The main organic components of saliva are:

- mucus, for lubrication in eating and speech
- amylase, which digests starch at alkaline pH
- lysozyme, which has antibacterial actions and protects against oral infection.

Secretion of saliva is reflexly stimulated via the salivary nuclei in the medulla oblongata in response to:

- stimulation of chemoreceptors and mechanoreceptors in the mouth
- activity in higher centres of the CNS, e.g., smelling, or thinking about food.

The efferent limb of the reflex is *parasympathetic* and follows cranial nerves VII and IX to reach the glands. Parasympathetic stimulation favours a rapid flow of enzyme-rich saliva. Drugs which block parasympathetic neurotransmission produce a dry mouth by inhibiting salivation.

Gastric secretion

Relevant structure

The inner surface of the stomach is thrown into a series of visible folds, known as *rugae*, which increase the surface area of the gastric mucosa. The mucosal surface consists of a flat layer of mucus-secreting, columnar epithelial cells, but this is interrupted by multiple *gastric pits*, each of which leads down into a number of tubular exocrine glands (Fig. 99). The mucosal cells lining the gastric glands are the source of the gastric secretions, with a number of specialist cells producing separate components. The upper

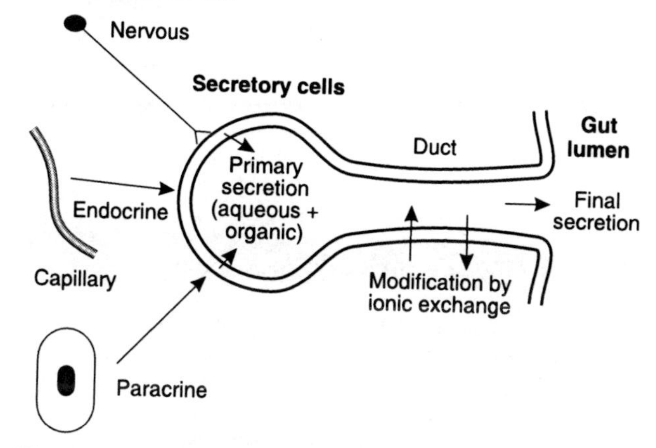

Fig. 98 The principles of exocrine secretion.

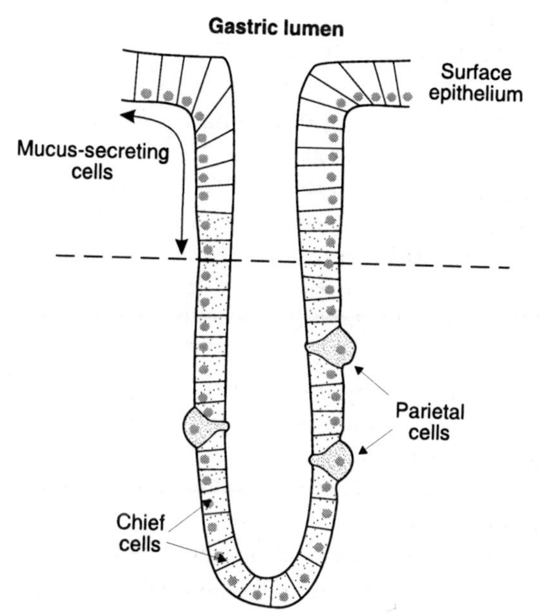

Fig. 99 A tubular gastric gland. Several glands open into each gastric pit. The surface epithelium secretes protective mucus and HCO_3^-.

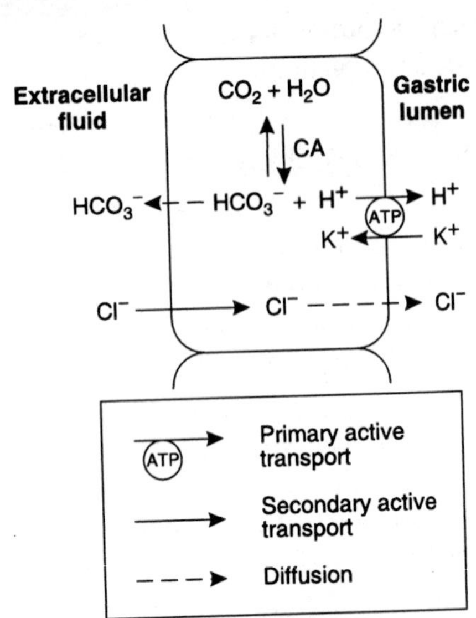

Fig. 100 Summary of the mechanism of HCl secretion in a gastric parietal cell. CA, carbonic anhydrase.

one-third to half is lined by *mucus*-secreting cells. The other two types of exocrine cell, which line the deeper regions of the gastric pits, are the acid-secreting *parietal cells* (or oxyntic cells) and the smaller, more numerous, cuboidal *chief cells*, which secrete the enzyme precursor pepsinogen. The gastric mucosa also contains cells which produce chemicals capable of stimulating acid secretion. The *G cells* produce the peptide hormone *gastrin*, while the *enterochromaffin-like cells* secrete *histamine*, which acts as a paracrine signal stimulating acid secretion in adjacent parietal cells.

Components of gastric secretion

Gastric acid. The parietal cells secrete HCl. This produces a pH of 2–3 within the stomach itself. Active transport of H^+ in exchange for K^+ by an ATP-dependent proton pump (H^+/K^+ ATPase) in the cell membrane facing the gastric lumen (the luminal membrane) is the central mechanism involved (Fig. 100). Proton production in the cell depends on dissociation of carbonic acid formed by the reaction of CO_2 with H_2O in a reaction catalysed by the enzyme carbonic anhydrase (Section 4). This also produces HCO_3^-, which is transported out of the parietal cell and into the blood. Because of this, gastric venous blood is more alkaline than systemic arterial blood.

Intrinsic factor. This is also secreted by parietal cells and binds to vitamin B_{12}, allowing it to be absorbed in the terminal ileum.

Pepsinogen. Pepsinogen is an enzyme precursor secreted by the chief cells. Its release is stimulated in parallel with acid secretion and pepsinogen is activated to pepsin by acid digestion in the stomach. Pepsin is an acid protease, i.e., it breaks dietary protein into peptides and amino acids at acid pH. Pepsin also promotes its own formation by breaking down pepsinogen.

Mucus. Mucus is secreted by the mucus cells of the gastric pits and the surface cells between the pits. This forms an acid-resistant layer over the stomach mucosa, the effectiveness of which is enhanced by the fact that HCO_3^- is secreted into the mucus, making it alkaline. As a result, the epithelial surface remains close to neutral while acid permeates through the mucin to the stomach lumen.

Control of acid secretion

Acid secretion is stimulated in three phases during a meal.

The cephalic phase. This depends on higher centres in the brain acting via the vagus nerve (parasympathetic). It is initiated by the thought, smell, sight or anticipation of food.

The gastric phase. This begins when food enters the stomach. Ingested food molecules, particularly proteins, stimulate both the intrinsic gastric nerves and release of the peptide hormone gastrin. This is secreted into the blood by G cells in the mucosa of the gastric antrum and acts as the main endocrine stimulus to gastric secretion.

The intestinal phase. This starts when chyme enters the intestine and probably depends on gastrin secretion from cells in the mucosa of the duodenum and small intestine.

Parietal cell stimulation At the level of the parietal cell itself, acid secretion can be stimulated by acetylcholine from parasympathetic and intrinsic nerves and by circulating gastrin. There is also paracrine regulation based on the stimulatory effect of histamine, secreted by cells in the gastric mucosa, which acts on H_2 receptors on the parietal cells.

Box 22 Clinical note: Peptic ulceration

An ulcer is a break in the mucosal epithelium and when this occurs in the stomach or duodenum it is referred to as a peptic ulcer. One of the main causes is believed to be mucosal damage by gastric acid but infection with the bacterium *Helicobacter pylori* may

Box 22 Clinical note: Peptic ulceration–cont'd

also contribute in many cases. Current drug treatments aim to neutralize or decrease acid secretion or increase mucosal protection. This may be combined with antibiotics to eradicate *H. pylori*, if present.

- The damaging effects of secreted acid can be reduced using weak alkali suspensions (antacids) or substances which form a protective mucosal coating, e.g., bismuth, alginates.
- The secretory mechanism in the parietal cell can be inhibited using H^+ ATPase inhibitors, e.g., omeprazole.
- Pathways which normally stimulate secretion can be inhibited using drugs which block appropriate receptors. Blockade of muscarinic receptors for acetylcholine do work but have many side-effects due to blockade of other muscarinic effects. Drugs which block H_2 receptors for histamine have been widely used for this purpose, e.g., cimetidine, ranitidine. Surgical division of the vagal branch to the stomach (vagotomy) was previously used to achieve a similar effect.
- Prostaglandin E_1 analogues, such as misoprostol, reduce acid secretion and increase protection of the gastric mucosa.

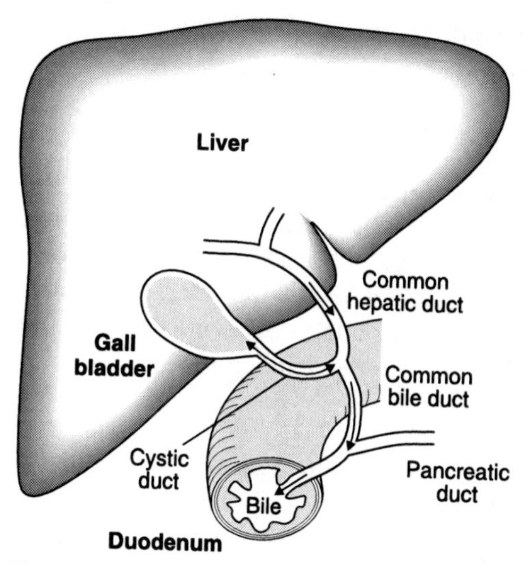

Fig. 101 The anatomy of the main pancreatic and biliary ducts.

Pancreatic secretion

The pancreatic secretions pass from the secretory cells, which are arranged in grape-like clumps termed *acini*, down branch ducts to the main *pancreatic duct*, which fuses with the common bile duct before opening into the duodenum (Fig. 101). The secretion is alkaline; therefore, it neutralizes the gastric acid entering the duodenum. The balance between H^+ secretion in the stomach and HCO_3^- secretion by the pancreas ensures that there is normally no overall change in arterial plasma pH following a meal. Pancreatic secretions sampled close to the acini contain the highest concentrations of HCO_3^-, but as these secretions pass along the pancreatic ducts there is absorption of HCO_3^- in exchange for Cl^-, and pH falls.

The pancreatic acini elaborate and secrete enzymes relevant to the digestion of each main food type.

Proteases and peptidases are secreted from the pancreas as inactive precursors (e.g., *chymotrypsinogen, trypsinogen* and *procarboxypeptidase*) which are only activated (to *chymotrypsin, trypsin* and *carboxypeptidase*) by enzymic degradation in the small intestine. Activation is catalysed by *enterokinase*, an enzyme secreted by the intestinal mucosa. Secretion in inactive form and the presence of trypsin inhibitor prevent autodigestion of the pancreas. The active enzymes catalyse the production of short peptides from proteins in the diet. This is adequate for normal protein digestion even in the absence of any gastric pepsin activity.

Pancreatic amylase digests carbohydrate under alkaline conditions, releasing oligosaccharides.

Lipase and cholesterol esterase produce monoglycerides, free fatty acids and cholesterol from triglycerides and cholesterol esters. Lipid digestion also depends on the emulsifying action of bile salts.

Control of pancreatic secretion

Pancreatic secretion is most strongly stimulated by food entering the small intestine, which triggers release of two intestinal peptide hormones.

Acid in the duodenum stimulates endocrine release of *secretin* which causes secretion of a large volume of HCO_3^--rich fluid.

Food molecules, particularly proteins and lipids, promote release of *cholecystokinin*, which activates enzyme production and secretion.

Parasympathetic nerve stimulation produces a scanty pancreatic secretion which is enzyme rich. This probably contributes to the cephalic and gastric phases of pancreatic secretion, which precede entry of food to the intestine (intestinal phase).

Bile

The aqueous component of bile contains $NaHCO_3$. The main organic components are:

- the bile salts or acids
- the bile pigments
- lecithin
- cholesterol.

This mixture is secreted into the bile canaliculi by the hepatocytes of the liver (see Fig. 113) and then stored and concentrated in the *gallbladder* (Fig. 101). Following a meal, the gallbladder is stimulated to contract by *cholecystokinin*, and bile is expelled through the common bile duct into the duodenum.

Bile salts

Bile salts have a steroid nucleus which is usually conjugated with an amino acid. This makes them *amphipathic* (part water soluble, part fat soluble), allowing them to stabilize fatty emulsions in aqueous conditions. As a result, dietary lipids form an intestinal emulsion of microscopic fat droplets, each droplet being held in its watery environment by a surface coat of bile salt. The bile salt molecules orient themselves with their fat-soluble side towards the lipid core of the emulsified droplets and their water-soluble aspect in contact with the external solution.

Bile salts are important in both the digestion and absorption of lipid.

Emulsification is necessary for efficient *fat digestion* since it maintains a large surface area for lipase action. This relies on the stabilizing effect of the bile salts. In their absence, the triglycerides would simply form a fatty layer floating on the surface, like an oil slick on water, and the rate of digestion would be insignificant.

Lipid absorption requires bile salts to stabilize the fatty acids and monoglycerides released by lipases in the form of even smaller molecular aggregates known as micelles. The products of lipid digestion diffuse out of these to be absorbed by the cells of the mucosal epithelium (Section 6.4).

Enterohepatic recirculation of bile salts Bile salts are reabsorbed in the small intestine, both by passive diffusion and by an active transport system located in the lower end of the ileum (the terminal ileum). Over 80% of the secreted bile salts are absorbed back into the blood in this way and as they pass through the liver they can be resecreted back into bile (Fig. 102). This reabsorption–secretion cycle is referred to as the enterohepatic recirculation of bile salts; it minimizes the amount of bile salt synthesis required.

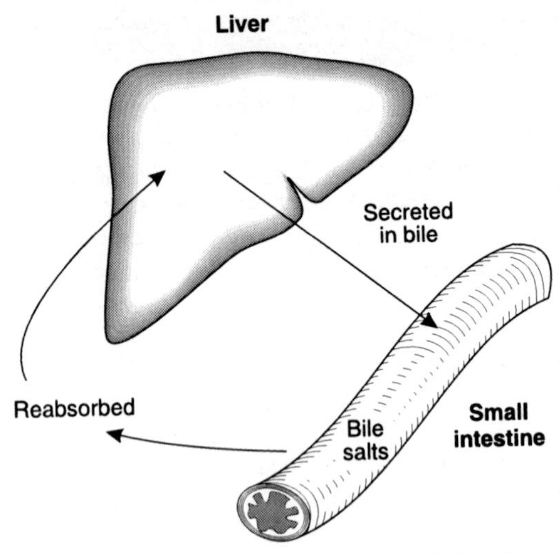

Fig. 102 The enterohepatic recirculation of bile salts.

Bile pigments: bilirubin

Bile pigments give bile its yellow/green colour. The main pigment is bilirubin. This is a breakdown product of the porphyrin subunit released from haemoglobin following lysis of ageing or damaged red cells by the reticuloendothelial system, particularly in the spleen (Section 2.3). Venous blood from the spleen enters the portal vein and so is carried directly to the liver (see Fig. 112). Bilirubin is relatively insoluble in water and has to be transported in plasma in association with albumin. Once within the hepatocytes, bilirubin is conjugated with *glucuronic acid* and this renders it water-soluble prior to its excretion via the bile canaliculi and ducts. Intestinal bacteria convert the bilirubin glucuronides to *urobilinogen*. Some of this is reabsorbed from the intestine and then recirculated to the bile, although a proportion is excreted in the urine (Fig. 103). The urobilinogen which is not reabsorbed is excreted in the faeces, in which form it is sometimes referred to as stercobilinogen.

Cell attached intestinal digestive enzymes

Normal digestion depends on secretion of enzymes into the lumen of the gut, notably pancreatic amylases, lipases and proteases, and gastric pepsin. The cells of the mucosal epithelium also have important digestive functions, particularly in the case of carbohydrate.

Oligosaccharidases. Oligosaccharides are produced in the small intestine through the action of amylase on longer-chain carbohydrate polymers. The intestine cannot absorb these, and further digestion must occur to produce monosaccharides. This is carried out by oligosaccharidases attached to the membranes of the intestinal epithelium (sometimes referred to as the epithelial brush border because of

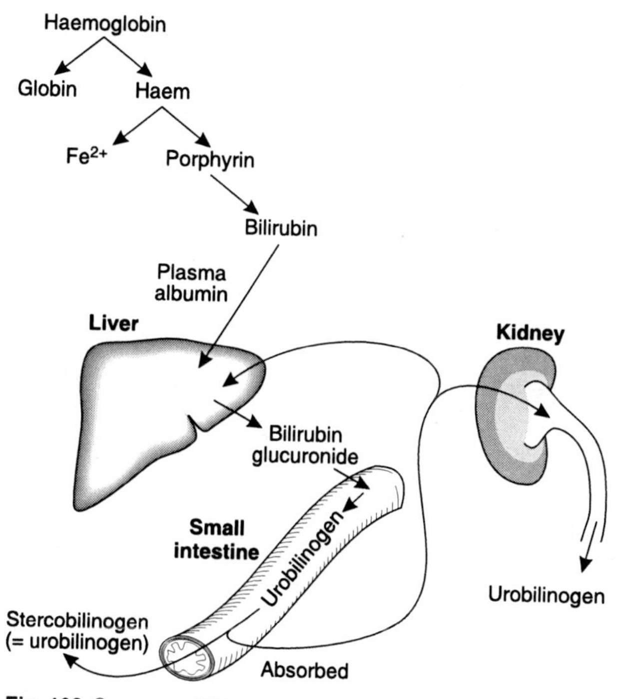

Fig. 103 Summary of bile pigment metabolism.

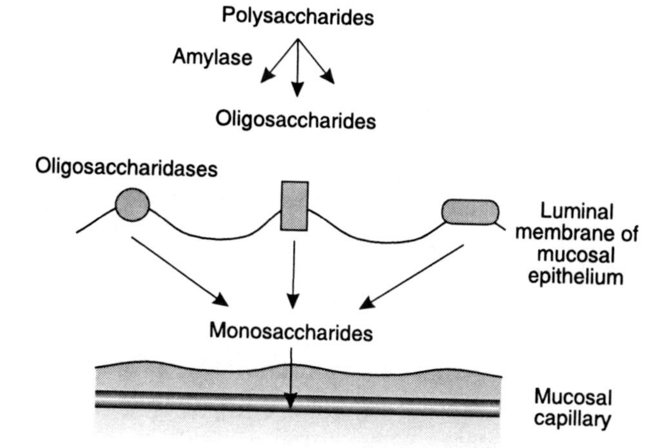

Fig. 104 Role of intestinal oligosaccharidases in carbohydrate digestion.

the multiple microvilli; Section 6.4, Fig. 106). The active sites of these digestive enzymes are exposed at the luminal surface of the mucosa and their action releases monosaccharides, which can then be absorbed (Fig. 104).

Peptidases. Although gastric and pancreatic digestion can convert protein to amino acids, many short peptides are actually digested by peptidases located both on the brush border membrane and within the epithelial cells of the small intestine (Fig. 105).

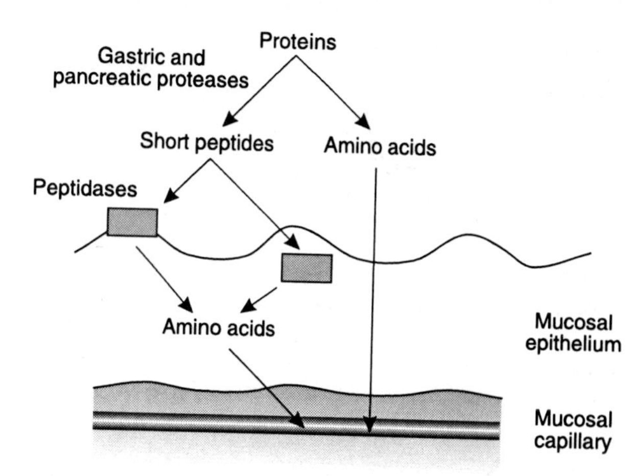

Fig. 105 Role of intestinal peptidases in protein digestion.

Box 23 Clinical note: Jaundice

An elevated plasma bilirubin level leads to a yellow discoloration of the skin, called jaundice. The causes may be classified as:

- pre-hepatic, caused by abnormal haemolysis and excess bilirubin production
- hepatic, caused by hepatocellular damage or defects
- post-hepatic or obstructive, caused by obstruction of the bile duct.

Bilirubin is fat soluble and accumulates in lipid-rich material, e.g., the sclera of the eyes, but has little toxicity in the adult. In the newborn, however, it may be deposited in cerebral neurones causing mental retardation (*kernicterus*). This is unlikely to occur after the first few months of life because of the protection provided by the maturing blood–brain barrier.

6.4 Absorption

Learning objectives

At the end of this section you should be able to:

- describe how the small intestine is specialized for absorption
- outline the absorption mechanisms for water and the major nutrients
- explain the causes and consequences of malabsorption.

Digestion is the enzymatic degradation of nutrient molecules into simpler component molecules. Absorption involves the transfer of these products of digestion, along with minerals, vitamins and ingested water, from the lumen of the gut into the blood. This mainly occurs in the small intestine,

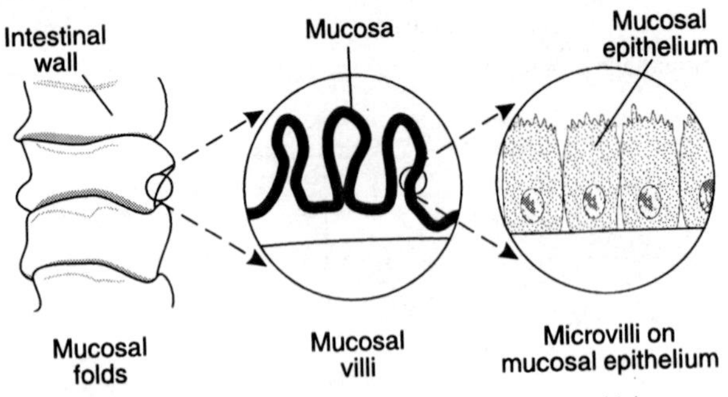

Fig. 106 Structural and ultrastructural specializations which increase the absorptive area in the small intestine.

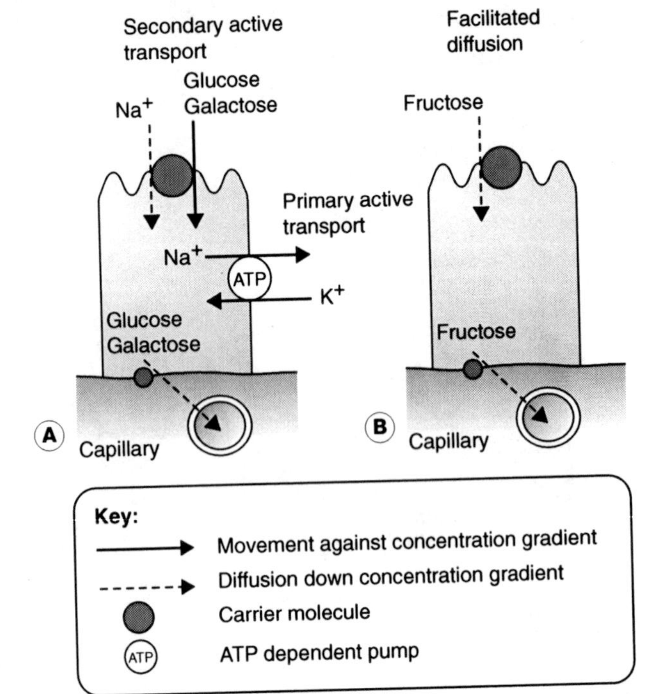

Fig. 107 Carbohydrate absorption mechanisms. (A) Glucose and galactose are absorbed by 2° active transport using Na^+ cotransport systems. Na^+ is removed using the Na^+/K^+ ATPase pump (1° active transport). (B) Fructose absorption is passive but relies on a carrier molecule, i.e., facilitated diffusion.

which is suitably adapted for the task by having both specific transport mechanisms and a very large absorptive surface.

Structural adaptations

The absorptive area of the small intestine is increased by macroscopic *mucosal folds*, microscopic mucosal projections called *intestinal villi* and folding of the epithelial cell membrane to produce *microvilli* (giving rise to the term *epithelial brush border*). The resultant surface area is increased several hundred fold over that of an equivalent plain cylindrical tube (Fig. 106).

Absorptive mechanisms

Carbohydrates

Carbohydrates in the diet are absorbed as glucose, galactose and fructose (monosaccharides) following digestion. Absorption occurs either by secondary active transport, which can carry sugars against a concentration gradient, or facilitated diffusion, which cannot (Section 1.3) (Fig. 107).

Secondary active transport uses diffusion of Na^+ into the cell as an energy source and depends on a *Na^+ cotransport*-carrier molecule in the luminal membrane of the mucosal epithelium. Both Na^+ and the sugar molecule (glucose or galactose) bind to the carrier and, as Na^+ moves from a region of high concentration outside the cell to low concentration inside it, the monosaccharide is carried inside with it. The Na^+/K^+ ATPase uses the energy available from ATP breakdown to pump Na^+ out again (primary active transport), maintaining a low Na^+ concentration inside the cell. The high intracellular concentrations of glucose/galactose allow these nutrients to diffuse out into the plasma in the mucosal capillaries.

Facilitated diffusion requires a diffusion gradient from the intestinal lumen to plasma, the carrier simply acting to decrease the diffusion barrier effect of the fatty cell membrane. Fructose absorption occurs by this mechanism. Both glucose and fructose cross the basolateral membrane by facilitated diffusion before entering the intestinal blood.

Protein

Proteins are broken down into amino acids and short peptides which are absorbed by secondary active transport, again using a *Na^+ cotransport* system (Fig. 108). Intact proteins may be absorbed through endocytosis at the luminal surface and exocytosis at the basal membrane, but this occurs at an insignificant rate after the first few months of life.

Lipids

The action of pancreatic lipase on dietary fat releases free fatty acids and monoglycerides which are initially contained within bile salt-coated micelles in the lumen of the gut. Eventually, these digestion products diffuse into the intestinal epithelium, where they are reconstituted into triglycerides by appropriate enzymes (Fig. 109). These are then packaged within a lipoprotein coat by the endoplasmic reticulum to form *chylomicrons*, which enter the submuco-

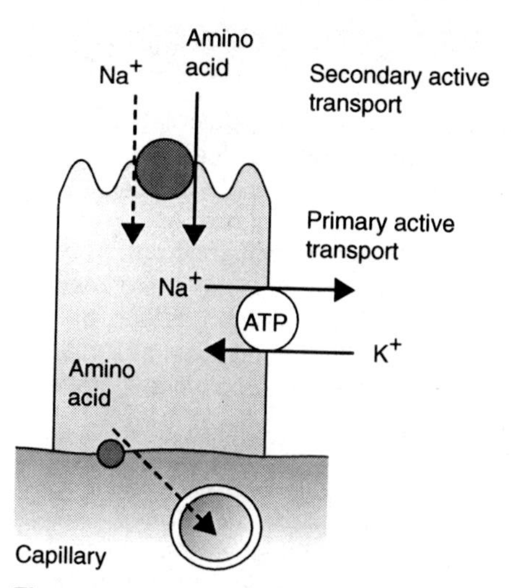

Fig. 108 Amino acids (and some short peptides) are absorbed by 2° active transport using Na^+ cotransport. (Symbols used have the same meaning as in Fig. 107.)

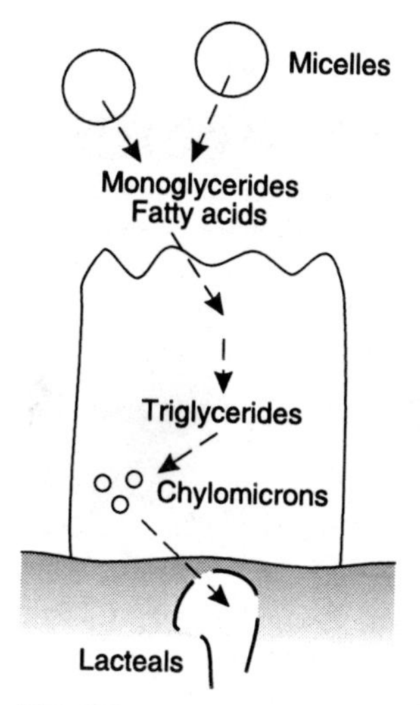

Fig. 109 Lipids are absorbed by diffusion after digestion to monoglycerides and fatty acids. These are reconstituted into triglycerides and packaged as chylomicrons within the cell before entering intravillous lymphatics (lacteals). These have an open endothelium (broken outline) and are, therefore, permeable to larger particles than are blood capillaries.

sal interstitium. Chylomicrons are too large to enter blood capillaries and pass into the *intravillous lymphatics*, or lacteals, instead. They are then returned to the blood via the lymphatic system.

Water and Na^+

Each day, 9 L of water enters the small intestine (1.5 L ingested, 7.5 L secreted). All but 0.5 L is absorbed prior to entry to the large intestine, where another 350 ml is also absorbed. This absorption depends on *osmotic* movement of water, much of which is secondary to absorption of NaCl.

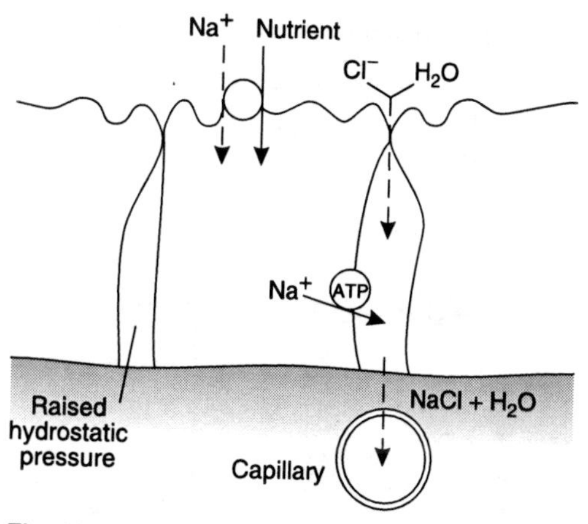

Fig. 110 Transport of Na^+ can drive absorption of Cl^- and osmosis of water from the intestine. This is accelerated in the presence of nutrient molecules (e.g., glucose). (Symbols have the same meaning as in Fig. 107).

When water enters the intestine, it reduces the osmolality of the gut contents and this favours osmotic absorption of water. Even isotonic solutions will be absorbed, however, since active transport of Na^+ out of the epithelial cells into the restricted spaces between adjacent cells is followed by electrostatic movement of Cl^- and osmotic movement of water (Fig. 110). This elevates the hydrostatic pressure in the paracellular space and fluid is forced out via the lowest resistance exit route into the interstitium. It is then absorbed by the blood capillaries. Epithelial cell Na^+ is replaced by diffusion from the intestinal lumen and, as this requires a carrier protein, Na^+ diffusion is enhanced during nutrient absorption involving one of the Na^+ cotransport systems.

The interdependence of water, Na^+ and nutrient absorption is very important in the *oral treatment of dehydration*, e.g., in cases of childhood diarrhoea and vomiting. Both Na^+ and water must be replaced, but Na^+ absorption is poor unless some nutrient is available for cotransport. Feeding appropriate mixtures of NaCl and sucrose dissolved in sterilized water can achieve life-saving results and is considerably more effective than feeding with an isotonic solution containing salt alone.

Vitamins and minerals

Fat-soluble vitamins (A, D, E and K) are absorbed with the lipids in which they are dissolved. Water-soluble vitamins generally diffuse freely across the intestinal mucosa, although B_{12} requires a special complexing protein (*intrinsic factor*) secreted by the parietal cells of the stomach. Absorption of the B_{12}-intrinsic factor complex also depends on a special carrier molecule

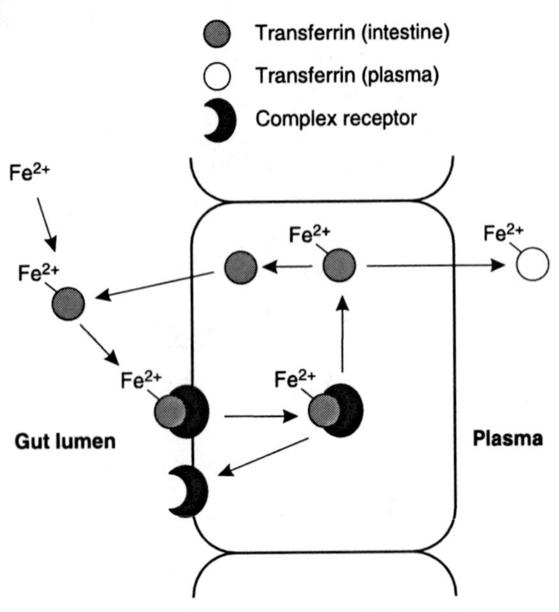

Fig. 111 Outline of iron absorption mechanisms in the upper small intestine.

in the terminal ileum, so that either gastric or ileal damage can cause vitamin B_{12} malabsorption.

Mineral absorption is usually proportional to dietary intake but absorption of iron and Ca^{2+} can be regulated depending on body needs. This is achieved through increases in the availability of carrier sites on specific transport molecules within the mucosa. About 1 mg of *iron*, which is necessary for haemoglobin synthesis, is absorbed each day, although 15–20 mg are ingested. Absorption occurs in the duodenum and jejunum, and ferrous (Fe^{2+}), rather than ferric (Fe^{3+}), ions are absorbed. Absorption relies on binding to a transport protein (*transferrin*) which is then taken into the cell by endocytosis (Fig. 111). This depends on a membrane receptor which attaches to the Fe^{2+}–transferrin complex. Eventually the iron is released across the basal membrane of the mucosal epithelium and is taken up by another form of transferrin in the plasma. In periods of iron deficiency or increased iron demand (e.g., following blood loss) the ability to absorb iron is increased. This is at least partly the result of an increase in the density of membrane receptors for the iron–transferrin complex.

Calcium absorption is dependent on Ca^{2+}-binding protein in the mucosal epithelial cells. The amount of this protein and, therefore, the rate of Ca^{2+} absorption is increased by the rise in vitamin D levels which occur in response to reductions in body Ca^{2+} levels (Section 8.4).

Box 24 Clinical note: Malabsorption

Failure to absorb any component of ingested food is referred to as malabsorption. This may lead to symptoms of malnutrition despite an adequate diet, with weight loss and specific deficiency states, e.g., anaemia or loss of bone mineral (osteomalacia). At the same time, the faecal bulk is increased and diarrhoea may result. Malabsorption may arise from defects of digestion as well as faults with the intestinal absorption process itself, since absorption of carbohydrate, protein and fat requires prior digestion.

Primary absorption defects

These may result from any condition which reduces the intestinal surface area available for absorption. One cause is *coeliac disease*, in which an abnormal sensitivity to the cereal protein gluten leads to destruction of the mucosal villi. Absorption of all food types is affected, leading to weight loss, weakness, anaemia and a wide range of other symptoms.

Digestion defects

These may occur because of a deficiency of digestive enzymes, e.g., because of pancreatic damage. Failure of bile salt secretion may also interfere with digestion, but in this case only the absorption of fat and fat-soluble vitamins is likely to be affected. Fat malabsorption is known as *steatorrhoea*.

6.5 Liver functions

Learning objectives

At the end of this section you should be able to:

- outline the main functions of the liver
- describe the changes in metabolism in fed and fasting states
- explain the hormonal control of metabolism.

Relevant structure

The liver is one of the most important metabolic organs in the body. It is responsible for conversion of nutrient-derived molecules into other substances, as dictated by body needs. This function is aided by a specialization of the vascular system, whereby blood is delivered to the liver not just from the *hepatic artery* but also by a *portal vein*, which transports nutrient-enriched blood from the capillaries of the intestine (Fig. 112). Absorbed molecules are, therefore, immediately transported to the liver where

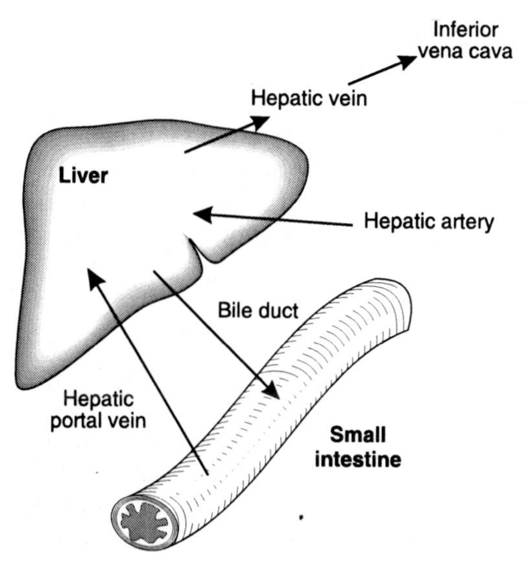

Fig. 112 Vascular and biliary connections of the liver. Note the dual blood supply: from the hepatic artery and the portal vein.

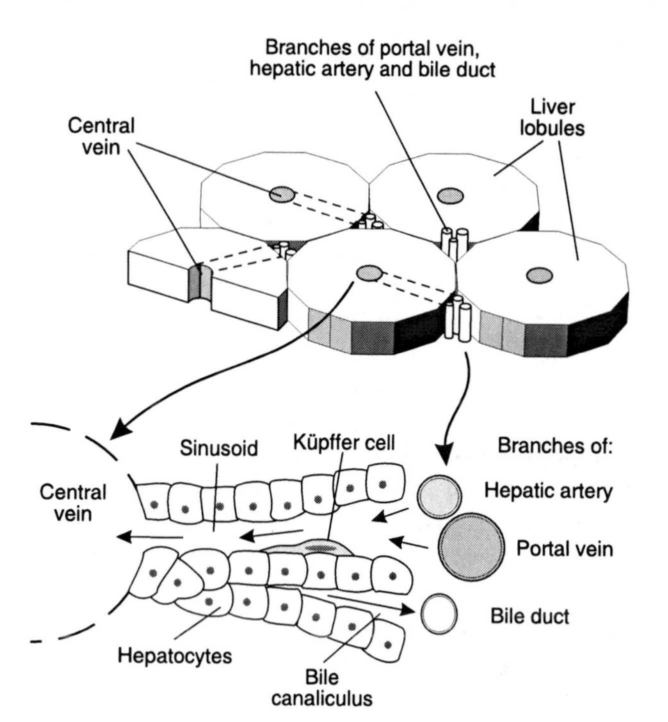

Fig. 113 Outline of hepatic structure. The expanded diagram shows a section across a lobule.

they come into contact with hepatocytes (the liver cells). Mixed hepatic arterial and portal venous blood passes along *hepatic sinusoids* to the *central vein* in each liver lobule, which eventually returns the blood to the inferior vena cava via the *hepatic vein*.

The functional hepatic subunit is the liver *lobule* (Fig. 113). Each lobule consists of cords of hepatocytes radiating out from the central vein and separated by the blood-filled hepatic sinusoids, which connect the central vein with branches of the hepatic artery and the hepatic portal vein. Another set of narrow channels, known as *bile canaliculi*, open into the *bile ducts*. It is by this route that the liver secretes bile into the biliary tree.

Main functions of the liver

- Secretion of bile
- Metabolic activities
- Plasma protein synthesis
- Vitamin D activation
- Detoxification
- Phagocytosis
- Vitamin and mineral store.

Secretion of bile

This is important since bile contains both the bile salts necessary for efficient lipid digestion and absorption and the bile pigments derived from the breakdown of haemoglobin (Section 6.3).

Metabolic functions

The main aspects of hepatic metabolism can be summarized by considering each of the metabolic substrate groups in turn.

Carbohydrate After a carbohydrate meal, there is a rich supply of absorbed glucose, which is converted into *glycogen* for short-term storage within the liver. This glycogen is equivalent to the energy needs for about 12 hours and is converted back to glucose to maintain blood glucose levels between meals. Thus, the liver is an important site of both *glycogen synthesis* and *breakdown*. Which process dominates depends on the hormonal environment, e.g., insulin promotes glycogen synthesis while glucagon promotes its breakdown (glycogenolysis). These hormones are considered in more detail below.

Protein Hepatocytes are involved in several aspects of amino acid and protein metabolism.

Gluconeogenesis. The liver is the main site of gluconeogenesis, i.e., the conversion of amino acids (and other non-carbohydrate molecules) to glucose. Again, this is hormonally controlled and is activated by glucagon some hours after a meal, when glucose levels tend to drop.

Urea production. Amino acid metabolism throughout the body results in the release of ammonia, which is normally converted to the innocuous substance urea in the liver. In liver failure, ammonia can accumulate producing encephalopathy, coma and, eventually, death.

Amino acid and protein synthesis. The liver is an important site of amino acid and protein synthesis.

Lipids A variety of lipid-related molecules are synthesized and broken down in the liver.

Free fatty acids. In the fed state, glucose is converted to free fatty acid within the liver. This is then transported to adipose tissue around the body where it is esterified with glycerol and stored as triglyceride. The fat store normally equates to the energy requirements for approximately 6–8 weeks. During fasting, fatty acids and glycerol are released from the fat stores. The fatty acids are oxidized in the liver, providing the ATP necessary for gluconeogenesis, while glycerol acts as a gluconeogenic substrate.

Lipoproteins and cholesterol. The liver acts as a major site for lipoprotein and cholesterol synthesis. Cholesterol is used in many synthetic pathways around the body, including bile acid production in the liver itself.

Hormonal control of metabolic function The summary above makes it clear that many biochemical reactions in the liver can be grouped into pairs, each being effectively the reverse of the other. We can observe either synthesis or breakdown of glycogen, amino acids and fatty acid. The role of hormones in determining which metabolic reactions will dominate at a given time will be considered in more detail here.

Body cells require a constant supply of energy and substrates for metabolism and synthesis. Our intake of these consumables is discontinuous, however, occurring in the form of meals spaced at intervals of 3–5 hours, or longer. The overall pattern of metabolism can be thought of as see-sawing between two extremes (Fig. 114). In the fed, or *absorptive,* state immediately after a meal, energy supply greatly exceeds demand and the excess is stored for future use in the form of glycogen and lipid. This is followed by the *post-absorptive* state some hours later, when nutrient absorption from the intestine has ceased and energy stores have to be mobilized to meet ongoing demands. This oscillation between periods of energy storage (*anabolism*) and periods of store mobilization (*catabolism*) is under hormonal control. The roles of the most important regulators of metabolic function in the liver, i.e., insulin, glucagon, the glucocorticoids, the catecholamines and growth hormone, will be outlined here. Further details will be found in the chapter on endocrine physiology (Ch. 8).

Insulin. The increased levels of blood glucose during a meal stimulate secretion of insulin by the *β cells* of the *islets of Langerhans*, regions of endocrine tissue in the pancreas. Insulin promotes the removal of glucose from the circulation by cellular uptake and storage as glycogen, or by conversion to fatty acids for transport to lipid stores. Therefore, insulin is an anabolic hormone.

Glucagon. Once glucose absorption from the intestine stops, blood glucose levels tend to drop and this stimulates secretion of glucagon, another pancreatic hormone produced by the *α cells* of the islets. Glucagon favours catabolism of glycogen and fat stores, releasing glucose into the circulation. This effect of glucagon also depends on the presence of normal levels of a group of steroid hormones secreted from the adrenal cortex and known as *glucocorticoids*. As the name suggests, these tend to elevate blood glucose levels and, although their levels do not usually rise in the post-absorptive state, they must be present in adequate amounts if the rising glucagon concentration is to be effective. They are said to have a *permissive role*, i.e., they permit glucagon to exert its physiological effects.

Catecholamines. The adrenal glands also secrete two other hormones which promote a rise in blood glucose, i.e., adrenaline (epinephrine) and noradrenaline (norepinephrine). These come from the adrenal medulla, not the cortex, and belong to a class of compounds known as catecholamines. They are

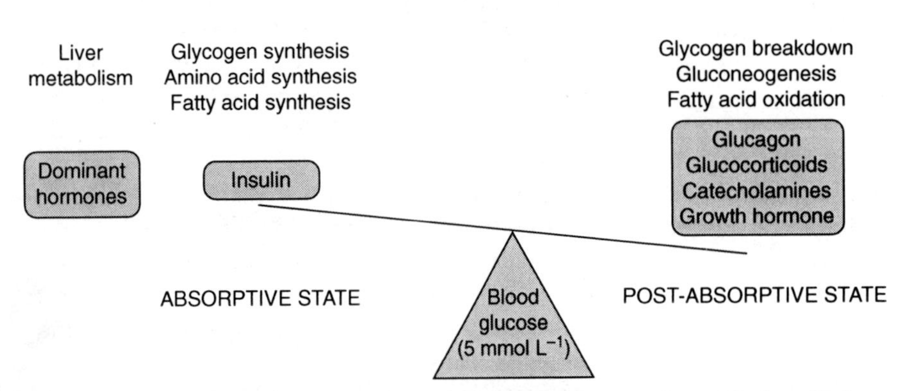

Fig. 114 The changing pattern of biochemistry and hormonal control as the metabolic profile 'see-saws' between the absorptive and post-absorptive states. Overall, these changes maintain a fairly stable blood glucose concentration both during and between meals.

secreted in increased amounts when the adrenal medulla is stimulated via sympathetic nerves. This occurs during physical and emotional stress but is also seen as a response to low levels of blood glucose. The effect of the released catecholamines is to increase glucose production and release in sites of fat and glycogen storage. This provides a backup control mechanism which can augment the normal glucagon/corticosteroid systems if necessary.

Growth hormone is secreted from the anterior pituitary in response to starvation or low blood glucose. Relevant actions include mobilization of fatty acids from lipid stores and the promotion of fatty acid oxidation for energy. There is a parallel decrease in glucose uptake and breakdown in fat and skeletal muscle (a glucose-sparing effect), while hepatic glucose production is increased. All these actions tend to raise the glucose concentration in the plasma.

Integrated hormone action. The overall effect of all these hormonal controls is to limit the variation in blood glucose concentration, which rarely exceeds 10 mmol L^{-1} in the absorptive state and is maintained close to 5 mmol L^{-1} in the fasting state. The latter observation is particularly important for the brain, since neurones are totally dependent on glucose as an energy source. Abnormally low blood glucose levels can lead to abnormal thinking and behaviour, followed by loss of consciousness and death.

Plasma protein synthesis

This includes production of albumin, which controls the plasma oncotic pressure (Section 3.8), as well as a number of other plasma proteins with specialist functions. For example, liver failure may cause bleeding problems because of reduced levels of certain clotting factors (Section 2.6).

Vitamin D activation

Vitamin D is important in regulating the uptake of calcium from the gut. It is absorbed as a fat-soluble vitamin from the diet or manufactured by the action of ultraviolet light on the skin. It then has to be activated in a two-stage hydroxylation process. The first step occurs in the liver and leads to the formation of 25-hydroxycholecalciferol. This is further hydroxylated to form calcitriol (1,25-dihydroxycholecalciferol) in the kidney (Section 8.5).

Detoxification

A number of natural molecules are inactivated in the liver, including steroid hormones and byproducts of protein metabolism. A variety of foreign molecules are also detoxified in the liver. This is a useful protective measure, since toxic substances absorbed from the intestine have to pass through the liver before gaining access to the rest of the body. It can be a problem in therapeutics, however, since drugs may also be degraded by the liver immediately after absorption from the intestine and before they exert any beneficial effect. Some drugs are designed to take advantage of this action, allowing an inactive precursor with reduced digestive side-effects to be taken orally and activated by liver enzymes after absorption.

Phagocytosis

The liver contributes to the removal of both ageing blood cells and foreign cells such as bacteria by means of the *Küpffer cells*. These are a type of phagocytic cell which lines the hepatic sinusoids (Fig. 113), forming part of the reticuloendothelial system (Section 2.4).

Vitamin and mineral storage

As well as glycogen, the liver acts as an important storage organ for a number of substances including iron, copper and vitamins, e.g., the amounts of vitamin B_{12} normally stored in the liver are adequate for approximately 1 year even in the absence of any dietary intake.

Self-assessment: questions

Multiple true/false questions

Each of the following statements consists of a stem followed by a number of possible endings. State whether each statement is True or False. For each stem, all, several or none of the statements may be true.

1. Weight gain could result from:
   - **a.** abnormal increase in output from the satiety centre
   - **b.** abnormal increases in output from the hunger centre
   - **c.** a block in transmission between the satiety centre and the hunger centre
   - **d.** emotional disturbances
   - **e.** a decrease in metabolic rate
   - **f.** damage to the hypothalamus

2. Gastric emptying:
   - **a.** propels chyme into the jejunum
   - **b.** increases when arterial blood pressure is abnormally reduced
   - **c.** is stimulated by parasympathetic nerves
   - **d.** is inhibited when acid enters the duodenum
   - **e.** is stimulated by secretin
   - **f.** can be slowed by muscarinic cholinergic receptor-blocking drugs

3. Faecal incontinence, i.e., loss of voluntary control of defecation, could result from:
   - **a.** damage to the spinal cord
   - **b.** damage to the cerebral cortex
   - **c.** damage to the lumbar sympathetic chain
   - **d.** damage to pelvic parasympathetic motor nerves
   - **e.** damage to pelvic somatic motor nerves

4. Acid secretion in the stomach:
   - **a.** relies on exocytosis
   - **b.** is a function of parietal cells
   - **c.** can be stimulated by gastrin
   - **d.** can be inhibited by histamine
   - **e.** causes the systemic arterial pH to rise significantly during a meal

5. Digestion:
   - **a.** of fat is aided by bile pigment
   - **b.** of carbohydrate requires an acid environment
   - **c.** tends to increase the osmolality of the gut contents
   - **d.** of fat is completed by lipases attached to the intestinal epithelial brush border
   - **e.** of protein largely depends on proteases which are initially secreted as inactive zymogens

6. Intestinal absorption:
   - **a.** of water is a primary active process
   - **b.** of glucose is coupled with sodium transport
   - **c.** of iron is assisted by the low pH in the stomach
   - **d.** is assisted by intestinal segmentation
   - **e.** results in most of the absorbed lipid directly entering the hepatic portal vein

7. Bile:
   - **a.** is produced by hepatocytes
   - **b.** contains urobilinogen
   - **c.** aids fat absorption through the action of bile salts
   - **d.** is concentrated in the gallbladder
   - **e.** is released from the gallbladder into the jejunum in response to the hormone cholecystokinin

8. Liver failure is likely to cause:
   - **a.** an increased tendency to bleed
   - **b.** elevated plasma urea levels
   - **c.** elevated plasma bilirubin levels
   - **d.** peripheral oedema
   - **e.** drowsiness and mental confusion

9. In the absorptive state:
   - **a.** plasma insulin levels are elevated
   - **b.** cell uptake of amino acids is increased
   - **c.** increased oxidation of fatty acids leads to a build-up of ketone bodies
   - **d.** liver glycogen content increases
   - **e.** there is an increased metabolic rate
   - **f.** insulin stimulates uptake of glucose by neurones

Single best answer questions

For each of the following questions, choose the single best answer.

1. Daily dietary requirements:
   - **a.** are usually defined in terms of total energy and protein requirements to ensure fat intake is minimized
   - **b.** are usually defined in terms of total carbohydrate, fat and protein requirements
   - **c.** are usually defined in terms of total energy and fat requirements as protein is usually plentiful in the diet
   - **d.** are usually defined in terms of total energy and protein requirements as dietary protein is the only source of essential amino acids
   - **e.** are usually defined in terms of total fat and protein requirements as this automatically ensures an adequate energy intake

2. Feeding behaviour:
   - **a.** is inhibited if the satiety centre is damaged
   - **b.** is promoted if the hunger centre is damaged
   - **c.** is inhibited by leptins secreted by muscle cells
   - **d.** is promoted by gastric distension
   - **e.** is inhibited by leptins secreted by adipocytes

3. Contraction of intestinal smooth muscle:
   - **a.** is triggered by spontaneous action potentials depending on Na^+ influx
   - **b.** is triggered by spontaneous slow waves depending on Ca^{2+} influx
   - **c.** is triggered by nervous stimulation of slow waves
   - **d.** is triggered by nervous stimulation of action potentials
   - **e.** is triggered by hormonal stimulation of slow waves

4. During swallowing:
   - **a.** a bolus of food is voluntarily pushed backwards with the tongue in the pharyngeal phase
   - **b.** oesophageal peristalsis is a feature of the pharyngeal phase
   - **c.** reflex motor activity is coordinated by a centre in the hypothalamus
   - **d.** the laryngeal opening is protected by upwards movement of the larynx and passive deflection of the epiglottis in the pharyngeal phase
   - **e.** the soft palate prevents food from entering the nasal passages in the oral phase

5. The defecation reflex:
   - **a.** is triggered by stretch of the sigmoid colon
   - **b.** depends on voluntary control of the external anal sphincter
   - **c.** depends on reflex inhibition of the internal anal sphincter by parasympathetic nerves
   - **d.** depends on intact ascending sensory pathways in the spinal cord
   - **e.** depends on reflex inhibition of the internal anal sphincter by somatic nerves

6. Which of the following is *not true* concerning gastric acid secretion:
   - **a.** it is stimulated by acetylcholine acting on muscarinic receptors
   - **b.** it is stimulated by histamine acting on H_1 receptors
   - **c.** it is stimulated by histamine acting on H_2 receptors
   - **d.** it is a function of the parietal cells
   - **e.** it is stimulated by the vagus nerve

7. Jaundice:
   - **a.** associated with haemolysis is caused by elevation of unconjugated bilirubin
   - **b.** associated with bile duct obstruction is caused by elevation of unconjugated bilirubin
   - **c.** associated with haemolysis leads to bilirubin in the urine
   - **d.** associated with bile duct obstruction leads to elevated unconjugated bilirubin in the urine
   - **e.** associated with hepatitis is caused by elevation of unconjugated bilirubin

8. Which of the following is absorbed by facilitated diffusion:
   - **a.** alanine
   - **b.** galactose
   - **c.** fatty acids
   - **d.** fructose
   - **e.** glucose

9. Absorption of fluid from the intestine:
   - **a.** depends on active H_2O transport
   - **b.** ceases when the intestinal fluid is isosmolar with body fluids

c. is increased in rate when both Na^+ and glucose are present, rather than Na^+ alone
d. is dependent on paracellular absorption pathways in the mucosal epithelium
e. is completed within the terminal ileum

10. Hepatic synthesis of glycogen:
 a. is stimulated by glucagon
 b. is inhibited by insulin
 c. is stimulated by growth hormone
 d. is stimulated by adrenaline
 e. is stimulated by cortisol

Matching item questions

Theme: Bile

Options

A. Lecithin
B. Cholesterol
C. Bile salts
D. Bile pigments
E. Bicarbonate
F. Urobilinogen

For each of the descriptions below choose the most appropriate option from the list above. Each option may be used once, more than once or not at all.

1. A phospholipid.
2. Makes bile alkaline.
3. Forms the most common kind of gallstone.
4. Manufactured by intestinal bacteria.
5. Usually secreted after conjugation with glycine or taurine.
6. Usually secreted after conjugation with glucuronic acid.

Theme: Control of gastrointestinal function

Options

A. Gastrin
B. Cholecystokinin
C. Secretin
D. Noradrenaline (norepinephrine)
E. Acetylcholine
F. Enterokinase

For each of the descriptions below choose the most appropriate option from the list above. Each option may be used once, more than once or not at all.

1. Release is stimulated by acid in the duodenum.
2. Promotes secretion of digestive enzymes by the pancreas.
3. Activates proteases by proteolytic cleavage of inactive proenzymes.
4. Stimulates secretion of saliva.
5. Relaxes the sphincter of Oddi.
6. Inhibits intestinal motility.

Short notes

Write short notes on the following:

a. recommended nutritional intakes
b. polysaccharide digestion
c. structure of the gastric glands
d. excretion of bilirubin
e. intestinal absorption of amino acids

Modified essay

Susan, a general practitioner, arrives home one evening. Her surgery ran late and it is long past dinner time so she is feeling extremely hungry.

Questions

1. What mechanisms are responsible for the sensation of hunger?
2. What metabolic reactions are likely to be dominant in the liver in someone who has not eaten for many hours? Which hormones stimulate these reactions?

Too tired to make a proper meal, Susan decides to have some tea and buttered toast before going to bed. She then sits down to eat this while watching TV.

Question

3. What mechanisms stimulate saliva secretion during eating and what benefits does this have?

Susan goes to bed as soon as she finishes her supper. She lies down and soon dozes off.

Questions

4. What will happen to the food in her stomach over the next hour or so?
5. How does the fat in the butter end up as lipid in the bloodstream?
6. What changes in gastrointestinal and pancreatic hormones are likely to occur after eating? What effects do these changes have?

After some time Susan is wakened by a burning pain behind her lower sternum. She recognizes this to be an episode of heartburn, to which she is rather prone since she has a hiatus hernia. She gets some extra pillows to lie against and eventually falls asleep once more.

Questions

7. Why should going to bed shortly after a meal tend to produce indigestion in a hiatus hernia patient?
8. What benefit is there in using extra pillows?

Data interpretation

In an experiment on the hormonal control of pancreatic secretion, a normal subject had a small tube inserted into the main pancreatic duct so that all pancreatic secretion could be collected and analyzed for its ionic and enzymatic content. The results have been graphed (Fig. 115) to show how (i) the rate of secretion, (ii) the bicarbonate concentration of the secretion, and (iii) the amylase concentration of the secretion varied with time following intravenous injections with the hormones secretin and cholecystokinin (CCK). Secretory rates and concentrations returned to their initial resting levels during the break in collections indicated by the cross lines. Table 10 shows the recorded values under control conditions and for the first sample after each injection.

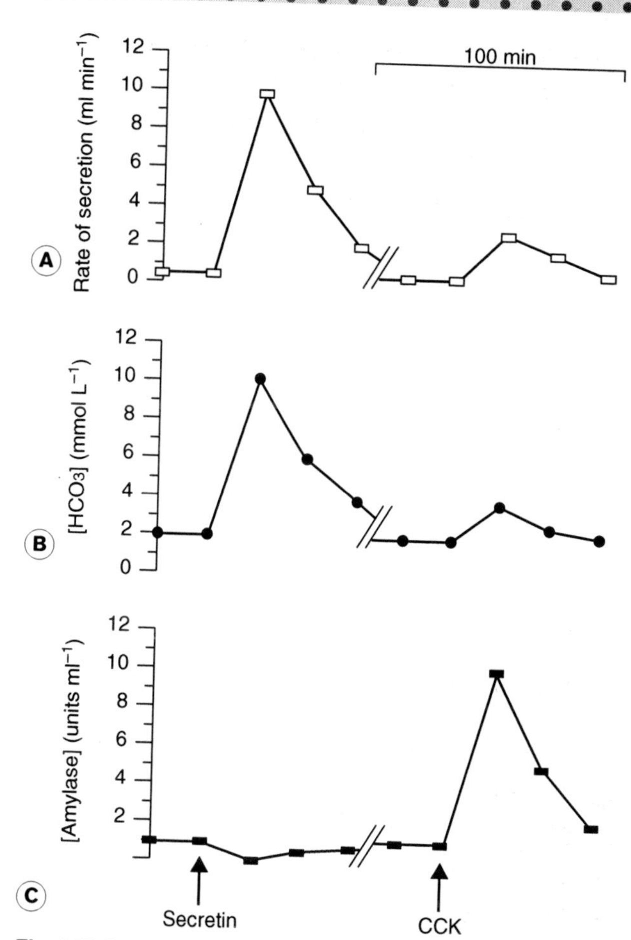

Fig. 115 An experiment was carried out on the control of pancreatic secretion of a normal individual. Changes in the total rate of secretion (A) and the HCO_3^- (B) and amylase (C) concentrations of the pancreatic secretion are shown following intravenous injections with secretion and CCK (marked with arrows). Some of these data have also been tabulated (Table 10).

Questions

1. What is the main gastrointestinal site of secretion for CCK and secretin? What stimulates this secretion?
2. What is the chemical nature of CCK and secretin? Why were they given intravenously, rather than orally, in this experiment?
3. What effects do the two hormones have on the volume of secretion? What about HCO_3^- secretion?
4. What conclusions can be drawn from the data given about the effects of these hormones on the rate of amylase secretion by the pancreatic acinar cells?
5. Discuss the functional benefits of the different responses described in your answers to Questions 3 and 4 in terms of the stimuli which lead to secretion of these two hormones.

Table 10 Values for flow rate, [HCO_3^-], and amylase concentration for pancreatic secretion under control conditions and for the first sample after injections of secretin and cholecystokinin (CCK). (See text and Fig. 115 for fuller details.)

Condition	Rate ($ml\,min^{-1}$)	[HCO_3^-] ($mmol\,L^{-1}$)	[Amylase] ($units\,ml^{-1}$)
Control	0.5	70	1
Post secretin	10	110	0.1
Post CCK	3	80	10

6. Which of the measured variables in this experiment appear to be positively correlated regardless of which hormone stimulates secretion?
7. If HCO_3^- is removed from the pancreatic secretions as they travel along the pancreatic duct, how might this correlation be explained?
8. What other major control of pancreatic secretion is there? What is its effect?

Clinical scenario

A 45-year-old woman comes to her doctor complaining of recent weight loss despite normal appetite. She has no other symptoms of hyperthyroidism. On questioning she reports that her stools have become very pale, float in the toilet, have an extremely bad smell, and are difficult to flush away. Physical examination reveals anaemia and widespread bruising. A tentative diagnosis of malabsorption is made.

Questions

1. Abnormalities of which two gastrointestinal functions may result in malabsorption?
2. List three main organs/processes, defects of which may give rise to such abnormalities. What effects would the defects described have in each case?
3. Explain the appearance of the stools in this case.
4. How might the anaemia have arisen?
5. Why do you think this woman bruises so easily?
6. What other possible complications of malabsorption should be tested for?

Viva questions

1. How is acid secreted in the stomach? What mechanisms control secretion?
2. Describe the main events that take place when you swallow food.
3. Tell me something about the roles parasympathetic nerves play in controlling gastrointestinal function.

Multiple true/false answers

1. a. **False.** This inhibits feeding.
   b. **True.**
   c. **True.** As a result of loss of inhibition of the hunger centre by the satiety centre.
   d. **True.** Can also cause weight loss.
   e. **True.** This reduces energy consumption.
   f. **True.** This is the site of the satiety and hunger centres and selective damage of these could lead to weight gain or weight loss, respectively.

2. a. **False.** The pylorus opens into the duodenum.
   b. **False.** The increased sympathetic nervous activity associated with low arterial pressure would reduce gastric contractility.
   c. **True.**
   d. **True.** Via intestinal hormones.
   e. **False.** Secretin and cholecystokinin inhibit emptying.
   f. **True.** Parasympathetic nerves stimulate muscarinic receptors (Section 7.6); blocking these will remove autonomic nervous stimulation of motility.

3. a. **True.** Spinal transmission of the sensory urge to defecate and voluntary motor signals controlling the external anal sphincter are blocked.
   b. **True.** The cortex is responsible for sensory awareness of the urge to defecate and initiating motor control of the sphincter.
   c. **False.** Sympathetic nerves are not involved in defecation.
   d. **False.** This would inhibit the defecation reflex itself, leading to constipation. It would not interfere with voluntary control.
   e. **True.** Results from loss of motor control of external anal sphincter.

4. a. **False.** Depends on transmembrane proton pumping.
   b. **True.**
   c. **True.** Secreted by gastric G cells.
   d. **False.** Stimulates acid secretion.
   e. **False.** pH rises in gastric veins but, overall, pancreatic HCO_3^- secretion balances gastric H^+ secretion after a meal.

5. a. **False.** Bile pigments are waste products; bile salts assist in fat emulsification and micelle formation.
   b. **False.** Pancreatic amylase requires alkaline conditions.
   c. **True.** Breakdown of complex molecules increases the total number of solute molecules.
   d. **False.** Intestinal digestive enzymes are only important for oligosaccharides and peptides.
   e. **True.** Inactive precursors are stored as zymogen granules in gastric parietal and pancreatic acinar cells.

6. a. **False.** It occurs passively as a result of osmotic forces developed secondary to active transport of sodium.
   b. **True.**
   c. **True.** Iron is absorbed from the duodenum in the reduced ferrous (Fe^{2+}) rather than the more oxidized ferric (Fe^{3+}) form. The acid conditions of the stomach promote reducing reactions.
   d. **True.** This promotes mixing.
   e. **False.** Carbohydrates and amino acids enter the portal vasculature but chylomicrons enter the lymphatic system.

7. a. **True.**
   b. **True.** This is resecreted by the liver after absorption into the plasma from the intestine.
   c. **True.**
   d. **True.**
   e. **False.** Cholecystokinin stimulates release but the common bile duct opens into the duodenum.

8. a. **True.** Reduced levels of clotting factors.
   b. **False.** Liver normally manufactures urea so plasma concentration tends to fall in hepatic failure.

c. **True.**

d. **True.** Reduced plasma protein concentration causing reduced plasma oncotic pressure.

e. **True.** This hepatic encephalopathy is caused by a build-up of toxic substances, e.g., ammonia.

9. a. **True.** Stimulated by the high levels of absorbed glucose.

b. **True.** This is one of insulin's actions (Section 8.7).

c. **False.** This is more typical of the post-absorptive state, especially during prolonged fasting, when insulin levels are low and glucagon levels are elevated.

d. **True.** Insulin stimulates hepatic glycogen synthesis from glucose.

e. **True.** This is called the specific dynamic action of food.

f. **False.** Insulin stimulates glucose uptake by muscle cells, but neuronal glucose uptake is independent of insulin.

Single best answers

1. d. Essential amino acids are those which cannot be synthesized by the body. 'High quality' dietary protein contains a high proportion of these amino acids.

2. e. This may provide a negative feedback system helping to limit the size of the body fat store.

3. b. This activity may be depend on specialized pacemaker cells known as interstitial cells of Cajal.

4. d. These activities are coordinated by the swallowing centre in the medulla oblongata.

5. c. This allows the faeces to be expelled from the rectum. Spinal sensory pathways and voluntary control of the external sphincter are important elements in continence of faeces but not for the defecation reflex itself.

6. b. H_1 receptors are stimulated by histamine released from mast cells in response to certain allergens and activate a variety of inflammatory responses. H_2 receptors on parietal cells stimulate acid secretion.

7. a. Haemolytic jaundice is caused by increased levels of unconjugated bilirubin but this is insoluble in water and so is not filtered into the urine. Obstructive jaundice causes an increase in the levels of conjugated bilirubin, which does appear in the urine. Hepatitis commonly causes an increase in levels of both conjugated and unconjugated bilirubin.

8. d. Fatty acids are absorbed by simple diffusion. Alanine (an amino acid), glucose and galactose are all absorbed by Na^+-cotransport mechanisms, and are examples of 2° active transport.

9. c. The rate of absorption of Na^+ via cotransporters is increased when a relevant substrate (glucose) is available. This generates an osmotic gradient across the epithelium, promoting H_2O absorption.

10. e. Cortisol, a glucocorticoid, promotes gluconeogenesis and glycogen synthesis, ensuring an adequate glycogen store. Insulin also promotes glycogen synthesis; the other hormones promote its breakdown.

Matching item answers

Theme: Bile

1. A. Lecithin and bile salts help prevent cholesterol from crystallizing out in bile.

2. E.

3. B. Forms stones if ratio of cholesterol to bile salts and lecithin is too high in gallbladder.

4. F. Produced by bacterial metabolism of bile pigment.

5. C. E.g., salts of glycocholic and taurocholic acid.

6. D. Mono- and diglucuronides of bilirubin. These are water soluble, unlike bilirubin itself.

Theme: Control of gastrointestinal function

1. C. Secretin promotes secretion of alkaline pancreatic juice which neutralizes the acid.

2. B. Cholecystokinin release is strongly activated by the presence of food molecules, particularly lipids, in the duodenum.

3. F. Activates chymotrypsinogen, trypsinogen and procarboxypeptidase.

4. E. Acetylcholine is the main postganglionic transmitter in parasympathetic nerves.

5. B. Cholecystokinin also stimulates gallbladder contraction. These effects combine to promote release of bile into the duodenum.
6. D. Noradrenaline (norepinephrine) is the main postganglionic neurotransmitter in sympathetic nerves. Sympathetic nervous activity inhibits activity in gastrointestinal smooth muscle.

Short note answers

a. These define the daily energy (about 40 kcal kg^{-1}) and protein (source of essential amino acids; about 30–40 g required in total) intake required by a healthy adult with average work load. Energy values are greatly increased in manual labourers and others carrying out high rates of physical exercise. Protein and energy needs increase in pregnancy. May also attempt to define needs for vitamins, e.g., C, D, K, B_{12}, folate, etc., and minerals, e.g., iron (for haemoglobin synthesis) and calcium (for bone mineral). Again, needs may alter with physiological state and there may be storage sites in the body, e.g., liver storage of B_{12} and iron.

b. Polysaccharides must be broken down into single sugar monomers prior to absorption. This depends on the alkaline amylases in saliva and pancreatic secretions, although the former probably acts for only a short time prior to swallowing since the acid in the stomach will inactivate it. Pancreatic amylase produces a number of short-chain oligosaccharides in the small intestine and these are broken down into monosaccharides by oligosaccharidases bound to the epithelial brush border.

c. A labelled diagram similar to Fig. 99 would be ideal. The three cell types (parietal, chief and mucus secreting) must be mentioned. Their functions may be listed but need not be discussed further; the question asks about *structure*, not function.

d. Water-insoluble bilirubin (from breakdown of porphyrin in haemoglobin) is transported by plasma albumin. It is taken into hepatocytes by a carrier protein and is then conjugated with glucuronic acid to form bilirubin mono- and diglucuronides. These water-soluble bile pigments are secreted into the bile canaliculi and then excreted into the intestine in bile. They are then converted into urobilinogen by bacteria. This is absorbed from the gut and excreted in urine (forms urobilin by oxidation on exposure to air) or secreted back into bile. Some enters the faeces, darkening them (referred to as stercobilinogen, which is oxidized to stercobilin on exposure to air).

e. This is an example of secondary active transport using a sodium cotransport mechanism. Active Na^+ extrusion by the Na^+/K^+ ATPase pump keeps intraepithelial $[Na^+]$ low, maintaining a diffusion gradient into the cell from the intestinal lumen. The energy available from this Na^+ diffusion is coupled to amino acid cotransport using specific carrier systems. This allows amino acids to be absorbed from the intestine even against an opposing concentration gradient. Amino acids then diffuse into the mucosal capillaries and are carried away in the portal venous blood. The large intestinal surface area (increased by villi and microvilli) and the constant mixing of intestinal contents by intestinal segmentation increase the efficiency of absorption.

Modified essay answers

1. Hunger depends on increased activity in the hunger centre of the hypothalamus linked with decreased inhibition of hunger by the satiety centre. One stimulus is reduced plasma glucose concentration, but higher centres also have important inputs which may be mediated through altered levels of endogenous opiates.
2. Catabolic reactions will dominate, e.g., glycogenolysis and gluconeogenesis from amino acids or glycerol, and oxidation of fatty acids for energy, thus sparing glucose. These reactions are stimulated by glucagon from the pancreas in the presence of glucocorticoids from the adrenal cortex. If blood glucose falls too low, adrenal catecholamine release may be stimulated following activation of the sympathetic nervous system along with growth hormone secretion from the anterior pituitary. These also have glucose raising actions.
3. Secretion of saliva is stimulated by the thought, sight, smell of food and the presence of food in the mouth acting via chemo- and mechanoreceptors to stimulate salivary nuclei

in the brain and activating parasympathetic outflow to glands. Saliva lubricates food, starts carbohydrate digestion (amylase) and provides protection against infection (lysozyme) and dental caries (alkali).

4. It will be mixed into chyme by waves of gastric peristalsis. Secretion of acid and pepsinogen will lead to acid digestion of protein in the bread, forming oligopeptides. Over the next hour or so, chyme will be delivered in regular quantities through the pyloric sphincter to the duodenum.
5. Fat is emulsified by bile salts and digested by pancreatic lipase in the small intestine. Free fatty acids and monoglycerides diffuse out of micelles into mucosal epithelium, are reconstituted as triglycerides and packaged with lipoprotein to form chylomicrons. These diffuse out into intestinal lacteals, travelling in the lymph through intestinal and thoracic lymph ducts to enter the bloodstream via the great veins in the thorax.
6. Increases in gastrin, cholecystokinin and secretin promote secretion into the gastrointestinal tract leading to gastric acid secretion, bile expulsion from the gallbladder and pancreatic secretion into the duodenum. As monosaccharide absorption commences after digestion of the carbohydrate in the toast, glucagon secretion will be depressed and insulin secretion stimulated. This favours glucose uptake and storage as glycogen and lipid, inhibits gluconeogenesis and fat breakdown, and limits the rise in blood glucose levels.
7. Gastric acid levels are high because of the stimulus to acid secretion provided by the meal. This continues after active eating has ceased because of the gastric and intestinal phases of acid stimulation. Lying horizontal makes it easier for gastric contents to enter the oesophagus.
8. This simply props the individual in a more vertical position so that gravity reduces acid reflux. Propping the head end of the bed also helps.

Data interpretation answers

1. They are secreted from the mucosa of the upper small intestine and their release is stimulated by food entering the duodenum. Secretin is particularly stimulated by acid from the stomach while CCK is more sensitive to fats and protein digestion products.
2. They are peptide hormones and have to be given by a route which avoids digestion (and thus inactivation) in the intestine. Intravenous injection allows the blood levels to be elevated rapidly, since no absorption is necessary as would be the case with a subcutaneous or intramuscular injection.
3. Both hormones increase the volumetric rate of secretion, but secretin is more effective. A similar pattern is seen in [HCO_3^-], so that the amount of HCO_3^- secreted in mmol min^{-1} is enormously accelerated by secretin.
4. Obvious conclusion is that CCK strongly stimulates secretion of amylase. Secretin's effect is less clear. The reduction in [amylase] following secretin does not necessarily indicate a true decrease in amylase secretion since the total volume of secretion is also greatly increased. Using volumetric secretion rate × [amylase] to calculate amylase secretion rate from the values in Table 10 we get 0.5 units min^{-1} under control conditions. This is slightly increased to a peak of 1 unit min^{-1} following secretin but peaks at 30 units min^{-1} following CCK. Thus CCK is generally regarded as strongly stimulating enzyme secretion by the acinar cells while secretin mainly affects volume and pH of pancreatic secretion.
5. Secretin is mainly stimulated by acid from the stomach and stimulates pancreatic secretion of a high volume of HCO_3^- rich fluid to neutralize this acid. CCK is more strongly stimulated by fat and protein products and stimulates secretion of digestive enzymes. Although amylase was measured here, lipase and protease secretion is similarly affected.
6. Rate of secretion and [HCO_3^-] appear to rise and fall in parallel. This could be better demonstrated by plotting these two variables on a scatter diagram.
7. The initial secretion has a high [HCO_3^-], which is reduced as it travels along the duct. The slower the rate of flow, the more time there is for HCO_3^- removal and the lower the final [HCO_3^-]. With faster secretion rates, there is less time for modification and the [HCO_3^-] is higher. Thus, the higher rate of flow following

secretin stimulation produces the more alkaline secretion.

8. Parasympathetic nerves from the vagus stimulate enzyme secretion in the acini, but the volume of secretion is low.

Clinical scenario answers

1. Malabsorption may result from abnormalities of digestion or of absorption itself.
2. Pancreatic insufficiency will lead to inadequate secretion of digestive enzymes for all food types. Abnormalities of the small intestine and its mucosal epithelium may interfere with absorption of any or all food types. Bile salt deficiency will reduce digestion and absorption of lipids.
3. Pale, smelly, floating stools suggest malabsorption of lipid, the fats making the stools less dense than water. This is known as steatorrhoea.
4. Malabsorption may give rise to deficiency of one or more of iron, folate and vitamin B_{12}, causing anaemia.
5. Bruising suggests clotting factor deficiency. This may result from vitamin K deficiency. Vitamin K is a fat-soluble vitamin and its absorption is likely to be impaired in cases of fat malabsorption.
6. Other possible deficiency states include deficiency of the other major fat-soluble vitamins, particularly vitamin D deficiency, which causes osteomalacia in adults, and vitamin A deficiency, which, rarely, causes night blindness. There may also be a generalized hypoproteinaemia, causing oedema due to reduced plasma oncotic pressure.

Viva answers

1. Your answer should cover the source (gastric glands), cell type (parietal) and mechanism of production (formation of carbonic acid from CO_2 and H_2O catalyzed by carbonic anhydrase, dissociation into H^+ and HCO_3^-, ATP-driven proton pump, and HCO_3^- diffusion into gastric capillaries). Secretion is stimulated by vagus (parasympathetic nerves secreting acetylcholine which acts on muscarinic receptors), gastrin (hormone secreted from G cells in stomach) and histamine (paracrine signal secreted from enterochromaffin cells in gastric mucosa). These signals result in cephalic, gastric and intestinal phases of secretion.
2. You need to emphasize the role of the swallowing reflex both in driving food down the oesophagus and in protecting the airway. Mechanical stimulation of the back of the pharynx activates the swallowing centre in the medulla oblongata, which sends out motor signals to stop breathing, raise the larynx and close the glottis (airway protection; assisted by presence of the epiglottis which passively deflects food away from laryngeal opening). It also activates oesophageal peristalsis and relaxes both oesophageal sphincters. Both somatic and parasympathetic nerves are involved.
3. Parasympathetic nerves are both motor (i.e., stimulate mechanical activity) and secretomotor (i.e., stimulate exocrine secretions) along the length of the gastrointestinal tract. Thus, they are involved in swallowing, receptive relaxation of the stomach, promotion of gastric contraction and emptying, rhythmic contraction of small and large intestine and the defecation reflex. They stimulate secretion by the salivary glands, the stomach, intestinal glands and the pancreas, although in the case of the pancreas their influence is less important than that of hormones. Further questions might include the source of the parasympathetic outflow to different areas (Xth cranial nerve to most of tract with the exceptions of VIIth and IXth cranial nerves to salivary glands and sacral parasympathetic outflow to lower large intestine), the neurotransmitter (acetylcholine) and receptor (muscarinic).

Neuromuscular physiology

Overview

The nervous system provides the most sophisticated signalling and control network within the body. This relies on rapid conduction of electrical signals along nerve axons and chemical transmission between nerves. Sensory receptors convert information about the external and internal environments into action potentials which are relayed to the central nervous system via sensory nerves. Sensory signals reflexly stimulate relevant motor outputs concerned with homeostatic and locomotor control, and many enter our conscious awareness in the form of sensory perceptions. Motor systems in the brain also allow us to control voluntary movements through links with the cranial and spinal motoneurones which activate skeletal muscle contraction. The brain is also responsible for consciousness, thought and our understanding and use of language, as well as the ability to learn from previous experiences. These higher functions are of particular interest, not least because they are poorly understood.

7.1 Relevant structure and organization of the nervous system

Learning objectives

At the end of this section you should be able to:

- list the main subdivisions of the nervous system
- describe the main structures of a neurone
- name the glial cells and outline their function
- describe the circulation of cerebrospinal fluid.

Subdivisions of the nervous system

Nerves are distributed throughout the body to form the nervous system. This can be subdivided into different elements based on a number of structural and functional considerations.

One simple anatomical division is that between the *central nervous system* (CNS), which includes all the structures within the brain and spinal cord, and the *peripheral nervous system* outside these regions. The tissue of the CNS may, in turn, be divided into grey matter, containing a high density of cell bodies, and white matter, containing many axons. Cerebral grey matter is found both in the convoluted outer layer of the brain, the *cerebral cortex*, and within a number of more centrally placed *cerebral nuclei*.

Peripheral nerves can be separated on a functional basis into *sensory*, or afferent, nerves bringing signals into the CNS, and *motor*, or efferent, nerves carrying information out.

An important distinction is made between the *somatic* nervous system, supplying the skin and musculoskeletal structures, and the *autonomic* nerves, which are distributed both to smooth muscle in the blood vessels and viscera of the body and to exocrine and certain endocrine glands.

These different classifications overlap, e.g., the somatic and autonomic nervous systems have elements in both the central and the peripheral nervous systems and contain sensory as well as motor components.

Neuronal structure

The main functional cell in the nervous system is the nerve cell or neurone, which is primarily engaged in signalling and control. Classically, each neurone is described as having three elements, with the *cell body* or soma connected to a large number of short projections, known as *dendrites*, and a single elongated projection called an *axon* (Fig. 116). Incoming signals affect the dendrites and soma, while outgoing signals are passed on via the axon. The connection between one nerve cell and the next is known as a *synapse* and there is often a swelling at the nerve ending known as a *synaptic bouton*. The membranes of adjacent nerve cells do not come into direct contact with each other at these sites, however, but are separated by a *synaptic cleft*. This separation has important consequences for signal transmission between cells.

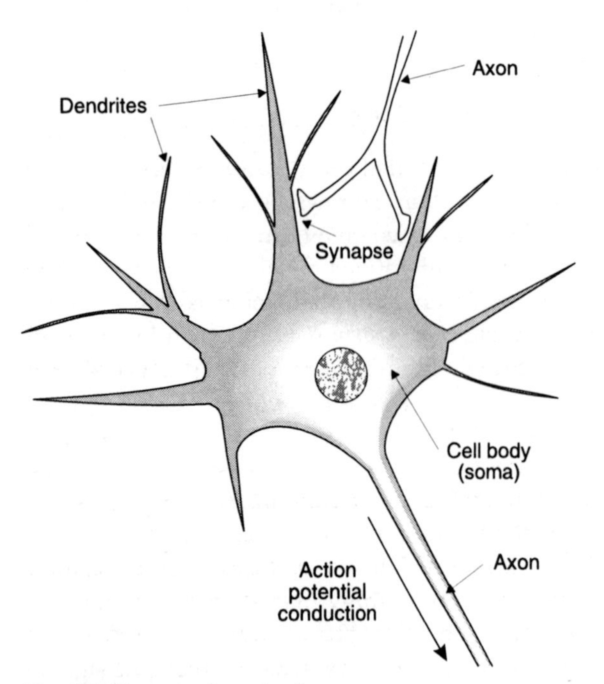

Fig. 116 The main elements of a neurone.

Glial cells in the nervous system

Although we will concentrate on the function of neurones in the rest of this chapter, it should be mentioned that the nervous system also contains a large number of nonexcitable, glial cells. In the CNS, these are referred to as neuroglial cells and consist of *macroglia* (astrocytes and oligodendrocytes), *ependymal cells*, lining the fluid-filled ventricles and spinal cord canal, and tissue macrophages called *microglia*. Astrocytes provide a supportive matrix around the neurones and form part of the blood–brain barrier, while oligodendrocytes provide an electrically insulating sheath of *myelin* around nerve axons. Myelination of peripheral nerves depends on *Schwann cells*, the only glial cells in the peripheral nervous system.

Cerebrospinal fluid

Cerebrospinal fluid (CSF) is a cell-free modified plasma containing low levels of protein which bathes the central nervous system. It acts as a buffer against mechanical injury, and its composition reflects that of the extracellular fluid within the CNS itself. Cerebrospinal fluid is continuously secreted by the choroid plexuses within the lateral, third and fourth ventricles of the brain, and circulates through the ventricular system before passing into the subarachnoid space from the fourth ventricle. The subarachnoid space lies between the arachnoid and pial layers of the meninges which cover the brain and spinal cord. Most of the CSF eventually flows through the arachnoid villi into the blood within the cranial venous sinuses.

Box 25 Clinical note: Hydrocephalus

If the circulation of CSF is blocked, the continuous secretion of additional fluid by the choroid plexuses leads to an increase in fluid pressure. The ventricular spaces become distended, compressing the surrounding cerebral tissue. The obstruction may be within the ventricular system itself, or at the level of the outflow from the fourth ventricle. This leads to non-communicating hydrocephalus, since the high pressure in the ventricular system is not 'communicated' to the subarachnoid space. In communicating CSF the obstruction occurs at the level of the arachnoid villi, and the pressure is elevated within the subarachnoid space as well as the ventricles.

7.2 Conduction and transmission in nerves

Learning objectives

At the end of this section you should be able to:

- compare and contrast single and compound action potentials
- explain how action potentials propagate along an axon
- identify factors which affect conduction velocity
- describe how action potentials cause transmitter release
- list, with examples, the main transmitter groups
- describe the properties of postsynaptic potentials and explain how they are generated
- explain presynaptic inhibition.

Information is conducted along nerves in the form of 'all or nothing' electrical signals known as *action potentials* (Fig. 117A). Intracellular recordings from axons show that the inside is normally negative with respect to the outside, i.e., there is a *resting membrane potential* (about –70 mV). If a stimulus decreases this potential (i.e., makes it less negative), so that the membrane reaches the *threshold potential*, then there is further, automatic and rapid *depolarization* which reverses the potential across the cell membrane. This is an active event and is followed by *repolarization* to resting conditions. The properties of this action potential, and the membrane mechanisms which generate it, have been considered in detail in Section 1.4. Little more need be said here other than to remind ourselves that depolarization reflects a voltage-dependent increase in the membrane permeability, or conductance, to Na^+, while repolarization is caused by the slower onset of a similar rise in K^+ conductance.

Compound action potentials

An anatomically identifiable peripheral nerve, such as the sciatic nerve in the pelvis and thigh, is made up of a large number of individual axons. The extracellularly recorded electrical activity in such a nerve reflects the total effect of all the individual action potentials in different axons and is called a compound action potential. When such a nerve is stimulated electrically, it can be shown that the amplitude of the compound action potential increases as the stimulus strength is increased (Fig. 117B). This observation appears to contravene the 'all or nothing' law for excitable cells but, in fact, it simply reflects the presence of axons with differing thresholds within the nerve. As the stimulus is increased, more and more axons reach threshold so that a greater proportion of the available fibres fire an action potential. Eventually all the axons will have been recruited in this way and further increases in the stimulus produce no further change in the compound action potential. These are referred to as supramaximal stimuli.

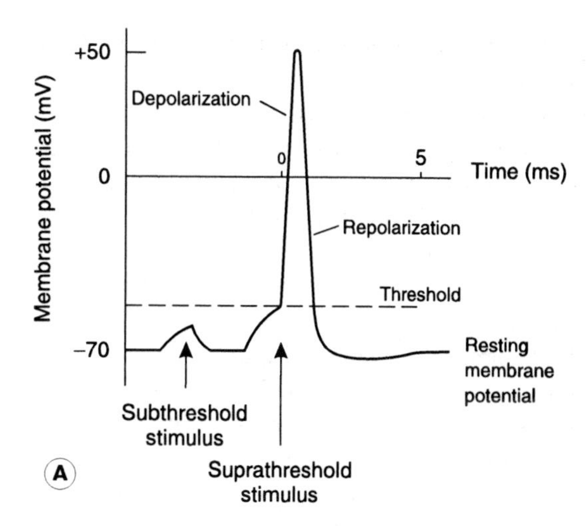

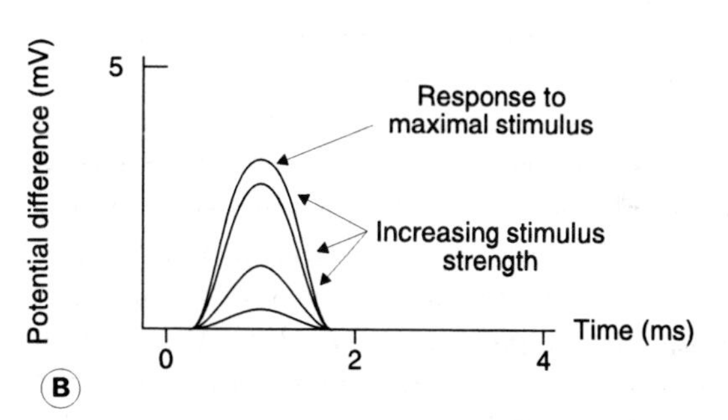

Fig. 117 Action potentials in nerves. (A) Intracellular recording from a single axon showing the 'all or nothing' response to a suprathreshold stimulus. (B) Extracellular recording of the compound action potential from a nerve containing many axons. The increasing response with increasing stimulus strength reflects recruitment of axons with higher thresholds.

Action potential conduction velocity

Once initiated, action potentials are propagated, or conducted along an axon by means of *local currents* (Section 1.4). Positive current spreads along the axon away from the site of an action potential and this depolarizes the adjacent membrane. Once threshold potential is reached, an action potential will be actively generated, and the whole process is repeated. Although the local currents are conducted equally in both directions along the axon, the fact that the nerve membrane remains refractory to further stimulation for a few milliseconds after firing an action potential ensures that the signal is conducted along the axon in one direction only.

The speed with which a signal can be conducted along an axon, i.e., the nerve conduction velocity, is determined by two aspects of nerve structure.

Axon diameter. Conduction velocity increases as axon diameter increases, so that rapid conduction is a feature of large nerve fibres. This reflects the reduction in the electrical resistance to the spread of current along the axoplasm which accompanies any increase in the cross-sectional area available for conduction. As a result, the voltage change resulting from an action potential generates a larger local current along the axon, causing a greater length of membrane to be depolarized to threshold, and so the action potential is propagated more rapidly along the nerve.

Myelination. Conduction velocity is greatly increased in myelinated, as opposed to unmyelinated, nerves. *Myelin sheaths* are produced by *Schwann cells* in peripheral nerves and *oligodendrocytes* in the CNS. These glial cells wrap their plasma membranes round and round the axon to produce a sheath of tightly stacked bilipid layers (Fig. 118). This myelin is not continuous, however, and there are gaps in the sheath, known as *nodes of Ranvier*, spaced at regular intervals along the axon. Action potentials are only generated at these sites, where the excitable membrane of the neurone is exposed. The spread of depolarizing current along the myelinated region to the next node is very rapid and the electrical insulation provided by the lipid myelin sheath helps to promote this rapid conduction by reducing the leakage of local current across the nerve membrane, leaving more available to depolarize the axon at the nodes. This type of conduction, in which the action potential effectively jumps from one node to the next along the axon, is referred to as *saltatory conduction* and is considerably more rapid than continuous conduction in unmyelinated nerves. Myelinated nerves conduct with velocities in the range 10–120 $m\,s^{-1}$, depending on their size, with the large somatic nerves to skeletal muscles conducting the fastest. Small, unmyelinated nerves, by comparison, may conduct at less than 1 $m\,s^{-1}$, e.g., slow pain fibres and autonomic motor nerves.

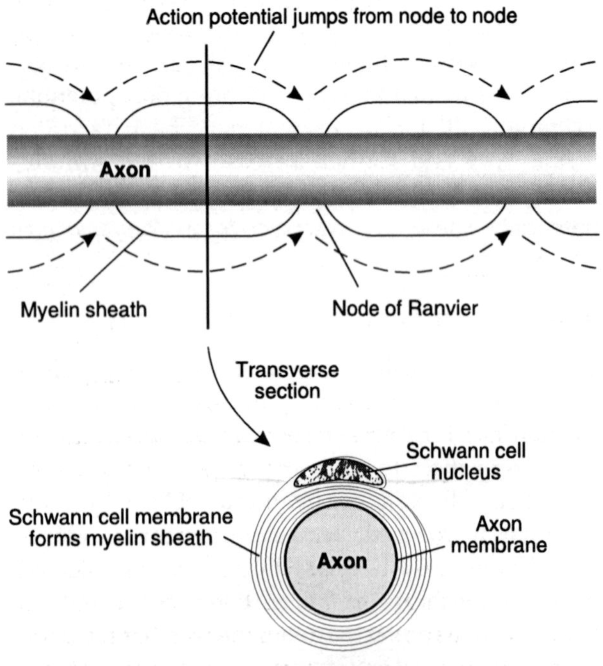

Fig. 118 Action potential conduction (saltatory conduction) in a myelinated axon.

Synaptic transmission

When an action potential reaches an axon ending, it must then influence the adjacent nerve if the signal is to be passed on. This occurs at the synapses between the axon of one neurone and the dendrites or cell body of the next (Fig. 116), and the process is referred to as synaptic transmission. Transmission usually depends on the release of a messenger molecule called a *neurotransmitter* from the nerve terminal, and this chemical affects the adjacent cell, tending either to excite or to inhibit action potential generation in that neurone. This is an example of chemical transmission.

Structure and function of the synapse

The nerve along which incoming signals arrive at the synapse is called the presynaptic nerve and this characteristically terminates in an expanded structure called a terminal bouton (Fig. 119A). Electron micrographs demonstrate that these boutons enclose a large number of membrane-bound vesicles containing neurotransmitter. A narrow gap (about 20 nm) known as the synaptic cleft separates the *presynaptic* membrane from that of the *postsynaptic* neurone. The existence of this separation between adjacent nerves suggests a chemical transmission mechanism, since electrical transmission (the possible alternative mode of transmission) requires electrical continuity between the cytoplasm of adjacent cells (e.g., the gap junctions between cardiac cells; Section 3.1).

The general sequence of events in synaptic transmission may be summarized as indicated by the numbered steps in Figure 119B.

1. Arrival of an action potential at a nerve terminal depolarizes the presynaptic membrane, increas-

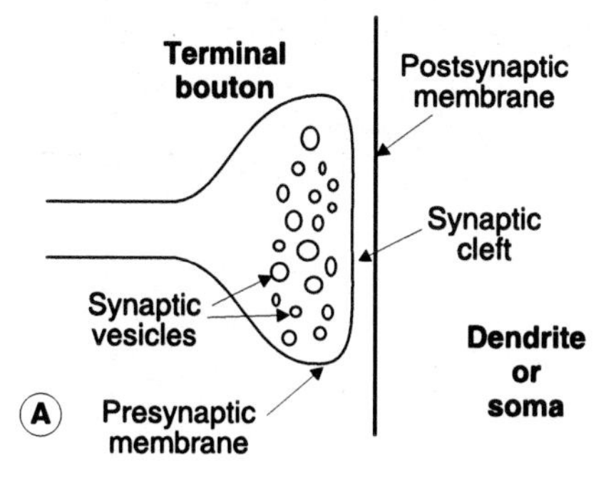

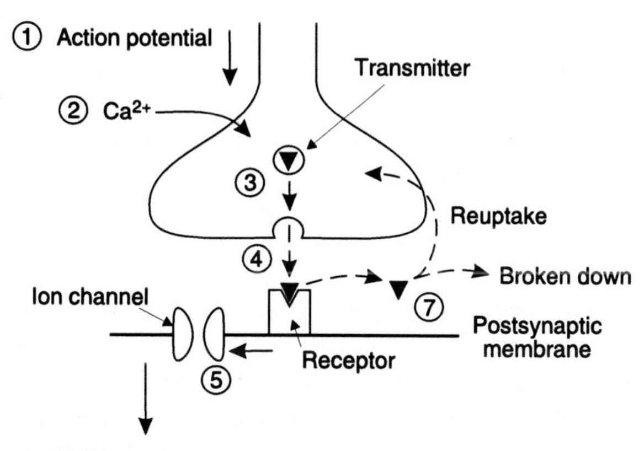

Fig. 119 Synaptic transmission. (A) Major structural features of a chemically transmitting synapse. (B) Main events in neurotransmission. See main text for full explanation of the numbered processes.

ing its permeability to Ca^{2+} by opening voltage-operated Ca^{2+} channels.

2. Calcium ions diffuse down their electrochemical gradient into the nerve terminal, since the $[Ca^{2+}]$ is higher in the extracellular fluid than in the cytoplasm.
3. The resulting increase in intracellular $[Ca^{2+}]$ triggers fusion of the synaptic vesicles with the presynaptic membrane and neurotransmitter is released into the synaptic cleft by exocytosis.
4. Transmitter diffuses across to the postsynaptic membrane where it binds with postsynaptic receptors.
5. Receptor-operated ion channels are opened as a result, generating a postsynaptic ion current.
6. Depending on the transmitter and receptor involved, this may either excite or inhibit the postsynaptic neurone.
7. The signal is terminated when the transmitter is released from the receptor. The transmitter is then broken down or taken up again by the nerves.

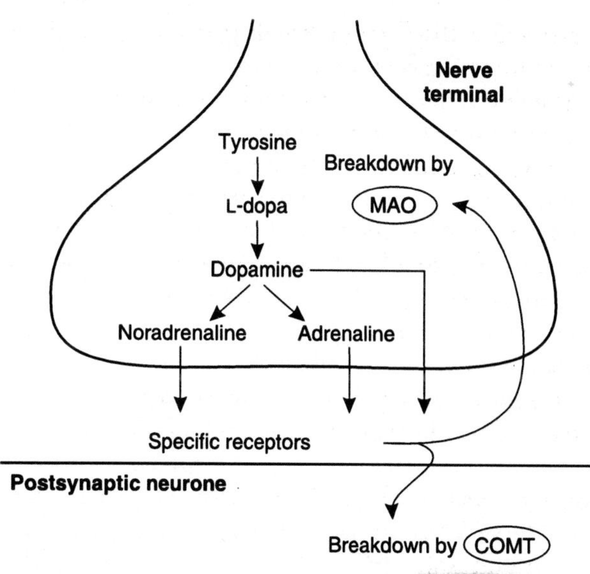

Fig. 120 Summary of synthesis and breakdown pathways for catecholamines. MAO, monoamine oxidase; COMT, catechol-O-methyl transferase.

Neurotransmitters

Although it is difficult to prove that a given substance acts as a neurotransmitter, a wide range of different substances have been identified as possible (putative) transmitters and some of the most important are listed below.

Acetylcholine is an important excitatory transmitter in the brain and spinal cord. It is also involved in ganglionic transmission in autonomic nerves, as well as being the peripheral transmitter at the skeletal neuromuscular junction (Section 7.5) and in postganglionic parasympathetic nerves (Section 7.6).

Biogenic amines include the catecholamines, 5-hydroxytryptamine (serotonin) and histamine. Catecholamines (*dopamine, noradrenaline* (norepinephrine) and *adrenaline* (epinephrine)) share a common synthetic pathway involving conversion of the amino acid tyrosine to L-dopa and then to dopamine (Fig. 120). This may then be further modified to form adrenaline (epinephrine) or noradrenaline (norepinephrine). After transmitter release, catecholamine action is terminated by reuptake into the neurones followed by breakdown through the action of monoamine oxidase (in the presynaptic neurone) or catechol-*O*-methyl transferase (on the postsynaptic membrane). Drugs which inhibit monoamine oxidase promote catecholamine-dependent transmission in the brain and are used in the treatment of clinical depression. Inadequate dopaminergic transmission is also believed to be involved in a motor disorder called Parkinson's disease (Section 7.5) and treatment with the dopamine precursor L-dopa is sometimes beneficial, presumably because it increases dopamine synthesis. Conversely, drugs which inhibit dopaminergic transmission by blocking dopamine receptors are used in the treatment of schizophrenia, a psychiatric disorder characterized by disrupted thought, hallucinations and delusions. One side-effect of this treatment can be the development of Parkinsonian type symptoms, which is consistent with the idea that dopamine deficiency is involved in causing Parkinson's disease.

Amino acids act as important neurotransmitters. Glycine inhibits spinal cord motoneurones while glutamate and aspartate are widespread excitatory transmitters in the brain. Glutamate may also be converted to γ-aminobutyric acid (GABA), an inhibitory transmitter in the brain.

Peptides. A wide range of peptides have now been identified as probable or possible neurotransmitters. Examples include substance P, which seems to play a role in the central transmission of pain signals, and endorphins, which inhibit pain pathways, perhaps by blocking release of substance P (Section 7.4). Some peptide transmitters may also be released along with a classical transmitter by a single neurone, e.g., acetylcholine and vasoactive intestinal polypeptide (VIP) coexist in some parasympathetic nerves. This phenomenon is known as cotransmission.

Other transmitters include *purines*, like ATP, which may also be coreleased with classical transmitters, and nitric oxide (NO).

Neuromodulation

It is now recognized that both classical transmitters and neuropeptides may act primarily to alter the amount of neurotransmitter released when a given nerve terminal is stimulated, rather than by having a direct effect on the postsynaptic cell. This mechanism for the regulation of transmission is referred to as neuromodulation.

Excitatory transmission

Excitatory neurotransmitters act on the postsynaptic membrane to produce depolarization, thus favouring the firing of an action potential. The electrical response recorded from the postsynaptic neurone consists of a brief depolarization followed by a slower decline to the resting potential (Fig. 121), and is known as an *excitatory postsynaptic potential* (EPSP). This differs from the action potential in two main ways.

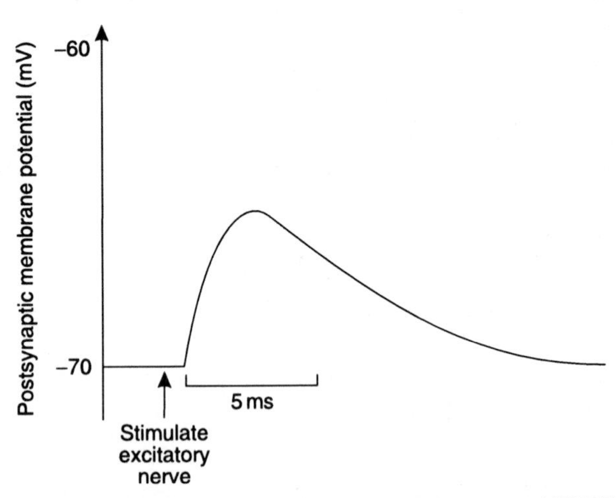

Fig. 121 Recording of an excitatory postsynaptic potential (EPSP) from the cell body of a neurone following stimulation by an excitatory nerve.

Passive conduction over the cell. The EPSP is not conducted over the cell in an active fashion. It does depolarize the adjacent membrane but this occurs passively because of the local currents which are set up. The further one gets from a synapse, however, the smaller the effect of the EPSP becomes. This is quite different from a propagated action potential, whose size remains constant as it is actively propagated along the membrane.

Graded response. The EPSP is a graded response, not an 'all or nothing' response like the action potential (Fig. 122A). The larger the number of excitatory nerves which are stimulated simultaneously, the larger is the resulting EPSP. This is called *spatial summation*, since the effects of excitatory neurotransmitter release at many different synapses on the postsynaptic neurone add together to give a larger total response. Similarly, repeated stimulation in a single excitatory nerve may increase the peak of the EPSP, since further depolarization occurs before the membrane returns to resting potential. This is *temporal summation* (Fig. 122B). If summation leads to a large enough depolarization (i.e., reaches threshold), the postsynaptic cell will generate an action potential which then travels down its axon.

Inhibitory transmission

The simplest form of inhibition (*postsynaptic inhibition*) is analogous to excitation in that synaptic transmission produces a change in postsynaptic membrane potential. In this case, however, hyperpolarization results (Fig. 123A), and this is termed the *inhibitory postsynaptic potential* (IPSP). By moving the membrane further away from the threshold potential, this makes it less likely that an action potential will be

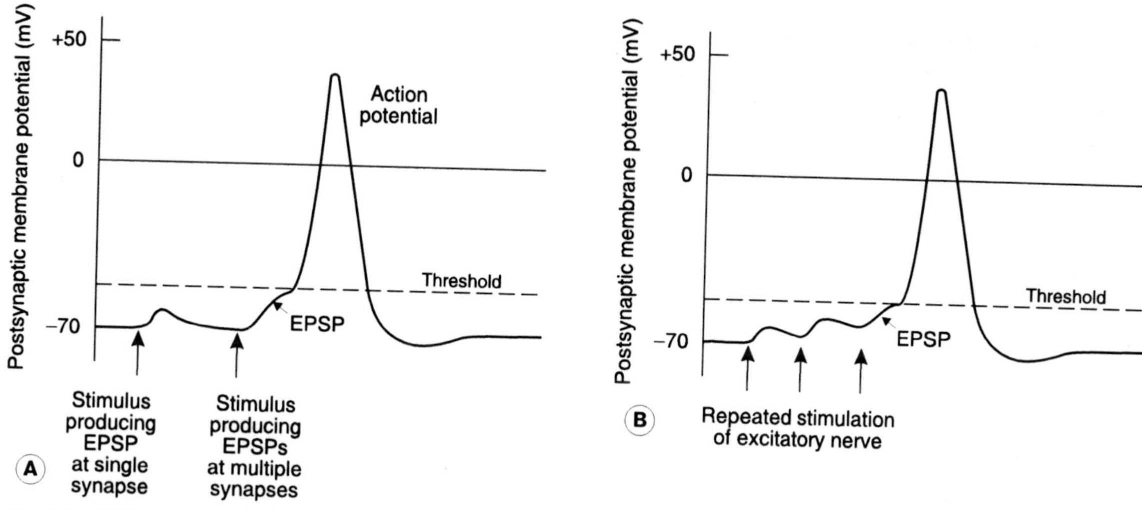

Fig. 122 EPSP summation may lead to an action potential if threshold is reached. (A) Spatial summation. (B) Temporal summation.

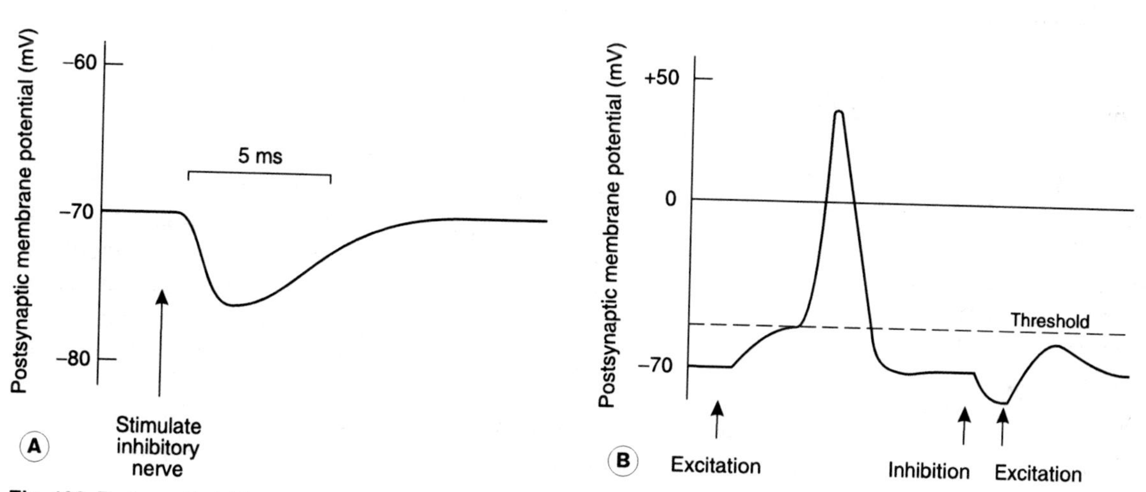

Fig. 123 Postsynaptic inhibitory effects. (A) Recording of an inhibitory postsynaptic potential (IPSP) following stimulation of an inhibitory nerve. (B) Recording of membrane potential showing how summation of an IPSP and an EPSP may prevent the firing of an action potential.

generated in response to other excitatory stimuli. In other words, IPSPs and EPSPs from different synapses acting on a given neurone will summate algebraically (Fig. 123B). The signals from the many different inputs to a cell are integrated in this way and an action potential will only be generated if the postsynaptic membrane in the region of the axon reaches threshold. Overall, the action potential frequency will reflect the balance between excitatory and inhibitory influences at any time.

Presynaptic inhibition

It is possible to reduce the size of the EPSP generated by excitatory stimulation through inhibitory nerves which do not produce any IPSP in the postsynaptic neurone. This presynaptic inhibition depends on inhibitory nerves which synapse with the incoming excitatory axon, rather than the postsynaptic cell itself (Fig. 124A). Activity in this inhibitory pathway reduces the amount of excitatory neurotransmitter released when the axon is stimulated. This depends on depolarization of the axon by the presynaptic neurotransmitter, which decreases the amplitude of any action potentials within that nerve terminal since the potential rises from an elevated baseline potential. The quantity of transmitter released is dependent on the size of the action potential and so the resulting EPSP is smaller than normal (Fig. 124B). Certain inhibitory interneurones in the spinal cord are believed to exert their influence on spinal motoneurones through presynaptic inhibition of this kind.

Ionic mechanisms of postsynaptic potentials

Neurotransmitters produce the changes in postsynaptic potential described in the preceding sections by

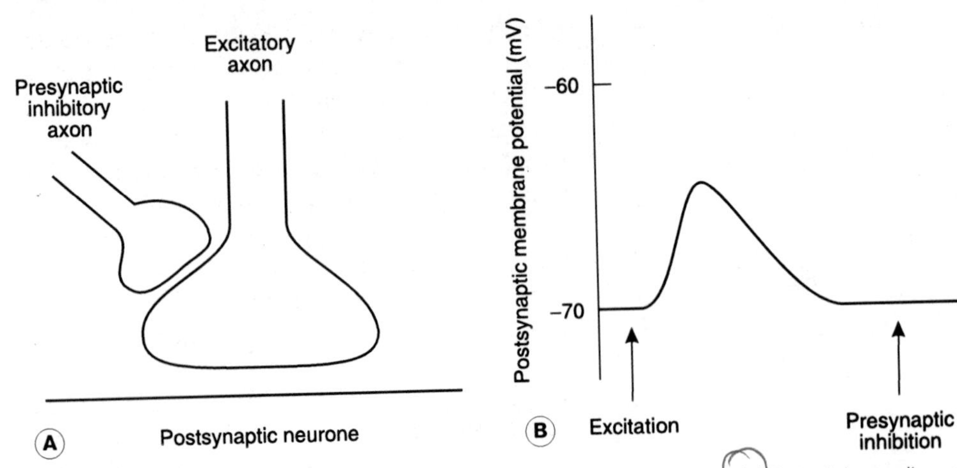

Fig. 124 Presynaptic inhibition. (A) Structure of a presynaptic inhibitory synapse. (B) Potential recordings show that stimulating the presynaptic inhibitory nerve has no direct effect on postsynaptic membrane potential but greatly reduces the size of a subsequently induced EPSP.

opening *receptor-operated ion channels*. In the case of spinal motoneurones it seems likely that:

EPSPs are produced in response to the excitatory transmitter *glutamate* which opens channels allowing both Na^+ and K^+ to flow through them (nonselective cation channels). Overall, this increases the permeability of the membrane to Na^+ relative to that to K^+ and, since E_{Na} (the equilibrium potential for sodium) is positive, this tends to depolarize the cell (Section 1.4).

IPSPs are generated in response to the inhibitory transmitter *glycine* as the result of opening of receptor-operated channels which allow Cl^- to flow through them. This shifts the membrane potential towards E_{Cl} which is slightly negative with respect to the resting potential. Hyperpolarisation.

7.3 General sensory mechanisms

Learning objectives

At the end of this section you should be able to:

- classify sensory receptors and explain how sensory modality is coded
- explain what is meant by a receptive field and why these vary in size
- describe a receptor potential and explain how stimulus strength is coded
- explain what is meant by sensory adaptation.

Sensory physiology deals with the mechanisms involved in the detection and interpretation of a wide variety of different stimuli. This relies on sensory receptors to convert these stimuli into electrical signals within the sensory nerves, which relay the information to the CNS. Within the CNS the sensory inputs are used both to control the movements of the body and to regulate the internal environment (*homeostasis*). This requires storage of reference information about normal conditions, e.g., the set points for different controlled variables (Section 1.1), and appropriate connections between the sensory (afferent) and motor (efferent) pathways so that compensatory adjustments can be made. Conscious awareness and interpretation of sensation is also a central nervous function, with particular areas of the cerebral cortex being devoted to processing activity from specific types of receptor.

Receptor types

Sensory receptors convert the energy in a sensory stimulus into action potentials within sensory nerves. Energy converting devices are known as transducers and so they are also called sensory transducers. These provide the first link in the communication chain which keeps the central nervous system informed about conditions within the body (the internal environment) and in the outside world (the external environment). Some receptors are structurally complex while others consist of nothing more than an exposed nerve ending.

Receptors may be classified in terms of the general class of stimulus to which they respond.

- *Mechanoreceptors* detect mechanical distortion or movement, e.g., pressure receptors in the skin and stretch receptors in muscles.
- *Photoreceptors*, like the rods and cones in the eye, detect electromagnetic radiation.
- *Chemoreceptors* are sensitive to chemical stimuli, e.g., taste receptors in the tongue or respiratory

chemoreceptors responsive to blood gases and pH.

- *Thermoreceptors* detect hot and cold.

Receptors may also be classified in terms of the purpose they serve. This is subtly different from a classification based on stimulus type. For example, *nociceptors* consist of all receptors responsible for painful sensations but this grouping contains a mixture of mechano-, thermo- and chemoreceptors (Section 7.4). Similarly, *proprioceptors* represent a specific subgroup of mechanoreceptors, all of which provide information about joint position.

Sensory modalities and the law of specific nerve energies

We may divide our conscious sensory experiences into a number of different sensory modalities, e.g., touch, taste, sight, hearing, etc. How the brain converts the indistinguishable action potentials conducted along different sensory nerves into these separate perceived sensations is not well understood. Receptors show a high sensitivity to only one type of stimulus, however, and activity in the sensory nerves connecting a receptor with the CNS will always be interpreted as being caused by that stimulus. For example, photoreceptors are sensitive to light and signals in the optic nerve from the retina are always interpreted in visual terms. This remains true, even if the receptors have been stimulated by a very strong stimulus of some other type, such as direct pressure over the eyeball. Equally, direct electrical stimulation of the optic nerve in the absence of any stimulus to the photoreceptors will evoke visual sensations. This demonstrates the general principle that a fixed interpretation is given to activity in a sensory nerve based on the nature of the receptor of origin, a rule referred to as the law of specific nerve energy.

Receptive fields

A receptive field may be defined as the region within which a stimulus will produce activity in a given sensory receptor. For a touch receptor, this will be determined by the area of skin overlying the branching sensory terminals; for a photoreceptor, it will depend on the target area presented by that receptor in the retina. Receptive fields can also be defined for more central neurones along the sensory pathway, and in this case the extent of each field will depend not just on effective receptor area but also on the degree of convergence of information between the periphery and the central sensory neurones. Thus, if a single neurone in the cerebral cortex receives inputs from many adjacent peripheral receptors (a highly convergent sensory pathway), its receptive field will be considerably larger than that of the individual receptors themselves. The receptive fields of the central neurones form a sensory map of the body, in which areas demonstrating a high degree of spatial resolution are represented by a large number of neurones, each with a small receptive field.

Differences of this kind can be demonstrated for touch using the two-point discrimination test. A subject is tested with two sharp points placed close to each other and asked whether they feel them as a single or separate stimuli. For example, the skin over the fingertips, which has a high receptor density and is represented centrally by a relatively extensive area of sensory cortex, can discriminate much more closely spaced stimuli than the skin of the arm or leg.

Receptor potentials

Sensory receptors convert a stimulus into an electrical response known as a *receptor potential*. This is often a graded depolarization that increases in amplitude as the stimulus strength increases (Fig. 125). Receptor potentials are localized to the receptor region itself and are not actively propagated along the sensory nerve. However, if the receptor potential depolarizes the nerve to threshold level, it stimulates action potentials that are then conducted along the sensory axon to the spinal cord. This relies on passive

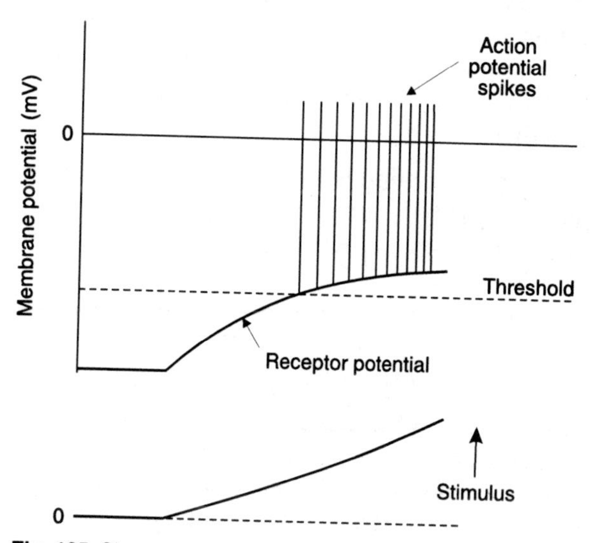

Fig. 125 Changes in receptor potentials of a slowly adapting receptor, in response to an increasing stimulus (recorded from a sensory axon adjacent to the receptor). Changes in membrane potential show a depolarizing receptor potential. If the receptor potential depolarizes the nerve to threshold level, this generates action potentials in the sensory nerve.

Seven

depolarization of the axon in receptors formed by adaptation of a sensory nerve ending, e.g., mechanoreceptors in the skin. In other cases, e.g., the hair cells in the inner ear, the receptor cells may be distinct from the relevant sensory axons. Here, the receptor potential controls transmitter release from the receptor itself, and this depolarizes the sensory axon, producing a graded *generator potential*. If this reaches threshold, propagating sensory action potentials result.

The mechanism underlying the receptor potential depends on the type of receptor. In the case of cutaneous sensation, the appropriate stimulus often leads to an increase in the Na^+ permeability of the receptor membrane. The resulting inward movement of Na^+ ions depolarizes the receptor and thus the sensory axon associated with it. When the relevant stimulus is applied at other points along the length of the sensory axon, however, it has no such effect.

Coding for stimulus strength in the sensory system

As well as identifying the nature or modality of a stimulus, we can also make some estimate of its strength. Information about stimulus amplitude must, therefore, be coded into both the receptor response and the signal in sensory nerves.

- Receptors code for stimulus strength through changes in the size of the graded receptor potential.
- Sensory nerves carry information in the form of action potentials which are constant in size. Information about stimulus strength is coded for by changes in the frequency of these action potentials.

Thus, increasing stimulus strength produces larger receptor potentials which generate a higher frequency of action potentials in the sensory axon (Fig. 125).

Sensory adaptation

If a constant stimulus is applied continuously to a receptor, the resulting receptor potential (and, therefore, the action potential frequency in the relevant sensory nerve) tends to decline as time passes (Fig. 126). This is referred to as adaptation, since the receptor has adapted itself to the new stimulus level. Receptors like the Pacinian corpuscle, a pressure receptor in the skin, adapt very rapidly and so are more sensitive to changes in stimulus strength than to a constant stimulus. Such rapidly adapting receptors are said to be *phasic*. *Tonic* receptors, by comparison, adapt very slowly and so continue to generate sensory action potentials during sustained stimuli. Some of the proprioceptors in the musculoskeletal system behave in this way, providing continuous information about joint position.

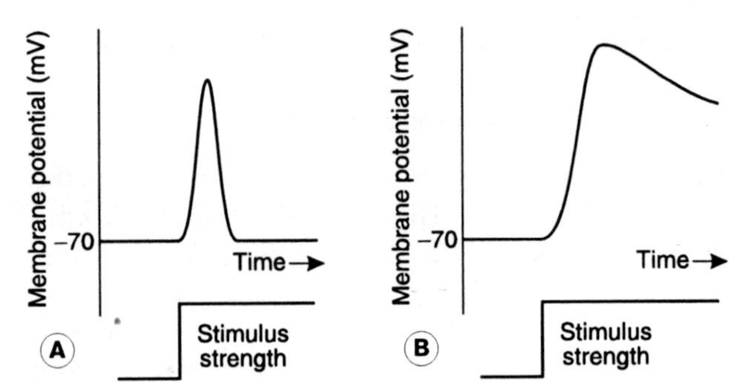

Fig. 126 Adaptation of sensory receptors. (A) Receptor potential response to a constant stimulus in a fast-adapting (phasic) receptor. Such receptors are sensitive to changes in stimulus strength. (B) Receptor potential during a constant stimulus in a slowly adapting (tonic) receptor.

7.4 Specific sensory systems

Learning objectives

At the end of this section you should be able to:

- describe the receptors, nerve pathways and cortical areas involved in somatosensory perception, taste, smell, hearing, balance and vision
- classify pain, explain how nociceptors work and outline how pain is modified in the CNS
- identify the main structures of the ear
- explain what is meant by auditory threshold and how it varies with frequency
- describe how sound is transmitted to the cochlea and transduced by hair cells
- explain how sound frequency and loudness are coded
- describe how the vestibular apparatus detects movement and body position
- draw a labelled diagram of the eye
- outline the optical properties of the eye and how these can be reflexly adjusted
- explain the principles of phototransduction, adaptation and colour vision
- describe some simple refractive and visual field defects.

Having considered the general properties of sensory systems, we can look at some specific details of individual modalities of sensation. In each case, we shall consider the receptors involved, a simple outline of the sensory pathways by which the information is relayed to the CNS and what is known about the processing of that information within the brain itself. Some of the more common sensory abnormalities will also be considered.

The somatosensory system

This is responsible for the sensations of touch, temperature, proprioception (joint position) and pain.

Somatosensory pathways

Information is relayed from the relevant peripheral receptors via unipolar sensory neurones which have their cell bodies in the dorsal root ganglia. The information enters the spinal cord through the dorsal nerve roots and is then transferred up to the brain (Fig. 127). Some sensory pathways (touch, pressure, temperature, pain) cross to the opposite (contralateral) side of the cord before ascending in the spinothalamic tracts. Others (proprioception, touch) travel up the dorsal columns on the same (ipsilateral) side of the cord before crossing the midline in the region of the medulla oblongata. Ultimately, therefore, all the sensory information from the right side of the body is routed through the left thalamus up to the left cerebral cortex. This explains why damage to the sensory pathways in the right cerebral hemisphere leads to sensory loss on the left side of the body. All sensory pathways also send inputs to the reticular-activating system in the brainstem area. This region is involved in controlling the level of consciousness, especially patterns of wakefulness and sleep. Increased sensory input increases awareness, making sleep less likely.

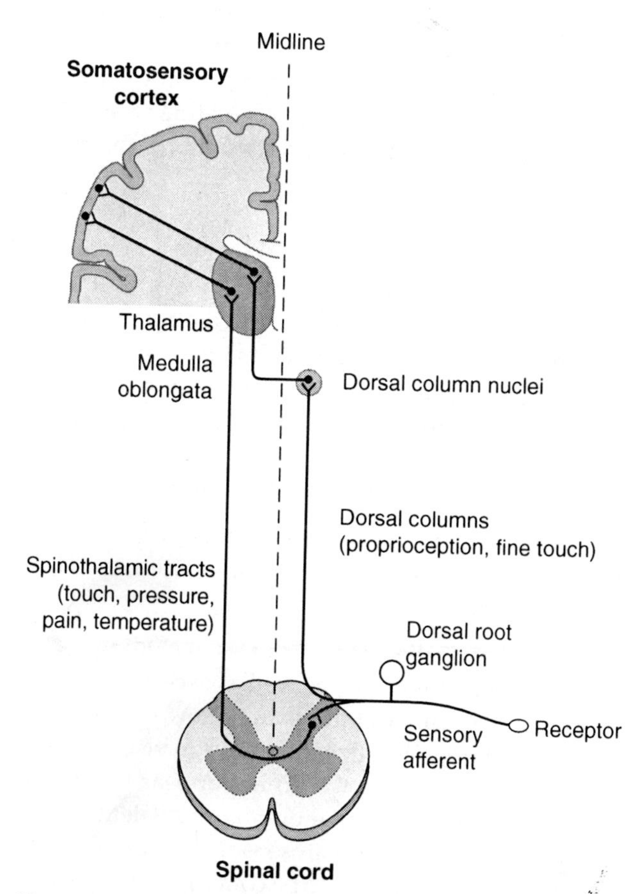

Fig. 127 The somatosensory pathways. Sensory information ascends the spinal cord in the ipsilateral dorsal columns and the contralateral spinothalamic tracts. These paths converge on the contralateral thalamus and thalamic neurones then connect to the somatosensory cortex in the postcentral gyrus of the parietal lobe.

Somatosensory cortex

Although we may become aware of some cutaneous sensations (particularly pain) at the level of the thalamus, conscious awareness and localization of sensory stimuli is mainly a function of the somatosensory cortex in the *postcentral gyrus* of the parietal lobe (Fig. 128). Different parts of the body are represented by distinct regions within the sensory cortex. The site of origin of any stimulus is thus coded for in the brain by the anatomical position of the activated cortical neurones, an example of *somatotopic representation*. As a result, one can plot out a cortical map of the body, called the somatosensory homunculus. In this, the area of cortex devoted to a given region reflects touch sensitivity rather than anatomical bulk, e.g., the cortex representing the skin over the fingertips is more extensive than that representing the skin of the leg or back. This can be illustrated by drawing the body with each part scaled in

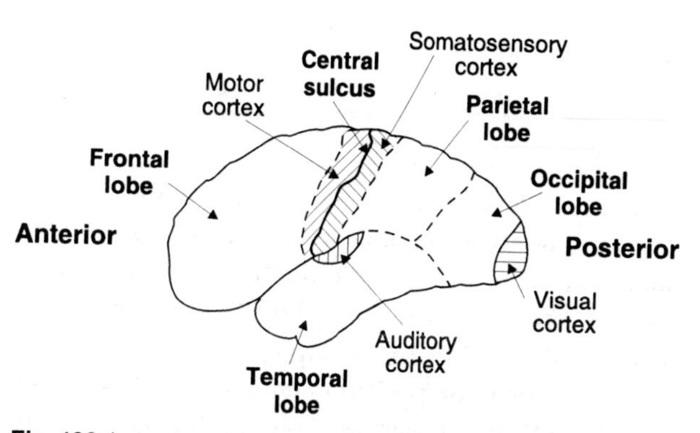

Fig. 128 Lateral view of the left cerebral hemisphere showing the four cerebral lobes and the areas of cortex involved in conscious awareness of sensation. Note the central sulcus which marks the boundary between the frontal and parietal lobes.

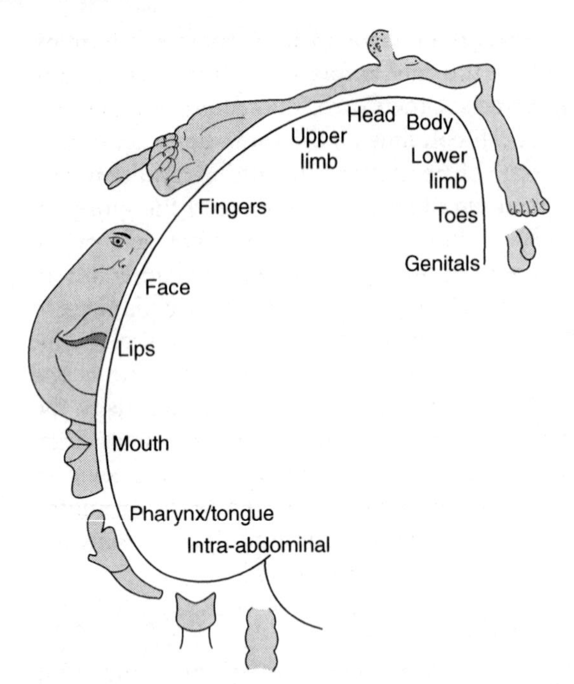

Fig. 129 Sensory homunculus for the left sensory cortex, which is responsible for sensation from the right side of the body.

proportion to the area of sensory cortex devoted to it—the so-called sensory homunculus (Fig. 129).

Somatosensory defects

Defects may arise because of a problem in a peripheral nerve, producing fairly localized symptoms, or because of defects in the spinal cord or brain, giving more widespread deficits.

Defective conduction in the primary sensory axons may result from trauma to a peripheral nerve, impaired conduction as part of a neuropathy, or damage to the spinal nerve roots as they enter the cord. The patient will typically complain of numbness (*anaesthesia*) or pins and needles (*paraesthesia*), and testing usually reveals a defect in all modalities, i.e., temperature, pain and joint position as well as touch. When the area of sensory loss is mapped out clinically, the nerve or nerve roots involved are often clear.

Damage to the ascending tracts in the spinal cord is normally bilateral, producing sensory loss below that level in all modalities and on both sides. If only one side of the spinal cord is involved, however, then the two sides of the body may be affected differently. A right-sided lesion, for example, would cause right-sided loss of joint position sense below that level (travels up the ipsilateral side of the cord) and left-sided loss of temperature and pain sensation (travel in contralateral tracts). Such defects are rare, however.

Somatosensory damage within the brain is frequently caused by a stroke involving the internal capsule, a region carrying ascending tracts from the thalamus to the sensory cortex. This always produces sensory loss involving the opposite side of the body. There are usually associated motor defects, since sensory and motor axons lie in close proximity within the inner capsule (Section 7.5).

Pain

The clinical significance of pain and its relief justifies further consideration of this somatosensory modality. Pain is an unpleasant sensation with a protective function and is normally generated by stimuli which are capable of causing tissue damage. The subjective perception of pain has a strong emotional, or affective, component and the fear and anxiety which are often experienced along with the pain itself contribute a lot to the unpleasantness of painful events. Pain is, therefore, a complex sensation which is influenced both by the peripheral transduction and conduction of information about damaging stimuli through nociceptors and sensory nerves, and by the central pathways which process this information. These central mechanisms include control loops which modify pain perception by regulating the ease with which signals are transmitted along the pain pathways.

Types of pain

Pain may be classified in several ways, but one general division is between pain coming from the skin, muscles, bones and joints (somatic pain), and pain originating in the viscera (visceral pain). Itch may be added to the list as a related, unpleasant sensation.

Somatic pain can be subdivided into:

- superficial pain, from the skin itself
- deep pain, from the underlying muscles, joints and bones.

Visceral pain has many causes including ischaemia and inflammation, as well as excess stretch or contraction of smooth muscle in hollow organs. It is often very hard to localize the source of visceral pain within the body, since we do not appear to have a neural map of our internal organs equivalent to the somatic map in the somatosensory cortex. In fact, visceral pain may appear to be coming from a part of the body anatomically distant from the true source, a phenomenon called *referred pain*. Ischaemic pain from the heart (angina), for example, may be experienced both as chest pain and as pain in the left arm. This seems to occur because sensory nerves from the heart enter the spinal cord at the same level as those from the arm. Similarly, irritation of the diaphragm produces pain referred to the shoulder tip since the sensory nerves from both diaphragm and shoulder enter the spinal cord in the cervical region. These are examples of the *dermatome rule*, i.e., pain from visceral afferents is often referred to the somatic dermatome whose sensory nerves enter the cord at the same nerve root level. Such considerations are important when seeking to interpret a patient's description of their pain in terms of a possible cause.

Responses to pain. Pain in general, and visceral pain in particular, is often associated with a variety of reflex autonomic responses. The nature of these is unpredictable, although sweating, nausea and vomiting are common, as is bradycardia caused by vagal slowing of the heart. This may reduce the blood pressure producing dizziness or fainting.

Fast and slow pain

Pain is also classified in terms of its speed of onset, with fast pain starting and stopping rapidly after the beginning and end of the painful stimulus itself. This type of pain is well localized and of a distinctive nature, so that the damaging stimulus can often be identified, e.g., fast pain from contact with a hot object is quite different from that caused by an incision. Fast pain is conducted by rapidly conducting, myelinated sensory axons and is only important in skin. Slow pain is much more diffuse, with an aching, burning or throbbing nature, and can persist for hours or days after the initial insult. This type of pain may originate in superficial or deep somatic structures, or from internal viscera, and is usually carried by slow conducting, unmyelinated nerves.

Itch

Itch is not strictly a form of pain but is a distinct, potentially very unpleasant sensation. Indeed, persistent and widespread itching may be just as distressing as chronic pain, e.g., in patients with severe atopic eczema. One important cause of itch is the release of histamine from mast cells, as occurs in skin allergies.

Pain receptors or nociceptors

Pain seems to be detected by specific receptors, called nociceptors, and is not simply the result of overstimulation of other types of receptor. Structurally, nociceptors are bare nerve endings and are distributed more densely within the skin than are other sensory receptors. Some are sensitive to specific damaging stimuli, e.g., mechanical or thermal nociceptors, but others respond to a range of different noxious stimuli (multimodal pain receptors). Many are chemoreceptors sensitive to substances released within damaged tissues, regardless of the cause of that damage. These chemicals include normal cell constituents, such as K^+ and ATP, which may be released following cell destruction, as well as inflammatory mediators like 5-hydroxytryptamine and bradykinin (Section 2.5).

Nociceptors do not adapt during prolonged stimulation, as anyone who has suffered from toothache for several days can testify. Their level of sensitivity can be modified, however, by other chemicals within the tissues. Prostaglandins are particularly important in this respect since they sensitize nociceptors, reducing the level of stimulation necessary to produce action potentials in the relevant sensory axons. This explains why drugs which inhibit prostaglandin synthesis (e.g., aspirin) have pain-relieving (*analgesic*) effects. It must be underlined that prostaglandins themselves do not directly stimulate nociceptor activity, they simply reduce the pain threshold in the presence of other damaging stimuli.

Central pathways in pain perception

Incoming afferents from nociceptors synapse within the dorsal horn of the spinal cord. There are a number of spinal cord reflexes which may be triggered directly at this level, e.g., withdrawal from the painful stimulus (Section 7.5), but there is no conscious awareness of pain at the spinal cord level. Axons cross the midline and ascend the cord in the anterolateral spinothalamic tract (Fig. 127). This makes important collateral connections with the reticular formation in the brainstem, which activates the autonomic nervous system and increases general cortical awareness (painful stimuli increase arousal). The thalamus relays the pain signals on to the cortex. The parietal cortex is important in localizing pain while associated frontal lobe activity contributes to the accompanying distress.

Regulation of pain by endogenous opiates: pain gate

The transmission of signals from neurone to neurone in the pain pathways can be inhibited by the activity of other, pain-regulating pathways. A group of peptides known as endogenous opioids or opiates, which include the *endorphins*, *enkephalins* and *dynorphins*, play an important role in this activity. They are naturally produced within the body (endogenous) and bind to the same cell receptors as opiate analgesic drugs like morphine. Conditions associated with increased endogenous opiate production, including exercise and certain forms of emotional stress, are also known to have analgesic influences. One proposed mechanism for this is release of enkephalin from a descending, inhibitory pathway, which blocks the transmission of the pain signal within the spinal cord. This may rely on presynaptic inhibition in which the enkephalin reduces the release of excitatory transmitter (possibly substance P) from the incoming pain fibre. It is sometimes said that there is a pain gate in the spinal cord, controlling the onward transmission of pain signals. In this model, the descending enkephalin pathways act to close the gate, reducing pain perception. Inputs from other peripheral sensory nerves may also inhibit the transmission of pain within the CNS, although not necessarily at the spinal cord level. This may explain why rubbing the site of an injury brings some relief; the increased sensory input inhibits the central transmission of pain signals from that region.

Box 26 Clinical note: Analgesia

Analgesics are treatments which are used to reduce pain. They may act at the level of the nociceptor (e.g., salicylates and other nonsteroidal anti-inflammatory drugs which inhibit prostaglandin synthesis, preventing receptor sensitization), the pain afferent axon (e.g., the use of local anaesthetics, like lidocaine (lignocaine), or local cooling to block nerve conduction) or within the CNS (e.g., the narcotic analgesics, such as codeine and morphine, which inhibit pain transmission centrally). Narcotic analgesics may also depress CNS activity in a nonspecific fashion, producing drowsiness and respiratory depression. Dissociative analgesics have little effect on the pain pathways themselves but greatly reduce the level of associated anxiety and distress, while general anaesthetics globally inhibit higher central nervous function, producing reversible loss of consciousness, e.g., during surgery.

Taste

Taste is dependent on chemoreceptors in *taste buds* within the mouth. These are particularly densely distributed along the sides of the papillae which cover the upper, or dorsal, surface of the tongue. The receptor consists of 40–50 chemosensitive cells which are exposed to the chemicals in their vicinity through an opening, or pore, in the upper surface of the taste bud (Fig. 130A). Microvilli, or *gustatory hairs*, project into this pore. Binding of a chemical to receptors on the microvilli causes release of neurotransmitter from the cell. This activates sensory nerves which

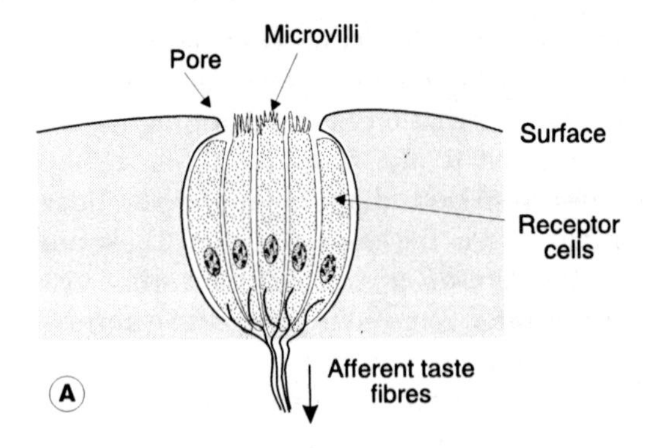

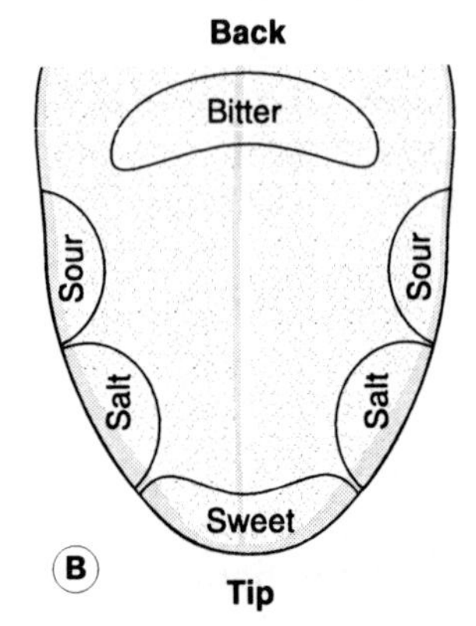

Fig. 130 Taste receptors. (A) A taste bud. (B) Areas of elemental taste sensitivity over the dorsal (upper) surface of the tongue.

connect the taste buds to the ipsilateral cortex (postcentral gyrus) via the thalamus, as well as sending relays to the brainstem and hypothalamus, regions involved in the control of feeding.

Taste is usually described in terms of one of four basic tastes, *sweet, salt, sour* (or acid) and *bitter*, and different regions of the tongue show varying sensitivities to each (Fig. 130B). These elementary tastes seem to reflect different overall patterns of taste bud stimulation rather than being caused by taste buds sensitive to only one type of chemical. Mammals are particularly sensitive to bitter tastes, which are often generated by chemicals in poisonous plants, presumably providing protection against their accidental ingestion.

Smell or olfaction

This depends on chemoreceptors in a small area of olfactory mucosa in the upper and posterior part of the nasal lining. The receptors are specialized neurones with microvilli projecting from their mucosal surface. Mucus-secreting cells support the receptor cells and molecules can only be detected once dissolved in the mucus layer. Axons emerging from the basal surface of the receptor neurones from the *olfactory nerve*, which passes through the bony cribriform plate and enters the olfactory bulb. This then connects directly with the olfactory cortex, which is part of the limbic system. The autonomic and behavioural controls exerted through hypothalamic and limbic mechanisms help explain the important emotional associations, and nervous and endocrine responses, which the sense of smell may evoke (Section 7.7).

Odour molecules can be detected at extremely low concentrations and produce a wide range of different perceived smells. These contribute greatly to the appreciation of food, for example, since much of what we term flavour is smell dependent; hence the loss of 'taste' when the nasal mucosa is inflamed during a common cold.

The auditory system

Hearing and balance both depend on specialized mechanoreceptors in the inner ear, with distinct but connected structures detecting sound (the auditory system) and the position and movement of the head (the vestibular system).

Auditory thresholds and audiometry

Sound is the perceived sensation generated in response to pressure waves travelling through the air to our ears. These sound waves may be characterized physically in terms of their frequency and amplitude, features which are perceived as *pitch* (high frequency, high pitch) and *loudness*, respectively. Although normal sounds are a complex mixture of frequencies and pressure amplitudes, hearing sensitivity is usually measured in terms of the *threshold amplitude* at which notes of a single frequency can just be detected by the subject (audiometry). The threshold pressure is measured for the different frequencies tested and plotted as an *audiogram*. Absolute sound pressure (P) is usually converted to a relative logarithmic scale measured in decibels (dB).

The sensitivity of hearing varies with the frequency of sound, being maximal around 2000–3000 Hz (the upper frequency range in speech) and falling off above and below this value (Fig. 131A). Clinical audiograms do not always make this point clear, however, since the threshold pressure is often plotted in terms of hearing loss relative to the expected threshold at any given frequency, giving a normal value of zero for all frequencies (Fig. 131B).

Structure of the ear

The ear is divided into the *outer*, *middle* and *inner ears* (Fig. 132). The first two structures conduct sound to

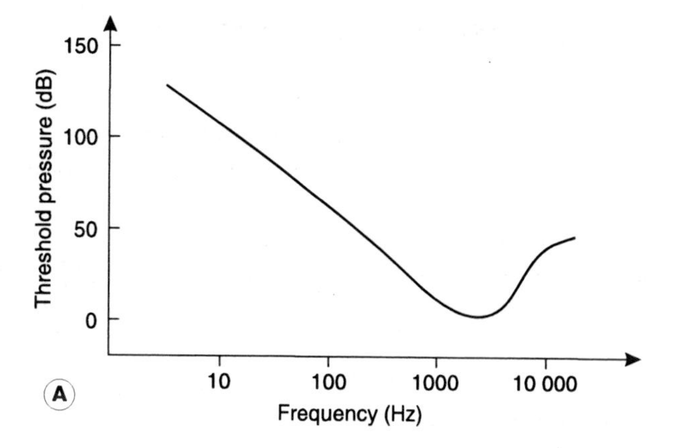

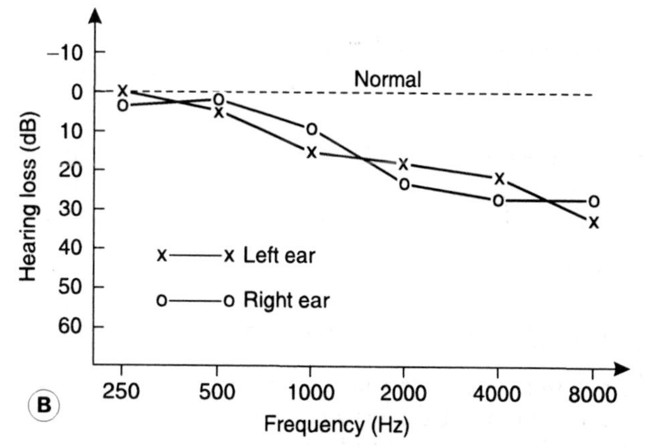

Fig. 131 Audiometry. (A) The pressure at which a sound can just be detected (i.e., the auditory threshold) is plotted against sound frequency. (B) Clinical audiograms compare the threshold pressure with the expected value and plot this in terms of hearing loss. This audiogram indicates a bilateral decline in sensitivity in the higher frequency range and is typical of noise-induced hearing loss and hearing loss with ageing (presbycousis).

the *cochlear* portion of the inner ear where it is detected by specialized receptors. The *vestibular apparatus*, which is important for balance control, is also located within the inner ear and will be considered in more detail below.

The outer ear consists of the visible ear, or pinna, and the *auditory canal*. Sound waves are channelled through this canal to the *tympanic membrane* at its inner end. This is set vibrating and these vibrations are conducted across the middle ear by movements of three bony *ossicles*, the *malleus*, *incus* and *stapes*. This rather elaborate construction allows vibration in the air to be converted into vibration in the fluid of the inner ear through the footplate of the stapes, which fits into the *oval window* of the cochlea. The direct transmission of sound waves between air and liquid would, in fact, be very inefficient, and the structures of the middle ear effectively act to amplify the pressure waves generated in the cochlear fluid by:

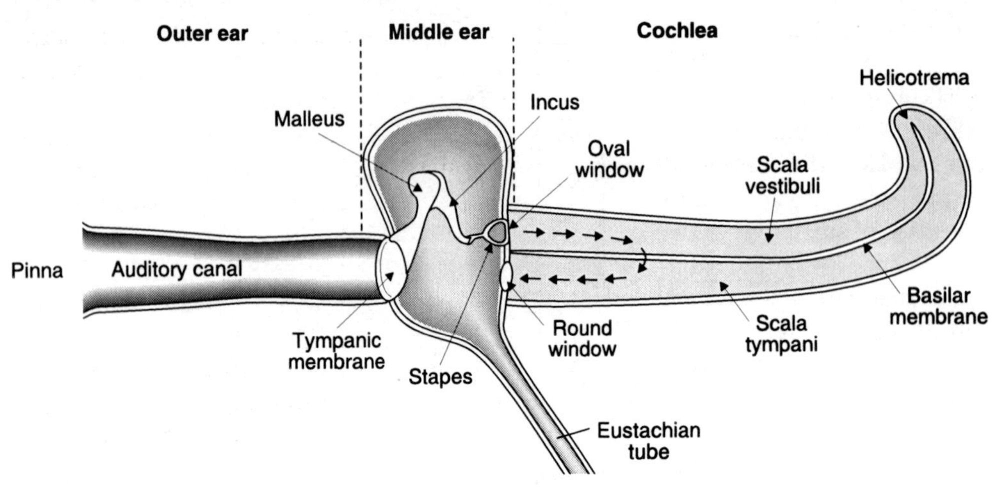

Fig. 132 The main auditory structures of the ear (the vestibular elements of the inner ear are not represented). The cochlea has been 'uncoiled' for clarity. The arrows represent sound vibrations transmitted from the oval window through the basilar membrane to the round window.

- transmitting the total force caused by vibration of the tympanic membrane to the much smaller area of the oval window, increasing the pressure generated
- providing a lever system which increases the force delivered to the foot of the stapes.

The structures of the middle ear are protected from damage caused by loud noises through reflex contraction of the *tensor tympani* and *stapedius* muscles in the middle ear. This restricts the movement of the tympanic membrane and ossicular chain. The middle ear is also connected to the nasopharynx by the *eustachian tube*, which is opened during swallowing and yawning. This allows the pressures inside and outside the tympanic membrane to be equalized if the external pressure changes, e.g., during ascent and descent in an aeroplane.

Cochlear structure and function

The cochlea consists of a coiled tube which forms part of the bony labyrinth. In cross-section, the cochlea is seen to be divided into three chambers by two membranes. *Reissner's membrane* (or the vestibular membrane) separates the *scala vestibuli* from the *scala media* (or cochlear duct), which is, in turn, divided from the *scala tympani* by the *basilar membrane* (Fig. 133). The scala vestibuli and scala tympani are filled with perilymph (similar in composition to cerebrospinal fluid) and are in continuity with each other via an opening at the very end of the basilar membrane known as the *helicotrema*. The scala media contains a K^+-rich fluid known as *endolymph* and forms part of the membranous labyrinth in continuity with the vestibular apparatus (see Fig. 134).

Vibrations in the stapes are communicated to the fluid of the scala vestibuli via the oval window. Very-low-frequency movements simply displace fluid from the scala vestibuli to the scala tympani through the helicotrema and are not detected as sound. Higher frequencies, however, set the basilar membrane in motion and this movement is detected by *hair cells* in the *organ of Corti*, which generate an auditory signal. Basilar membrane vibration is possible because movement of the membranous *round window*, which

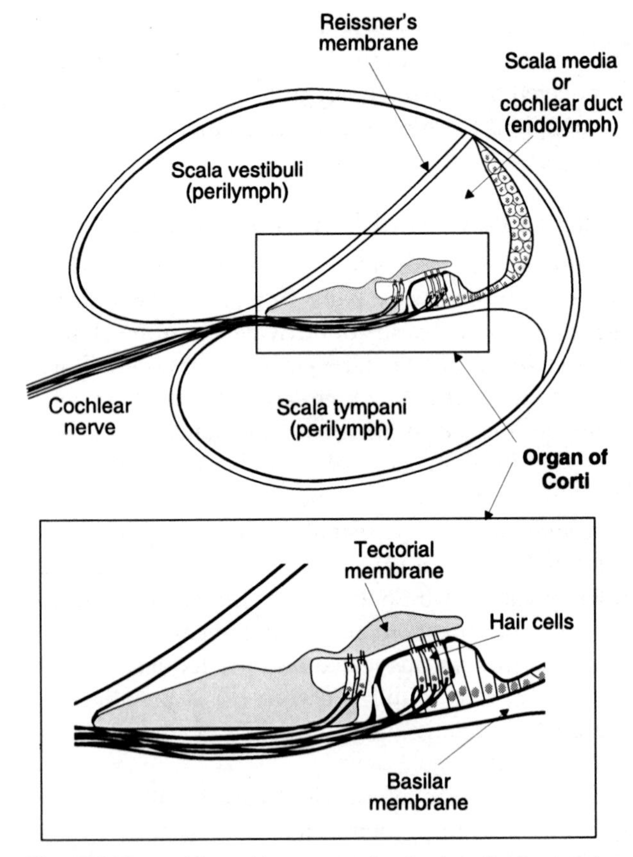

Fig. 133 The cochlea cut in cross-section to show the three main chambers and the membranes which divide them.

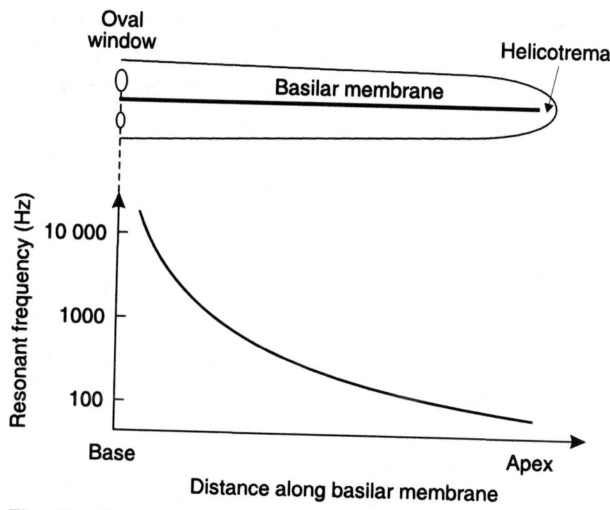

Fig. 134 The resonant frequency of the basilar membrane (shown in the diagram of the 'uncoiled' cochlea in the upper part of the figure) decreases with distance away from the oval window (at the base) towards the helicotrema (at the apex).

opens back onto the middle ear, accommodates the displacement of fluid in the scala tympani.

Organ of Corti The auditory receptor is the organ of Corti which lies within the scala media (Fig. 133). It sits on the basilar membrane and contains a number of hair cells, each projecting a series of *stereocilia* (the 'hairs' in question) from its upper pole. The ends of these cilia are embedded in the overlying *tectorial membrane* so that sound-induced vibration of the basilar membrane produces shearing movements between the cilia and the body of the hair cells. It is this distortion which sets up the receptor potentials in these specialized mechanoreceptors. Hair cells synapse with the neurones of the cochlear nerve, releasing excitatory transmitter when activated. The sensory nerve fibres carry auditory information centrally to the brain.

Pitch determination One important aspect of hearing is our ability to distinguish between sounds of differing frequency, differences which we perceive as changes in pitch. This requires that the auditory system should be able to code for sound frequency in some way. Two different coding methods are used for low- and high-frequency sounds, one based on frequency of action potential discharge and the other based on the position of the activated hair cells.

Phase locking: For low-frequency sounds up to 4000 Hz, the frequency of the sound can be directly coded for by the pattern of action potentials in the afferent axons, which discharge in phase with the stimulus frequency. This is referred to as phase locking.

Tonotopic coding: At higher sound frequencies, the frequency is coded for by the site along the basilar membrane at which hair cells are maximally stimulated, i.e., coding by place, rather than the pattern of action potentials in the auditory axons. Mechanical properties, such as the width and thickness of the basilar membrane, vary along its length. As a result, its resonant frequency also changes, being high (20 000 Hz) close to the oval window, and low (100 Hz) near the helicotrema (Fig. 134). Sound waves generate the largest amplitude of vibration at the point along the basilar membrane where the resonant frequency matches the frequency of the incident sound, and hair cell stimulation will be maximal at that point. Thus, the position of an activated hair cell along the length of the basilar membrane can be used to code for the frequency of a sound. Activity from receptors close to the base of the cochlea (the oval window end) is interpreted as being caused by a high-pitched sound.

Central auditory pathways

The *cochlear nerves* form part of the vestibulocochlear nerves (VIIIth cranial nerve, sometimes referred to as the auditory nerve) and travel to the cochlear nuclei in the medulla oblongata. From here the auditory pathways ascend to the medial geniculate nuclei in both the contralateral and the ipsilateral thalamus, sending collaterals to the reticular activating system en route. The thalamus, in turn, connects to the auditory cortex in the temporal lobe (Fig. 128).

Box 27 Clinical note: Deafness

Impaired hearing may result from defects in the conduction of sound from the air to the cochlea (conductive deafness) or from lesions in the transduction–transmission systems of the cochlea and cochlear nerve (sensorineural deafness).

Conductive deafness may arise because of blockage of the auditory canal (e.g., by accumulated wax), or damage to the middle ear through mechanical disruption (e.g., perforation of the tympanic membrane) or infection (otitis media).

Sensorineural deafness can be caused by damage to the cochlear hair cells following repeated exposure to loud noise. This usually produces more marked hearing loss at higher frequencies and may appear many years after sound exposure has ceased. A similar pattern of high-frequency deafness is also seen in many elderly subjects, a condition known as *presbycousis* (see Fig. 131B). Other causes of sensorineural defects include intrauterine infections such as rubella, leading to congenital deafness, and acquired deafness secondary to meningitis. Childhood deafness is a particularly severe handicap since it interferes with the acquisition of spoken language.

The vestibular system

The second major sensory function of the inner ear is to provide information about the position and acceleration of the head. This is particularly important in the maintenance of balance and control of movement and depends on the vestibular apparatus, which is located within the bony labyrinth adjacent to the cochlea. The inner, membranous labyrinth includes five vestibular components (three *semicircular canals*, the *utricle* and the *saccule*, Fig. 135), in addition to the cochlear duct of the auditory system. It is filled with *viscous endolymph*, while the surrounding space between it and the bony labyrinth contains *perilymph*. The sensory transducers are hair cells which synapse chemically with the fibres of the *vestibular nerve*.

Structure and function of the semicircular canals

The three semicircular canals are oriented approximately at right angles to one another and each detects any *rotation* in its own plane. The sensory receptor, known as the *crista*, is located within the distended *ampulla* of each canal and consists of hair cells whose *stereocilia* project into a gelatinous dome known as the *cupula* (Fig. 136A). When the head starts to rotate in the plane of the canal, the inertia of the viscous endolymph causes it to be displaced, deflecting the cupula in the opposite direction to that of the rotation (Fig. 136B). The stereocilia bend, producing a change in the membrane potential of the hair cells. On stopping rotation, the inertia of the endolymph deflects the cupula in the reverse direction. This produces the opposite effect on membrane potential to the initial deflection, so that if the hair cells stimulate activity in the vestibular nerves at the onset of the rotation, for example, they will reduce it when rotation stops. It can be seen that maximal stimulation of the receptors is produced by a change in rotation, i.e., an angular acceleration, rather than during constant rotation (Fig. 136B). In normal life, however, prolonged periods of constant rotation are rare.

Since there are three sets of hair cells in differently oriented canals (Fig. 135), the actual direction, or axis, of any body rotation can be identified from the overall pattern of receptor activity. This is analogous to the mathematical definition of a point's position

Fig. 135 Simplified diagram of the membranous labyrinth of the left ear. The utricle, saccule and ampullary swellings in the semicircular canals are shown, along with their receptor organs, but all other communicating ducts (as well as the cochlear duct) are represented by single lines for simplicity.

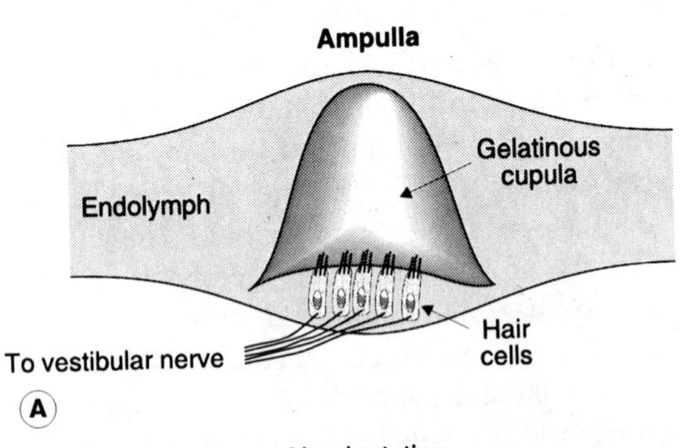

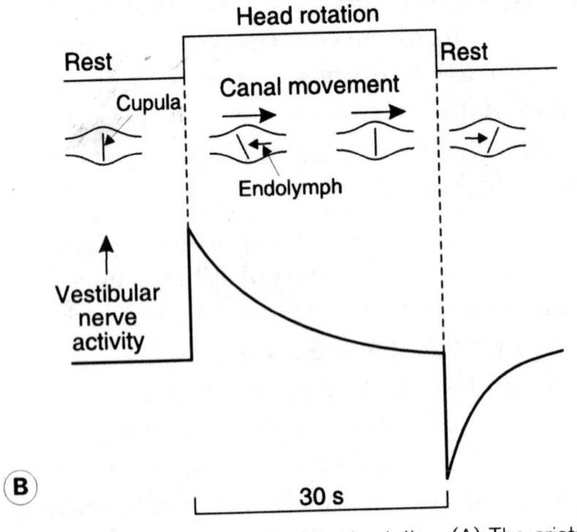

Fig. 136 Semicircular canals detect rotation. (A) The crista inside the ampulla of a semicircular canal. (B) Effects of head rotation on the cupula and vestibular nerve activity. The onset of rotation deflects the hair cell cilia because the inertia of the endolymph acts as a brake resisting the movement of the cupula. The resulting hair cell receptor potential increases vestibular nerve activity. During constant rotation, however, the cupula returns to its original position, and receptor stimulation ceases. When rotation stops, the inertia of the endolymph deflects the cupula in the opposite direction, reversing the receptor potential and reducing vestibular nerve activity.

in space using three different coordinate axes (x, y and z).

Structure and function of the utricle and saccule

The utricle and saccule detect *linear acceleration* and changes in *head position*. The receptor region is known as the *macula* and consists of fixed hair cells whose stereocilia are embedded in an overlying gelatinous membrane (Fig. 137A). This is known as the *otolithic membrane* because it contains tiny crystals of calcium carbonate, or *otoliths*, which increase its inertia and weight. If the otolithic membrane exerts a force at right angles to the attached stereocilia, receptor potentials are produced in the hair cells, altering the activity in the vestibular nerve.

Because of their different orientations in the head, the utricle and saccule provide slightly different information.

Horizontal acceleration. Both utricle and saccule are sensitive to horizontal accelerations. These distort the cilia because of the inertia of the attached otolithic membrane.

Body position. The utricle is particularly useful in supplying information about the position of the body relative to the vertical, e.g., during bending movements. When upright, the utricular macula lies close to the horizontal plane (Fig. 137B). In this position, gravity acts vertically along the length of the hair cells and has little tendency to bend the cilia sideways. As a result, receptor output is minimal. Bending of the waist or neck, however, allows a component of otolithic membrane weight to act at right angles to the receptor cells, and this activates them.

Vertical acceleration. The saccular macula is oriented vertically in a parasagittal plane (Fig. 135) and is, therefore, sensitive to vertical acceleration, e.g., when jumping or riding in a lift.

Central vestibular pathways

Vestibular information is carried in the vestibular portion of the vestibulocochlear nerve (the VIIIth cranial nerve) to the vestibular nuclei in the brainstem. These connect with:

- areas of the cerebellum and spinal cord involved in the control of motor activity, posture and balance
- the oculomotor centre which controls eye movements; this helps maintain gaze fixation during movement and rotation
- the thalamus, which relays the input to the postcentral gyrus in the temporal lobe of the cerebral cortex. This region is involved in the conscious awareness of position and acceleration, and the voluntary control of balance.

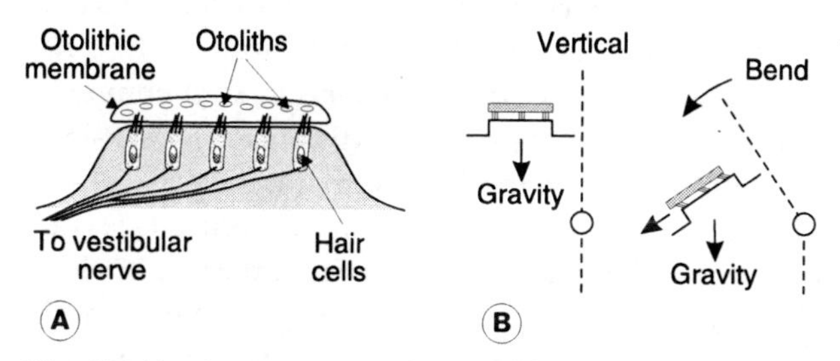

Fig. 137 Macular structure and function. (A) The main components of the macula. (B) Detection of head position relative to the vertical is one function of the utricular macula.

Box 28 Clinical note: Vestibular defects

Abnormal vestibular stimulation (e.g., during motion sickness) and vestibular damage can produce loss of balance, with a profound sense of dizziness known as ***vertigo***, and associated nausea. There may also be repetitive involuntary eye movements in which the eyes drift to one side before rapidly jerking back to the central position. Vestibular *nystagmus*, as this is called, is also seen in normal individuals on starting and stopping vigorous rotation.

The visual system

Vision depends on a complex receptor organ, the eye, which contains important optical elements as well as the photoreceptors themselves. These receptors are located within the retina, which converts the incoming visual signal into an electrical signal in the optic nerve. This carries the information centrally for further processing in the brain.

Structure of the eye

The outer coat of the eye consists of a tough, connective tissue layer known as the *sclera*, part of which is normally visible as the white of the eye (Fig. 138). Anteriorly this is continuous with the cornea, a transparent window through which light enters the eye. The cornea is continuously washed and lubricated by a film of tear fluid, which is renewed and dispersed by blinking of the eyelids. The space between the back of the cornea and the lens of the eye is filled with a clear liquid called *aqueous humour*. This region is divided into anterior and posterior chambers by the *iris*, which controls the diameter of the pupil through contraction of radial and longitudinal smooth muscle fibres. The *lens* itself is suspended from the *ciliary body* by the *suspensory*

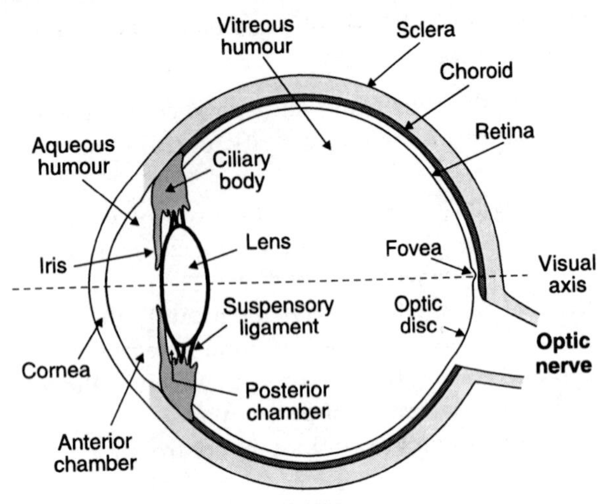

Fig. 138 The main structures of the eye.

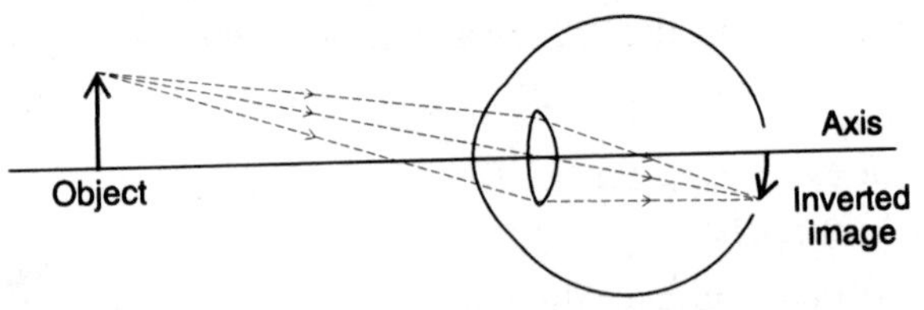

Fig. 139 The optics of the eye focus all the diverging rays of light from a given point on an object onto a fixed point on the retina. The result is a sharply focused, but inverted, retinal image.

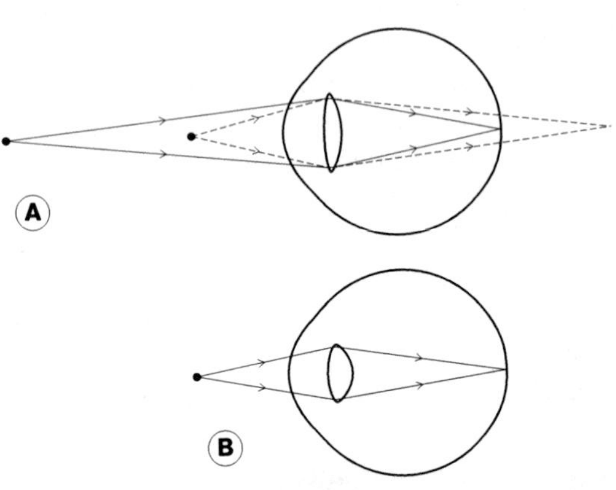

Fig. 140 Accommodation for near vision. (A) In the relaxed state, the lens is pulled outwards into a relatively flat structure with limited ability to converge incoming light rays. This is appropriate for distant vision (solid line ray paths) but more divergent light from near objects (broken lines) will only be brought to a focus behind the retina. (B) Contraction of the ciliary muscle for near vision produces a more convex lens with increased focusing power.

ligament (or zonule). Contraction of the ciliary smooth muscle affects the shape of the lens, allowing the eye to alter its focal distance to accommodate vision to objects at different distances.

The chamber behind the lens is filled with gelatinous *vitreous humour* and is lined by the retina. The retina contains the photoreceptors and a large number of neurones involved in the processing of visual signals. Retinal blood vessels ramify over the inner surface of the retina and may be observed using an ophthalmoscope. Other visible features include the pale *optic disc*, where the retinal axons enter the optic nerve, and the *macula* (or macula lutea), a region close to the visual axis of the eye which is free from large retinal vessels. In the middle of the macula is a central depression called the *fovea* (or fovea centralis). This is specialized to produce the sharpest vision, i.e., it provides the highest level of visual resolution or acuity.

The vascular *choroid* layer lies between the retina and sclera, and supplies most of the nutritional needs of the photoreceptors. Choroidal melanin, in conjunction with the pigment layer of the retina, probably also helps to absorb stray light, preventing it from being scattered back into the photoreceptor layer where it would blur the visual image.

Visual optics

The optical properties of the eye result in the formation of a sharp image of external objects on the retina. This depends on refraction of light by the cornea and lens of the eye. Rays of light are bent, or refracted, at the interface between regions with differing refractive indices, e.g., on passing from air into the cornea. Because of the convex outer surface of the cornea and the biconvex shape of the lens, the overall effect is to provide a convergent optical system. Diverging rays of light reaching the eye from each point on an object can, therefore, be brought back to a focus in the plane of the retina (Fig. 139). The resulting image is inverted both vertically and laterally, so that parts of the object in the top right of the visual field produce an image on the bottom left of the retina.

Accommodation The lens of the eye contributes a smaller fraction of the total focusing power of the eye than the cornea. It makes an important contribution to visual performance, however, because its focal length can be altered by altering its shape, allowing us to focus on objects close to the eyes. Accommodation, as this is known, is controlled through the ciliary muscles. When these are relaxed, the suspensory ligaments pull outwards on the lens, flattening it and reducing its tendency to converge light. This is appropriate for distance vision, since the light entering the eye from far objects is less divergent and so has to be bent, or focused, less to produce a sharp retinal image than that from close objects (Fig. 140A). For near vision, parasympathetic nerves stimulate the ciliary muscle to contract, moving the ciliary body towards the lens and reducing the tension on the lens attachment. This allows the natural elasticity of the lens to pull it into a more convex structure with greater focusing power (Fig. 140B). The closest point at which objects can be brought into focus is known as the *near point* and is normally <10 cm from the eye in a young adult. With ageing, the near point moves further and further from the eye (*presbyopia*) as the elasticity of the lens decreases, diminishing its ability to form a highly convex structure. Because of this, many older people require spectacles for reading.

Pupil size and depth of focus The iris controls the size of the pupil. Sympathetic nerves stimulate contraction of the radially disposed smooth muscle, producing pupillary dilatation, while parasympathetic activation of the circumferential muscle leads to constriction. Reflex constriction of the pupil in response to both light and accommodation (the *pupillary reflexes*) is routinely tested in the clinical examination of the nervous system. Pupillary constriction has two important consequences for vision.

Levels of light. Constriction reduces the amount of light entering the eye in bright conditions. This is analogous to reducing the aperture on a camera and helps ensure that the level of photoreceptor stimulation is optimal for image contrast.

Depth of focus. Constriction increases the depth of focus of the eye. This is most easily understood by thinking of photography again. If a camera is focused on some distant object, such as a mountain range, then most of the objects in the photograph will also appear acceptably sharp, even though they may be quite a long way in front of the true focal plane. We say that there is a large depth of focus under these circumstances. In a close-up shot of a face, however, the depth of focus is small, so that nearly everything other than the subject itself appears blurred. Reducing the aperture of the camera improves the depth of focus by cutting out peripheral rays of light, since these produce more blurring than light close to the optical axis. In the eye, the depth of focus is also diminished during near vision, but reflex constriction of the pupil during accommodation helps limit this effect.

Retinal structure and function

The retina consists of several layers of cells, with the photoreceptors (the *rods* and *cones*) located towards the external side (furthest from the vitreous humour). This means that light must normally pass through the other, more internally situated cell layers before stimulating the receptors (Fig. 141). In the central area of the macula, however, the retinal layers are displaced to either side of a depression called the

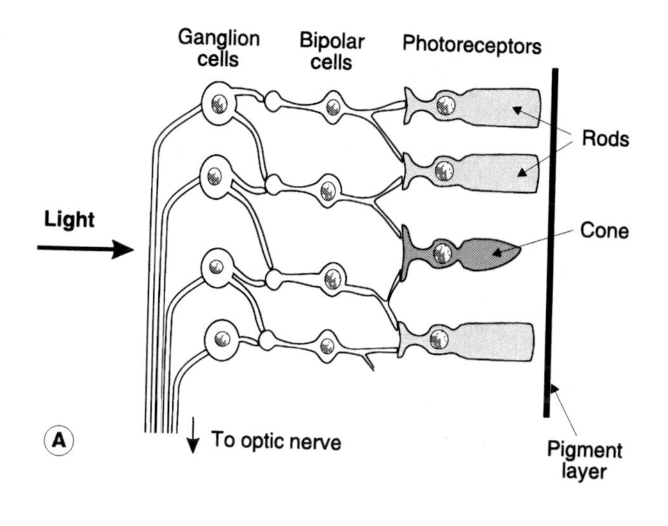

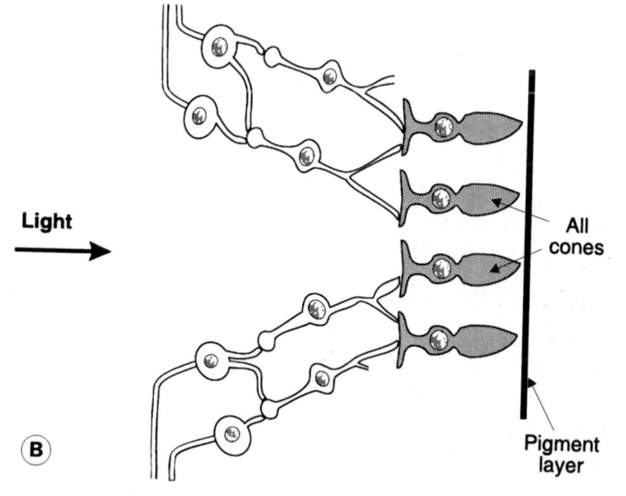

Fig. 141 Retinal structure. (A) The peripheral retina. (B) In the foveal region, the internal cell layers are displaced sideways so that light is minimally scattered before it reaches the photoreceptors, which are all cones.

fovea. At this point, which lies on the visual axis of the eye, there is minimal scattering of incoming light before it reaches the photoreceptors, which are exclusively cones. When looking directly at an object the image falls on the fovea and this region gives the highest level of *visual acuity* of any part of the retina.

Photoreceptors

The two types of receptor cell, the rods and cones, are distinguished by the shape of their membranous outer segments (Fig. 142). They also show functional differences, since rods only detect light intensity and give no information about its wavelength or colour. Colour vision depends on the cones, which also provide a more highly resolved (sharper) image than that from the rods. Rods have a greater sensitivity to stimulation at low light levels, however, generating

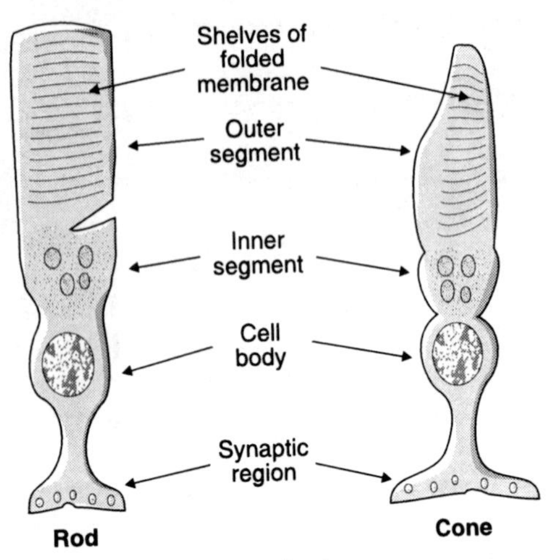

Fig. 142 Rod and cone structure.

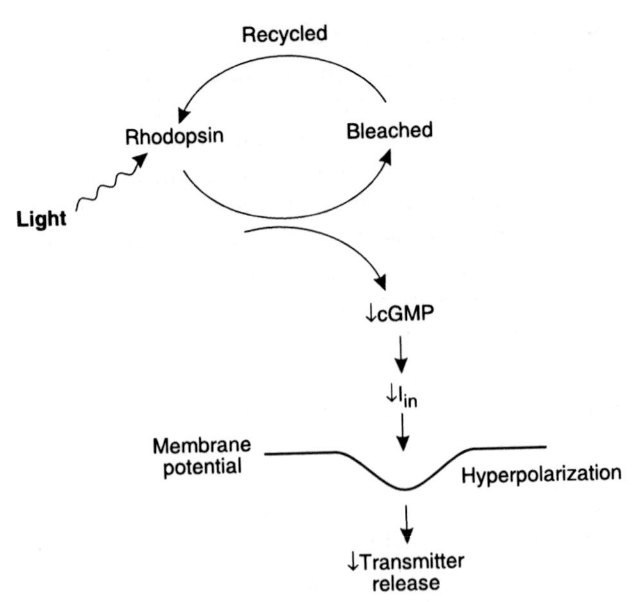

Fig. 143 Summary of transduction mechanisms in a rod.

an image of the world painted in shades of grey when conditions are too dim to activate the cones. The distribution of the two receptors varies across the retina, with many more rods than cones in the peripheral areas and an increased density of cones in the macula. The fovea contains only cones, and these are more tightly packed and have a smaller diameter than elsewhere. This helps explain the high visual acuity in this area, since retinal resolution of two points of light requires that they stimulate separate receptors. The smaller the receptor diameter, the smaller the receptive field (Section 7.3), and the more closely stimuli can be spaced without loss of visual discrimination.

The *outer segment* of the receptors, which contains a series of folded membrane shelves stacked on top of each other, functions as the specialized receptor region in both rods and cones (Fig. 142). The *inner segment* contains mitochondria and other organelles and is continuous with the cell body of the receptor. At its internal end, each receptor synapses with *bipolar cells* and these connect in turn with the *ganglion cells* of the retina. It is the ganglion cell axons which eventually leave the retina at the optic disc to form the *optic nerve*.

Receptor mechanisms Phototransduction (the conversion of light to an electrical signal) relies on *photosensitive pigments* which coat the membrane shelves of the outer segment in rods and cones. These pigments differ chemically in different receptors, but the general principles of the light response remain the same in each case. The photopigment in rods is called *rhodopsin* and consists of an opsin protein attached to a vitamin A derivative called *cis*-retinal. Absorption of light leads to bleaching of the rhodopsin, in which the retinal changes shape, converting to the *trans*-retinal isomer, and then splits off the opsin. These events regulate the intracellular levels of the messenger molecule cyclic guanosine monophosphate (cGMP), and result in *membrane hyperpolarization* (Fig. 143). When unstimulated, the relatively high level of cGMP opens membrane ion channels, allowing a Na^+ current to flow into the outer segment of the rod (the *dark current*). This normally tends to depolarize the membrane, but exposure of the photopigment to light causes a reduction in the concentration of cGMP, switching off this inward current. As a result, the membrane potential becomes more negative than normal. This light-induced, hyperpolarizing receptor potential contrasts with the depolarization seen in most other receptors but is equally effective as a means of controlling the activity in the optic nerve, the final pathway to the brain for visual information. Bleached photopigment, which is light insensitive, is eventually recycled to form active rhodopsin again.

Adaptation

The light sensitivity of the rods and cones increases as the intensity of light decreases, a phenomenon known as *dark adaptation*. Sensitivity mainly depends on the concentration of pigment in the receptors, which is reduced in bright light because of the rapid rate of bleaching. In the dark, bleaching is greatly reduced and pigment concentration rises. Adaptation of the rods can increase their sensitivity by a factor of 105 and takes about 40 minutes to develop fully. Reversal of this process in bright light (*light adaptation*), however, is complete within seconds. Rods are particularly important for night vision since they demonstrate a sensitivity to light 100 times greater than that of cones when both are fully dark adapted.

In certain circumstances, e.g., at night on the bridge of a ship, it is desirable to provide a relatively high level of illumination while maintaining dark adaptation of rods. This double goal can be achieved by using red light, since the photopigment in rods is insensitive to bleaching by long wavelengths (Fig. 144). Cones provide for high-acuity vision under these conditions but the rods remain fully dark adapted, and so high-sensitivity vision at shorter wavelengths is retained.

Colour vision

The *trichromatic theory* explains colour vision in terms of three classes of cone (red, green and blue), which absorb light maximally in different parts of the visual spectrum because of differences in the opsin proteins of their photopigments (Fig. 144). When a given wavelength of light strikes the retina, it produces a pattern of activity in the different cones which depends on how well the pigment each contains can absorb that light. Yellow light with a single wavelength of 560 nm, for example, will stimulate the red and green cones almost equally, but produce no activity in the blue cones. Equivalent stimulation of the cones can, however, be achieved by shining a balanced mixture of red and green light (two different wavelengths) at the retina. The trichromatic theory would predict that this should be visually indistinguishable from the monochromatic yellow light and, indeed, a mixture of red and green light is perceived as yellow. If all three cones are stimulated equally, the image appears white, a 'colour' with no equivalent wavelength.

The importance of multiple cone types for colour discrimination is emphasized by the observation that *colour blindness* is associated with the inherited absence of one or more class of cone. As an *X-linked, recessive* condition, this is more common in men than women and results in an inability to distinguish between certain colours, e.g., red and green. The trichromatic theory cannot account for all aspects of colour vision, since it has been shown that perceived colour is not solely dependent on the mixture of wavelengths entering the eye. An object's colour, for example, appears relatively constant under a wide range of varying illumination conditions, e.g., in both daylight and artificial light. Since the pattern of reflected wavelengths is usually very different under these two conditions, *colour constancy* cannot be simply explained on the basis of the absorption spectra of the cones. The mechanisms involved, however, are not known.

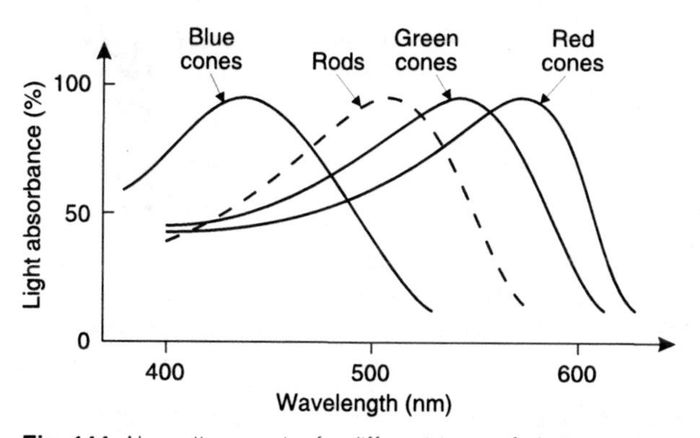

Fig. 144 Absorption spectra for different types of photoreceptor.

Retinal connections

Rods and cones are linked to the ganglion cells of the retina by bipolar cells in the inner nuclear lamina (Fig. 141). Light-induced hyperpolarization reduces neurotransmitter release from the receptors and this produces graded potential changes (not action potentials) in the bipolar cells. Some cells are depolarized while others are hyperpolarized, depending on whether they are inhibited or excited by the transmitter, and this regulates release of excitatory transmitter at the synapses between bipolar and ganglion cells in the retina. Thus, changes in incident light, as detected by the photoreceptors, control the action potential output in the ganglion cell axons within the optic nerve.

Retinal processing of visual information The complex connections between receptors, bipolar cells and ganglion cells in the retina allow for considerable processing of visual information before it is passed to the brain. At the receptor level, strongly stimulated cones inhibit more weakly stimulated neighbours in a process called *lateral inhibition*. Ganglion cell output is also sensitive to local variations in stimulus levels, with *on-centre* ganglion cells responding most strongly when the light is brightest in the centre of the relevant receptive field while *off-centre* cells respond best when the edge of the receptive field is bright. These mechanisms tend to enhance visual contrast, making edges and boundaries in the image easier to identify.

Visual acuity

Visual acuity refers to an observer's ability to resolve the detail in an image and is often assessed using the familiar Snellen chart, with its rows of different sized letters. Each row is labelled with the maximum distance in metres from which someone with normal vision should be able to read letters of that size. The chart is viewed from 6 m and the distance label for the smallest size of letter which can be read is used as a measure of acuity. For a row which equates to a distance N, the visual acuity is said to be $6/N$. Normal acuity equates to 6/6, while $N > 6$ (i.e., the smallest letters identified should be visible from a distance of greater than 6 m) indicates impairment.

Visual fields

The visual field may be defined as the region of space within which a visual stimulus can be detected without moving the eye. Each eye is tested separately, the observer being asked to fix their gaze on the centre of a target and then report when they first become aware of a stimulus brought from beyond their line of sight. The field extends further in the lateral or temporal direction than on the medial, or nasal side, where the nose acts as a physical obstruction (Fig. 145). There is also a *blind spot* on the temporal side of the field for each eye. Images from this area fall on the optic disc on the nasal side of the retina, where there are no photoreceptors. The visual fields of the two eyes overlap extensively allowing for *binocular* or stereoscopic vision, in which the slightly different images from the two eyes are combined within the brain. This is an important mechanism in depth perception and three-dimensional vision.

Central visual pathways

All the visual information from each eye passes backwards in the optic nerve until it reaches the optic chiasma immediately below the hypothalamus. At this point, axons cross the midline so that the signals from the left visual field of each eye, i.e., from the right half of each retina, are brought together in the optic tract on the right side of the brain, and vice versa (Fig. 146). These tracts synapse in the lateral geniculate nucleus of the thalamus on each side, from where the optic radiations extend out to the visual cortex in the occipital lobes of the brain (Fig. 128). Relays from the optic tracts also pass to the superior colliculi, which are involved in the control of eye movement.

Cortical processing of visual information is complex. Not only is the spatial distribution of the retinal receptors represented in the cortical map, but neurones in different parts of the cortex respond to specific visual features. Some are stimulated by lines, others detect corners, while others may only be stimulated by movement in a certain direction. Eventually these responses are integrated into a single image.

Visual development and neural plasticity

Normal development of the lateral geniculate nucleus and visual cortex in early life is dependent on normal visual experiences so that a defect in the eye itself can actually disrupt the growth of brain areas involved in the analysis of the incoming visual information. The ability of sensory experience to mould the anatomical development of sensory pathways in this way is one example of neural plasticity, in which the structure and function of the nervous system is modified by the signals it carries. Susceptibility to these effects is maximal during early life,

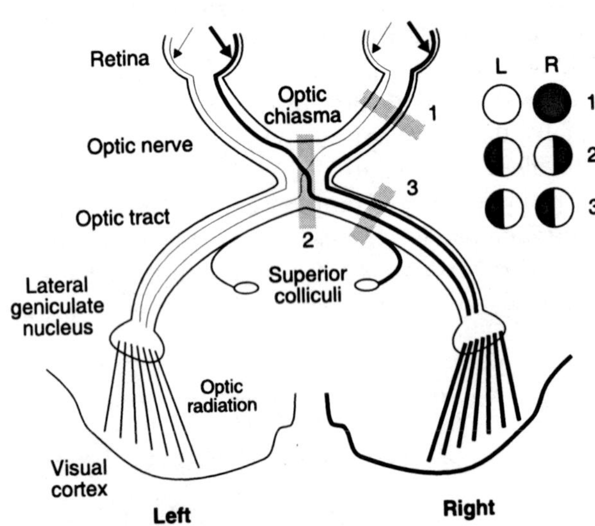

Fig. 146 The visual pathways. All information from the right visual field is carried via the left optic tract and radiation to the left occipital cortex (light lines) and from the left visual field via the right tracts (heavy lines). The visual field defects produced by damage at different points along the pathway are indicated on the field diagrams to the right. 1: right monocular blindness caused by right optic nerve damage; 2: bitemporal hemianopia caused by optic chiasma damage; 3: left homonymous hemianopia caused by right optic tract damage.

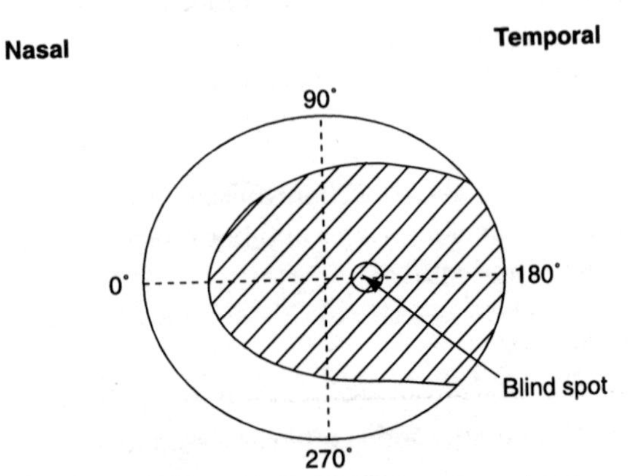

Fig. 145 The visual field of the right eye (shaded). The eye is staring fixedly at the intersection of the vertical and horizontal axes so light from this point falls on the fovea.

so that permanent visual deficiencies can occur as a result of any childhood defect which is allowed to persist through a critical period up to the age of 5 or 10 years. Later correction of the initial problem, e.g., realignment of a squinting eye (*strabismus*), is likely to be of limited benefit because of a permanent reduction in the visual acuity of that eye resulting from abnormal neuronal development. Visual impairment secondary to visual deprivation of this kind is referred to as *amblyopia*.

Box 29 Clinical note: Visual defects

Reduced visual acuity

Visual defects often present as blurring of vision, i.e., as a loss of visual acuity. This can arise because of a problem in the optical pathway, a defect in the retina or a fault anywhere along the visual pathways in the brain. Probably the most common cause, however, is abnormal refraction, in which the image is focused in front of or behind, but not on, the retina.

Abnormal refraction

There are two main faults, myopia and hypermetropia (Fig. 147).

Myopia

The myopic, or short-sighted, eyeball is effectively too long for the refraction system when the lens is unaccommodated (minimal focusing power), so that light rays from distant objects converge in front of the retina even though the lens is minimally convex. Distance vision is blurred, although closer objects, which require increased convergence, can be focused clearly by appropriate accommodation. This defect can be corrected using concave lenses to diverge the light.

Hypermetropia

The hypermetropic, or long-sighted, eye is effectively too short, so that when the lens is fully accommodated for near vision the ability to converge the diverging rays of light is inadequate and the image falls behind the retina. Distance vision is fine, but near vision is blurred. This may be corrected by using convex lenses to increase the total focusing power of the system.

Visual field defects

The visual field in one or both eyes may be abnormally restricted because of defects at different points along the visual pathways (see Fig. 150), producing different patterns of field loss. A brain tumour in the region of the optic chiasma, for example, damages fibres carrying information from the temporal half of the visual field of both eyes. The resulting field defect, in which peripheral vision is lost to both sides, is referred to as *bitemporal hemianopia*. Damage to the right optic tract, however, will produce a defect in the left half of the visual field for each eye, a left *homonymous hemianopia*.

Glaucoma

The underlying defect in glaucoma is an elevated pressure within the eye caused by an excess accumulation of aqueous humour. This is secreted into the posterior chamber of the eye by the ciliary body and is drained from the irido-corneal angle of the anterior chamber through the *canal of Schlemm*. If drainage is impaired, *intraocular pressure* rises and this may eventually damage the retina causing blindness.

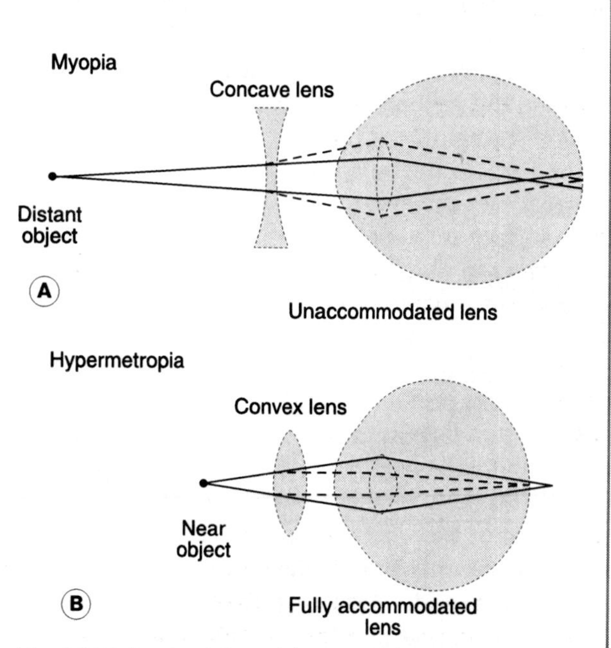

Fig. 147 Refraction defects. (A) Myopia. (B) Hypermetropia. Solid lines denote light from the object, broken lines the effect of the correcting lens.

7.5 Control of motor function

Learning objectives

At the end of this section you should be able to:

- explain what a motor unit is and how contractile force is controlled
- outline the major electrical and chemical events in neuromuscular transmission
- describe the main spinal motor reflexes
- outline how the brain controls motor function
- explain the physiological basis of major motor defects.

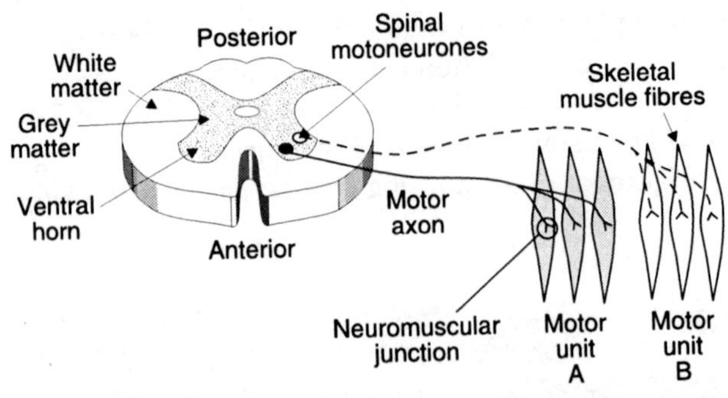

Fig. 148 The components of two motor units (A and B) in a single muscle. The α-motoneurone cell bodies lie in the ventral horn of the spinal grey matter. Each supplies several muscle fibres, whereas a single fibre is only innervated by one motor axon.

Motor control involves a variety of mechanisms which regulate skeletal muscle activity. We will begin by considering the spinal motoneurones, which act as the final common pathway in the initiation of muscle contraction, regardless of the original source of the motor signals.

Spinal motoneurones and motor units

Contraction in skeletal muscle is controlled by means of nerves distributed from the spinal cord to the muscle fibres themselves. These are known as *α-motoneurones* and, like all nerves, they transmit information in the form of action potentials, which are rapidly conducted along myelinated motor axons. In this case, the cell bodies lie in the ventral, or anterior, horn of the spinal cord grey matter. Axons pass out of the cord in the anterior, or ventral, roots of the spinal nerves and eventually terminate on the muscle fibre surface in a modified synapse known as the *neuromuscular junction* (Fig. 148). Each motoneurone supplies several muscle fibres. Since all of these fibres contract each time the relevant axon is excited, a motoneurone plus the muscle cells it innervates constitutes an important functional unit, called a *motor unit*. In muscles requiring very fine motor control, such as the extraocular muscles of the eye, each motor unit involves only a few muscle fibres, whereas a single motor axon may supply several hundred different muscle cells in large limb muscles.

Although each nerve supplies several muscle cells, each muscle fibre is only innervated by one motor axon. Since action potentials cannot spread from one muscle fibre to another (unlike cardiac muscle and visceral smooth muscle), skeletal muscle excitation is completely dependent on the activity in the relevant motoneurone. There is a one-to-one relationship between action potentials in the motoneurone and in the innervated muscle fibre.

Force regulation in skeletal muscle

The amplitude of the action potentials in both motoneurones and skeletal muscle fibres is fixed (the 'all or none' rule of excitation). Action potential size cannot be varied, therefore, in order to control the force of muscle contraction. Contractile force may be regulated in the intact muscle in two ways:

- *motor unit recruitment*, in which increasing the number of activated motoneurones also increases the number of contracting muscle fibres.
- *action potential frequency*, since at higher frequencies of motoneurone activity the interval between consecutive muscle action potentials is shortened. This increases the tension generated by each fibre up to a maximum which is determined by muscle type and resting fibre length (Section 1.6). Stimulation frequency can affect tension because the action potential in the muscle is much shorter in duration than the resulting mechanical response (or twitch), allowing for *summation* of consecutive twitches (Fig. 149). At still higher frequencies of stimulation, a continuous contraction, or *tetany*, results.

Neuromuscular transmission

This heading covers the mechanisms whereby an action potential in a motoneurone generates an action potential in the muscle fibre. At the neuromuscular junction, motor axons lose their myelin sheath before terminating in grooves in the muscle membrane known as synaptic gutters (Fig. 150). The specialized postsynaptic membrane of the muscle fibre is known as the *motor end-plate* and is separated from the presynaptic membrane of the axon terminal by a *neuromuscular cleft*, about 50 nm wide. This gap

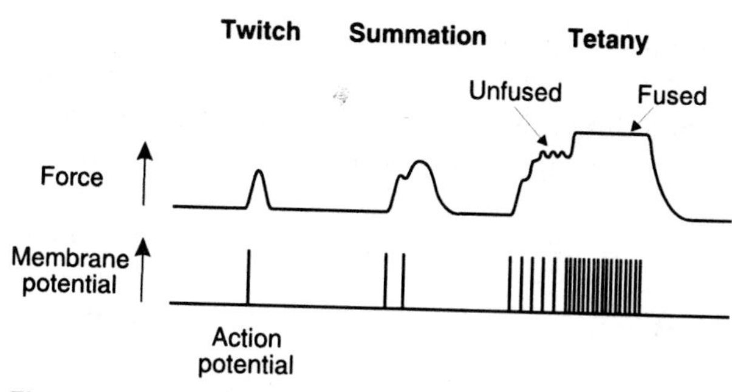

Fig. 149 The force of skeletal muscle contraction can be controlled by changing the frequency of stimulation.

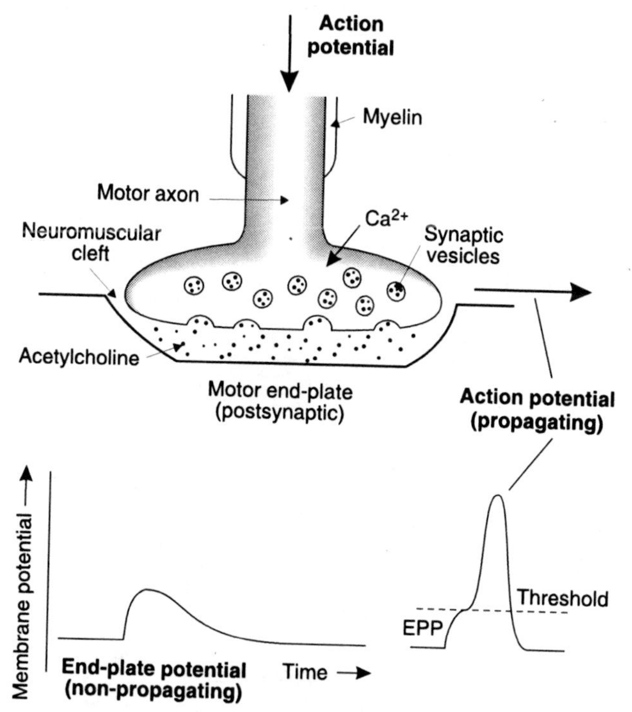

Fig. 150 The structure of the neuromuscular junction and the main events in neuromuscular transmission. Motor axon action potentials trigger release of acetylcholine leading to a postynaptic end-plate potential (EPP). This triggers an action potential in the muscle fibre.

between nerve and muscle suggests that chemical transmission is likely to be involved in any communication between them, since electrical transmission would require direct continuity from one cell to the next. In fact, the basic mechanisms underlying neuromuscular transmission are similar to those involved in synaptic transmission between one nerve and another (Section 7.2).

When an action potential reaches the end of the nerve it depolarizes the nerve ending, or terminal. This makes the presynaptic plasma membrane more permeable to Ca^{2+}, which diffuses into the nerve terminal. As a result, the *secretory vesicles* in the terminal fuse with the plasma membrane, releasing the neurotransmitter *acetylcholine* into the neuromuscular cleft. This diffuses across the cleft and binds to specialized receptors on the motor end-plate. There are several types of cholinergic receptor in different parts of the body, but the receptors on skeletal muscle fibres are referred to as *nicotinic receptors* because they also respond to the drug nicotine.

Once the receptors have been activated by binding with acetylcholine, they open receptor-operated channels permeable to Na^+ and K^+, and the net inward current produces a local depolarization, which is referred to as the *end-plate potential* (EPP) (Fig. 150). This EPP does not propagate over the muscle fibre, but it sets up local currents which depolarize the muscle membrane adjacent to the motor end-plate. An action potential is initiated when threshold is reached, and it is conduction of this action potential across the plasma membrane which acts as the signal for muscle contraction through the excitation–contraction coupling mechanisms of the cell (Section 1.6).

The transmission process is brought to an end by the release of acetylcholine from its receptor, which allows the end-plate to repolarize back to its resting potential. The transmitter is rapidly broken down into acetate and choline by the enzyme *acetylcholinesterase*, which is present at high concentrations within the neuromuscular cleft. A considerable fraction of the choline is reabsorbed into the presynaptic terminal and is used for further synthesis of acetylcholine.

Box 30 Clinical note: Therapeutic manipulation of neuromuscular transmission

The nicotinic cholinergic receptors on the motor end-plate can be blocked by several drugs which induce paralysis by inhibiting neuromuscular transmission. One such agent is a naturally occurring poison known as *curare*, derivatives of which are used by anaesthetists to promote muscle relaxation during surgery. Transmission can be promoted by inhibiting the action of acetylcholinesterase with drugs such as *neostigmine*. This effectively increases the concentration of transmitter in the end-plate region and is used to treat a condition called myasthenia gravis which causes muscle fatigue because of defective neuromuscular transmission. Anticholinesterases are also used in anaesthetic practice to reverse the effects of muscle relaxants.

Spinal cord reflexes

Skeletal muscles are often termed voluntary muscles because of our ability to consciously control their actions. It would be wrong to suppose, however, that all skeletal muscle activity is voluntary. Many reflex responses occur, in which specific sensory stimuli automatically produce a fixed pattern of motor response. This reflex muscle action is important in the control of posture and muscle tone, providing a background level of muscle activity appropriate for effective voluntary actions.

Each reflex involves a sensory detector and sensory nerve (the afferent limb) which is linked to the relevant motor nerve (the efferent limb) within the CNS. This set of connected elements is referred to as the *reflex arc*.

The muscle stretch reflex

This is the simplest type of reflex in the body since it involves only one synapse, i.e., it has a *monosynaptic reflex arc* (Fig. 151). It is a classical example of a spinal reflex, i.e., one which is coordinated within the spinal cord. The stretch reflex is activated by passive stretching of a muscle, which causes the muscle to contract actively in response. This type of reflex is important in controlling the resting tone in muscles, especially the extensor muscles of the lower limbs which resist gravity. Any tendency for the knee to buckle (flex), for example, will be resisted by the stretch this puts on the extensor (quadriceps) muscles of the upper thigh, which reflexly contract, keeping the leg in its normal, extended position. Stretch reflexes are also commonly tested by clinicians, e.g., when they elicit a knee jerk.

The receptor for the stretch reflex is the *muscle spindle*, a structure about 1 cm long which contains specialized fibres (intrafusal fibres) arranged in parallel with the normal (extrafusal) fibres of the surrounding muscle. Spindle fibres have a centrally located receptor region, which activates the associated sensory nerve endings, as well as contractile regions at either end of the cell (Fig. 152A). These areas are innervated by *γ-motoneurones* from the ventral horn area of the spinal cord. When a muscle spindle is stretched, the receptor region depolarizes, thus generating action potentials in the associated sensory (afferent) nerve. This signal contains information both about the absolute length of the muscle (the static response) and the rate of length change

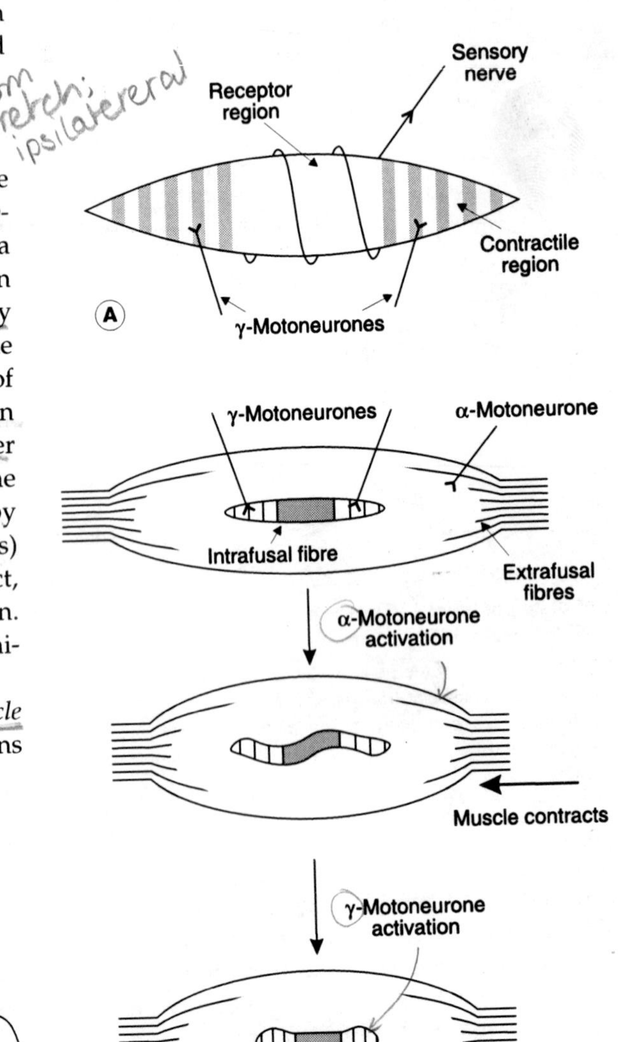

Fig. 152 Muscle spindle function. (A) A muscle spindle fibre. (B) The degree of stretch of the receptor area (shaded) controls its sensitivity: voluntary contraction of the extrafusal fibres reduces sensory output while stimulation of the intrafusal (spindle) bundles by the γ-motoneurones stretches the receptor region once more and restores sensitivity.

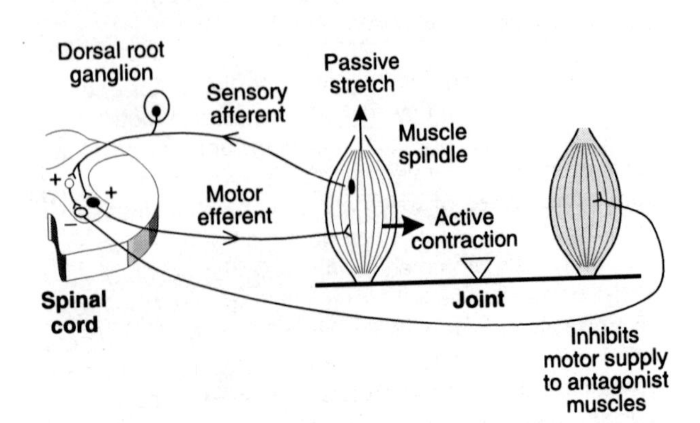

Fig. 151 The stretch reflex depends on a monosynaptic reflex arc. The motor supply to antagonist muscles is inhibited via interneurones. Excitatory synapse (+) and inhibitory synapse (–).

(the dynamic response). The afferent axon, which has its cell body in the dorsal root ganglion, enters the spinal cord through the dorsal spinal nerve root and continues into the ventral horn where it makes excitatory synapses directly onto *α-motoneurones* supplying the stretched muscle. The resulting action potentials in the motor axon trigger muscle contraction to oppose the stretch stimulus.

The sensory afferents from the muscle spindle also inhibit contraction in any muscles which oppose the action of the stretched muscle (Fig. 151). This *reciprocal inhibition* depends on excitation of inhibitory interneurones (Renshaw cells) which synapse with α-motoneurones supplying antagonist muscles. This reduces antagonist resistance to shortening of the stretched muscle.

Regulation of muscle spindle sensitivity by γ-motoneurones

The output from the muscle spindle is directly dependent on the degree of stretch of the receptor region. This can be increased by stimulating contraction of the ends of the spindle fibres through the γ-motoneurones (Fig. 152A), even if the overall length of the muscle does not change. Functionally, this is important in at least two respects.

Control of muscle tone. Since the stretch reflex is an important determinant of a muscle's ability to resist passive changes in length, i.e., its tone, γ-motoneurone-induced changes in spindle sensitivity provide an important mechanism for the control of general muscle tone.

Control of voluntary contractions. Being able to regulate the sensitivity of the spindle in this way is beneficial during voluntary contractions. As the muscle actively shortens, the spindle fibres tend to become slack, thus disabling the sensory output relating to muscle length and velocity of shortening (Fig. 152B). To prevent this, γ-motoneurones are normally *coactivated* along with the α-motoneurones supplying the extrafusal fibres during voluntary motor control. Contraction at either end of the spindle stretches the central receptor region, and sensory output is maintained.

Golgi tendon organ reflex

The Golgi tendon organs are another set of musculoskeletal stretch receptors but, in this case, they are attached in series with the muscle fibres, being located, as their name suggests, within the relevant tendon. They are sensitive to muscle tension, rather than length, and initiate an inhibitory response. The sensory afferents act on the α-motoneurones supplying the contracting muscle through inhibitory interneurones within the spinal cord, and so reduce active contraction. This reflex seems to be protective, limiting tension so as to avoid permanent damage to muscle or tendon.

Withdrawal and crossed extensor reflexes

This pattern of reflex activity is induced by painful stimuli and is sometimes referred to as the pain reflex. It is more complex than the reflexes described above, involving muscles on both sides of the body and not just on the side of the injury.

Any painful stimulus causes a reflex withdrawal of the affected part of the body from the damaging agent. On touching a hot object, for example, the hand is reflexly pulled away because of excitation of flexor muscles in that arm by polysynaptic connections between the sensory nerves from the pain receptors and the relevant α-motoneurones (Fig. 153). Other branches of the sensory nerve make connections via inhibitory interneurones to the α-motoneurones supplying the extensor muscles in the arm, and these relax, leaving flexion unopposed. Taken together, these responses constitute the flexor or withdrawal reflex.

Further spinal cord connections are made to the muscles in the opposite arm and give rise to an additional, crossed extensor reflex. This is a mirror image of the flexor reflex, with extensors being activated and flexors inhibited. The result is a rapid straightening of the uninjured arm, which tends to push the whole body away from any source of danger. Similar reflexes are elicited by painful stimuli affecting other parts of the body.

The pain reflex provides a good example of how conscious control can modify reflex actions. If you pick up a hot dish your reflex response is to drop it, flexing your arms. If the dish contains your dinner, however, you may be able to consciously inhibit the reflex, allowing you to transport the food safely to the table.

Brain areas involved in motor control

Many areas of the brain are involved in the control of motor function. Axons descend from the cortex and brainstem to synapse with spinal motoneurones, and these motor pathways are essential for normal voluntary movement. In addition, the basal ganglia and cerebellum play important roles in controlling skeletal muscle activity even though they have no direct output to spinal neurones. All these areas will be considered in turn.

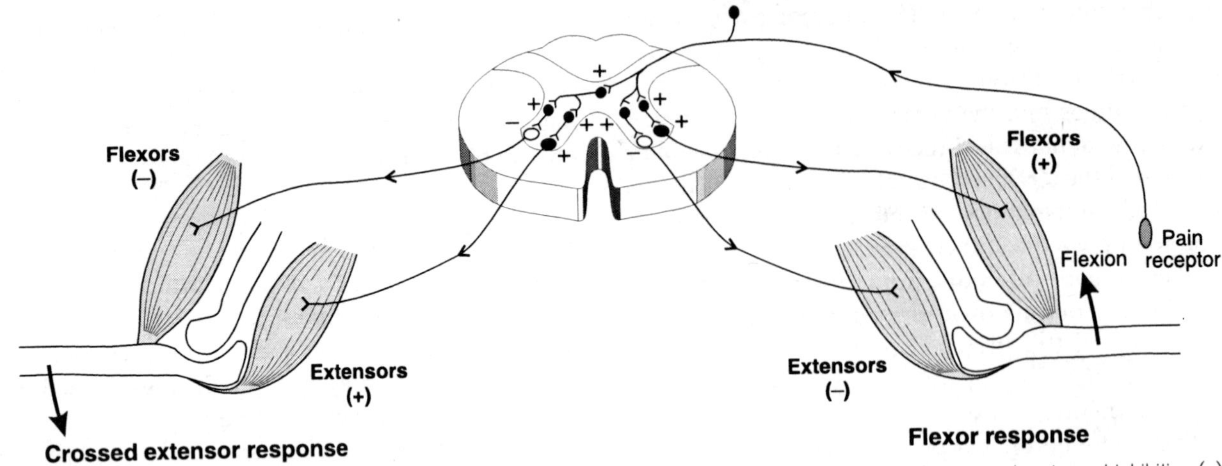

Fig. 153 Spinal reflex arcs involved in pain responses. Polysynaptic pathways lead to excitation (+) of the flexors and reciprocal inhibition (–) of extensors on the side of the injury (flexor, or withdrawal response). This pattern is reversed on the opposite side of the body (crossed extensor response).

Motor cortex and corticospinal tracts

The motor cortex lies in the *precentral gyrus* of the frontal lobe (Fig. 128). Axons from this region descend to synapse directly with motoneurones, i.e., there are no interneurones in the pathway. *Corticobulbar tracts* carry axons to the motor nuclei of the cranial nerves, while the fibres innervating spinal motoneurones descend as the *corticospinal tracts* within the cord itself (Fig. 154). Traditionally, these are known as the *pyramidal* tracts, since the axons pass through the pyramids in the medulla oblongata. Although the functional distinction between pyramidal and extrapyramidal motor tracts is now known to be spurious, these terms are still used in clinical practice, as described later.

All the descending fibres from the motor cortex cross the midline, either in the medulla or within the spinal cord itself, so the left side of the body is controlled from the right cerebral hemisphere (the *contralateral hemisphere*), and vice versa. Different parts of the body are represented by different regions of the motor cortex, in a fashion analogous with the organization of the somatosensory cortex (Section 7.4). The resulting body map, or homunculus, is anatomically distorted, however, since the cortical area devoted to a given set of muscles does not reflect their bulk. The muscles of the hand and face receive a particularly extensive cortical innervation, emphasizing the significance of the motor cortex for voluntary control of fine hand movement (via spinal motoneurones) and facial expression (via cranial nerves). Indeed, the motor homunculus shares many features with the sensory homunculus in the postcentral gyrus (Fig. 129).

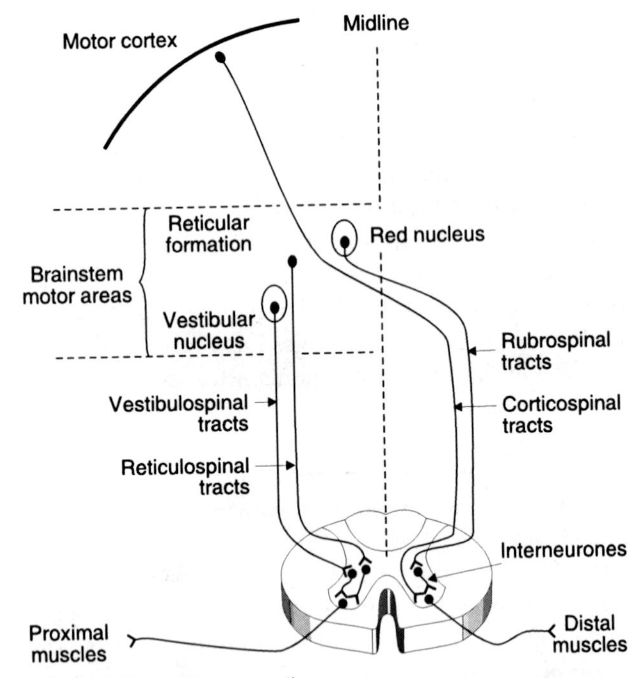

Fig. 154 The main motor pathways.

Motor areas in the brainstem

The other motor pathways innervating the spinal motoneurones originate in the brainstem. These include the vestibulospinal tracts from the vestibular nuclei, the rubrospinal tracts from the red nuclei and the reticulospinal tracts from the reticular formation (Fig. 154). The term *extrapyramidal* pathways is sometimes used to refer to all of these tracts collectively, but this implies a unity of structure and function which they do not possess. It is now believed that the rubrospinal and corticospinal tracts may be more sensibly classified together as the *lateral motor pathways*, as they terminate on neurones in the lateral part of the spinal cord grey matter (i.e., in the ventral horns). In this classification, the vestibulospinal and reticulospinal tracts constitute the *medial motor pathways*, terminating in the medial layers of the ventral horns.

The rubrospinal tract. Like the corticospinal tract, this crosses the midline to descend contralaterally. It is mainly distributed to spinal motoneurones supplying distal limb muscles, e.g., the muscles controlling finger movement. Thus, the corticospinal and rubrospinal tracts (the lateral pathways) are both involved in fine motor control. Selective experimental damage to the rubrospinal or, more particularly, the corticospinal tract interferes with finger and toe movement but has little effect on posture and balance.

The vestibulospinal and reticulospinal tract. These descend to supply motoneurones on the same side (ipsilateral) of the body as the brainstem regions from which they originate. They innervate proximal muscles involved in the control of balance and locomotion. The tracts of the medial motor pathways are important in the postural reflexes initiated by signals from the vestibular apparatus through the vestibular nuclei, and are also involved in cerebellar motor control. Interrupting the vestibulospinal or reticulospinal tract reduces tone in the extensor muscles of the proximal limbs and trunk, making it very difficult to walk or remain upright.

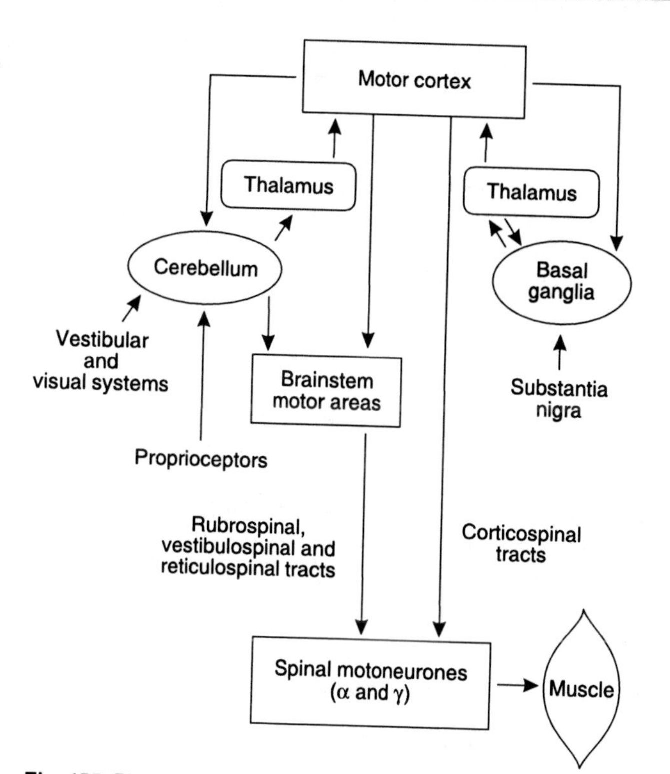

Fig. 155 Block diagram summarizing the main motor control areas with relevant inputs and outputs. Only the motor cortex and brainstem motor areas make connections with the spinal motoneurones.

The cerebellum

The cerebellum plays a significant role in motor control, even though no descending pathways connect it directly to the spinal motoneurones (Fig. 155). The cerebellum receives *inputs* from a wide variety of sources including:

- the vestibular apparatus
- peripheral sensory receptors, especially muscle spindles and proprioceptors in joints
- the visual and auditory systems
- the corticospinal pathways.

Cerebellar *outputs* are directed to the motor cortex (via the thalamus) and brainstem motor areas. By comparing information about intended motion (as signalled by the relays from the corticospinal neurones) with the actual motion (as detected by sensory receptors) the cerebellum can detect motor deviations and correct them through its outputs to other motor regions. This produces smooth, coordinated movement. The cerebellum also plays an important role in the maintenance of balance, by regulating motor responses to vestibular, visual and proprioceptive stimuli.

The basal ganglia

These are major subcortical nuclei within the forebrain. Like the cerebellum they make no direct motor connections with spinal motoneurones. The basal ganglia receive *inputs* from:

- wide areas of cortex, including motor cortex
- the thalamus, a relay station for sensory pathways
- the substantia nigra in the midbrain.

The major *output* is to the motor cortex via the thalamus (Fig. 155). This suggests that the basal ganglia are involved in the control of voluntary movement through their influence on corticospinal output from the motor cortex. They seem to be particularly important in the initiation of movement and in controlling body posture to allow for fine movements, e.g., by ensuring the shoulder and upper arm are positioned appropriately for hand movement.

Box 31 Clinical note: Defects of motor control

Abnormalities at different levels in the motor control system produce different patterns of motor deficit.

Damage to the spinal motoneurones themselves, either in the anterior horn or along peripheral motor axons, leads to paralysis and wasting of the muscles supplied. Motor tone is reduced (flaccid paralysis) and the stretch reflex is absent, since there is no efferent pathway for muscle stimulation. Lesions of this type are classified clinically as *lower motor neurone* defects.

Bilateral spinal cord injuries are generally the result of accidental trauma and cause paralysis in all the muscles below the level of the defect (*paraparesis*) because of transection of all descending motor pathways from the brain. After an initial period of flaccidity (spinal shock), tone is generally increased (spastic paralysis) and muscle stretch reflexes become exaggerated. The mechanism of these increases in tone and reflexes remains unclear but in clinical practice these signs are regarded as typical of an *upper motor neurone* lesion.

Global damage above the brainstem, as opposed to selective damage to higher motor centres, leads to increased tone, with the limbs and trunk in a fully extended position (*decerebrate rigidity* or *opisthotonos*). This reflects increased facilitation of extensor motoneurone activity by the brainstem nuclei, following the loss of inhibitory inputs to the vestibulospinal and reticulospinal tracts from higher brain centres.

Cerebellar damage is characterized by loss of balance and an unsteady gait (*ataxia*), along with an inability to perform rapidly alternating movements smoothly. Muscle tone is decreased since cerebellar facilitation of the motor areas in the cortex and brainstem is lost.

Basal ganglia defects are characterized by involuntary movements (*dyskinesia*), slowness in initiating voluntary movement (*bradykinesia*) and changes in muscle tone. Clinical deficits of this kind are often referred to as extrapyramidal disorders even though this is almost certainly a misnomer, since the main motor influence of the basal ganglia is exerted through connections with the corticospinal (pyramidal) system. The most common example of such a condition is Parkinson's disease, which produces tremor, rigidity and poverty of movement. This seems to result from a deficiency of dopamine-secreting neurones in the substantia nigra, one of the regions with a major input to the basal ganglia. Some of the symptoms of this condition can be relieved by treatment with L-dopa, a dopamine precursor (Fig. 120).

Defects of cerebral control of motor function are commonly seen in stroke victims who have suffered from cerebral ischaemia or haemorrhage involving corticospinal neurones. This produces paralysis down the opposite side of the body (*hemiplegia*), with increased muscle tone and pronounced stretch reflexes, i.e., the classical signs of an upper motor neurone defect. The fact that experimental lesions involving only the corticospinal tracts do not increase muscle tone or reflexes suggests that the classical clinical picture actually reflects more widespread damage, possibly also involving destruction of inhibitory cortical outputs to the brainstem nuclei.

7.6 Autonomic nervous system

Learning objectives

At the end of this section you should be able to:

- describe the organization of sympathetic and parasympathetic nerves
- name the main autonomic transmitters and receptors and where they are found
- explain the role of the adrenal medulla in sympathetic responses
- describe the 'fight or flight' response
- give examples of the important classes of autonomic drugs, and describe how they act.

Autonomic nerves control a wide range of systems within the body, regulating cardiac and smooth muscle activity, endocrine and exocrine gland secretion, and the metabolism of certain cells. They are not generally under our conscious control but are important in many crucial functions, e.g., the digestion of food and the maintenance of blood pressure.

Relevant structure

Anatomically, autonomic nerves may be divided into two divisions: *sympathetic* and *parasympathetic*. In each case, the connection between the CNS and the target cell depends on two nerves in series which synapse in a ganglion. This ganglion lies outside the CNS and the neurones which enter and leave it are known as the *preganglionic* and *postganglionic* neurones, respectively. Within this overall pattern, however, there are several features which distin-

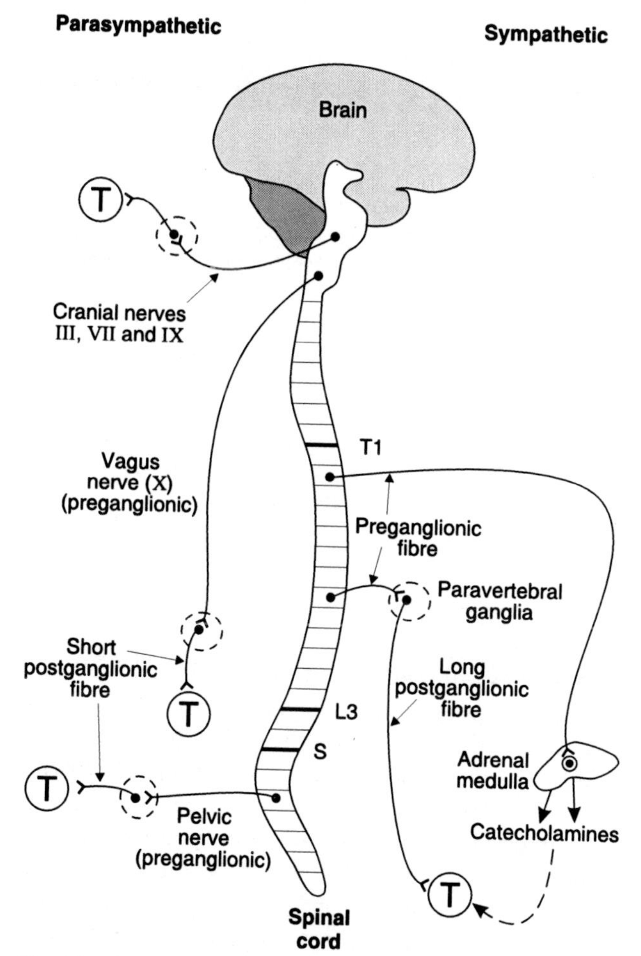

Fig. 156 The main anatomical features of the autonomic nerves. Sympathetic fibres synapse in ganglia (dotted circles), releasing acetylcholine as transmitter to activate nicotinic receptors on postganglionic fibres. These activate targets (T) by release of noradrenaline (norepinephrine). The adrenal medulla is also innervated by preganglionic sympathetic fibres which stimulate endocrine secretion of catecholamines. Both preganglionic and postganglionic transmission in parasympathetic nerves are cholinergic, acting on nicotinic and muscarinic receptors, respectively.

guish sympathetic from parasympathetic nerves (Fig. 156).

Preganglionic sympathetic neurones have their cell bodies in the grey matter of the thoracic and upper lumbar spinal cord (T1–L3). Their axons pass out in the ventral ramus to enter the paravertebral sympathetic chain, which runs parallel with the vertebral column. Sympathetic ganglia are located along this chain, or within the major autonomic nerve plexuses, e.g., the coeliac ganglion. Long postganglionic nerve fibres are distributed to their targets within the spinal nerves or along the walls of blood vessels.

Parasympathetic preganglionic fibres arise from one of a number of nuclei within the brain, or from the sacral region of the spinal cord. They exit the CNS via cranial nerves or pelvic splanchnic nerves, e.g., the vagus nerve (cranial nerve X) supplies the heart, airways and the gastrointestinal tract as far as the transverse colon, while pelvic nerves supply the rest of the gut and the bladder. These preganglionic axons synapse with short postganglionic fibres in, or close to, the target organ itself.

Autonomic transmitters and receptors

Ganglionic transmission is cholinergic in both sympathetic and parasympathetic nerves, with acetylcholine acting on postganglionic nicotinic receptors (i.e., receptors which can be activated by nicotine). Postganglionic neurotransmitters, however, differ in the two divisions of the autonomic system.

Noradrenaline (norepinephrine) release by sympathetic nerves. The major postganglionic transmitter in sympathetic nerves is noradrenaline (norepinephrine). This acts on target cell membrane receptors of two main types, the α-and β-adrenoceptors. It is difficult to generalize about the actions of each class of adrenoceptor since this varies from tissue to tissue, e.g., β-adrenoceptors stimulate cardiac contraction but generally inhibit smooth muscle. Receptor classification for a given tissue is based on pharmacological studies, i.e., drugs which are known to bind to one type of adrenoceptor rather than the other are used in an attempt to mimic or inhibit the effects of sympathetic stimulation. Some important, adrenergically mediated sympathetic effects are summarized in Table 11. It should be noted, however, that a small minority of postganglionic sympathetic nerves release acetylcholine rather than noradrenaline (norepinephrine), e.g., sympathetic nerves which stimulate sweating.

Acetylcholine release by parasympathetic nerves. Postganglionic parasympathetic nerves release acetylcholine, which acts at *muscarinic* receptors (i.e., cholinergic receptors activated by the drug muscarine) on the target cell membrane. This should be contrasted with the nicotinic receptors found in autonomic ganglia. Parasympathetic nerves are less widely distributed than sympathetic nerves and have little effect on the vasculature or metabolism. Some major parasympathetic effects are listed in Table 11.

Autonomic cotransmitters. Although noradrenaline (norepinephrine) and acetylcholine are the dominant transmitters, it is well recognized that postganglionic autonomic nerves may simultaneously release additional transmitters, a process known as cotransmission. Examples of cotransmit-

Table 11 Some important autonomic effects

Sympathetic effects (α-receptors)	Sympathetic effects (β-receptors)	Parasympathetic effects
Vasoconstriction	Vasodilatation	
	Increased heart rate	Decreased heart rate
	Increased cardiac contractility	
	Intestinal relaxation	Intestinal contraction
Intestinal sphincter contraction		Intestinal sphincter relaxation
	Uterine relaxation	
	Bronchodilatation	Bronchoconstriction
Bladder sphincter contraction	Bladder relaxation	Bladder contraction
Pupillary dilatation		Pupillary constriction
		Lens accommodation
Scant mucoid exocrine secretion		Enzyme-rich exocrine secretion
Contraction of vas deferens and epididymis during ejaculation		Penile erection
	Metabolic stimulation	

ters include adenosine triphosphate (ATP) for some sympathetic nerves, and vasoactive intestinal polypeptide in certain parasympathetic nerves.

The adrenal medulla

The central region of the adrenal gland is known as the adrenal medulla and it is regulated by preganglionic sympathetic nerves (Fig. 156). Transmission is through nicotinic cholinergic receptors, as in the sympathetic ganglia, but instead of exciting postganglionic nerves, preganglionic activity stimulates adrenal secretion of catecholamines (*noradrenaline* and *adrenaline*) into the circulation. Noradrenaline (norepinephrine) is particularly potent as an α-adrenoceptor agonist, while adrenaline (epinephrine) shows similar potency at both α- and β-receptors. This helps explain why different cardiovascular effects are seen when each of these substances is injected into a vein. Noradrenaline (norepinephrine) is a potent vasoconstrictor (an α effect), leading to a rise in arterial pressure. The baroreceptor response to this inhibits the heart rate by stimulating parasympathetic activity (Section 3.6) and this more than outweighs the rather weak, direct effect of noradrenaline (norepinephrine) on the cardiac β-receptors. The overall response is a rise in peripheral resistance and blood pressure, with a fall in heart rate. Adrenaline (epinephrine), on the other hand, strongly stimulates the heart, raising heart rate and stroke volume. Blood pressure rises because of this but peripheral resistance falls, both through reflex inhibition of sympathetic vasoconstrictor nerves and because of the direct, β-adrenoceptor-mediated vasodilatation in skeletal blood vessels.

Physiological stimulation of the adrenal medulla leads to release of a mixture of both catecholamines in a ratio of about 80% adrenaline (epinephrine) to 20% noradrenaline (norepinephrine). This produces similar effects in most target organs to those elicited by direct stimulation through sympathetic nerves. Although slightly slower in onset, adrenal-dependent responses are of longer duration than those mediated by nerves alone, persisting until all the circulating hormone is taken up and metabolized (Section 7.2). There is no equivalent endocrine extension of the parasympathetic nerves.

Autonomic reflexes, 'fight or flight'

Many homeostatic mechanisms and control systems depend on autonomic reflexes, e.g., blood pressure control, regulation of gut and bladder motility, and regulation of digestive secretions. These are considered elsewhere in the context of the relevant systems. One less specific reflex is the 'fight or flight' response to anxiety or rage. This is believed to prepare the body for vigorous exercise and involves a generalized increase in sympathetic nervous activity and secretion of adrenal catecholamines. Cardiovascular effects predominate, with increases in heart rate, blood pressure and skeletal muscle blood flow. Pallor results from cutaneous vasoconstriction. Metabolism is stimulated and sweating increases but gastrointestinal blood flow and motility are reduced. This may produce nausea. Salivary secretion is inhibited, making the mouth dry and sticky, and there may be a fine, skeletal muscle tremor.

Box 32 Clinical note: Autonomic drugs

A variety of agents exist which can interfere with autonomic transmission. These may act at one of several sites along the nerve pathway.

Ganglionic transmission may either be stimulated, e.g., by nicotine, or blocked, e.g., by hexamethonium. In either case, postganglionic activity is affected in both divisions of the autonomic system. For example, hexamethonium reduces blood pressure (sympathetic blockade) and inhibits gastrointestinal secretion and motility (parasympathetic blockade).

Postganglionic transmission in the parasympathetic system may be selectively mimicked by muscarinic agonists, e.g., pilocarpine, or blocked by muscarinic antagonists, e.g., atropine.

Postganglionic transmission in sympathetic nerves may be mimicked and blocked in a receptor-selective fashion. For example, α-adrenoceptors are stimulated by phenylephrine and inhibited by phentolamine, while β-adrenoceptors are stimulated by isoprenaline and inhibited by propranolol.

Many of these agents have important therapeutic uses, e.g., β-blockers are widely prescribed for hypertension and ischaemic heart disease.

7.7 Other functions of the brain

Learning objectives

At the end of this section you should be able to:

- outline the homeostatic and behavioural functions of the hypothalamus
- describe the role of the brain in speech, learning and memory
- explain the EEG in principle and describe some activity patterns
- outline factors controlling consciousness
- outline frontal lobe functions.

The brain is responsible for our thoughts and emotions, allows us to store and retrieve memories, generates our sense of personal identity and provides us with the ability to communicate. These higher cerebral functions are poorly understood but some will be considered briefly.

Hypothalamic and limbic functions

The hypothalamus lies adjacent to the third ventricle in the forebrain and acts as a homeostatic control centre for autonomic and endocrine functions. It is also linked to the cerebral cortex and thalamus by the limbic system, and these pathways seem to be important in determining certain aspects of behaviour, especially those with a significance for survival and reproduction.

Homeostatic controls. These require information concerning the state of the internal environment, which is supplied to the hypothalamus both from peripheral sensory receptors (via the thalamus and limbic system) and from specialized receptor cells within the hypothalamus itself, e.g., hypothalamic osmoreceptors. The hypothalamus can activate appropriate responses to these signals directly through its connections with autonomic and somatic nerves, e.g., it promotes sympathetically driven vasoconstriction and somatically stimulated shivering in response to the cold (Section 1.1). In addition, the hypothalamus exerts indirect control over many endocrine aspects of homeostasis through regulation of pituitary function. Thus, neural and hormonal control systems are integrated through hypothalamic mechanisms, a function considered in more detail elsewhere (Section 8.2).

Behavioural controls. These appear to involve both the hypothalamus and the limbic system, which is often regarded as being important in determining emotional states.

Eating and drinking. The hypothalamus contains localized feeding and satiety centres as well as a thirst centre. These control eating and drinking.

Fear and aggression. These seem to reflect activity both in the hypothalamus and in the amygdala within the limbic system.

Sex drive, or libido. This is influenced strongly by the hypothalamic–limbic system.

Reward and punishment. Such systems seem to exist within the brain, involving a number of areas in the hypothalamus and the frontal lobe. Stimuli which activate the reward system lead to feelings of well-being and associated behaviour tends to be reinforced. Activities which stimulate the punishment centre have the opposite effects and tend to be avoided.

Language skills

The ability to understand and generate speech resides in the left cerebral hemisphere in 90% of right-handed people and 70% of left-handed people. This is said to be the *dominant hemisphere* in these individuals. Two cortical regions are of particular importance:

- *Broca's area* is located in the frontal lobe and lies close to the motor cortex controlling the face. It is particularly important in speech production.
- *Wernicke's area* is found in the temporal lobe, close to the auditory cortex and allows us to understand spoken language.

Box 33 Clinical note: Aphasia

Damage to the dominant hemisphere, e.g., as a result of a stroke, may lead to disruption of language, known as *aphasia*. This may be subdivided into expressive and receptive aphasias, although most aphasic individuals demonstrate elements of both.

Expressive or motor aphasia is characterized by an inability to produce speech while retaining comprehension of language. It is a specific defect caused by damage to Broca's area and not simply the result of a failure of phonation caused by muscle weakness, e.g., the sufferer may still be able to hum a melody quite accurately.

Receptive or sensory aphasia results from damage to Wernicke's area and is characterized by an inability to understand language even though hearing is normal. Words may be articulated clearly in these individuals but they are not generally put together in an intelligible sequence, presumably because their meaning is obscure to the speaker as a result of the comprehension defect.

Learning and memory

Learning may be defined as any change in behaviour that results from previous experience. This relies on memories, i.e., information stored during the learning experience which is used to direct the new behaviour. This information is often described as passing first into *short-term memory*, which retains small units of information for seconds or minutes, before it is transferred to *long-term memory*, which can hold an enormous amount of information for periods lasting from days to decades. Memory storage and retrieval is greatly enhanced by the repeated *rehearsal* of information, as evidenced by the benefits of practising motor skills or revising previously studied material. The mechanisms underlying this are unclear, but seem to involve changes in the efficiency of neurotransmission within pathways which are stimulated frequently.

- Transmitter release may be enhanced, making future activation of that pathway easier.
- There may eventually be anatomical changes within the relevant pathway, perhaps involving an increase in the number or size of synapses. This kind of structural modification in response to nerve stimulation is referred to as *neural plasticity* and may explain some aspects of long-term memory.

The electroencephalogram (EEG)

The electrical activity within cerebral neurones leads to surface potentials, which may be detected using scalp electrodes. The resulting recording is called an electroencephalogram (EEG) and is analogous to the electrocardiogram (ECG) for the heart. The EEG is much more complex than the ECG, however, reflecting varying levels of activity in neurones in different parts of the cortex under different conditions, e.g., as sensory stimuli change. In general, as cortical neuronal activity increases, the EEG becomes more irregular (desynchronized), with a rise in the frequency and a fall in the amplitude, or voltage, of the recorded potentials. Diagnostically, the EEG is most used in possible cases of epilepsy, which produces characteristic changes in the EEG pattern.

Consciousness and sleep

Consciousness is defined clinically in terms of our awareness of sensory stimuli. Someone who answers questions promptly is clearly conscious while an individual who does not respond to pain is deeply unconscious. The mechanisms underlying consciousness are unknown but it does seem that the reticular formation in the brainstem plays a role in regulating the general level of awareness. The *reticular activating system* within the reticular formation, which receives inputs from most sensory and motor systems and sends outputs to cortical and other areas of the CNS, seems to be particularly important in this regard.

Sleep

Sleep is an episodic and spontaneous loss of consciousness which is easily reversed. Normal sleep consists of prolonged periods of deep, or *slow wave* sleep, interrupted every 90 minutes or so by episodes of paradoxical or *rapid eye movement* (REM) sleep lasting 5–20 minutes.

Slow wave sleep is so named because of the associated, low-frequency delta waves on the electroencephalogram (EEG). It typically leads to reductions in blood pressure, respiration, metabolic rate and gastrointestinal activity.

Dreaming occurs during REM sleep, which is associated with a variable heart rate and respiration, irregular muscle movements and a general reduction in muscle tone. The EEG shows high-frequency beta waves, indistinguishable from those in an alert individual; this is the paradoxical aspect of REM sleep.

Control of sleep. The alternation between sleep and wakefulness seems to be controlled from the reticular activating system. Both sensory and motor activity tend to excite this region, reducing drowsiness. Sleep deprivation eventually leads to extreme disturbances of thinking and emotion, but why sleep should be so important for normal cerebral function is unknown.

Frontal lobe function

In addition to the motor and premotor areas, the frontal lobe includes considerable areas of cortex of rather uncertain function. A few suggestions can be made regarding possible roles of this *prefrontal cortex*, based on the observed effects of damage to this region. No obvious sensory or motor defects result, but there may be significant emotional and behavioural changes. There is a general deterioration in concentration and forward planning so that following instructions, or organizing and carrying out a series of simple tasks to achieve a required goal, becomes difficult. In particular, there is a tendency to repeat behavioural responses after they have ceased to be appropriate (*perseveration*). For example, a patient with a frontal lobe lesion may correctly draw a circle if asked but is likely to produce another circle when subsequently asked to draw a square. These features more than offset any gains from possible parallel reductions in aggressive behaviour, so that treatment of psychiatric patients by surgical destruction of the prefrontal area (prefrontal lobotomy) is no longer justified.

Parietal lobe function

In addition to the somatosensory cortex, the parietal lobe contains the parietal association area, which helps process both visual and somatosensory information, as well as supplementary motor areas, which participate in the planning and execution of complex motor movements. Parietal lobe damage may result in defects of sensory function, or of complex motor control, in the absence of any specific sensory or power loss. One specific sign of a parietal lobe related problem is known as *anosognosia*, in which the sufferer ignores or neglects half of the sensory environment, usually half of the visual field, or even half of their own body. This most commonly affects the non-dominant (usually the right) parietal lobe, a major site for spatial awareness.

Self-assessment: questions

Multiple true/false questions

Each of the following statements consists of a stem followed by a number of possible endings. State whether each statement is True or False. For each stem, all, several or none of the statements may be true.

1. Neurones:
   - **a.** synthesize neurotransmitters
   - **b.** can be inhibited by agents which block Na^+ channels
   - **c.** may be myelinated by microglial cells
   - **d.** outnumber any other cell type in the brain
   - **e.** may demonstrate graded changes in membrane potential in response to neurotransmitters

2. Axonal action potentials:
   - **a.** are propagated unidirectionally as a result of a unidirectional flow of local current
   - **b.** tend to travel more quickly along large axons
   - **c.** are propagated discontinuously in myelinated axons
   - **d.** are generated by every EPSP produced in the dendrites or cell body of a neurone
   - **e.** increase the Ca^{2+} permeability of the axon terminal

3. Chemical neurotransmission:
   - **a.** requires presynaptic receptors
   - **b.** is potentiated by inhibitors of transmitter reuptake
   - **c.** may be inhibited presynaptically
   - **d.** results from a change in presynaptic membrane potential
   - **e.** involves exocytosis

4. The neurotransmitter:
   - **a.** acetylcholine activates the same class of receptor on both skeletal and cardiac muscle
   - **b.** dopamine is a precursor for noradrenaline (norepinephrine) synthesis
   - **c.** noradrenaline (norepinephrine) can be broken down by monoamine oxidase
   - **d.** glycine excites spinal motoneurones
   - **e.** γ-aminobutyric acid (GABA) is generally inhibitory

5. Inhibitory transmission mechanisms:
   - **a.** are always postsynaptic
   - **b.** may involve a postsynaptic increase in Cl^- conductance (g_{Cl^-})
   - **c.** may depolarize the postsynaptic membrane
   - **d.** may depolarize the presynaptic membrane
   - **e.** tend to reduce the action potential frequency in the postsynaptic nerve

6. A sensory receptor:
   - **a.** produces 'all or nothing' receptor potentials
   - **b.** is only stimulated by stimuli within its receptive field
   - **c.** typically responds to a broad range of stimuli
   - **d.** codes for stimulus strength by changes in receptor potential amplitude
   - **e.** which adapts rapidly is well suited to the monitoring of absolute stimulus strength

7. Somatosensory signals:
   - **a.** are all generated by touch receptors
   - **b.** enter the spinal cord via the ventral nerve root
   - **c.** all ascend in spinal tracts contralateral to the receptor
   - **d.** affect the somatosensory cortex contralateral to the receptor
   - **e.** all reach the level of conscious awareness

8. Pain:
   - **a.** signals from the viscera are usually conducted through myelinated nerves
   - **b.** normally results from activation of nociceptors
   - **c.** fibres travel in the spinothalamic tracts
   - **d.** transmission may be inhibited by enkephalin
   - **e.** thresholds are usually reduced in patients suffering from a peripheral neuropathy, in which nerve conduction is inhibited or slowed

9. The cochlea:
   - **a.** makes use of mechanoreceptor hair cells to detect sound

b. contains the scala tympani and scala vestibuli which are filled with K^+-rich perilymph
c. contains the basilar membrane whose resonant frequency increases as one moves towards the helicotrema
d. contains the scala media, which is part of the membranous labyrinth
e. in the intact ear demonstrates maximal sensitivity to sounds of a frequency over 10^4 Hz

10. In the vestibular system:
a. only the utricular macula is sensitive to horizontal acceleration
b. the semicircular canals detect rotational acceleration
c. otoliths in the cupula enhance the sensitivity of the crista to endolymph displacement
d. the saccular macula is sensitive to vertical acceleration
e. outputs from the vestibular nuclei go to the nuclei of the oculomotor nerves (cranial nerve III)

11. In the eye:
a. the fovea lies on the visual axis
b. aqueous humour is secreted into the anterior chamber
c. dark adaptation is only seen in rods
d. light reflexly activates the parasympathetic nerves supplying smooth muscle in the iris
e. most of the focusing power resides in the cornea

12. Photoreceptors:
a. contain pigments which are broken down, or bleached, by light
b. synapse directly with ganglion cells in the retina
c. generate hyperpolarizing potentials in response to light
d. all demonstrate the same absorbance spectrum
e. in the fovea are all cones

13. At the skeletal neuromuscular junction:
a. each motor axon action potential results in a muscle action potential
b. a depolarizing end-plate potential (EPP) is generated as a result of opening of cation-permeable channels
c. there are muscarinic cholinergic receptors
d. gap junctions bridge the space between pre- and postsynaptic membranes
e. intracellular acetylcholinesterase degrades the transmitter

14. The stretch reflex in skeletal muscle:
a. is promoted by stimulation of γ-motoneurones
b. depends on muscle spindles sensitive to muscle tension
c. is associated with reciprocal inhibition of antagonist muscles
d. is brisk in individuals suffering from a lower motor neurone lesion
e. involves a slowly conducting sensory afferent

15. Spinal motoneurones:
a. receive an input from the vestibular nuclei which mainly influences distal muscles
b. receive an excitatory input from the contralateral motor cortex
c. have their cell bodies in the ventral horn grey matter
d. may be inhibited by neurones which release glycine
e. innervate a single muscle fibre

16. The cerebellum:
a. sends outputs to spinal motoneurones via the spinocerebellar tracts
b. receives inputs from the motor cortex
c. sends outputs to the motor cortex
d. causes symptoms on the same side as the lesion, if damaged
e. is part of the brainstem

17. Sympathetic neurones:
a. innervating the adrenal medulla are postganglionic
b. which release noradrenaline (norepinephrine) are preganglionic
c. should function normally following blockade of all nicotinic receptors
d. stimulate ciliary muscle contraction in the eye
e. act on β-adrenoceptors in the heart

18. In the brain:
a. the frontal lobes contain cortical areas which affect personality and behaviour

b. the frontal lobe contains the motor cortex
c. the right hemisphere is dominant in the majority of left-handed people
d. the area in the temporal lobe known as Wernicke's area is important for understanding speech
e. the reticular activating system plays a role in controlling patterns of wakefulness and sleep

Single best answer questions

For each of the following questions choose the single best answer.

1. Myelinated nerves conduct more rapidly than unmyelinated nerves because:
   a. they have larger action potentials
   b. they have a larger diameter
   c. myelin reduces transmembrane current leak at the nodes of Ranvier
   d. myelin reduces transmembrane current leak over the internodal region
   e. they are found in the peripheral, rather than the central, nervous system.

2. Synaptic transmission:
   a. is dependent on an increase in postsynaptic permeability to Ca^{2+}
   b. generates an action potential in the postsynaptic cell
   c. typically depends on direct electrical transmission
   d. generates an action potential in the presynaptic cell
   e. is dependent on an increase in presynaptic permeability to Ca^{2+}

3. Sensory receptors:
   a. are always depolarized by the relevant stimulus
   b. are always hyperpolarized by the relevant stimulus.
   c. code for stimulus strength by the amplitude of the receptor potential
   d. code for stimulus strength by the frequency of the receptor potential
   e. send impulses to the CNS via efferent nerves

4. Damage to one side of the spinal cord:
   a. may inhibit ipsilateral pain sensation
   b. may inhibit contralateral joint sensation
   c. may produce contralateral muscle weakness
   d. may produce ipsilateral muscle weakness
   e. none of the above

5. Which of the following are *not* true of the cerebellum?
   a. it receives inputs from the vestibular system
   b. it receives inputs from the visual system
   c. it innervates spinal motoneurones
   d. it sends outputs to the thalamus
   e. it forms part of the brainstem

6. Damage to the basal ganglia:
   a. increases the level of voluntary movement
   b. reduces motor power
   c. reduces muscle tone
   d. promotes involuntary muscle activity
   e. impairs proprioception

7. Autonomic transmission:
   a. by preganglionic sympathetic nerves depends on muscarinic cholinergic transmission
   b. by preganglionic parasympathetic nerves depends on muscarinic cholinergic transmission
   c. by postganglionic sympathetic nerves depends on muscarinic cholinergic transmission
   d. by postganglionic parasympathetic nerves depends on muscarinic cholinergic transmission
   e. within autonomic ganglia depends on α-adrenergic transmission

8. Adrenaline (epinephrine):
   a. is released by cells in the adrenal cortex
   b. directly stimulates heart rate more potently than noradrenaline (norepinephrine)
   c. promotes vasoconstriction more potently than noradrenaline (norepinephrine)
   d. promotes contraction of the radial muscle in the iris more potently than noradrenaline (norepinephrine)
   e. is a steroid hormone

9. In the cerebral cortex:
   a. speech is a function of right parietal cortex
   b. the frontal lobe contains the somatosensory cortex
   c. the parietal lobe contains the primary motor cortex

d. spatial awareness is usually a function of the right parietal cortex
e. hearing is a function of the occipital cortex

10. Which of the following are not a feature of slow wave sleep:
   a. reduced blood pressure
   b. reduced metabolic rate
   c. dreaming
   d. δ-waves in the EEG
   e. reduced ventilation

Matching item questions

Theme: Autonomic control

Options

A. α-adrenoceptors
B. β_1-adrenoceptors
C. β_2-adrenoceptors
D. Muscarinic receptors
E. Adrenal medullary cells
F. Adrenal cortical cells

For each of the descriptions below choose the most appropriate option from the list above. Each option may be used once, more than once or not at all.

1. Slow the heart when activated.
2. Increase cardiac contractility when activated.
3. Release catecholamines in response to activation of nicotinic receptors.
4. Cause vasoconstriction when activated.
5. Cause bronchodilatation when activated.
6. Lead to accommodation for near vision when activated.

Theme: Neuromuscular transmission

Options

A. Presynaptic depolarization
B. End-plate potential
C. Muscle action potential
D. Vesicular exocytosis
E. Nicotinic receptors
F. Fast Na^+ channels

For each of the descriptions below choose the most appropriate option from the list above. Each option may be used once, more than once or not at all.

1. Propagates down T-tubules.
2. Ligand-gated ion channels.
3. Occurs in response to Ca^{2+} influx.
4. Open during axonal action potential.
5. Nonpropagating electrical event.
6. Releases transmitter into the neuromuscular cleft.

Short notes

Write short notes on the following:

a. postsynaptic mechanisms in neurotransmission
b. adaptation in sensory receptors
c. coding for pitch in the ear
d. normal and abnormal motor control from the basal ganglia
e. cortical control of language

Modified essay

The maintenance of posture and balance may be considered as an example of an integrated control system involving sensory inputs, central integration and motor outputs.

Questions

1. What are the major sensory receptors in this system? Which stimuli do each detect?
2. What connections are made between these sensory inputs and centres within the CNS which are relevant to balance?
3. What are the main descending motor pathways involved in control of balance? What are the main inputs to these pathways and what is their termination?
4. Draw a block diagram summarizing the main systems involved in posture control and their interconnections.

Data interpretation

The recording shown (Fig. 157) represents an electromyogram (EMG) from a normal subject. The recording electrode was placed over the hypothenar muscles and the complex waveform observed represents the surface potentials developed as electrical activity spreads through the underlying skeletal muscle following their activation by motor fibres in the ulnar nerve. Stimulating electrodes were used to stimulate this nerve at the wrist and elbow, giving traces A and B, respectively.

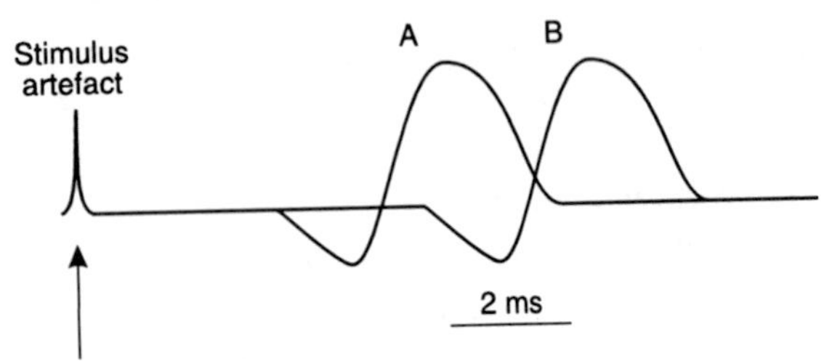

Fig. 157 Electromyographic recordings from muscles of the hypothenar eminence following stimulation of the ulnar nerve at a point close to the wrist (record A) and one near the elbow (record B). The stimulation sites (A and B) were 25 cm apart.

Questions

1. Increasing the stimulus voltage applied to the nerve from a low value produced EMG recordings of constant shape, but their amplitude increased up to a maximum response which could not be augmented by further increments in stimulus voltage. How can this finding be explained in the light of the 'all or nothing' law?
2. There is a delay between the stimulus and the beginning of the recorded response, known as the latency. What events may contribute to this delay?

EMG recordings of this sort may be used to calculate the nerve conduction velocity. The distance between the two stimulus points is divided by the difference in latencies for two records to give the estimate.

Questions

3. Why is this approach preferable to simply dividing the distance between the stimulating and recording electrodes by the latency?
4. Use the time base given to calculate the latency difference in this case. If the stimulus positions for A and B were 25 cm apart on the arm, what is the nerve conduction velocity in this case?

Clinical scenario

A 75-year-old gentleman was admitted to hospital after suddenly losing consciousness at home. The symptoms and signs suggest that he had suffered from a stroke. A week later he has recovered consciousness but is having difficulty speaking and cannot move the right side of his body. The muscle stretch reflexes are found to be exaggerated on the affected side of the body and muscle tone is elevated.

Questions

1. Which cerebral hemisphere do you think has been affected by the stroke? What features of the victim's condition suggest this?
2. Name the elements of the stretch reflex arc. What activates the reflex?
3. How may the sensitivity of the stretch reflex be controlled? If muscle tone is increased does this suggest increased or decreased stretch sensitivity?
4. What role does reciprocal inhibition play in the normal stretch reflex and how is it mediated?
5. If the stroke also involved the postcentral gyrus in the parietal lobe, what signs and symptoms might you expect?

Viva questions

1. Explain how light leads to vision.
2. How is an action potential propagated (conducted) along an axon?
3. How is voluntary movement controlled?

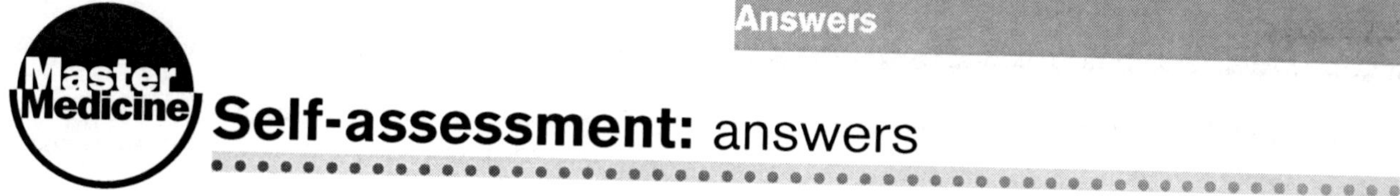

Self-assessment: answers

Multiple true/false answers

1. a. **True.** Classical transmitters are synthesized in the nerve terminal; peptide transmitters are transported down the axon from the cell body.
   b. **True.** Blocks action potential production.
   c. **False.** Microglia are macrophage-type cells.
   d. **False.** There are 10× more glial cells.
   e. **True.** Excitatory and inhibitory postsynaptic potentials.

2. a. **False.** Local currents spread bidirectionally; unidirectional propagation relies on the refractoriness of recently stimulated axon.
   b. **True.** Decreased axonal resistance to local current.
   c. **True.** Rapid, saltatory conduction.
   d. **False.** It requires spatial or temporal summation of multiple EPSPs to produce an action potential.
   e. **True.** Ca^{2+} entry is an important step in neurotransmitter release.

3. a. **False.** Requires postsynaptic receptors. Presynaptic receptors may be involved in neuromodulation but are not requisite for transmission.
   b. **True.** These prolong transmitter action, e.g., cocaine prolongs the action of catecholamines.
   c. **True.** Reduces transmitter release, a form of neuromodulation.
   d. **True.** Depolarization leads to transmitter release.
   e. **True.** Exocytosis of vesicle-bound transmitter from presynaptic terminal.

4. a. **False.** Nicotinic receptors at the skeletal neuromuscular junction, muscarinic receptors at the postganglionic parasympathetic nerve endings in the heart.
   b. **True.** See Section 7.2.
   c. **True.** After reuptake into presynaptic neurones.
   d. **False.** It inhibits them.
   e. **True.**

5. a. **False.** May be presynaptic.
   b. **True.** The likely basis of the IPSP.
   c. **False.** IPSPs are hyperpolarizing.
   d. **True.** This reduces neurotransmitter release and is the basis of presynaptic inhibition.
   e. **True.** By making it more difficult for excitatory stimuli to depolarize the nerve to threshold.

6. a. **False.** Receptor potentials are graded.
   b. **True.** The receptive field is defined as the region within which applied stimuli activate the receptor.
   c. **False.** Receptors are selective for one type of stimulus.
   d. **True.** This is converted into variations in action potential frequency in sensory nerves.
   e. **False.** Nonadapting, or slowly adapting receptors can provide continuous information about absolute stimulus strength; rapidly adapting receptors detect changes in stimulus strength.

7. a. **False.** Include pressure, temperature.
   b. **False.** Enter via dorsal root.
   c. **False.** Dorsal column tracts ascend ipsilaterally, carrying proprioceptive information.
   d. **True.**
   e. **False.** Not all proprioceptive signals reach the conscious level, e.g., muscle spindle output.

8. a. **False.** Visceral pain is usually 'slow' because of slow conduction through unmyelinated pain fibres.
   b. **True.**
   c. **True.**
   d. **True.** Probably inhibits transmitter release from pain fibres in the spinal cord.
   e. **False.** Poor pain conduction in neuropathic sensory nerves leads to pain insensitivity, raising the pain threshold. This results in accidental, traumatic damage to the peripheries.

9. a. **True.** These are the receptor cells of the organ of Corti.
   b. **False.** Endolymph is K^+ rich, not perilymph.
   c. **False.** Resonant frequency decreases as one moves towards the helicotrema. This provides the basis for frequency coding by the cochlea.
   d. **True.** Also called the cochlear duct; contains endolymph.
   e. **False.** Maximal sensitivity of hearing is at about 2000–3000 Hz.

10. a. **False.** Even though it is oriented vertically, the saccular macula is also sensitive to horizontal acceleration because it lies in a sagittal plane.
    b. **True.**
    c. **False.** Otoliths are found in the macula, not the crista.
    d. **True.**
    e. **True.** This helps explain the link between rotation and nystagmus since the oculomotor nerves are involved in controlling eye movements.

11. a. **True.** This is the region of the retina employed when directly gazing at an object.
    b. **False.** Secreted into the posterior chamber, drained from the anterior chamber.
    c. **False.** Cones also dark adapt but their maximal sensitivity is lower.
    d. **True.** Leads to reflex constriction of the pupil.
    e. **True.** Cornea provides majority of total focusing power; only the lens can be regulated during accommodation, however.

12. a. **True.** The initial step in phototransduction.
    b. **False.** They synapse with bipolar cells.
    c. **False.** This decreases transmitter release from the receptors.
    d. **False.** This differs for rods and different types of cone.
    e. **True.** Provides maximum resolution or visual acuity.

13. a. **True.** The EPP is normally suprathreshold.
    b. **True.** These channels are permeable to Na^+ and K^+.
    c. **False.** They are nicotinic.
    d. **False.** Gap junctions are associated with electrical transmission, e.g., between cardiac muscle cells.
    e. **False.** Acetylcholinesterase is located extracellularly in the neuromuscular cleft.

14. a. **True.** This stimulates the peripheral, contractile portions of the spindle fibres to contract stretching the central, sensory region.
    b. **False.** Sensitive to length.
    c. **True.** Through inhibitory interneurones.
    d. **False.** Brisk stretch reflexes are a feature of upper motor neurone lesions.
    e. **False.** These are very rapidly conducting, myelinated nerves.

15. a. **False.** Vestibulospinal tracts mainly innervate motoneurones supplying proximal muscles involved in posture and balance.
    b. **True.** Via the corticospinal tracts.
    c. **True.**
    d. **True.** Glycine is an important inhibitory neurotransmitter in the spinal cord.
    e. **False.** Each neurone supplies several muscle fibres to form a motor unit.

16. a. **False.** There are no cerebellar outputs to the spinal motoneurones; the spinocerebellar tracts carry proprioceptive information up to the cerebellum.
    b. **True.** Relays information about the intended motor outcome.
    c. **True.** Helps correct errors in motor control.
    d. **True.** This contrasts with damage to the motor cortex which affects contralateral muscles.
    e. **False.**

17. a. **False.** They are preganglionic.
    b. **False.** Noradrenaline (norepinephrine) release is a feature of postganglionic sympathetic fibres.
    c. **False.** This would block cholinergic transmission in sympathetic ganglia.
    d. **False.** This is a parasympathetic action and leads to accommodation for near vision in the eye.
    e. **True.** Accelerates heart and increases contractility.

18. a. **True.** The prefrontal association areas.
    b. **True.** In the precentral gyrus.
    c. **False.** Left hemisphere is dominant in about 70% of left-handed people.
    d. **True.** Damage here produces a sensory, or receptive, aphasia.
    e. **True.** Controls general level of awareness.

Single best answers

1. d. This promotes the passive spread of depolarization along the axon, speeding conduction. Myelin is absent from the nodal areas, where the action potentials are fired. Increasing the axonal diameter also increases conduction velocity, but this is not relevant to myelination, per se.
2. e. This activates exocytosis of neurotransmitter. Synaptic transmission may be inhibitory as well as excitatory. An increase in postsynaptic Ca^{2+} permeability may result but is not essential for synaptic transmission.
3. c. Although most receptors are depolarized during stimulation, some, e.g., photoreceptors, are hyperpolarized. The graded receptor potential results in changes in the frequency of action potentials in the afferent, sensory nerves.
4. d. The majority of neurones in descending motor tracts cross the midline in the brain and so spinal damage to these will produce weakness in muscles on the same side. Pain sensation ascends in spinothalamic tracts on the opposite side of the cord from the receptors, proprioceptive tracts on the same side, so the effects for these modalities would be the reverse of those stated.
5. c. Neither the cerebellum nor the basal ganglia innervate the spinal motoneurones directly. They exert their effects indirectly through cortical and brainstem motor areas.
6. d.
7. a. Transmission in autonomic ganglia depends on nicotinic cholinergic transmission for both divisions of the autonomic nervous system. Postganglionic sympathetic nerves mainly release noradrenaline (norepinephrine), which activates α- and β-adrenoceptors.
8. b. This is a β-adrenoceptor action, and adrenaline (epinephrine) is more potent an activator of β-receptors than noradrenaline (norepinephrine). The other effects are mediated by α-adrenoceptors, which are more potently activated by noradrenaline (norepinephrine). Note that contraction of the radial muscle in the iris leads to pupillary dilatation.
9. d.
10. c. This is a feature of REM sleep.

Matching item answers

Theme: Autonomic control

1. D. Vagal slowing of the heart is a muscarinic cholinergic effect.
2. B. β_1-adrenoceptors are the main sympathetic receptors in the heart.
3. E. Sympathetic preganglionic fibres activate adrenal medullary cells via nicotinic cholinergic receptors. Similar transmission mechanisms are found in the autonomic ganglia.
4. A. Sympathetic vasoconstrictor nerves control peripheral resistance through α-adrenoceptors.
5. C. The different adrenoceptor subtypes found in the heart and smooth muscle allows relatively selective drugs to be produced.
6. D. Another important parasympathetic action.

Theme: Neuromuscular transmission

1. C. This rapidly propagates the electrical signal to the centre of the muscle fibre.
2. E. Binding of acetylcholine opens these receptor/channel units, thus generating the end-plate potential.
3. D. A rise in intraneuronal $[Ca^{2+}]$ is a crucial signal for neurotransmitter release.
4. F. Causes rapid depolarization.
5. B. End-plate potentials do not propagate, unlike action potentials.
6. D. Vesicular exocytosis is the final step in transmitter release.

Short note answers

a. The main events are generation of nonpropagating EPSPs and IPSPs in response to excitatory and inhibitory neurotransmitters. These transmitters bind to the relevant receptors, leading to the opening of ion channels. Thus, the excitatory transmitter glutamate causes activation of cation-permeable channels which carry a net inward current, producing a depolarizing EPSP. Inhibitory transmission by glycine, on the other hand, opens Cl^- channels, producing an outward current which causes a hyperpolarizing IPSP. EPSPs and IPSPs are integrated by the mechanisms of temporal and spatial summation, with the membrane potential of the postsynaptic cell body being determined by the algebraic sum of the inputs. If this reaches the threshold level, a propagating action potential is triggered in the axon.

b. Sensory adaptation refers to a declining receptor potential output in the face of a constant stimulus. Rapidly adapting receptors (phasic receptors) are essentially responsive to changes in stimulus strength, e.g., the Pacinian corpuscles which act as pressure receptors in the skin. In this case adaptation is achieved through the multilayered, fluid-filled structure around the nerve ending which quickly allows a redistribution of force when pressure is applied and so axon deformation is short lived. Tonic receptors adapt much more slowly and so can provide information about the absolute strength of a maintained stimulus, e.g., some proprioceptors and cardiovascular baroreceptors. Some adaptation occurs even in these cases, e.g., baroreceptors adapt to a new pressure range if blood pressure remains elevated for several days.

c. Pitch is mainly determined by the frequency of sound, since this determines how far sound travels within the cochlea before its energy is converted into vibration of the basilar membrane. The resonant frequency of this structure falls progressively as one moves away from the oval window towards the helicotrema, so high-frequency sounds cause vibration close to the oval window, while low-frequency sounds are detected much closer to the cochlear apex. Basilar membrane vibrations set up shearing motions which activate the auditory receptor hair cells in the organ of Corti, and the position of a stimulated hair cell along this membrane provides a code for the frequency of the incident sound. This is then interpreted as the pitch of the perceived sound by the auditory cortex.

d. The basal ganglia are involved in initiating and regulating voluntary movement and in controlling body posture and muscle tone. This relies on inputs on sensory pathways, the motor cortex and the substantia nigra, and outputs to the motor cortex. Abnormalities of basal ganglia function (e.g., secondary to dopamine deficiency in the substantia nigra in Parkinson's disease) are typified by a lack of voluntary movement, involuntary movements of various kinds and abnormal muscle tone.

e. Language is a function of the dominant cerebral hemisphere, usually the left. Two cortical areas have especial importance. Broca's area in the frontal lobe is important for the generation of speech; it is sometimes called the motor speech area. Wernicke's area in the temporal lobe is important in understanding speech; it is sometimes called the sensory speech area. Damage in either area may cause speech difficulties (aphasia). These may be primarily motor (expressive aphasia) or sensory (receptive aphasia) but are usually mixed.

Modified essay answers

1. The vestibular apparatus detects head tilt (utricle), horizontal acceleration (utricle and saccule), vertical acceleration (saccule) and rotation (semicircular canals). Peripheral proprioceptors detect joint position and muscle spindles detect changes in muscle length. The eyes provide visual cues relating to body orientation and movement.
2. Vestibular inputs connect to the vestibular nuclei in the brainstem and are relayed to the cerebellum. The cerebellum also receives proprioceptive inputs from ascending spinocerebellar tracts. Sensory afferents from the muscle spindles make direct connections with α-motoneurones in the ventral horn grey matter of the spinal cord.

3. Vestibulospinal and reticulospinal tracts descend from the brainstem to innervate spinal motoneurones. Their activity is controlled by inputs from the vestibular apparatus, the cerebellum and also premotor areas of frontal cortex. They terminate on motoneurones supplying the muscles of the body trunk and proximal limbs, i.e., the main postural muscles.
4. See Fig. 158.

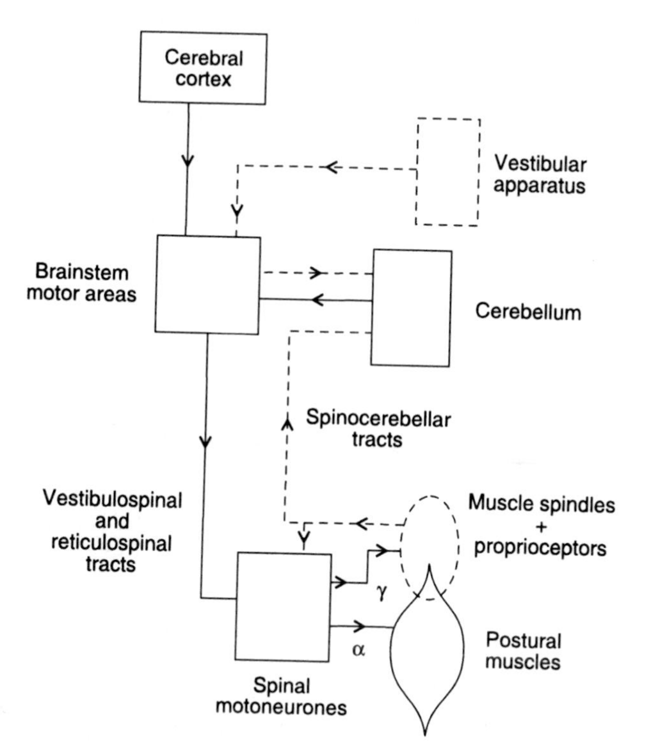

Fig. 158 Block diagram showing some of the main postural control systems. Sensory elements are indicated with broken lines, motor elements with solid lines. Note the involvement of both brainstem and spinal cord reflexes.

Data interpretation answers

1. EMG voltage will reflect the proportion of the muscle fibres in the hypothenar eminence which are activated. This depends on the number of motor axons which are stimulated and, as the stimulus voltage increases, more axons reach threshold and so fire an action potential. Individual action potentials are 'all or nothing' in nature but the total number of motor units activated increases with stimulus voltage until all available axons are recruited (a maximal stimulus).
2. The latency is contributed to by the time required for conduction of the action potential along the nerve axon and the transmission process at the neuromuscular junction. These events must be completed before the muscle action potentials, which are responsible for the EMG record, can be generated.
3. Such a result would include the time taken for neuromuscular transmission since this also contributes to the latency. The difference in latencies for two stimulation sites avoids this possible error, and should only depend on the conduction velocity.
4. Difference in latencies = 3.5 ms

   Velocity = Distance/Time

   $= 0.25\,\mathrm{m}/3.5 \times 10^{-3}\,\mathrm{s} = 71.4\,\mathrm{m\,s^{-1}}$.

Clinical scenario answers

1. The left hemisphere has been damaged, as suggested by motor difficulties on the right side of the body (which is controlled from the contralateral motor cortex) and the difficulties with speech. The speech cortex areas are on the left side in the majority of people.
2. The stretch reflex involves a muscle spindle and a sensory afferent nerve, which synapses with an α-motoneurone. This innervates the extrafusal fibres of the skeletal muscle. The reflex is activated by increases in muscle length.
3. Stretch sensitivity may be increased by stimulation of γ-motoneurones, which innervate the contractile portions of the spindle (intrafusal) fibres. Increased tone suggests increased stretch sensitivity.
4. Reciprocal inhibition refers to inhibition of muscles which act antagonistically to the stretched muscle and is mediated via relays from the sensory afferent through inhibitory interneurones to the relevant α-motoneurones.
5. Loss of any or all of the somatosensory modalities on the right side of the body, producing insensitivity to touch, temperature, pain and joint position.

Viva answers

1. Vision is a very extensive subject and you could never cover it all in a viva. Rather general questions are sometimes asked, however, and you should use them to cover the relevant material with which you are most

comfortable unless the examiner follows up with more specific points. Vision depends on formation of a focused image on the retina. Convergent refraction by the cornea and lens are relevant and accommodation for near vision may be mentioned. The examiner is most likely to be interested in transduction, however, and you should move on to this quickly unless subsequent questioning emphasizes the optical properties of the eye. The key topics to be mentioned include photobleaching in rods and cones leading to reduced cGMP and reduced inward membrane current (dark current). This causes hyperpolarization, reducing transmitter release. The signal is transmitted via bipolar cells to the ganglion cells whose axons form the optic nerve which carries the information into the brain. Information from each half of the visual field is relayed via the thalamus to the contralateral occipital cortex where it is converted into an image. Possible side issues include colour vision and dark adaptation.

2. The important concept here is the passive spread of excitation by local currents due to the potential differences which exist between depolarized membrane (e.g., at the peak of the action potential) and adjacent areas of axon which are at the resting potential. You will probably be asked to draw a diagram like Fig. 12 to illustrate what this means and in what direction current would flow. It is then easy to show that adjacent axonal membrane will be passively depolarized by the local currents until they reach threshold, at which point it fires an active action potential and the process continues. refractoriness ensures unidirectional propagation. Follow-up questions may well centre on factors which speed conduction velocity, e.g., reduced axoplasmic resistance to local currents in large axons and reduced loss of current across the membrane in myelinated axons which demonstrate saltatory conduction.
3. This is another huge topic. One way to start is by considering the role of the motor cortex in the prefrontal gyrus which regulates skilled movements, particularly in distal limb muscles, e.g., hand movements. Control is exercised by links to the α-motoneurones which directly innervate the muscles themselves, whether these are found in the cranial nerve nuclei (corticobulbar tracts) or the ventral horn of the spinal cord (corticospinal tracts). Cortical areas also influence the brainstem motor areas. These are linked to α-motoneurones by the rubrospinal tracts (which help supply the distal muscles) and vestibulo- and reticulospinal tracts (which are more involved in proximal limb and trunk muscles, controlling tone and posture). Skilled movements also require appropriate control from the basal ganglia (which help initiate movement) and the cerebellum (which monitors and fine tunes movement as it is occurring).

Endocrine physiology

Overview

Endocrine regulation is an important element of homeostatic control. Integration of endocrine and nervous controls is a major function of the hypothalamus. These neuroendocrine links will be considered in some detail, along with the actions of various hormones which regulate cell metabolism or some aspects of fluid and electrolyte balance. Other functions which are also under endocrine control are considered elsewhere, e.g., cardiovascular performance (Ch. 3), renal function (Ch. 5), gastrointestinal motility and secretion (Ch. 6) and reproductive physiology (Ch. 9).

8.1 Principles of endocrine function

Learning objectives

At the end of this section you should be able to:

- describe the main features of hormonal control
- outline how hormone release is controlled.

Endocrine control involves secretion of a *hormone* into the circulation. This messenger molecule binds to any cells which carry the relevant receptors, initiating the observed response (Section 1.5). As far as the target cell is concerned this is similar to other forms of chemical signalling, e.g., chemical neurotransmission. Indeed, a given substance may act both as a neurotransmitter and as a hormone at different sites. For example, noradrenaline (norepinephrine) is a postganglionic sympathetic neurotransmitter but is also released into the circulation as a hormone from the adrenal medulla (Section 7.6). Because of the large diffusion distances and circulation delays involved, however, hormonal responses are generally slower in onset than those mediated by nerves. They are also more persistent, since removal of the hormone from the bloodstream may take some time after secretion has stopped.

The release of endocrine substances is controlled in three ways:

Control by a regulated solute. Changes in a controlled variable may directly trigger changes in hormone levels; e.g., the cells which secrete parathormone, which acts to reduce the concentration of Ca^{2+} in the extracellular fluid, are themselves directly sensitive to $[Ca^{2+}]$ (Section 8.4). This is an example of a classical negative feedback mechanism.

Control by a different hormone. Secretion may be regulated by another hormone, e.g., several pituitary hormones stimulate secretion from other endocrine glands. This is also often under negative feedback control, with inhibition of pituitary secretion by the product hormone. Positive feedback may occur, however, leading to a rapid rise in hormone levels, e.g., prior to ovulation in the female reproductive cycle (Section 9.4).

Nervous control. Secretion may be under nervous control, and this provides a mechanism for the integration of nervous and endocrine responses. For example, hormone release may be altered by changing sensory stimuli or emotional states. The hypothalamus plays a central role in linking various endocrine systems with the nervous system in this way.

8.2 Hypothalamic and pituitary function

Learning objectives

At the end of this section you should be able to:
- outline the development of the pituitary
- draw a labelled diagram of the links between hypothalamus and pituitary
- explain how pituitary function is controlled and the role negative feedback at the hypothalamic and pituitary level plays in this
- list the anterior and posterior pituitary hormones and outline their actions
- describe the main consequences of abnormal pituitary function.

The hypothalamus regulates both autonomic nervous activity (Section 7.7) and several aspects of endocrine function. The latter role is fulfilled through its links with the pituitary gland, which secretes a wide range of hormones. Some of these regulate human physiology directly, while others control the activity of various endocrine glands around the body. The result is a multitiered endocrine control system which can be influenced through the nervous inputs to the hypothalamus.

Relevant structure

The hypothalamus is located adjacent to the third ventricle in the forebrain and is connected by the *hypophyseal stalk* to the pituitary gland (*hypophysis cerebri*) immediately inferior to it (Fig. 159). This has two embryologically distinct components.

- The *anterior pituitary*, or *adenohypophysis*, is a classical endocrine gland containing secretory epithelial cells. It is derived from an outpouching of the developing pharyngeal epithelium (*Rathke's pouch*). The anterior pituitary receives its blood supply from a series of *hypophyseal portal vessels* linking it with the hypothalamus.
- The *posterior pituitary*, or *neurohypophysis*, develops as a downgrowth from the brain and is continuous with the hypothalamus itself. It is made up of adapted axons which release hormones into the blood from their terminals. The cell bodies of these specialized neurones lie within the *paraventricular* and *supraoptic* nuclei of the hypothalamus.

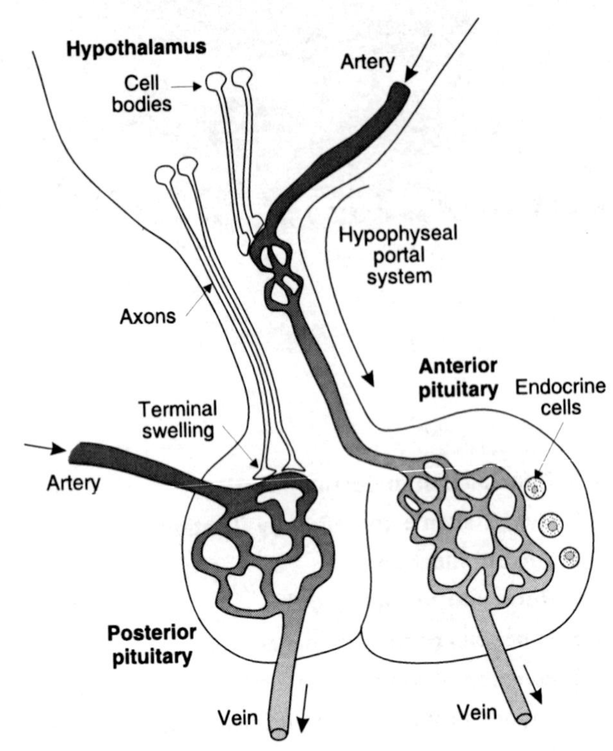

Fig. 159 The links between the hypothalamus and the pituitary gland.

Hypothalamic control of pituitary function

This is different for the anterior and posterior lobes of the pituitary.

Anterior pituitary

The endocrine activity of the anterior pituitary is regulated by hormones secreted from the hypothalamus and transported in the hypophyseal portal blood. Each of these factors selectively promotes or inhibits the secretion of a specific pituitary hormone and so they are referred to as *releasing* or *inhibiting hormones*, respectively. They are generally oligopeptides, although prolactin-inhibiting hormone is probably the transmitter substance dopamine. Hypothalamic secretion of these pituitary-regulating hormones is influenced by neurological inputs and feedback control. Both the pituitary hormones themselves and circulating products of the endocrine systems controlled by those hormones can regulate hypothalamic secretion, usually through feedback inhibition (Fig. 160).

Posterior pituitary

The posterior pituitary secretes two peptides which are manufactured in the cell bodies of hypothalamic neurones (Fig. 159). These hormones are packaged in vesicles and transported along the axons into the posterior pituitary itself to be stored in the axon

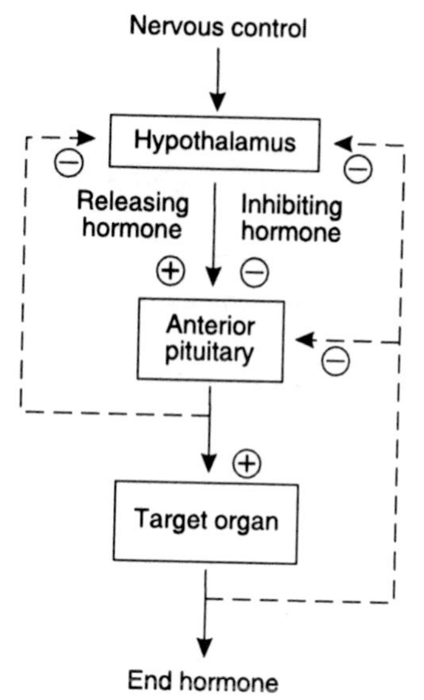

Fig. 160 Summary of the hypothalamic–pituitary control system. Note that secretion of both anterior pituitary and hypothalamic hormones may be inhibited by the circulating hormones whose release they stimulate. This provides negative feedback regulation of hormone levels.

terminals. Appropriate sensory stimuli lead to activation of the hypothalamic cells and the release of the hormones from the posterior pituitary into the bloodstream. This adaptation of neurotransmission to fulfil an endocrine role is sometimes called *neurocrine* control.

Anterior pituitary hormones

The anterior pituitary secretes six peptide hormones with well-defined functions in humans. These include hormones controlling the endocrine functions of the adrenal cortex, the thyroid and the gonads, as well as prolactin and growth hormone.

Adrenocorticotrophic hormone (ACTH) is secreted in response to *corticotrophin-releasing hormone* (CRH) from the hypothalamus. It stimulates the release of glucocorticoids from the adrenal cortex. A number of other peptides are synthesized and secreted from the anterior pituitary along with ACTH. These include the endogenous opiate β-endorphin, and precursors of *melanocyte-stimulating hormone*. Their roles are unclear.

Thyroid-stimulating hormone (TSH), or thyrotrophin, is secreted in response to *thyrotrophin-releasing hormone* (TRH) from the hypothalamus. It stimulates thyroid hormone secretion.

Follicle-stimulating hormone (FSH) and luteinizing hormone (LH) are secreted in response to *gonadotrophin-releasing hormone* (GnRH) from the hypothalamus. These gonadotrophins stimulate the male and female gonads (Sections 9.2 and 9.4).

Prolactin secretion may be influenced by a *prolactin-releasing hormone* (PRH) but is mainly controlled by a *prolactin-inhibiting hormone* (PIH), (this is probably dopamine). Stimuli which reduce PIH release from the hypothalamus raise prolactin levels. This occurs during pregnancy, favouring development of the breasts ready for lactation, and as part of the suckling reflex during breast feeding. The resultant prolactin peaks stimulate milk production (Section 9.5). Prolactin also inhibits GnRH secretion from the hypothalamus, thus inhibiting the reproductive cycle during lactation.

Human growth hormone (hGH), or somatotrophin, is controlled by two antagonistic hypothalamic hormones. *Growth hormone-releasing hormone* (GHRH) is believed to be more important than the *growth hormone-inhibiting hormone* (GHIH), which is also known as somatostatin. Growth hormone is a major anabolic hormone which stimulates cell division and growth in both bony and soft tissues around the body. These effects are probably indirect, being mediated by growth promoters, known as *somatomedins*, produced by the liver in response to hGH stimulation. Growth hormone is particularly important during childhood and adolescence, but although it can no longer stimulate long bone growth after the epiphyses have fused, hGH continues to have important metabolic functions throughout life. These favour protein synthesis in most body tissues, while promoting the breakdown of fat for energy use. This has a carbohydrate-sparing effect, so that glycogen stores increase and blood glucose levels tend to rise. Low blood glucose stimulates hypothalamic GHRH and inhibits secretion of somatostatin, both from the hypothalamus and the δ cells of the pancreatic islets. The resulting increase in hGH helps raise glucose concentrations back towards normal. Growth hormone secretion is also stimulated by trauma and stress, through neural control of the hypothalamus.

Posterior pituitary hormones

The posterior lobe of the pituitary secretes two short chain peptide hormones, oxytocin and antidiuretic hormone.

Oxytocin is produced by cells in the paraventricular and supraoptic nuclei of the hypothalamus. These are reflexly activated by sensory inputs from mechanoreceptors in the breast during suckling, and oxytocin is released from the posterior pituitary in response. This stimulates the ejection of milk through contraction of the myoepithelial cells surrounding the milk ducts (Section 9.5). Oxytocin also stimulates uterine contraction during labour (Section 9.5).

Antidiuretic hormone (ADH) is also produced in the paraventricular and supraoptic nuclei of the hypothalamus. Sensory inputs, both from local osmoreceptors and from stretch receptors in the cardiovascular system (the cardiopulmonary and systemic arterial baroreceptors), stimulate these cells whenever the osmolality of the extracellular fluid rises, or if blood volume or blood pressure falls. The resulting increase in ADH promotes water reabsorption from the collecting ducts and distal convoluted tubules of the kidney (Section 5.4), and so tends to reduce the osmolality and expand the volume of the extracellular fluids by promoting the production of a small volume of concentrated urine. At the same time, peripheral resistance is increased through arteriolar constriction, which also helps to maintain arterial pressure (Section 3.6). This pressor effect, though less important physiologically, explains the derivation of ADH's alternative name, which is vasopressin.

8.3 Thyroid function

Learning objectives

At the end of this section you should be able to:

- name the main cell types in the thyroid
- outline how thyroid hormones are synthesized
- explain how thyroid secretion is controlled
- explain the significance of protein transport of thyroid hormones
- list the main actions of thyroid hormones
- describe the main consequences of abnormal thyroid function.

The thyroid hormones are key metabolic regulators and are particularly important in determining metabolic rate and heat production.

Relevant structure

The thyroid gland is located in the neck, in front of and just below the level of the larynx, and consists of two lobes joined by a central isthmus. Histologically, it consists of numerous spherical *follicles*, each with an outer layer of cuboidal epithelium and filled

Box 34 Clinical note: Abnormal pituitary function

This may be classified into problems arising from deficient pituitary secretion and those caused by excess hormone.

Deficient secretion

Deficiencies resulting from insults to the entire gland, e.g., because of a local tumour or its surgical treatment, may involve all the pituitary hormones, a condition known as *panhypopituitarism*. Alternatively, there may be a deficit in a single hormone caused by a defect in the relevant pituitary cells or in the hypothalamic cells which regulate their function. The main problems which can result are:

- corticosteroid deficiency caused by lack of ACTH (Section 8.5)
- thyroid deficiency caused by lack of TSH
- failure of sexual function caused by lack of FSH/LH
- dwarfism caused by lack of hGH in childhood; this may arise as a congenital defect and can be treated with hGH supplements
- diabetes insipidus caused by lack of ADH; this results in an inability to concentrate urine and so the patient passes 8–10 L of dilute urine per day and has to drink a commensurate volume.

Excess secretion

The following conditions resulting from oversecretion of a single pituitary hormone are observed in clinical practice:

- corticosteroid excess caused by elevated ACTH (Section 8.5)
- impaired reproductive function caused by elevated prolactin. *Hyperprolactinaemia* is now recognized as an important cause of menstrual failure and infertility. This seems to be caused by prolactin's ability to block GnRH production by the hypothalamus
- abnormal growth caused by elevated hGH. This stimulates long bone growth in children, causing *gigantism*. If hGH levels become raised in adult life, after epiphyseal fusion, there is no increase in height, but soft tissue and body organ growth are stimulated. This is known as *acromegaly*. A persistent excess of hGH may eventually lead to elevated blood glucose levels, i.e., diabetes mellitus results.
- fluid retention and low plasma osmolality caused by elevated ADH. This is particularly associated with cancer of the lung and is known as *inappropriate ADH secretion*.

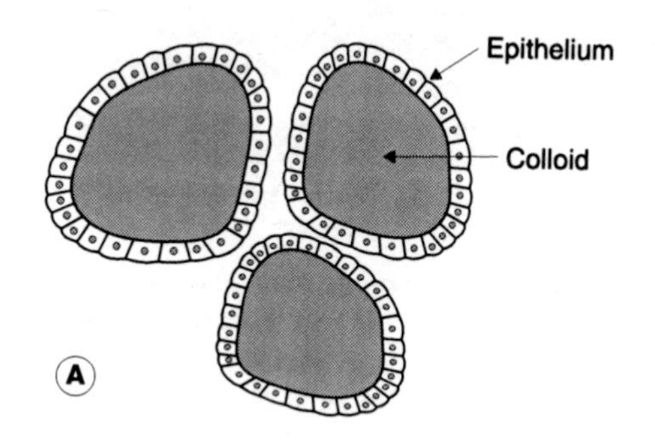

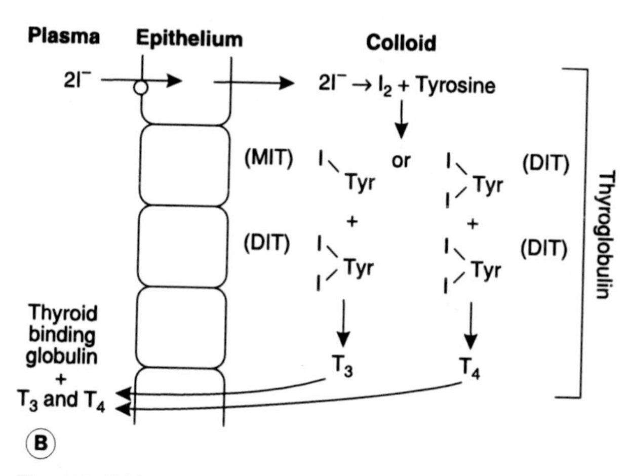

Fig. 161 (A) Thyroid follicles. (B) Main steps in thyroid hormone synthesis. Iodination of tyrosine produces monoiodotyrosine (MIT) and diiodotyrosine (DIT). These are then combined to form triiodothyronine (T_3) and thyroxine (T_4).

with proteinaceous *colloid* (Fig. 161A). These follicles represent the functional subunits of the gland, responsible for synthesis, storage and release of the thyroid hormones. The thyroid also contains *parafollicular C cells*, which secrete calcitonin. These will be considered along with other aspects of body calcium regulation (Section 8.4).

Hormone synthesis

The first step in the synthesis of the thyroid hormones involves active pumping of *iodide ions* (I^-) from the extracellular space into the follicular epithelium (Fig. 161B). The trapped I^- enters the colloid and is oxidized to iodine, which then combines with *tyrosine* molecules attached to a colloid-binding protein known as *thyroglobulin*. Monoiodotyrosine (MIT) and diiodotyrosine (DIT) are generated as a result. These then combine to produce the hormonal products, *triiodothyronine* (1 MIT + 1 DIT) and *thyroxine* (2 × DIT). These thyroid hormones may be stored in the colloid for some months but are eventually detached from the thyroglobulin and released into the bloodstream. The quantity of thyroxine (T_4) produced greatly exceeds that of triiodothyronine (T_3), but the latter is more biologically active.

Regulation of thyroid secretion

Thyroid activity is controlled by the hypothalamus and anterior pituitary and provides a classical example of the feedback loops typical of such regulation (Section 8.2; Fig. 160). The hypothalamus releases thyrotrophin-releasing hormone (TRH) into the hypophyseal portal blood and this stimulates secretion of thyroid-stimulating hormone (TSH) from the anterior pituitary. Thyroid hormone synthesis and secretion are both stimulated by TSH. Normal levels of circulating T_3 and T_4 are maintained through their negative feedback effects on TRH and TSH secretion. Other stimuli may also influence secretion, e.g., cold conditions may stimulate hypothalamic TRH release as part of temperature homeostasis (Section 1.1).

Protein transport of thyroid hormones

Most of the thyroid hormone in the blood is bound to plasma proteins, particularly *thyroid-binding globulin* (TBG). This transport is necessary as T_3 and T_4 are lipophilic molecules which would not dissolve readily in the aqueous plasma. Protein bound hormone also provides a reservoir which helps to maintain steady tissue levels. Only free (unbound) hormone is biologically active, however, since only it can diffuse out of the capillaries and into body cells. Increasing the concentration of plasma binding proteins may increase the total concentrations of T_3 and T_4 without affecting the levels of free hormone and this may have to be considered when interpreting hormone assays, e.g., the normal values for total T_4 concentrations in plasma increase during pregnancy as a result of increased protein binding, but this is not associated with a rise in the concentration of free hormone.

Intracellular actions of thyroid hormones

Because they are lipid-soluble, both T_3 and T_4 can diffuse across the plasma membrane and act intracellularly. A large proportion of the T_4 is converted to T_3 within the target cell and thyroid hormone receptors on the nucleus bind T_3 with a higher affinity than T_4. Receptor binding stimulates increased DNA transcription and, therefore, promotes protein synthesis. This leads to increased enzyme activity, elevating metabolic rate and promoting growth.

Whole body actions of thyroid hormones

These are many and varied but may be summarized briefly under the headings of metabolic, systemic and developmental effects.

Box 35 Clinical note: Abnormal thyroid function

This leads to inadequate or excess amounts of circulating thyroid hormone.

Deficient secretion

Inadequate thyroid secretion is known as hypothyroidism. This may be subdivided into primary hypothyroidism and hypothyroidism secondary to deficient TSH secretion.

In *primary hypothyroidism* the fault lies within the thyroid gland itself. Causes include iodide deficiency, congenital thyroid enzyme deficiencies and inflammation of the gland (thyroiditis). The low T_3 and T_4 levels lead to elevated TSH levels (reduced negative feedback) and this stimulates enlargement of the thyroid, i.e., goitre formation.

In *secondary hypothyroidism* there is deficient TSH secretion from the anterior pituitary. This leads to thyroid atrophy rather than goitre.

The major symptoms and signs of hypothyroidism can be related to the normal metabolic, systemic and developmental actions of the hormone.

- **Metabolic consequences:**
  - –a reduced basal metabolic rate leading to cold intolerance with constricted peripheries under fairly normal temperature conditions; appetite is reduced but weight increases
  - –reduced blood sugar levels
  - –elevated blood lipids, which may eventually lead to atherosclerosis
- **Systemic consequences:**
  - –low resting heart rate
  - –decreased gut motility leading to constipation
  - –slowed thinking and somnolence; muscle stretch reflexes may be delayed
- **Developmental consequences**: the most important developmental consequence is irreversible mental retardation in cases of congenital hypothyroidism resulting from inadequate brain development in the early months of life. This condition, known as *cretinism*, can be prevented by screening newborn children and providing immediate thyroid replacement therapy for those who require it.

Excess secretion

Hyperthyroidism, or *thyrotoxicosis*, can arise in two main ways.

- *Abnormal autoantibodies* may bind to the TSH receptors in the thyroid and thereby stimulate the gland, which enlarges, producing a goitre, and secretes excessive amounts of T_3 and T_4. Increased negative feedback reduces the circulating levels of TSH.
- A *hormone-secreting tumour* (an adenoma) may develop within the thyroid. This functions independently of TSH control and so the normal regulatory mechanism is lost. TSH levels will be low so that areas of normal gland tissue will atrophy.

The major symptoms and signs of hyperthyroidism may again be summarized using the metabolic, systemic, developmental headings.

- **Metabolic consequences:**
  - –a dramatic elevation of basal metabolic rate with increased heat production, heat intolerance, vasodilatation and sweating; appetite increases but weight declines.
- **Systemic consequences:**
  - –rapid heart rates and arrhythmias, especially atrial fibrillation; the pulse is bounding, because of the large pulse pressure; persistent elevation of the cardiac output in response to the increased metabolic needs of the body may eventually lead to high-output cardiac failure
  - –abnormal breathlessness during exercise
  - –increased gut motility leading to diarrhoea
  - –mental overactivity and insomnia; muscle reflexes are brisk but the muscles are often weak; sympathetic nervous system symptoms include anxiety and muscle tremor
- **Developmental consequences**: the main result of excess thyroid secretion in childhood is early closure of the epiphyseal plates in long bones so that, although early growth may be accelerated, adult stature is reduced.

- **Metabolic effects** involve nearly all the tissues of the body and include:
  - —elevation of basal metabolic rate with increased O_2 consumption and heat production (Section 4.7)
  - —stimulation of carbohydrate metabolism with increases in gastrointestinal and cellular absorption of glucose, glycolysis and gluconeogenesis
  - —increased catabolism of free fatty acids with depletion of fat stores and reduction in blood lipids
  - —increased protein synthesis and breakdown
- Systemic effects:
  - —direct stimulation of heart rate and an indirect reduction of peripheral resistance caused by the increased metabolic demands of the tissues; cardiac output and pulse pressure tend to rise, but mean arterial pressure is little affected
  - —increased ventilation
  - —increased gastrointestinal motility and secretion
  - —increased CNS activity and alertness
- Developmental effects:
  - —stimulation of skeletal growth in childhood
  - —promotion of normal brain development in the early postnatal period.

8.4 Hormonal control of Ca^{2+}

Learning objectives

At the end of this section you should be able to:

- identify the main Ca^{2+} pools in the body
- explain why extracellular $[Ca^{2+}]$ is tightly regulated
- describe the source and actions of parathormone
- describe the sources and actions of vitamin D
- explain how phosphate levels are controlled
- describe the main consequences of abnormal Ca^{2+} control.

Calcium enters the body by absorption of dietary Ca^{2+} from the gut and is lost mainly through urinary excretion. Within the body itself, Ca^{2+} is found in a number of functionally distinct reservoirs, or 'pools'. A variety of hormones act to regulate Ca^{2+} levels by controlling both the exchange between these Ca^{2+} pools and the rate of absorption and excretion of Ca^{2+} from the body.

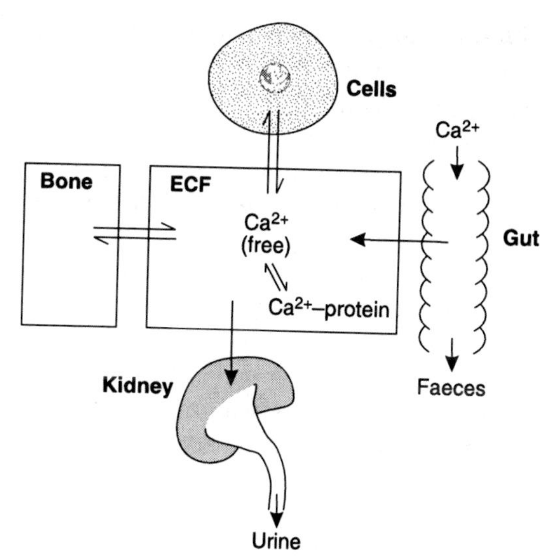

Fig. 162 The main Ca^{2+} pools in the body. ECF, extracellular fluid.

Calcium pools in the body

There are three main body calcium pools (Fig. 162):

- bony skeleton
- intracellular pool
- extracellular pool.

Bony skeleton. About 1 kg of Ca^{2+} is deposited in the mineral component of the bony skeleton. Calcium phosphate crystals form within the connective tissue framework (*osteoid*) laid down by *osteoblasts*. This bone mineral provides impressive resistance to compressive loads. Resorption of bone by *osteoclasts* breaks down the bony matrix, releasing Ca^{2+} and phosphate back into the extracellular fluid. Bone also contains *osteocytes*, which can transfer Ca^{2+} to the extracellular fluid without destroying the bone structure. This provides a *rapidly exchangeable pool* of bone Ca^{2+} which acts as a functionally distinct fraction of total bone Ca^{2+}.

Intracellular pool. The size of the intracellular Ca^{2+} pool is hard to estimate since much of it is normally bound to proteins within stores of unknown capacity. It is clear, however, that changes in cytoplasmic $[Ca^{2+}]$, whether caused by release of stores, as in skeletal muscle (Section 1.6), or by entry of Ca^{2+} from the extracellular fluid, as at the axon terminal (Section 7.2), act as important intracellular signals in controlling a wide variety of cell functions.

Extracellular pool. Calcium in the extracellular fluid is involved in continuous exchange with that in bone and body cells (Fig. 162). Absorption from the gut and urinary excretion also occur directly into and out of the extracellular space. All these processes are affected by the extracellular [Ca^{2+}]. Even more importantly, extracellular Ca^{2+} is crucial in determining the ease with which excitable cells can be stimulated and so plasma [Ca^{2+}] must be closely regulated. The normal value is about 2.5 mmol L^{-1} but approximately half of this is bound to protein. The remainder consists of Ca^{2+} free in solution, and it is only this fraction which is biologically active.

Physiological functions of extracellular Ca^{2+}

Several vital processes are dependent on the maintenance of an appropriate Ca^{2+} concentration in the extracellular space.

The excitability of nerve and muscle is increased by a fall in plasma [Ca^{2+}] (*hypocalcaemia*). This may result in spontaneous skeletal muscle contraction (which can be fatal because of laryngeal spasm and respiratory arrest), cardiac arrhythmias and abnormal sensations caused by spontaneous sensory nerve activity (*paraesthesia*). The mechanism underlying these effects is an increase in ion channel permeability to Na^+ in the presence of low extracellular [Ca^{2+}]. This promotes inward Na^+ currents so that the membrane depolarizes towards the threshold for action potential production. Conversely, a high extracellular [Ca^{2+}] depresses nerve and muscle activity by reducing the Na^+ current (Section 1.4).

Excitation–contraction coupling in muscle depends on an increase in intracellular [Ca^{2+}] (Section 1.6). Smaller decreases in extracellular [Ca^{2+}] than those which stimulate spontaneous excitability may reduce the strength of contraction, both by decreasing the influx of Ca^{2+} during the action potential and by depleting the intracellular stores within the sarcoplasmic reticulum.

Stimulus–secretion coupling refers to mechanisms leading to release of cell products following stimulation of nerve terminals and both endocrine and exocrine glands. These secretory processes are often triggered by an increase in intracellular [Ca^{2+}], either due to release of intracellular stores or Ca^{2+} entry from outside the cell.

Blood clotting is dependent on plasma Ca^{2+}, which acts as an essential clotting factor (Section 2.6).

Regulation of body Ca^{2+}

Calcium control mechanisms act to regulate two main variables, the free [Ca^{2+}] in extracellular fluid (in the short term), and the total body Ca^{2+} content (in the medium to long term). Assuming that the dietary intake is adequate, regulation of plasma [Ca^{2+}] mainly depends on two hormones, parathormone and vitamin D, with a less important contribution from a third, calcitonin.

Parathormone (PTH)

This is a protein hormone secreted from four parathyroid glands located posterior to the lobes of the thyroid gland in the neck. It is responsible for the tight control of free [Ca^{2+}] in the extracellular fluid and is essential for life. Parathormone secretion is directly stimulated by a fall in plasma [Ca^{2+}] and acts to elevate [Ca^{2+}] in several ways, thereby providing negative feedback control. It has four main actions: one in bone and three in the kidney.

- Stimulation of Ca^{2+} release from bone. Initially osteocytes mobilize Ca^{2+} from the rapidly exchangeable bone pool, transferring it into the extracellular fluid and thus raising plasma [Ca^{2+}]. In the longer term, parathormone promotes a continued but slower release of bone Ca^{2+} from the osteoclastic breakdown of the bone matrix itself.
- Stimulation of the rate of Ca^{2+} reabsorption from the renal tubules so that less is lost in the urine. This helps conserve plasma Ca^{2+}.
- Stimulation of urinary phosphate excretion, which reduces plasma phosphate concentration. This is important for Ca^{2+} regulation since it reduces the tendency for Ca^{2+} and phosphate ions to react, forming a solid precipitate of calcium phosphate.
- Stimulation of the rate at which vitamin D is converted to its most biologically active form, calcitriol (also known as 1,25-dihydroxycholecalciferol) within the kidney. Thus, parathormone indirectly stimulates the uptake of both calcium and phosphate from the gut by promoting the actions of vitamin D.

Vitamin D

Vitamin D is a fat-soluble vitamin which comes from two main sources, the diet (vitamin D_2 or ergocalciferol) and the skin (vitamin D_3 or *cholecalciferol*). These two forms differ slightly in structure but fulfil identical functions. The main dietary sources are fish, liver and ultraviolet (UV) irradiated milk, and D_2 is absorbed along with lipid in the small intestine. Alternatively, UV radiation from sunlight can convert a cholesterol derivative into vitamin D_3 in the skin. The relative importance of these sources in a given individual largely depends on the local climate.

Once in the circulation, *cholecalciferol* (D_3) undergoes two further activation steps. It is converted to *25-hydroxycholecalciferol* (25-(OH)-D_3) in the liver and further hydroxylated to form calcitriol (*1,25-dihydroxycholecalciferol,* 1,25-$(OH)_2$-D_3) in the kidney. This is the most active metabolite and is responsible for much of vitamin D's activity in the body. Its formation is stimulated both by parathormone, secreted in response to low plasma [Ca^{2+}], and by low plasma phosphate concentrations.

Vitamin D acts to elevate plasma levels of both Ca^{2+} and phosphate. It achieves this by:

- increasing the rate of active Ca^{2+} uptake from the gut. This helps replace calcium lost in the urine and is an important mechanism in the maintenance of total body calcium
- stimulating phosphate absorption from the gut
- stimulating Ca^{2+} and phosphate reabsorption in the kidney
- stimulating osteoclastic bone resorption, which releases both Ca^{2+} and phosphate into the extracellular fluid. This effect is only seen with very high levels of activated vitamin D.

Vitamin D also promotes mineralization of newly formed osteoid, which requires Ca^{2+} and phosphate, and adequate supplies are especially important during the periods of rapid skeletal growth in childhood and adolescence.

Calcitonin

This hormone is secreted by the parafollicular C cells of the thyroid gland. It acts on bone to reduce the rate of release of Ca^{2+} to the extracellular fluid and, therefore, tends to lower plasma [Ca^{2+}]. It appears to be secreted during episodes of *hypercalcaemia* (elevated [Ca^{2+}]) and may be protective against such abnormal rises. It does not seem to play any significant role in normal Ca^{2+} control.

Regulation of plasma phosphate concentration

The hormones involved in Ca^{2+} control also regulate phosphate, which is present as calcium phosphate in bone mineral and free in solution within the extracellular fluid. Phosphate is also a crucial intracellular ion, acting both as an enzyme cofactor and a substrate for phosphorylation reactions. Parathormone reduces plasma phosphate levels, while calcitriol increases it. Low phosphate concentrations stimulate renal activation of vitamin D and this provides the main feedback system for phosphate control. Increased vitamin D activity will also tend to elevate [Ca^{2+}], but this is prevented by the resultant inhibition of parathormone secretion. Under these circumstances, the reduced levels of parathormone make it easier to raise the phosphate concentration since parathormone promotes renal excretion of phosphate.

Box 36 Clinical note: Abnormal Ca^{2+} control

The wide range of causes of low (hypocalcaemia) and high (hypercalcaemia) plasma concentrations of Ca^{2+} will not be considered here. This section will restrict itself to problems resulting from deficient and excess levels of parathormone and vitamin D.

Deficiencies

Hypoparathyroidism may be caused by parathyroid autoantibodies or accidental damage during thyroid surgery. The main outcome is hypocalcaemia with elevated plasma phosphate levels. Increased neuromuscular excitability results and this may lead to muscle tetany, laryngeal spasm and convulsions.

Deficient vitamin D activity may result from an inadequate diet or lipid malabsorption (vitamin D is a fat-soluble vitamin) coupled with low levels of UV exposure. Failure of vitamin D activation may also cause problems in patients with chronic renal failure. There is only likely to be a mild hypocalcaemia, because of the compensatory stimulation of parathormone secretion, but plasma phosphate levels are reduced. The most obvious effects often reflect demineralization of bone, leading to osteomalacia with bone pain in adults and skeletal deformities (rickets) in children.

Excesses

Primary hyperparathyroidism results from failure of the normal negative feedback of plasma [Ca^{2+}] on parathormone secretion, often caused by a parathormone-secreting tumour. Plasma [Ca^{2+}] is elevated (hypercalcaemia) while phosphate is reduced. Bone destruction causes erosions and cyst formation while calcium may be deposited elsewhere in the body, e.g., in the urine (as renal calculi) and kidneys.

Hypervitaminosis D may result from excessive use of vitamin D supplements. The main effects are those of hypercalcaemia. Plasma phosphate levels will be high.

8.5 Functions of the adrenal cortex

Learning objectives

At the end of this section you should be able to:

- outline the structure and biochemistry relevant to adrenal corticosteroid secretion
- outline the cellular action of steroid hormones
- describe how aldosterone release is regulated
- describe the main actions of aldosterone
- describe how glucocorticoid release is regulated
- describe the main actions of glucocorticoids
- describe the main consequences of abnormal adrenocortical function.

The adrenal glands are situated at the upper poles of each kidney and consist of an outer cortex surrounding the central medulla. These regions are embryologically and functionally distinct and will be considered separately.

Relevant structure and biochemistry

The adrenal cortex secretes a variety of steroid hormones and can be subdivided into three histological zones. The outermost *zona glomerulosa* is responsible for aldosterone secretion. Beneath it lies the *zona fasciculata*, while the *zona reticularis* is immediately adjacent to the medulla. These last two regions are capable of secreting both glucocorticoids (mainly from the zona fasciculata) and adrenal androgens (mainly from the zona reticularis).

Hormone synthesis and excretion

Mitochondrial conversion of *cholesterol* to *pregnenolone* provides a common biochemical platform leading to the production of *aldosterone* (the main mineralocorticoid), *cortisol* and *corticosterone* (the main glucocorticoids), and adrenal *androgens* (Fig. 163). Following release, a considerable fraction of these lipophilic (hydrophobic) adrenal steroids is bound to plasma proteins. Eventually the circulating hormones are broken down in the liver and the metabolites are conjugated with glucuronic acid prior to excretion in the faeces and urine. This explains why conditions associated with excess production of adrenal corticosteroids may be usefully diagnosed by measuring urinary excretion of specific metabolites.

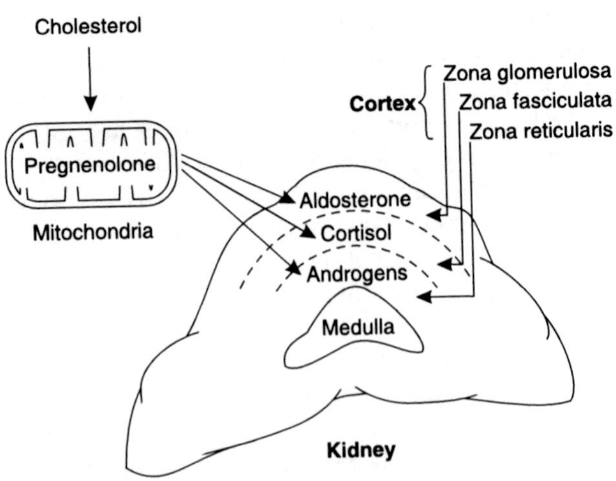

Fig. 163 Outline of adrenal structure and function.

Intracellular actions of adrenal steroids

Being lipophilic, these steroid hormones cross the plasma membrane of target cells easily and then bind to intracellular receptors. The receptor–hormone complexes enter the nucleus where they modify DNA transcription. This leads to increased production of enzymes and other functional proteins and is presumed to explain the changes in cell function seen. The specific effects of any given hormone will depend on exactly which genes it switches on.

Mineralocorticoids

Mineralocorticoid function is primarily concerned with the regulation of extracellular fluid volume through renal reabsorption of Na^+. *Aldosterone* accounts for about 95% of mineralocorticoid action but glucocorticoids also exert a small mineralocorticoid effect.

Regulation of aldosterone secretion

The secretory activity of the zona glomerulosa is controlled by a number of factors, but the dominant stimulus is the extracellular fluid volume, as detected by the kidney.

The renin–angiotensin system is activated by reductions in extracellular fluid volume, arterial blood pressure or plasma $[Na^+]$, as detected by the juxtaglomerular apparatus in the kidney (Section 5.3). The resulting increases in circulating levels of *angiotensin II* stimulate aldosterone secretion.

Increases in plasma K^+ concentration directly stimulate synthesis and secretion of aldosterone from the zona glomerulosa of the adrenal cortex.

Adrenocorticotrophic hormone (ACTH) from the anterior pituitary can also stimulate aldosterone secretion. This is not important in the normal regulation of aldosterone levels, however, and providing there is an adequate background level of ACTH, the actual rate of secretion is controlled from minute to minute by the plasma levels of angiotensin II and K^+.

Mineralocorticoid actions

Aldosterone and other hormones with mineralocorticoid actions are primarily involved in the control of the electrolyte and fluid content of the extracellular space. Their main site of action is the kidney. Similar effects are seen at sites other than the kidney, however, including the sweat glands, salivary glands, and secretory cells and glands along the gastrointestinal tract. In each case aldosterone promotes retention of Na^+ and loss of K^+.

Reabsorption of Na^+ The dominant mineralocorticoid effect is to stimulate active reabsorption of Na^+ from the distal convoluted tubule, and this leads to osmotic reabsorption of water (Section 5.4). The overall effect is to expand the extracellular fluid and plasma volumes by increasing the total Na^+ content of the extracellular space. Since the Na^+ is effectively absorbed in an isotonic solution, however, this has little effect on plasma $[Na^+]$. Although reabsorption of only a small fraction of the total Na^+ filtered by the kidneys is under aldosterone control (about 3%), the extremely high filtration rate in the kidneys means that this still potentially represents a large quantity of fluid each day (over 35% of the total extracellular volume). This mechanism helps to maintain normal hydration, since any reduction in extracellular volume leads to a parallel fall in blood volume which stimulates aldosterone release through the renin–angiotensin system.

Secretion of K^+ Mineralocorticoids stimulate active secretion of K^+ from the plasma into the distal convoluted tubule and thus promote its urinary excretion. There is no parallel movement of water and so plasma $[K^+]$ falls. This provides the main negative feedback mechanism for the control of $[K^+]$, since a rise in $[K^+]$ directly stimulates aldosterone release.

Secretion of H^+ Mineralocorticoids stimulate secretion of H^+ into the distal convoluted tubule. This effect is less pronounced than their action on K^+.

Glucocorticoids

These hormones play a major role in the control of blood glucose and other aspects of metabolism, and are also essential elements in the normal response to stress, increasing the body's ability to withstand trauma, infection and abnormal environmental conditions. Glucocorticoids are essential for life, and deficiency states require very careful management. *Cortisol* (hydrocortisone) is the dominant glucocorticoid, although smaller amounts of corticosterone and other hormones are also secreted by the zona fasciculata of the adrenal cortex.

Regulation of glucocorticoid secretion

The rate of glucocorticoid secretion is directly dependent on the plasma level of *adrenocorticotrophic hormone* (ACTH). This anterior pituitary hormone binds to membrane-bound adrenal receptors, stimulating synthesis and secretion of cortisol. ACTH release is promoted by hypothalamic release of corticotrophin-releasing hormone (CRH) into the hypophyseal portal vessels (Section 8.2). The system is under negative feedback control with inhibition of both CRH and ACTH release by circulating cortisol.

There are two important physiological factors which modulate glucocorticoid secretion:

- normal circadian rhythm
- stress.

Cortisol levels do not remain constant during a normal 24-hour cycle but demonstrate a *circadian rhythm*, with a peak in the early part of the morning (about 7–9 a.m.) and a sustained trough in the middle of the night (about 1–3 a.m.). This pattern is driven by a parallel, diurnal variation in ACTH secretion and can be influenced by a number of events, e.g., changing patterns of sleep and wakefulness can change the timing of the peak and trough. This suggests that nervous inputs play an important part in the regulation of glucocorticoid secretion, presumably through their influence on hypothalamic release of CRH. A further, practical consequence of this variation is that blood samples for cortisol estimation should be taken at the same time each day if serial measurements are to be compared.

As well as varying cyclically between morning and night, ACTH and cortisol secretion may rise acutely (within 10 minutes) and dramatically (increasing up to 20 fold) in response to a wide range of *stressful conditions*. These include:

- trauma, whether accidental or surgical
- burns
- infection and fever
- anxiety

Eight

- heavy exercise
- low blood glucose levels (hypoglycaemia).

Glucocorticoid actions

These are wide ranging.

Metabolic effects

Cortisol is an important catabolic hormone. It promotes:

- protein breakdown (favouring negative nitrogen balance)
- glucose synthesis from amino acids (gluconeogenesis)
- glycogen storage
- mobilization of fat stores, allowing oxidation of fatty acids for energy use and so preserving cell glucose.

These effects are particularly important in the prevention of hypoglycaemia during the postabsorptive state (Section 6.5), particularly if there is a prolonged fast. Glucocorticoid levels do not rise when fasting unless hypoglycaemia itself develops, and this does not usually occur even after many days of fasting. The increases in amino acid mobilization and gluconeogenesis normally seen in fasted individuals do not occur in the absence of normal levels of cortisol, however, and, since this reduces glycogen stores, stimulation of glycogenolysis by glucagon and adrenaline (epinephrine), an important component of the normal response to fasting (Section 6.5), has limited benefit when prior exposure to glucocorticoids is reduced. This is an example of a *permissive effect*, in which the actions of other hormones depend on the presence of normal levels of cortisol. Glucocorticoids also stimulate appetite, an effect which is relevant to their metabolic actions since it augments the energy supply to the body.

Stress resistance

The increased glucocorticoid levels stimulated by stressful conditions, such as trauma and infection, are vital to the body's ability to withstand these insults. The mechanism of this protection is unclear but the following actions may be involved.

Availability of glucose. Cortisol promotes a rapid supply of glucose to the body tissues through its metabolic effects.

Cardiovascular effects. Cortisol plays a permissive role in the cardiovascular responses to stress. Increased sympathetic activity is reflexly activated by stresses such as exercise, anxiety, hypoglycaemia and blood loss, and this causes vascular constriction, which helps to maintain arterial pressure and directs blood towards the most vital organs (Sections 3.6 and 3.7). Glucocorticoids are essential for these pressor effects because vascular smooth muscle becomes unresponsive to catecholamines in the absence of glucocorticoids.

Suppression of inflammation and immunity

Although glucocorticoids play an essential metabolic role in the normal response to infection, they also act as immunosuppressive hormones, especially at high concentration.

Inflammation is a nonspecific form of immune response in which tissue prostaglandins and chemotaxins promote local vasodilatation, oedema and the accumulation of phagocytic leucocytes at any site of injury or infection (Section 2.5). Cortisol is a potent anti-inflammatory agent, and synthetic glucocorticoid analogues are used clinically to suppress the undesirable effects of inflammation, e.g., reducing airways swelling and obstruction in asthma, or limiting pain and inflammatory joint damage in arthritis.

Specific immune responses depend on antigen-specific B and T lymphocytes. Glucocorticoids inhibit lymphocyte activation and this is often exploited in transplant recipients, who are treated with cortisol with a view to reducing the risk of organ rejection. This suppresses all specific immune responses, however, leaving these patients at a high risk of infection.

Adrenal androgens

Secretion of androgens from the zona reticularis of the adrenal cortex is stimulated by ACTH. This is insignificant in comparison with testicular testosterone production in males but is the main source of androgen in the female.

8.6 Function of the adrenal medulla

Learning objectives

At the end of this section you should be able to:

- outline the endocrine function of the adrenal medulla
- describe the main consequences of abnormal adrenal medullary function.

Relevant structure and regulation of hormone release

The central, medullary region of the adrenal gland (Fig. 163) acts as an endocrine extension of the sympathetic nervous system (Section 7.6). It contains

Box 37 Clinical note: Abnormal adrenocortical function

Deficient secretion

This is referred to as hypoadrenalism or *Addison's disease*. It may be caused by a primary reduction in adrenal gland function or arise secondary to an anterior pituitary defect.

Primary hypoadrenalism is caused by failure of the adrenal cortex itself, e.g., following destruction by autoantibodies, tumour invasion or surgery. Reduced feedback inhibition from circulating cortisol leads to a rise in ACTH and related pituitary peptides which stimulate melanocyte activity leading to a characteristic pigmentation of the skin and buccal mucosa. A test dose of ACTH has no effect on steroid levels in the blood or urine.

Secondary hypoadrenalism results from reduced stimulation of the gland by ACTH following anterior pituitary damage. A test dose of ACTH leads to a rapid adrenal response.

The problems resulting from hypoadrenalism reflect the combined effects of deficiencies in both mineralocorticoid and glucocorticoid action.

Aldosterone deficiency leads to:

- increased loss of Na^+ and water in the urine causing dehydration, plasma depletion and hypotension; if untreated, this may be fatal within a few days
- renal retention of K^+ and an elevated plasma $[K^+]$ (*hyperkalaemia*): this increases cardiac excitability by depolarizing the plasma membrane (Section 1.4) and can cause ventricular fibrillation, a fatal cardiac arrhythmia
- renal retention of H^+ producing a *metabolic acidosis*.

Glucocorticoid deficiency leads to:

- loss of weight and appetite, with poor exercise tolerance and muscle weakness; hypoglycaemia may occur during periods of prolonged fasting
- reduced resistance to trauma and infection so that relatively minor insults can lead to physical collapse.

Patients on regular glucocorticoid replacement therapy require increased dosages during periods of illness or prior to surgery so as to mimic the physiological rise in secretion normally stimulated by such events.

Excess secretion

Oversecretion of any of the three main types of adrenocortical hormone may arise as an isolated condition, producing three separate patterns of disease.

Primary hyperaldosteronism

Primary hyperaldosteronism reflects the activity of a tumour in the adrenal cortex which secretes aldosterone. The main effects are:

- renal Na^+ and water retention leading to an increased blood volume and hypertension; this reduces renin secretion from the juxtaglomerular apparatus so angiotensin levels are low
- renal K^+ loss leading to reduced plasma $[K^+]$; this *hypokalaemia* can cause muscle weakness
- renal H^+ loss leading to *metabolic alkalosis*.

Excess glucocorticoids

Cushing's syndrome results from increased glucocorticoid activity. This may be caused by excess secretion of ACTH, a glucocorticoid-secreting tumour in the adrenal cortex or prolonged periods of treatment with glucocorticoids, e.g., in transplant patients. The main effects are:

- an increase in blood glucose which is resistant to insulin treatment (Section 8.7)
- increased protein breakdown leading to muscle wasting, osteoporosis in bone and the appearance of stretch lines known as striae in the skin; growth is reduced in children
- weight gain caused by the stimulation of appetite, with abnormal fat deposits over the upper back (buffalo hump) and in the face (moon face)
- hypertension caused by the mineralocorticoid action of cortisol
- hirsutism and acne caused by the androgenic actions of cortisol.

Excess androgens

Adrenogenital syndrome is a genetically determined biochemical defect which results in increased secretion of androgens from the adrenals. The effects depend on the sex of the individual affected.

- The secretion of excess adrenal androgens in childhood leads to rapid growth and early (precocious) puberty in young boys. Their adult height is often reduced, however, as a result of early epiphyseal fusion.
- Developing females may become masculinized, with a male body build and hair distribution. The external genitalia may be virilized to such a degree that it is possible to 'misdiagnose' a genetic female as a male.

cells which lack axons but possess many of the other properties of postganglionic sympathetic neurones. These *chromaffin cells* release the contents of their intracellular vesicles into the bloodstream in response to stimulation by the preganglionic sympathetic nerves which innervate the adrenal medulla. As a result, activation of sympathetic nerves leads to an increase in the circulating levels of the catecholamines, i.e., adrenaline (epinephrine)and noradrenaline (norepinephrine). As in neuronal synthesis of noradrenaline (norepinephrine) (Section 7.3), these hormones are manufactured from the amino acid tyrosine in the chromaffin cells. However, in contrast with sympathetic neurones which release noradrenaline (norepinephrine), the adrenal medulla secretes about 80% adrenaline (epinephrine) to 20% noradrenaline (norepinephrine). After release, they are broken down again within minutes in the liver and kidney in reactions involving catechol-*O*-methyl transferase and monoamine oxidase, so their actions can be switched off rapidly. Catecholamine metabolites, such as vanillylmandelic acid, and small quantities of free hormone are excreted in the urine.

Cellular actions of the adrenal catecholamines

These hormones act in the same way as catecholamine neurotransmitters, i.e., they bind to α- and β-adrenoceptors on the target cell membrane (Section 7.6). Noradrenaline (norepinephrine) is more effective than adrenaline (epinephrine) at α-receptors, while adrenaline (epinephrine) is the more potent agonist at β-receptors. Activation of these receptors modulates the intracellular concentrations of a variety of *second messengers*, e.g., levels of cyclic adenosine monophosphate (cAMP) may rise or fall (Section 1.5). The pattern of second messenger response actually observed varies from cell to cell and depends on the receptor subtype involved. These intracellular signals alter cell function, often through second messenger-dependent kinases (cAMP promotes the activity of protein kinase A, for example) which phosphorylate important functional proteins and so alter their biochemical activities.

Metabolic actions of the adrenal catecholamines

The major cardiovascular and general systemic effects of the adrenal catecholamines have already been considered in the context of the autonomic nervous system (Section 7.6; Table 11). Here we will simply outline the metabolic actions. The main features are:

- increased metabolism and heat production
- increased glycogenolysis
- increased gluconeogenesis
- decreased glucose uptake and utilization in fat and muscle
- increased fat breakdown by hormone-sensitive lipase, with increased fatty acid oxidation and ketone production.

Therefore, activation of the adrenal medulla, either in response to stress (the 'fight or flight' response) or hypoglycaemia, favours a rise in blood glucose. Catecholamines also stimulate release of pancreatic glucagon, whose actions also tend to elevate blood glucose levels (Section 8.7).

Box 38 Clinical note: Abnormal adrenomedullary function

The only condition of clinical significance related to the adrenal medulla is oversecretion of catecholamines by a chromaffin cell tumour known as a *phaeochromocytoma*. The dominant feature is systemic hypertension, reflecting the pressor effects of adrenaline (epinephrine) and noradrenaline (norepinephrine) in the cardiovascular system (Sections 3.6 and 7.6). There may also be hyperglycaemia due to the metabolic actions of these hormones. Diagnosis depends on detection of increased levels of the catecholamines and their metabolites in blood and urine.

8.7 Endocrine functions of the pancreas

Learning objectives

At the end of this section you should be able to:

- name the cell types and hormones secreted by the islets of Langerhans
- describe how insulin secretion is regulated
- describe the metabolic actions of insulin
- describe how glucagon release is regulated and summarize its main actions
- describe the main consequences of abnormal function of the endocrine pancreas.

Synthesis and release of digestive enzymes by the exocrine pancreas is an important component of gastrointestinal function (Section 6.3). The pancreas also plays a crucial endocrine role, however, secreting hormones which are particularly important in the control of plasma glucose concentration. These help regulate the changes in metabolism necessitated by the continuous oscillation between conditions of rich nutrient supply in the absorptive state immediately after a meal, and postabsorptive periods of relative nutrient deficiency, when the body's energy stores have to be mobilized (Section 6.5).

Relevant structure

The greatest bulk of the pancreas consists of the acini and ducts involved in the secretion of enzymes into the duodenum. Between these exocrine structures, however, there are small groups of cells specialized for endocrine secretion into the blood, which form the *islets of Langerhans*. There are three main cell types within these islets, each producing a different hormone.

- insulin is secreted by β cells
- glucagon is secreted by α cells
- somatostatin is secreted by δ cells.

Insulin

This hormone consists of two peptide chains linked by two disulphide bridges. It survives for a relatively short time in the circulation and is broken down after approximately 10 minutes, mainly by the liver and kidney. This rapid turnover ensures that insulin levels drop rapidly if β cell secretion ceases. Insulin's chief role is to regulate cellular absorption and utilization of glucose, and so it is an important determinant of plasma glucose concentration. Glucose levels are normally maintained close to 5 $mmol L^{-1}$ under fasting conditions but may temporarily rise as high as 8 $mmol L^{-1}$ after a meal.

Regulation of insulin secretion

A number of factors help regulate insulin secretion but plasma glucose is the most important.

Elevation of the plasma glucose concentration directly stimulates a rapid rise in insulin secretion. Insulin levels are low at normal fasting glucose concentrations but rise as glucose is absorbed immediately following a meal. Since insulin reduces the amount of glucose in the extracellular fluid, this acts as a negative feedback system which limits any increase in plasma glucose. Secretion by β cells decrease as glucose levels decline, so insulin falls back to a low concentration in the postabsorptive state 3–4 hours after a meal (Section 6.5).

Increased amino acid concentrations also stimulate insulin secretion and this promotes their cellular uptake and use for protein synthesis.

Somatostatin from islet δ cells inhibits insulin release.

Other nerves and hormones can influence insulin release, e.g., parasympathetic nerves from the vagus stimulate release and may contribute to the rise in insulin at the start of a meal.

Cellular mechanisms of insulin action

Insulin binds to membrane receptors on the cell surface and this triggers important changes in the activity of both membrane transport proteins (for glucose and amino acids) and intracellular enzymes. A range of cell mechanisms seem to be involved, including intracellular second messenger systems and activation of inactive proteins by cell kinases (Section 1.5). Responses of this type can be very rapid in onset, e.g., the rate of cellular glucose uptake increases in less than a minute after insulin exposure. Insulin can also stimulate gene transcription, increasing the concentration of key enzymes, but the resulting effects may take minutes to hours to develop.

Metabolic actions of insulin

Insulin is a major *anabolic* hormone, affecting carbohydrate, protein and lipid metabolism in a wide range of body tissues.

Carbohydrate metabolism

Insulin stimulates glucose uptake in most body tissues. The hydrophilic glucose molecules are transported across the lipid cell membrane by facilitated diffusion (Section 1.3) dependent on glucose transporter (GLUT) complexes. Promotion of glucose influx by insulin is particularly important in the liver, skeletal muscle and adipose tissue, but the mechanisms responsible are different. Resting skeletal muscle cells and fat cells have a very low permeability to glucose but insulin stimulates uptake by increasing the density of GLUT4 transporters in the plasma membrane. Hepatocytes, in contrast, are glucose permeable even in the absence of insulin (GLUT2 transporter), but absorption is normally limited by rapid equilibration of [glucose] inside and outside the cell. Insulin activates a kinase which phosphorylates intracellular glucose, keeping the levels of free cytoplasmic glucose low and allowing further diffusion to occur (Fig. 164). Transport into brain cells is not insulin dependent and relies on GLUT1, which maintains a relatively fixed level of glucose uptake at all physiological concentrations of glucose, although uptake will decrease during hypoglycaemia. Cerebral metabolism is highly dependent on glucose, so maintenance of an adequate supply of plasma glucose is crucial for normal brain function.

Insulin promotes glycogen storage through increased glycogen synthesis (glycogenesis) and decreased glycogen breakdown (glycogenolysis). This is an important mechanism allowing carbohydrate storage during absorption of a meal. Liver glycogen can be converted back to free glucose to maintain plasma levels in the postabsorptive state. Because muscle lacks the phosphatase enzyme necessary to release free glucose following glycogenolysis, muscle glycogen can only be used for glycolysis within the muscle cells themselves (Fig. 164).

Insulin stimulates the use of glucose for energy by stimulating glycolysis.

Protein metabolism

Insulin promotes protein accumulation within cells. It does this by:

- stimulating amino acid uptake, particularly in muscle
- stimulating protein synthesis and inhibiting protein breakdown
- inhibiting conversion of amino acids to glucose (gluconeogenesis) and thus increasing substrate availability for protein synthesis.

Lipid metabolism

Insulin promotes the deposition of triglycerides (fatty acids esterified with glycerol) in body lipid stores. It does this by:

- inhibiting breakdown of stored lipid by hormone-sensitive lipase
- stimulating fatty acid synthesis from glucose, particularly in the liver
- stimulating release of fatty acids from triglycerides in circulating lipoprotein complexes by stimulating lipoprotein lipase in adipose tissue. The fatty acids can then be absorbed by lipid cells and resynthesized into triglycerides for storage
- promoting glycerol synthesis from glucose in lipid cells. This is used for triglyceride synthesis
- promoting carbohydrate metabolism so that fat is spared, e.g., resting muscle normally oxidizes fatty acids for energy but immediately after a meal, when glucose and insulin levels are high, it switches to carbohydrate.

Glucagon

This catabolic peptide hormone tends to raise plasma sugar levels. Its secretion from the α cells of the pancreatic islets is directly stimulated by:

- low plasma glucose concentrations: glucagon levels rise in the postabsorptive state some hours after a meal
- high levels of circulating amino acids. This may be important in the maintenance of normal

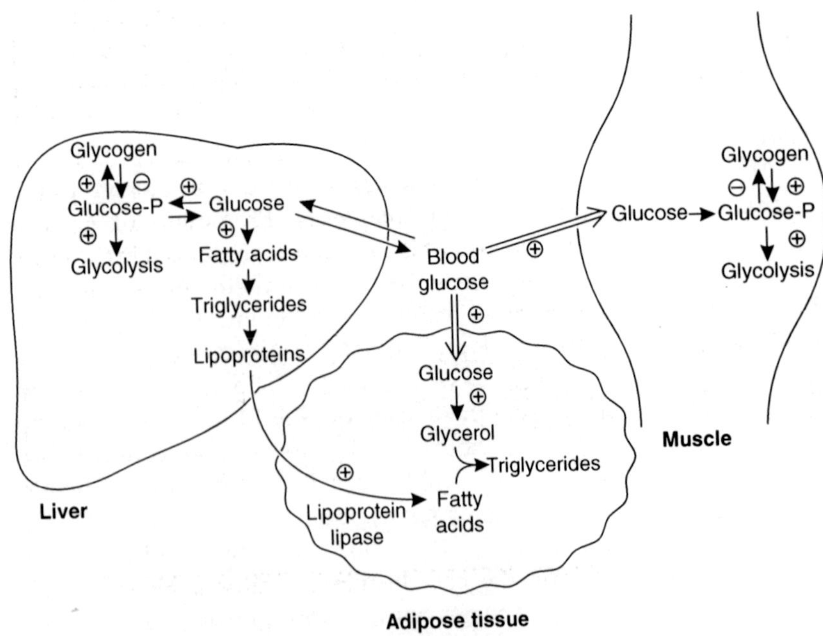

Fig. 164 Summary of insulin's effects on carbohydrate metabolism in liver, muscle and adipose tissue. (+) indicates steps which are directly stimulated by insulin, (–) those which are inhibited.

Box 39 Clinical note: Abnormal insulin function

Although secretion of any of the three islet hormones may go awry in theory, only abnormalities of insulin secretion, particularly insulin deficiency, are of major clinical significance.

Insulin deficiency

Inadequate insulin effect leads to the clinical syndrome of *diabetes mellitus*. This may arise in three ways:

- A primary deficiency of insulin is classified as Type I, insulin dependent, or juvenile onset diabetes. The underlying trigger for the β-cell damage leading to failure of insulin secretion is unknown, although an abnormal immune reaction against islet cells in the pancreas seems to play a role (autoimmune destruction of the β cells).
- In other individuals insulin secretion is relatively unimpaired but insulin's metabolic effects on the cells of the body are inhibited. This condition, known as insulin resistance, is seen in more elderly, overweight patients and gives rise to Type 2, non-insulin dependent, or maturity onset diabetes.
- Occasionally, diabetes mellitus is not due to a defect in insulin action but is secondary to excess secretion of a diabetogenic hormone, e.g., cortisol, growth hormone or, very rarely, glucagon.

Many of the symptoms and biochemical disturbances in insulin dependent diabetes mellitus can be explained on the basis of the normal metabolic effects of insulin.

Problems caused by abnormal carbohydrate metabolism

Hyperglycaemia. Decreased glucose uptake and utilization in body cells leads to high plasma glucose levels.

Glycosuria. Plasma glucose is freely filtered in the kidney but is normally completely reabsorbed within the convoluted tubules (Section 5.4). Abnormally high levels of plasma glucose, however, lead to saturation of the T_{max}-limited renal transport mechanisms, and the excess is then excreted in the urine (glycosuria). This occurs when the plasma glucose rises above the normal *renal threshold* concentration for glucose excretion; about 11 mmol L^{-1}.

Polyuria. The elevated glucose within the renal tubules causes osmotic retention of water and so there is an increased rate of urine production (polyuria). This is referred to as an *osmotic diuresis*.

Polydipsia. Thirst and drinking increase (polydipsia) in response to the dehydration resulting from renal fluid loss.

Problems caused by abnormal protein metabolism

Reductions in amino acid uptake and protein synthesis lead to negative nitrogen balance, with impaired growth in children and weight loss with muscle wasting in adults.

Problems caused by abnormal lipid metabolism

Lipid mobilization from fat stores. This contributes to weight loss and leads to increased plasma fatty acid levels.

Ketosis. With low levels of glucose uptake and utilization, β-oxidation of fatty acids is increased and this leads to formation of ketone bodies (ketosis), causing a metabolic acidosis.

Increased ventilation. Respiratory compensation for this acidosis leads to increased ventilation (Section 5.9).

Treatment of insulin deficiency

If untreated, diabetes mellitus will eventually lead to a ketotic coma and death from dehydration and acidosis. Careful restriction of carbohydrate intake along with regular insulin injections can lead to good control of blood glucose levels and a near normal life expectancy. In some cases, in which glycosuria and hyperglycaemia are associated with obesity, diet alone may be adequate.

Excess insulin

An insulin-secreting pancreatic tumour (an ***insulinoma***) may cause abnormally high levels of blood insulin. A relative insulin excess more commonly results from a mismatch between carbohydrate intake and the dose of injected insulin in a diabetic individual, e.g., when a meal is missed. The significant outcome is a reduction in plasma glucose levels (***hypoglycaemia***) and the symptoms reflect the dependence of cerebral metabolism on an adequate supply of glucose from the circulation. Low blood glucose directly activates the sympathetic nervous system leading to tremor, sweating and an increased heart rate. The patient may then become anxious or aggressive before hypoglycaemic coma results. This must be reversed as rapidly as possible using intravenous glucose or glucagon if permanent brain damage is to be avoided.

plasma glucose levels during absorption of a protein meal, since amino acids also stimulate insulin secretion and this might cause hypoglycaemia in the absence of any opposing action on carbohydrate metabolism.

Metabolic effects of glucagon

Glucagon's effects oppose those of insulin and are most pronounced in the liver.

Carbohydrate metabolism

Hepatic output of glucose to the plasma is promoted by:

- increased glycogenolysis
- increased formation of glucose from amino acids and glycerol (gluconeogenesis)
- glucose sparing through preferential β-oxidation of fatty acids for energy. This leads to formation of ketone bodies (acetone, acetoacetate and β-hydroxybutyric acid).

Lipid metabolism

Glucagon favours an increase in plasma levels of fatty acids and glycerol through activation of hormone-sensitive lipase.

Somatostatin

Somatostatin is a peptide hormone released by δ cells in the pancreas in response to increases in plasma glucose, amino acids and fatty acids. It slows down gastrointestinal function, decreasing motility, secretion and absorption, and so protects against very rapid increases in plasma nutrients during the absorptive phase. Somatostatin also inhibits release of both insulin and glucagon.

Self-assessment: questions

Multiple true/false questions

Each of the following statements consists of a stem followed by a number of possible endings. State whether each statement is True or False. For each stem, all, several or none of the statements may be true.

1. Endocrine control:
   - **a.** is important in the regulation of the extracellular fluid volume
   - **b.** of the anterior pituitary is an important hypothalamic function
   - **c.** always depends on exocytosis of a hormone by an endocrine cell
   - **d.** frequently involves negative feedback mechanisms
   - **e.** may be influenced by nervous controls

2. The pituitary gland:
   - **a.** is derived entirely from oropharyngeal ectoderm
   - **b.** releases prolactin from its posterior lobe
   - **c.** contains nerve axons
   - **d.** lies close to the optic chiasma
   - **e.** secretes a hormone which regulates parathyroid function

3. A patient with a low level of thyroid hormone and a high level of thyroid-stimulating hormone (TSH):
   - **a.** is likely to have a reduced metabolic rate
   - **b.** is likely to have a low peripheral resistance
   - **c.** will prefer a warm environment
   - **d.** suffers from primary hypothyroidism (i.e., not caused by a hypothalamic or pituitary fault)
   - **e.** would be expected to respond to an injection of thyrotrophin-releasing hormone (TRH) with increased TSH secretion

4. Triiodothyronine (T_3):
   - **a.** is the only biologically active form of thyroid hormone
   - **b.** is all protein bound in the circulation
   - **c.** is formed from thyroxine (T_4) inside target cells
   - **d.** binds to plasma membrane receptors
   - **e.** promotes glycolysis

5. Parathormone secretion:
   - **a.** is directly stimulated by Ca^{2+} in the extracellular fluid
   - **b.** promotes negative phosphate balance
   - **c.** may promote phosphate release from bone
   - **d.** is a function of the parafollicular C cells in the thyroid
   - **e.** is lower than normal in all hypercalcaemic states

6. The plasma Ca^{2+} concentration:
   - **a.** is increased by 1,25-dihydroxycholecalciferol
   - **b.** is increased by parathormone
   - **c.** directly regulates renal activation of vitamin D
   - **d.** may be raised by stimulating osteocytic Ca^{2+} mobilization
   - **e.** is an important determinant of neuronal excitability

7. Aldosterone:
   - **a.** generates a maximal renal response in minutes rather than hours
   - **b.** secretion is increased in response to hyperkalaemia (high plasma [K^+])
   - **c.** usually raises plasma [Na^+] significantly
   - **d.** excess may led to hypokalaemia (low plasma [K^+])
   - **e.** secretion is primarily regulated by ACTH in normal individuals

8. Cortisol:
   - **a.** promotes renal reabsorption of Na^+
   - **b.** is an anabolic steroid
   - **c.** leads to a negative nitrogen balance in patients who have undergone major surgery
   - **d.** levels are normally low in the middle of the night
   - **e.** promotes ACTH release from the anterior pituitary

9. Circulating adrenaline (epinephrine):
   - a. is a peptide hormone
   - b. tends to raise blood glucose levels
   - c. inhibits glucagon secretion
   - d. can cause vasoconstriction
   - e. can cause vasodilatation
   - f. is more likely to directly increase the heart rate than circulating noradrenaline (norepinephrine)

10. Insulin:
    - a. is secreted by the α cells of the pancreatic islets
    - b. promotes protein breakdown in muscle cells
    - c. stimulates fatty acid oxidation
    - d. deficiency tends to increase plasma osmolality
    - e. is largely protein bound in the circulation

11. Insulin deficiency (diabetes mellitus) may lead to:
    - a. abnormally high plasma glucose concentrations after a meal
    - b. a respiratory acidosis
    - c. an increased rate of glucose filtration in the kidney
    - d. reduced glucose uptake in the brain
    - e. weight loss

Single best answer questions

For each of the following questions choose the single best answer.

1. Endocrine control:
   - a. depends on the presence of relevant receptors on the plasma membrane
   - b. is independent of neuronal control at the whole body level
   - c. requires that the messenger be highly water soluble
   - d. will favour a longer latency between signal and response than paracrine control
   - e. is generally regulated through changes in the rate of breakdown of the hormone concerned

2. During a period of physiological stress, the concentration of adrenocorticotrophic hormone (ACTH) would be expected to be highest in:
   - a. the hypophyseal portal vessels
   - b. venous blood draining the posterior pituitary
   - c. venous blood draining the anterior pituitary
   - d. systemic arterial blood
   - e. venous blood draining the adrenal glands

3. The posterior pituitary:
   - a. manufactures and secretes antidiuretic hormone
   - b. manufactures and secretes oxytocin
   - c. manufactures and secretes vasopressin
   - d. manufactures and secretes antidiuretic hormone and oxytocin
   - e. secretes antidiuretic hormone and oxytocin

4. Enlargement of the thyroid gland:
   - a. occurs in response to iodine deficiency
   - b. occurs during thyrotoxicosis
   - c. is associated with elevated levels of thyroid stimulating hormone (TSH)
   - d. is associated with reduced levels of thyroid stimulating hormone (TSH)
   - e. occurs in response to excess stimulation of TSH receptors in the thyroid gland

5. Parathormone:
   - a. is under negative feedback control by free $[Ca^{2+}]$ in the plasma
   - b. is under negative feedback control by total $[Ca^{2+}]$ in the plasma
   - c. is under negative feedback control by total body Ca^{2+}
   - d. is likely to be decreased in concentration in response to hypophosphataemia
   - e. is likely to be decreased in concentration in response to hypocalcaemia

6. Calcitriol (1,25 dihydroxycholecalciferol):
   - a. promotes phosphate secretion in the kidney
   - b. promotes phosphate reabsorption in the kidney
   - c. promotes release of parathormone
   - d. is likely to be decreased in concentration in response to hypophosphataemia
   - e. is likely to be decreased in concentration in response to hypocalcaemia

7. Aldosterone:
   - a. is secreted in response to low plasma $[K^+]$
   - b. regulates absorption of up to 20% of the fluid filtered by the kidneys

c. stimulates H^+ absorption
d. has a larger effect on plasma osmolality than vasopressin
e. is secreted in response to high plasma $[K^+]$.

8. Cortisol:
   a. is an anabolic steroid hormone
   b. promotes weight loss
   c. is released from the zona glomerulosa of the adrenal cortex
   d. promotes gluconeogenesis
   e. promotes peripheral vasodilataion

9. Glucagon:
   a. secretion is stimulated by high levels of plasma amino-acids
   b. is a catabolic steroid hormone
   c. promotes glycogen synthesis
   d. inhibits hormone-sensitive lipase
   e. is secreted by the δ-cells of the islets of Langerhans

10. Insulin deficiency:
    a. causes hypoglycaemia
    b. elevates circulating free fatty acid levels
    c. favours metabolic alkalosis
    d. promotes cellular uptake of K^+ uptake
    e. is more characteristic of Type 2 than Type 1 diabetes mellitus

Matching item questions

Theme: Diabetes mellitus

Options

A. Weight loss
B. Ketoacidosis
C. Obesity
D. Hypoglycaemia
E. Glycosuria
F. Polydipsia

For each of the descriptions below choose the most appropriate option from the list above. Each option may be used once, more than once or not at all.

1. Causes an osmotic diuresis.
2. May be reversed with glucagon.
3. Is a feature of Type 2 rather than Type 1 diabetes mellitus.
4. Results from the diuresis.
5. May cause hyperventilation.
6. Is not a feature of untreated diabetes mellitus.

Theme: Metabolic control

Options

A. Thyroid hormone
B. Glucocorticoids
C. Insulin
D. Glucagon
E. Adrenaline (epinephrine)
F. Somatostatin

For each of the descriptions below choose the most appropriate option from the list above. Each option may be used once, more than once or not at all.

1. Is converted into its most active form intracellularly.
2. Promotes protein breakdown.
3. Inhibits hormone-sensitive lipase.
4. Acts on cell membrane receptors to increase metabolism.
5. Secretion is stimulated by acetycholine acting on nicotinic receptors.
6. Synthesis requires oxidation of iodide ions.

Short notes

Write short notes on the following:

a. the actions of human growth hormone
b. synthesis of thyroid hormones
c. hormone binding by plasma proteins
d. the circadian rhythm of glucocorticoid secretion
e. the actions of glucagon

Modified essay

The thyroid gland plays a key role in the regulation of body metabolism.

Questions

1. Outline how thyroid hormone affects the metabolism of each of the three main metabolic substrates in the body.

2. How is thyroid activity regulated?
3. What are the main steps in the synthesis of thyroid hormone?
4. How is thyroid hormone transported in the blood? Explain two advantages that result from this.
5. Outline the cellular mechanism of action of thyroid hormones.
6. Explain how hypothyroidism is classified on the basis of its cause.
7. How may these causes be distinguished using a simple blood test?

Data interpretation

Weight and the daily urinary excretion of Na^+ and K^+ were measured in a normal individual for 3 days before and 3 days after the initiation of treatment with aldosterone supplements. The daily intake of Na^+ and K^+ were kept constant throughout the experiment. Use the results (Fig. 165) to attempt to answer the following questions.

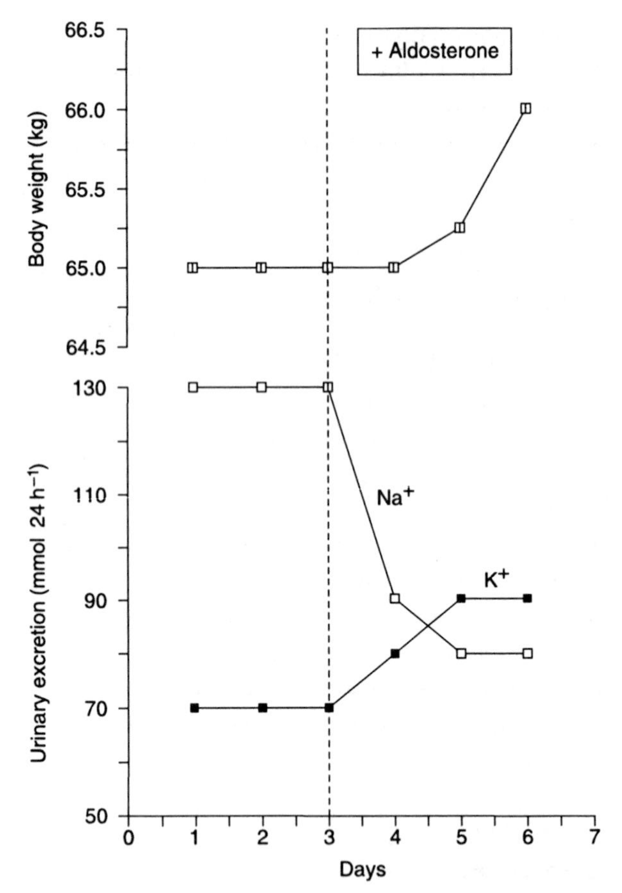

Fig. 165 Measurements of body weight and urinary electrolyte excretion rates in a normal individual over a 3-day control period and during 3 days of treatment with aldosterone supplements. Electrolyte intake rates remained constant throughout.

Questions

1. What is your best guess as to the daily intake of Na^+ in mmol? What was that for K^+? On what do you base your answers and what assumptions have you made?
2. Why do you think body weight increases during aldosterone treatment? What might the underlying mechanism for this be, given the observed changes in electrolyte excretion?
3. What do you notice about the relative time to onset of the changes in body mass and electrolyte excretion? Is this consistent with your suggestion as to the mechanism underlying the increase in weight?
4. Based on the measurements given, what can you say about the likely changes in total body K^+ and plasma $[K^+]$ in the presence of additional aldosterone?
5. Based on the measurements given, what can you say about the likely changes in total Na^+ and plasma $[Na^+]$ during aldosterone treatment?
6. Relate these results to what you know about the renal actions of aldosterone.

Clinical scenario

A young woman attends her general practitioner complaining that she has been drinking a lot of water recently and is passing a lot more urine than normal. Otherwise she is feeling healthy, although she has lost some weight even though her appetite is increased.

Questions

1. Name two hormones, deficiencies of which could lead to increased urine volume and drinking.
2. What mechanisms produce these symptoms in each of these two abnormalities?
3. Based on the symptoms above, which of these causes is most likely, and why?
4. What two simple measurements would help confirm the diagnosis? What values would you

expect normally? The diagnosis remains in some doubt after these procedures have been carried out and so a glucose tolerance test is performed. In this, the subject fasts overnight and then takes a glucose meal (75 g of glucose). Plasma glucose levels are measured immediately before, and at regular intervals after this test meal. The resulting values are plotted as a graph (Fig. 166, solid line). A control curve from a normal female of similar age and weight is plotted for comparison (broken line).

5. Identify at least three features of the test result in this individual which are consistent with the diagnosis you suggested in Answer 3. Briefly explain how these features arise on the basis of the normal function of the relevant hormone.

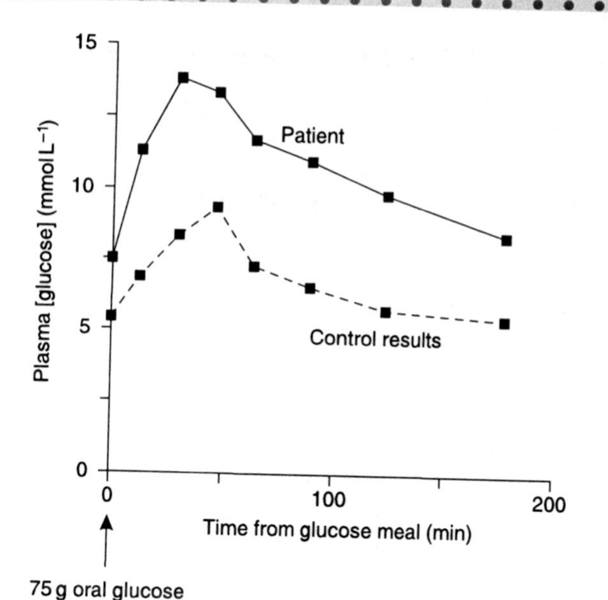

Fig. 166 Changes in plasma glucose concentration following rapid ingestion of 75 g of glucose (a glucose tolerance test).

Viva questions

1. How does the body regulate the plasma level of thyroid hormone?
2. Explain how plasma calcium concentration is controlled.
3. Tell me something about the hormonal responses to stress.

Self-assessment: answers

Multiple true/false answers

1. a. **True.** For example, aldosterone and antidiuretic hormone.
   b. **True.** Through hypothalamic secretion of releasing and inhibiting hormones into the hypophyseal portal system.
   c. **False.** This is generally true for protein and peptide hormones but steroids can simply diffuse out of the cell.
   d. **True.** Central to hormonal homeostatic mechanisms.
   e. **True.** For example, through the hypothalamus (posterior pituitary) or autonomic nerves (adrenal medulla).

2. a. **False.** Anterior pituitary only, posterior pituitary is derived from neural tissue.
   b. **False.** From the anterior lobe.
   c. **True.** Running into posterior lobe from hypothalamus.
   d. **True.** Pituitary tumours may cause bitemporal visual field defects (Section 7.4).
   e. **False.** Parathormone secretion is directly regulated by [Ca^{2+}].

3. a. **True.** Because of low thyroid levels.
   b. **False.** This is likely to be normal or high; peripheral resistance is reduced in hyperthyroidism.
   c. **True.** Hypothyroidism is associated with intolerance of cold conditions.
   d. **True.** TSH levels would be low in secondary hypothyroidism.
   e. **True.** This response is exaggerated in primary hypothyroidism because of the reduced negative feedback by thyroid hormones on TSH release.

4. a. **False.** T_4 is also active, although less potent.
   b. **False.** Large proportion is protein bound but only the free hormone is active.
   c. **True.**
   d. **False.** Binds to nuclear receptors.
   e. **True.** Also promotes glucose transport and gluconeogenesis.

5. a. **False.** Ca^{2+} inhibits its secretion.
   b. **True.** Lowers plasma phosphate by increasing urinary excretion.
   c. **True.** Phosphate is released along with Ca^{2+} from bone mineral.
   d. **False.** C cells produce calcitonin; parathormone is secreted by four parathyroid glands.
   e. **False.** In primary hyperparathyroidism the rise in [Ca^{2+}] is caused by increased parathormone.

6. a. **True.** Through actions on the gut.
   b. **True.** Through actions on kidney and bone.
   c. **False.** The effect of [Ca^{2+}] is indirect, via changes in parathormone secretion. Low phosphate concentrations directly stimulate activation.
   d. **True.** Mobilizes the rapidly exchangeable bone pool.
   e. **True.** Low [Ca^{2+}] increases excitability.

7. a. **False.** Requires 1–2 hours, reflecting time required for increased DNA transcription.
   b. **True.** Potassium ions directly stimulate aldosterone release.
   c. **False.** Since Na^+ and water are absorbed isotonically, [Na^+] is little affected, although total Na^+ increases. ADH controls plasma [Na^+].
   d. **True.** By stimulating renal K^+ secretion.
   e. **False.** Renin–angiotensin system and [K^+] are most important controls.

8. a. **True.** Cortisol has some mineralocorticoid effect, although this is much weaker than that of aldosterone.
   b. **False.** It is catabolic.
   c. **True.** Surgical stress stimulates cortisol release leading to protein breakdown.
   d. **True.** Trough in circadian cycle.
   e. **False.** Feedback inhibition of ACTH release by cortisol.

9. a. **False.** It is a catecholamine derived from tyrosine.

b. **True.** Increases glucose release and gluconeogenesis and decreases glucose utilization.

c. **False.** Catecholamines promote glucagon secretion; this contributes to their hyperglycaemic effect.

d. **True.** Through its action on α-adrenoceptors in regions where these are predominant, e.g., in skin and abdominal organs.

e. **True.** Through its action on β-adrenoceptors in regions where these are predominant, e.g., in skeletal muscle.

f. **True.** Cardioacceleration is a β-adrenoceptor action and adrenaline (epinephrine) is a more potent β-agonist than is noradrenaline (norepinephrine).

10. a. **False.** Secreted by β cells.

 b. **False.** Anabolic hormone, favouring protein synthesis.

 c. **False.** Promotes carbohydrate metabolism by stimulating glucose uptake and glycolysis.

 d. **True.** Through an increase in plasma glucose concentration.

 e. **False.** As a peptide hormone it is soluble in plasma.

11. a. **True.** Decreased glucose uptake by liver and muscle during absorptive phase.

 b. **False.** Metabolic acidosis caused by ketone bodies may lead to respiratory compensation and reduced arterial P_{CO_2}.

 c. **True.** Through increased plasma glucose concentration.

 d. **False.** Neuronal uptake of glucose will tend to rise since it is determined by plasma [glucose] and is not dependent on insulin.

 e. **True.** Loss of normal anabolic effect on protein synthesis.

Single best answers

1. d. The increased delay reflects the time taken for circulation of the hormone round the body as well as the diffusional delays within the tissues. Paracrine control only involves local diffusion. Hormone receptors may be intracellular as well as membrane bound (e.g., for lipophilic thyroid and steroid hormones, which are not very water-soluble). Hormonal control is influenced by neuronal inputs, especially at the hypothalamic–pituitary level. Hormone levels are modified by changes in secretion rather than changes in breakdown.

2. c. ACTH is secreted by the anterior pituitary and will reach the highest concentrations in the veins draining that region.

3. e. The posterior pituitary secretes both of these hormones but they are synthesized in the supraoptic and paraventricular nuclei of the hypothalamus.

4. e. All five answers can be true, but e. is the best answer in my judgement, since it covers both the situation of iodide deficiency, in which TSH is elevated, and the situation in which TSH-receptor antibodies drive the hyperplasia, causing thyrotoxicosis.

5. a. Parathormone secretion provides the primary mechanism for regulating free [Ca^{2+}], increasing when [Ca^{2+}] falls, whereas Vitamin D, particularly the most active form calcitriol, can be thought of as regulating total body Ca^{2+}. Total plasma [Ca^{2+}] is a function of free [Ca^{2+}] and plasma [protein]. Calcitriol controls plasma [phosphate].

6. b. Calcitriol levels are elevated in response to low [phosphate], increasing intestinal and renal absorption of phosphate. Parathormone levels are determined by free [Ca^{2+}], not vitamin D, but may fall if high levels of vitamin D activity lead to a rise in [Ca^{2+}]. Since parathormone promotes activation of vitamin D, however, calcitriol levels may rise during hypocalcaemia.

7. e. This is the main mechanism controlling plasma [K^+], since aldosterone promotes renal K^+ loss. H^+ secretion is also stimulated. The increased Na^+/H_2O absorption affects no more than 3% of the filtered load, and there is relatively little effect on osmolality. H^+ secretion is stimulated.

8. d. This favours the deposition of glycogen stores and results from the catabolic breakdown of protein. Cortisol is mainly secreted from the zona fasciculata, with some secretion from the zona reticularis. It favours weight gain, through increased appetite with fluid retention at high concentrations (a mineralocorticoid effect), and plays a permissive role

during sympathetically mediated vasoconstriction.

9. a. Glucagon is a catabolic peptide hormone which mobilizes glycogen stores and fatty acid release from adipose tissue. It is secreted by the α cells.

10. b. Insulin promotes the deposition of fatty acids as triglycerides in lipid store. Insulin deficiency is typical of Type 1 diabetes mellitus in which hyperglycaemia and metabolic acidosis are observed. Insulin itself promotes cell uptake of K^+. This has important clinical consequences for the treatment of diabetic ketoacidosis. The patient's $[K^+]$ may drop following insulin treatment, and intravenous K^+ supplementation may have to be used to prevent this.

Matching item answers

Theme: Diabetes mellitus

1. E. The glucose acts osmotically to retain fluid within the nephron, increasing the urinary volume.
2. D. Glucagon acts to raise blood glucose.
3. C. Weight loss characterizes Type 1 diabetes mellitus, but insulin resistance, which is the underlying problem in Type 2 diabetes mellitus, is strongly correlated with obesity. Weight loss improves glucose control, and diet alone may provide adequate control.
4. F. Diuresis leads to dehydration which promotes thirst and drinking.
5. B. Hyperventilation blows off CO_2, providing respiratory compensation for the metabolic acidosis; see Section 5.9.
6. D. Diabetes mellitus causes hyperglycaemia; hypoglycaemia is only seen in insulin dependent diabetics if the insulin dose is too large for the amount of carbohydrate eaten, or if unexpected exercise reduces insulin needs.

Theme: Metabolic control

1. A. T_4 is converted to T_3.
2. B. This is the catabolic action of glucocorticoids and favours amino acid release for gluconeogenesis.
3. C. This promotes energy storage in lipid stores.
4. E. Thyroid hormone acts intracellularly.
5. E. The preganglionic sympathetic nerves release acetylcholine which stimulate nicotinic receptors on the chromaffin cells, which behave as modified sympathetic postganglionic neurones.
6. A.

Short note answers

a. Growth hormone is anabolic, stimulating hepatic production of growth promoters known as somatomedins which act on bone and soft tissues. Protein synthesis is increased and fat stores are degraded and used for energy. This spares carbohydrate from metabolic utilization, raising blood glucose.

b. Iodide is trapped by the follicular epithelium, transported into the colloid and oxidized to iodine. Tyrosine residues within a colloidal-binding protein known as thyroglobulin are then iodinated to form mono- and diiodotyrosine (MIT and DIT). These precursors combine, producing the thyroid hormones triiodothyronine ($MIT + DIT = T_3$) and thyroxine ($2 \times DIT = T_4$). This series of reactions is driven by the thyroid peroxidase enzyme. When required, T_3 and T_4 are released from the colloid and cross the epithelium to enter the blood. Both synthesis and release are stimulated by thyroid-stimulating hormone.

c. Hormones may become reversibly attached to plasma proteins. This may involve either specific binding proteins, e.g., the transport of thyroid hormones by thyroid-binding protein and thyroid-binding prealbumin, or nonspecialized proteins such as albumin, e.g., steroid hormones. Such binding confers water solubility on steroid hormones (e.g., aldosterone, cortisol, sex steroids) and thyroid hormones; these are naturally hydrophobic molecules. Only unbound (free) hormones are biologically active. The total plasma concentrations of protein-bound hormones may be altered by changes in protein levels, e.g., causing a rise in total thyroid hormone concentration during pregnancy, but this will not affect the concentration of free hormone. Protein binding is not a feature of catecholamine and peptide/protein hormone physiology.

d. The plasma levels of glucocorticoids such as cortisol vary cyclically over each 24-hour period in a circadian rhythm. Average levels are high in the early part of the morning and low in the middle of the night. This cycle is caused by a rise and fall in release of adrenocorticotrophic hormone (ACTH) from the anterior pituitary since this controls glucocorticoid secretion by the adrenal cortex. ACTH release is, in turn, regulated by corticotrophic-releasing hormone (CRH) from the hypothalamus, and it is presumably cyclical changes in CRH release which dictate the pattern of diurnal variation. This may be altered by external stimuli, e.g., changing patterns of darkness and light, or changes in sleep habit, presumably through nervous influences on the hypothalamus.

e. Glucagon's main actions are catabolic and hyperglycaemic. It stimulates glucose release from glycogen and its synthesis from noncarbohydrate substrates (gluconeogenesis). It reduces the use of glucose as an energy source by stimulating the β-oxidation of fatty acids (glucose-sparing effect). These are made available in increased amounts through stimulation of hormone-sensitive lipase, which releases glycerol and fatty acids from lipid stores. Fatty acid metabolism also leads to increased formation of ketone bodies (acetone, acetoacetic acid and β-hydroxybutyric acid), i.e., glucagon is a ketogenic hormone.

Modified essay answers

1. Thyroid hormone increases metabolic rate. Protein: synthesis and breakdown are both increased (increased protein turnover). Carbohydrate: glucose uptake, glucose utilization and gluconeogenesis are increased. Lipid: fatty acid oxidation is increased, and fat stores and blood lipids are reduced.
2. Thyroid gland activity is stimulated by thyroid-stimulating hormone (TSH) from the anterior pituitary, which is regulated by thyrotrophin-releasing hormone (TRH) carried to the anterior pituitary from the hypothalamus in the hypophyseal portal blood. There is negative feedback through inhibition of TSH and TRH release by thyroid hormones, and of TRH secretion by TSH. This maintains a relatively constant level of thyroid activity but this may be modified by neural inputs to the hypothalamus, e.g., in cold conditions, thyroid activity may be increased.
3. Iodide (I^-) is transported into the colloid of the thyroid follicles by the epithelium. I^- is oxidized to iodine (I) and combined with tyrosine residues in the thyroglobulin of the colloid. These reactions result in the formation of mono- and diiodotyrosine (MIT and DIT) and are catalysed by the thyroid peroxidase enzyme. These combine to form the hormones triiodothyronine (T_3) and tetra-iodothyronine (thyroxine, T_4). These are stored in the colloid until they are secreted by the follicular epithelium following localized proteolysis of thyroglobulin.
4. Thyroid hormones are mainly transported in association with plasma proteins such as thyroid binding globulin (TBG) and thyroid binding prealbumin. This:
   - confers water solubility (thyroid hormones are lipophilic)
   - increases the circulating reservoir of hormone, buffering concentration changes.
5. Thyroid hormones diffuse across cell membranes and act intracellularly (they are lipophilic). Inside the cell:
   - T_4 is converted to the more active T_3
   - T_3 binds to its nuclear receptor
   - transcription of DNA to mRNA increases
   - expression of a range of metabolic and other enzymes/proteins increases
   - cell metabolism increases.
6. Hypothyroidism may be:
   - primary, due to a defect at the level of the thyroid gland itself
   - secondary, due to a deficiency of TSH from the anterior pituitary.
7. In primary hypothyroidism, TSH levels are high (reduced negative feedback by thyroid hormone), in secondary hypothyroidism they are low (defect is in the anterior pituitary).

Data interpretation answers

1. If we assume that steady-state conditions prevail during the control period then intake and loss must be equal for Na^+ and K^+ (and for

water). The existence of a steady state is suggested (but not proved) by the constancy of both weight and urinary electrolyte losses. Since there was a urinary Na^+ loss of 130 mmol day^{-1} and a urinary K^+ loss of 70 mmol day^{-1} prior to aldosterone supplementation, it is likely that these values also represent the daily intakes. This involves the further assumption that renal excretion accounts for nearly all losses of these ions, which is not always the case, e.g., K^+ and Na^+ can both be lost from the gastrointestinal tract in appreciable quantities, especially in cases of diarrhoea, and large quantities of Na^+ may be lost in sweat.

2. The increase in body weight is probably caused by accumulation of extracellular fluid. This might be explained on the basis of Na^+ accumulation secondary to the measured decrease in Na^+ excretion since total body Na^+ content is the main osmotic determinant of extracellular fluid volume.
3. The decline in Na^+ excretion precedes any measurable increase in weight. This time course is consistent with increased Na^+ accumulation as the mechanism underlying the weight gain since 'cause' must always precede 'effect'.
4. Increased K^+ excretion will lead to a decrease in total body K^+ assuming that intake is constant (as originally stated). If the increase in body weight reflects an increase in extracellular fluid volume then there will be a reduced amount of K^+ dissolved in a larger volume of extracellular fluid, so plasma $[K^+]$ should decrease.
5. Decreased Na^+ excretion will increase total body Na^+ but the change in plasma $[Na^+]$ is difficult to predict since this extra Na^+ is dissolved in an expanded extracellular volume, as indicated by the increased body weight. We can actually make a rough estimate of the likely direction of change in $[Na^+]$ in the following way:

   Total Na^+ accumulation over the 3 days of aldosterone treatment = (130 – 90) + (130 – 80) + (130 – 80) = 140 mmol
   Total fluid accumulation = 1 L (assuming a fluid density of 1 kg L^{-1})

   This is equivalent to fluid expansion with 1 L of 140 mmol L^{-1} Na^+ solution, i.e., a solution approximately isotonic with normal extracellular fluid. This should not alter the overall $[Na^+]$ at all. (Note: this assumes that all the additional fluid and Na^+ is in the extracellular space; if appreciable proportions enter the intracellular space the calculation is invalid.)
6. All these findings are consistent with the stimulation of Na^+/water reabsorption and K^+ secretion in the distal convoluted tubule of the nephron by aldosterone. This would be expected to decrease urinary Na^+ and water loss, leading to an increased extracellular volume. This explains the weight gain. At the same time K^+ excretion increases, leading to a fall in plasma $[K^+]$.

   Note: the situation shown in Fig. 165 could not continue indefinitely as Na^+ and H_2O would accumulate, leading to overhydration. Na^+ excretion must eventually fall back to match intake, but with a new, expanded ECF volume.)

Clinical scenario answers

1. Deficient secretion of either antidiuretic hormone (ADH) from the posterior pituitary or insulin from the pancreatic β cells could increase urinary excretion and thirst.
2. ADH normally promotes reabsorption of water from the collecting ducts, so abnormally low levels lead to increased excretion of dilute urine. This reduces extracellular volume and increases plasma osmolality, both of which effects tend to promote thirst through the hypothalamus. Insulin normally reduces plasma [glucose], and the high levels in its absence lead to incomplete reabsorption of glucose in the convoluted tubules of the kidney, which causes an osmotic diuresis. Dehydration and increased osmolality caused by high [glucose] promote drinking.
3. Insulin deficiency (diabetes mellitus) is the more likely cause since there is weight loss, a classic symptom of this condition. This reflects the loss of insulin's anabolic effects on protein and lipid stores. ADH has no significant metabolic actions, although ADH deficiency (diabetes insipidus) will produce a more dramatic increase in urine volume than insulin deficiency.

4. Urinary and plasma glucose concentrations should be measured in an attempt to confirm the diagnosis. These should be zero and about 5 mmol L^{-1}, respectively, but will be raised in diabetes mellitus.
5. The higher fasting glucose level, more rapid rise to a higher peak value and slower decline towards resting levels are consistent with insulin deficiency. These all reflect decreased insulin secretion with a consequent reduction in cell uptake and utilization of glucose from the plasma, particularly in liver and muscle.

Viva answers

1. You could either begin at the thyroid and work back, or start with the hypothalamus and go on from there. Taking the first option, you should explain that synthesis and release of hormone from the thyroid gland is regulated by thyroid-stimulating hormone (TSH) which binds to appropriate receptors on the follicular cells. This TSH comes from the anterior pituitary where the relevant cells (thyrotrophs) are activated by thyrotrophin-releasing hormone carried from the hypothalamus in the hypophyseal portal blood. Relatively stable hormone levels are maintained through negative feedback since thyroid hormone inhibits both TSH and TRH release (long loop feedback). TSH also inhibits TRH release (short loop feedback). Some discussion of the main actions of the thyroid hormones is likely to follow but only move on to this when asked by the examiner.
2. There is a lot to be covered here but you should start with parathormone if plasma concentration (rather than total body Ca^{2+}) is asked about. The important points are the location of the parathyroid glands and the nature and actions of parathormone. Remember PTH (a peptide hormone) acts on bone to mobilize Ca^{2+} from the bony pool (rapidly exchangeable Ca^{2+} is transferred to ECF by osteocytes; slower breakdown of bone by osteoclasts) and on the kidney to promote Ca^{2+} reabsorption and PO_4^{3-} excretion (this helps prevent precipitation of reabsorbed Ca^{2+}). Activation of vitamin D is also stimulated in the kidney, promoting Ca^{2+} uptake from the gut (an indirect effect of PTH). Normal levels of vitamin D (from UV exposure or diet) favour increased bone mineralization through increased intestinal absorption of Ca^{2+} and PO_4^{3-}.
3. Stressors such as trauma, surgery, burns, infection, hypoglycaemia and exercise favour release of a group of hormones which include glucocorticoids, catecholamines, glucagon and growth hormone. Glucocorticoids are the most important and you should begin by commenting on how stress favours rapid and dramatic increases in ACTH secretion by the anterior pituitary leading to a marked rise in cortisol secretion from the adrenal cortex (zona fasciculata). Cortisol is a catabolic hormone which promotes gluconeogenesis at the expense of protein and lipid breakdown. It is very important in stress resistance. Catecholamines from the adrenal medulla help maintain blood pressure and glucocorticoids play a permissive role since vasoconstrictor responses are poor in the absence of cortisol. Catecholamines, growth hormone and glucagon all help to elevate blood glucose levels and this seems to be important in stress. Growth hormone is anabolic, the rest are catabolic.

Reproductive physiology

Overview

This chapter focuses on the male and female reproductive systems and the control mechanisms which regulate their function. The primary reproductive organs, or gonads, produce the male and female gametes (spermatozoa and ova) and also act as important endocrine organs, secreting sex hormones. Reproduction requires fusion of one sperm with one ovum (fertilization), followed by implantation of the developing blastocyst in the endometrial lining of the uterus. During most of pregnancy, the fetus relies on maternal homeostatic systems and is sustained by gas and nutrient exchange across the placenta. Eventually, however, it is mature enough to survive life outside the uterus, and parturition (birth of the baby) can take place.

9.1 Sexual development

Learning objectives

At the end of this section you should be able to:

- identify the factors which control sexual development in utero
- outline the changes which occur at puberty.

Sexual development occurs in two main steps, the development of male or female sex organs during fetal life and the onset of full reproductive function at puberty.

Sex determination in utero

The sex of an individual may be considered in terms of *genetic* sex, *chromosomal* sex, *gonadal* sex and *genital* sex. The gene for maleness is carried on the Y chromosome so that embryos with the XY sex chromosome pair become males. The primary effect of this gene is to cause the primitive gonads to form testes in utero, producing a gonadal male. The testes secrete testosterone, the male sex hormone, and this, in turn, leads to the development of the rest of the internal and external male sex organs. The testes also secrete an inhibitory substance (antimüllerian hormone), which actually blocks the formation of the uterus and vagina from the embryonic müllerian duct. Therefore, the development of the male genitalia relies on the endocrine activity of the testes in utero. In the absence of the relevant gene on the Y chromosome, the primitive gonads appear to be programmed to develop into ovaries (producing a gonadal female) and the other relevant embryonic structures form female sex organs (a genital female).

Puberty

Puberty is the term given to the development of full reproductive function and normally commences in the early teenage years. On average, its onset is 1–2 years earlier in females than males. Pubertal changes extend over a period of years and there is an associated spurt in growth. At the same time, the *secondary sex characteristics* emerge. These include the development of an adult distribution of pubic hair in both sexes, with breast development in females. In males the penis and testes enlarge, the voice 'breaks', and facial and body hair appears. Puberty may be delayed in girls of low body weight (athletes and eating disorders).

The trigger for the onset of puberty is unknown, but some change affecting the *hypothalamus* seems to be crucial. In both sexes, one of the first detectable developments is an increase in the plasma concentrations of the gonadotrophins, i.e., *follicle-*

stimulating hormone (FSH) and *luteinizing hormone* (LH). These are protein hormones secreted by the anterior pituitary gland, but their secretion is controlled by gonadotrophin-releasing hormone (GnRH), a peptide secreted by the hypothalamus (Section 8.2). Why GnRH secretion should rise at puberty is not known, but the resulting increase in gonadotrophin levels stimulates the previously dormant testes or ovaries. This initiates spermatogenesis in the male and cyclical follicular development and ovulation in the female. At the same time, the endocrine activity of the gonads causes an increase in the levels of the sex steroids—testosterone in the male and oestrogens in the female—promoting the secondary sex characteristics.

Box 40 Clinical note: Pseudohermaphroditism

Pseudohermaphroditism is the name given to any condition in which an individual has the genes and gonads of one sex but develops the external genitalia of the other. Such an outcome may result if a female fetus is exposed to abnormally high levels of androgen, e.g., because of abnormal secretion from the adrenal glands. These male sex hormones can cause enlargement of the clitoris, producing a female pseudohermaphrodite. Male pseudohermaphroditism, on the other hand, results from genetic failure of the testes to form adequate testosterone so that the fetus develops female external genitalia. In many cases male pseudohermaphrodites are brought up as girls and their true sex only becomes clear at puberty, often following investigation of failure of menstruation. These rare problems underline the significance of the male sex hormones in determining the path of normal sexual development in utero.

9.2 Male reproductive function

Learning objectives

At the end of this section you should be able to:

- outline the structure of the male reproductive system
- describe the process of spermatogenesis
- summarize the functions of the Sertoli cell
- describe the endocrine control of male reproductive function
- list the functions of testosterone.

Relevant structure

The male gonad is the *testis* (Fig. 167). This is the source of the male gametes, the *spermatozoa*, and produces the male sex hormone *testosterone*. These functions rely on two different structures in the testis. Spermatogenesis occurs in the *seminiferous tubules*, while testosterone is secreted by specialized cells in the interstitium between these tubules known as the *interstitial cells of Leydig* (Fig. 168).

During the last months of fetal development, the testes descend through the inguinal canals from their initial position in the body cavity, so that at birth they lie within the *scrotum*. Like other aspects of male sexual development, this migration is stimulated by fetal testosterone. Normal sperm production (spermatogenesis) only occurs 2–3°C below

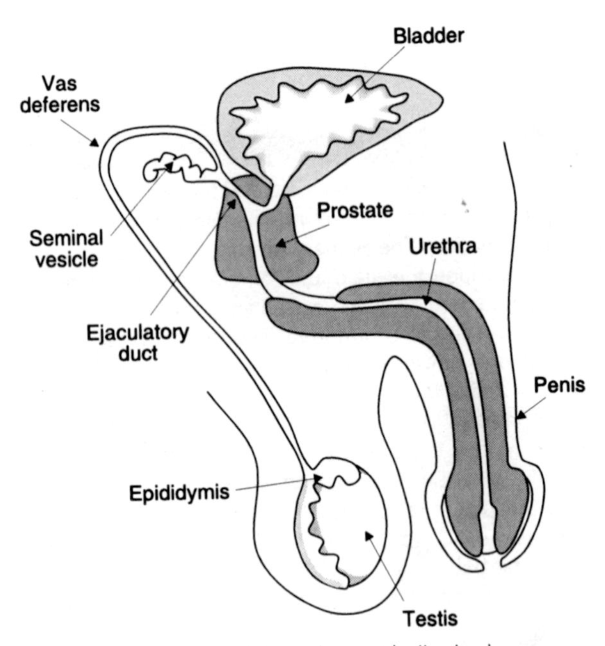

Fig. 167 Lateral view of the male reproductive tract.

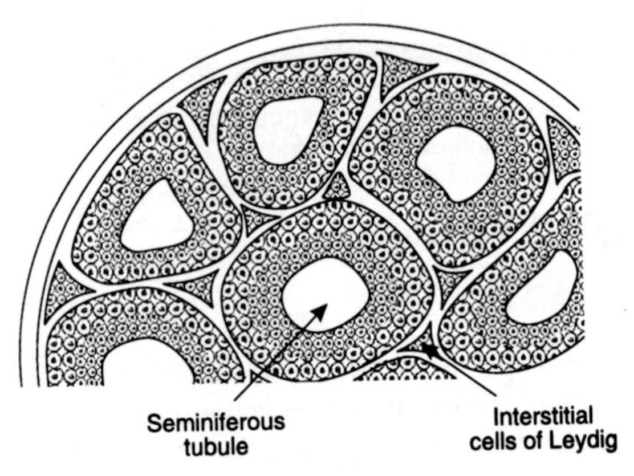

Fig. 168 A cross-section of the testis.

body core temperature, and if a testis fails to descend into the scrotal sac it will not function properly. Testicular temperature can actually be regulated using the cremaster muscle around the spermatic cord and the dartos muscle in the wall of the scrotum. In cold weather these contract, pulling the testis and scrotum towards the body and thus conserving heat.

Once structurally mature, spermatozoa are flushed into the *epididymis* which encloses the posterior pole of the testis (Fig. 167). Removal of fluid leaves a condensed mass of sperm which reach functional maturity within the epididymis, becoming motile. They then move on into the *vas deferens*, a muscular tube which acts as a storage reservoir for sperm. At its distal end, the vas fuses with the duct from the *seminal vesicle*, a reproductive exocrine gland, to form the ejaculatory duct. This passes through the *prostate gland* at the base of the bladder and several ducts from the prostate empty into the ejaculatory duct before it opens into the urethra.

Spermatogenesis

Sperm production continues throughout life from puberty onwards. The development of spermatozoa from *spermatogonia*, which are distributed around the periphery of the seminiferous tubules in the testes, is called spermatogenesis (Fig. 169). Spermatogonia first divide mitotically and then mature into *primary* (1°) *spermatocytes*. Since only some of the cells resulting from division develop further, however, leaving others in reserve for future replication, the total number of spermatogonia is not depleted. This allows continuous production of gametes from puberty onwards, and contrasts with oogenesis in the female, in which no new primary oocytes are formed after birth (Section 9.4).

Next, the primary spermatocytes (*diploid* cells containing 23 homologous chromosome pairs) undergo *meiosis* (Fig. 169). The DNA is replicated prior to the first meiotic division, with each chromosome forming two sister chromatids. There is then exchange of genetic material (*crossing over*) between the partner chromosomes in each chromosome pair, and this leads to new gene combinations in the daughter cells. Separation of the chromosome pairs followed by cell division completes the first meiotic division, producing two *secondary* (2°) *spermatocytes*. Each of these contains half the normal complement of chromosomes (one chromosome from each pair), i.e., these are *haploid* cells. It is important to realize, however, that each chromosome still consists of two connected sister chromatids, so 2° spermatocytes actually contain the normal number of genes. There is no further replication of DNA prior to the second meiotic division, however, during which the chromatids separate. This produces four haploid *spermatids*, each containing one chromosome from each chromosome pair and one gene from each gene pair.

There is no further cell division, but the spermatids must still undergo a final maturation process to produce spermatozoa. This is sometimes referred to as spermiogenesis and results in the head, midpiece and tail structures of the mature spermatozoon (Fig. 170). The head consists of the chromosomal material covered by an enzyme-filled sac called the *acrosome*. The midpiece contains mitochondria, which provide the ATP necessary for sperm motility. This motility is the result of the swimming action of the tail, which is driven by the contractile activity of its microtubular core.

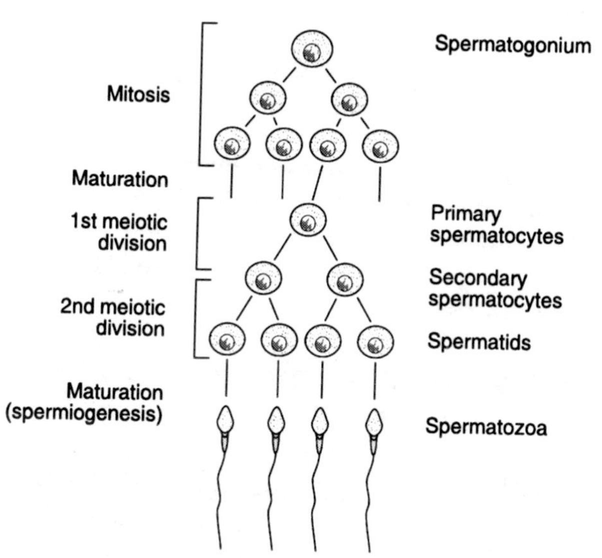

Fig. 169 Summary of the sequence of events involved in spermatogenesis.

Sertoli cell function

At each stage of spermatogenesis, there is a close association between the developing sperm and large cells which line the seminiferous tubules, known as Sertoli cells (Fig. 171). These fulfil the following functions:

- nutrition of the sperm
- phagocytosis of dead or damaged cells
- protection of the sperm from blood-borne toxins (the blood–testis barrier)
- formation of seminiferous tubule fluid; this flushes the structurally mature, but immotile, spermatozoa into the epididymis.

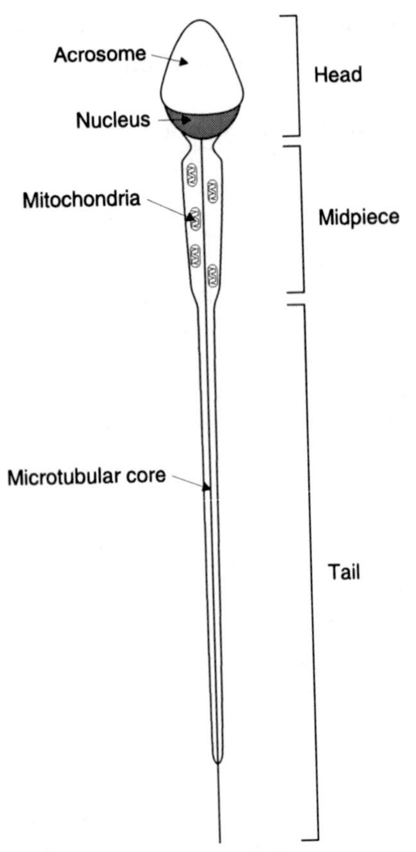

Fig. 170 Structural components of a mature spermatozoon.

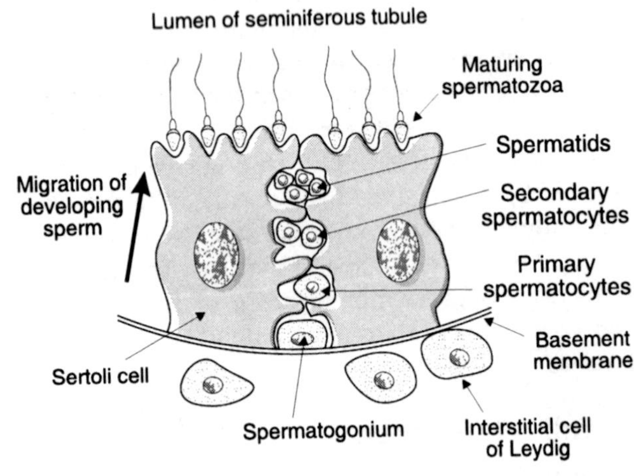

Fig. 171 Cross-section of a seminiferous tubule.

Hormonal control of male reproduction

Control of male reproductive function depends on a series of hormones whose secretion is regulated through negative feedback loops (Fig. 172). The overall pattern has several similarities to that seen in control of the female reproductive system (see Fig. 176). The hypothalamus produces GnRH which is carried in the hypophyseal portal system to the anterior pituitary (Section 8.2). This stimulates

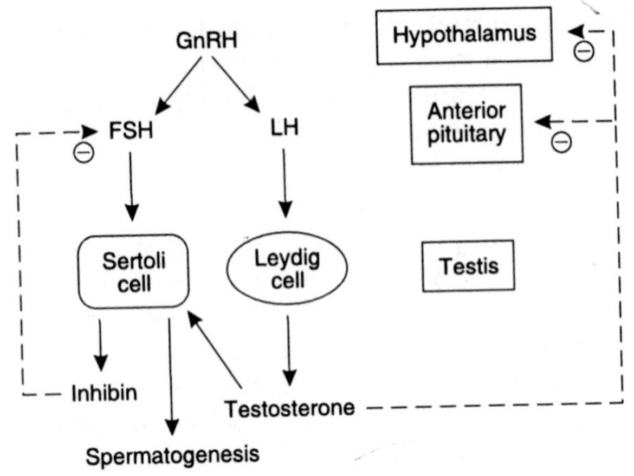

Fig. 172 Endocrine control of testicular function. Note the negative feedback loops (–).

release of the gonadotrophins, FSH and LH, which are carried in the circulation to the testes, their site of action. Each acts on a different target cell, with FSH stimulating spermatogenesis through its action on the Sertoli cell, while LH promotes testosterone secretion from the interstitial cells of Leydig. This results in a high tissue concentration of *testosterone* in the testis, which is necessary for spermatogenesis. Circulating testosterone exerts a negative feedback effect by inhibiting hypothalamic and anterior pituitary secretion (Fig. 172). Sertoli cells also secrete a peptide hormone called *inhibin*, which has a relatively specific effect in reducing FSH secretion.

Functions of testosterone

Testosterone is the dominant male sex steroid. Its actions include:

- stimulation of the differentiation and growth of the male reproductive system, and stimulation of spermatogenesis
- development of the male secondary sex characteristics at puberty
- promotion of adolescent skeletal growth and subsequent fusion of the epiphyses
- promotion of skeletal muscle development
- acceleration of the basal metabolic rate
- behavioural effects, including increased libido and aggression.

9.3 Sexual intercourse

Learning objectives

At the end of this section you should be able to:

- describe the reflexes responsible for erection and ejaculation.

Sexual intercourse makes fertilization possible by introducing sperm to the female genital tract. It involves erection of the penis, allowing penetration of the vagina to occur, followed by ejaculation of *semen*, a mixture of sperm and secretions from the glands of the male reproductive tract.

Erection

Erection results from engorgement of the vascular spaces in the penis. This may be initiated by erotic cognitive stimuli and is reinforced by stimulation of mechanoreceptors on the glans penis. These activate a spinal cord reflex causing dilatation of the penile arterioles. This depends both on activation of *parasympathetic dilator nerves* and inhibition of *sympathetic constrictor nerves*. As engorgement progresses, the veins draining the penis are mechanically compressed, further promoting erection.

Ejaculation

Ejaculation is a two-stage process.

- **Emission.** During the emission phase, *sympathetic nerves* stimulate contraction of the smooth muscle in the walls of the epididymis and vas deferens as well as the accessory reproductive glands. This causes sperm and secretions to be mixed together to produce semen which is then forced into the urethra.
- **Expulsion.** The expulsion phase relies on rhythmical contraction of skeletal muscle around the base of the penis, which compresses the urethra, discharging its contents into the vagina. This activity depends on reflex stimulation of *somatic motor nerves*.

Normally about 3 ml of semen is released. Male fertility requires both adequate numbers and quality of sperm, as measured in terms of the sperm count (typically around $120 \times 10^6 ml^{-1}$) and the percentage of abnormally shaped or immotile spermatozoa. Semen contains secretions from the seminal vesicles, which provide fructose, an energy source for sperm, along with prostaglandins and other hormones. Prostatic secretion is alkaline and helps to neutralize vaginal acid. It also contains a mixture of clotting factors and fibrinolysin. Lubricating mucus is added by the bulbourethral glands.

9.4 Female reproductive function

Learning objectives

At the end of this section you should be able to:

- outline the structure of the female reproductive system
- describe the timing and process of oogenesis
- describe the main events in the ovarian cycle
- describe the endocrine control of the ovarian and uterine cycles
- list the functions of the female sex hormones
- outline the causes and consequences of the menopause.

Relevant structure

The female gonad is the *ovary* which produces *ova* and secretes the female sex steroids, i.e., *oestrogens* (particularly estradiol) and *progesterone*. The ovaries are loosely enfolded in the frond-like fimbriae which guide ova into the *Fallopian*, or uterine, tubes (Fig. 173). These are connected to the *uterus*, which opens through the *cervix* into the *vagina* below.

Oogenesis

Oogenesis refers to the development of mature ova from their primitive precursors, the *oogonia* (Fig. 174). Initially, the oogonia multiply by mitotic division and then undergo a maturation process to form

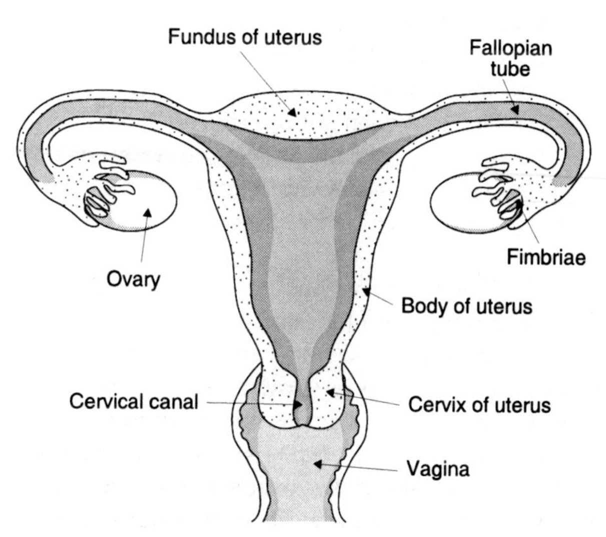

Fig. 173 Anatomy of the female reproductive tract.

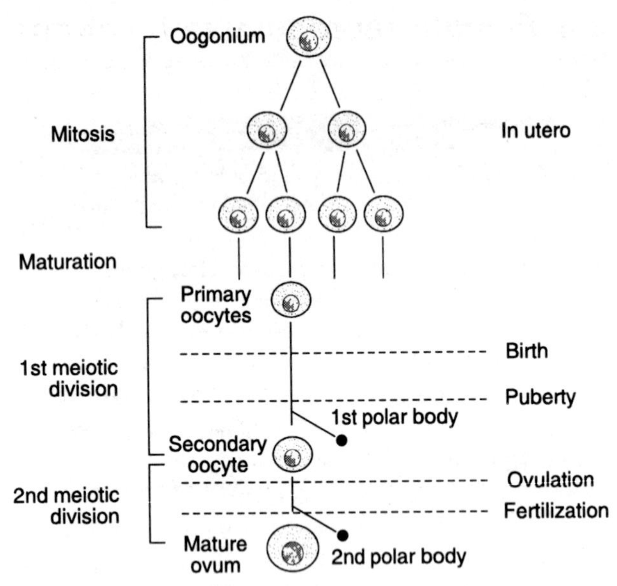

Fig. 174 Summary of the sequence of steps involved in the development of an ovum (oogenesis).

the *primary oocytes* within the ovary. Multiplication only occurs during fetal life, however, and the 1–2 × 10^6 primary oocytes present at birth are the source for all the mature ova which will develop during reproductive life. This should be contrasted with the continuous replication of spermatogonia in postpubertal males. Only 400–500 primary oocytes ever reach maturity, while the rest simply degenerate.

The primary oocytes (diploid cells) begin, but do not complete, the first meiotic division before birth (Fig. 174). All further development is arrested until after puberty, when the first meiotic division may finally be concluded. This produces a *secondary oocyte* (a haploid cell), which contains nearly all the available cytoplasm and a small *polar body*, which consists largely of chromosomal material. Formation of the secondary oocyte is delayed until just prior to ovulation, and the secondary oocyte only completes the second meiotic division if it is fertilized. Under these circumstances, a further uneven division of cell resources occurs, producing the mature *ovum* (containing 22 autosomes and a single sex chromosome) and a second polar body.

Since no new primary oocytes form during reproductive life, the age of a woman's ova increases as she grows older. This may explain why certain genetic defects, such as Down's syndrome, occur with a higher frequency in the children of older mothers. Presumably, the risk of chromosomal abnormalities caused by problems during meiosis increases as the age of the primary oocyte increases.

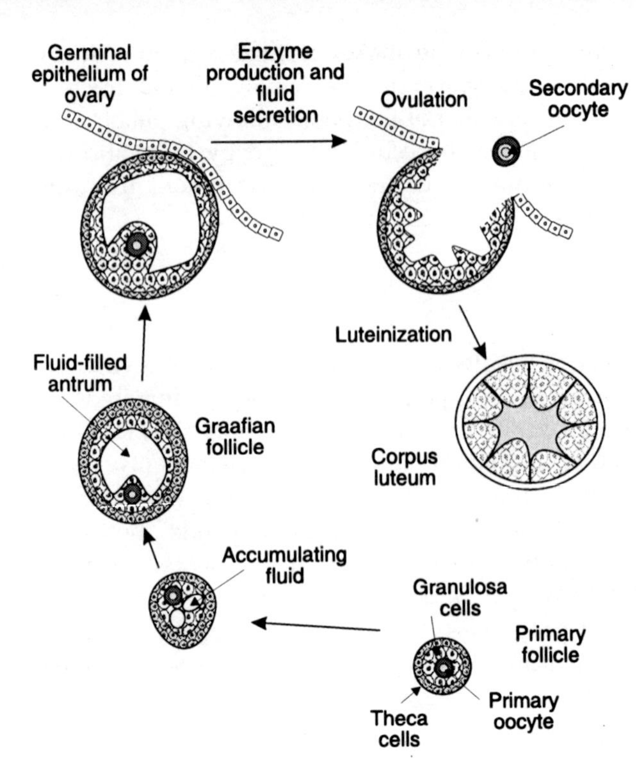

Fig. 175 Main structural changes during the ovarian cycle.

The female reproductive cycle

During reproductive life the female reproductive system develops and regresses in a cyclical fashion, each cycle lasting approximately 28 days. This involves changes in the ovary (the *ovarian cycle*) and the endometrial lining of the uterus (the *uterine cycle*). A single ovum is released around the middle of each cycle at a time when the endometrium is thickened and nutrient rich in preparation for pregnancy, should fertilization occur. If it does not, the endometrial lining is shed during *menstruation*, and the cycle is repeated.

Ovarian cycle and follicular development

Ova develop within structures known as follicles. *Primordial follicles* form in utero and consist of a primary oocyte surrounded by the basal lamina and a single layer of *granulosa cells*. At any time from late intrauterine life onwards, primordial follicles may mature further to form *primary follicles*. Granulosa cells multiply and lay down a jelly-like, mucopolysaccharide layer, the *zona pellucida*, around the oocyte, and a layer of cells outside the basal lamina differentiates to form *theca cells* (Fig. 175).

Although many primary follicles simply regress during life, a proportion will go on to develop further during the ovarian cycles of reproductive life. These cycles commence at puberty and each cycle is normally 28 days long, consisting of a 14-day follicular phase followed by ovulation and a 14-day luteal phase.

Follicular phase From puberty onwards, about 5–15 primary follicles begin to develop at the start of each reproductive cycle. Granulosa cells multiply and secrete fluid which eventually pools to form the *antrum*, thus producing an antral or *Graafian follicle* (Fig. 175). The theca cells also multiply and differentiate to form the theca interna surrounded by the theca externa. After 7–10 days, one follicle becomes dominant and enlarges further, while the rest become atretic. By 14 days, the developing follicle actually distorts the ovarian surface. Rapid fluid accumulation increases the intrafollicular pressure and proteolytic enzymes digest the overlying tissues. Ovulation follows as the follicle ruptures, releasing the ovum (a secondary oocyte) surrounded by the zona pellucida and adherent granulosa cells (the corona radiata) into the Fallopian tube (see Fig. 178). Some women report that they can feel the rupture of the follicle, which may be painful.

Luteal phase Following ovulation, the follicular remains seal over and become luteinized, forming the *corpus luteum*. Granulosa and thecal cells become swollen with fatty droplets, giving them a yellow appearance (hence the name corpus luteum, meaning yellow body) and the structure becomes more vascular. The resulting increase in blood supply serves to promote the endocrine function of the corpus luteum. If a fertilized ovum fails to implant in the endometrium, however, the corpus luteum degenerates after a further 10 days, i.e., about day 24 of a 28-day cycle. A new batch of primary follicles then starts to develop.

Hormonal control of the ovarian cycle

This is similar to control of male reproduction (Fig. 172), except for the monthly cycling of hormone levels seen in women. As in males, the hypothalamus produces GnRH, which causes release of the gonadotrophins FSH and LH from the anterior pituitary gland (Fig. 176). As its name suggests, FSH stimulates follicular development by activating granulosa cell division and secretion of oestrogens (e.g., estradiol). During the follicular phase, LH plays a supporting role as it stimulates *androgen secretion* by the theca cells. These androgens diffuse into the granulosa cells, which use them as precursors for the synthesis of oestrogens. LH is also responsible for triggering ovulation once follicular development is complete, and it then stimulates *luteinization* of the follicular remnants to form the corpus luteum. This secretes both progesterone and oestrogen.

Gonadotrophin secretion is under feedback control from both oestrogens and progesterone. Their effect is normally inhibitory, but very high oestrogen levels can actually exert positive feedback on LH secretion. This may explain the sudden surge in LH levels immediately preceding ovulation.

Using the ideas outlined above, we can attempt to explain the normal pattern of hormonal changes seen during a single, 28-day, reproductive cycle (Fig. 177). At the beginning of the cycle, oestrogen and progesterone levels are low, so they exert little negative feedback and FSH levels are high. This stimulates follicular development, leading to a rise in oestrogen levels with resultant inhibition of gonadotrophin release. FSH and LH concentrations fall to a minimum about day 10–12. At this stage, the very high oestrogen concentrations appear to generate a positive feedback effect, however, leading to a dramatic peak in LH concentrations. This *LH surge* stimulates *ovulation*, which occurs about day 14. Thereafter, increasing production of progesterone and oestrogen by the corpus luteum leads to a steady decline in LH and FSH. At about day 22–24, the corpus luteum undergoes programmed regression, or atresia, causing a rapid decline in the levels of ovarian hormone. The resulting reduction in feedback inhibition allows gonadotrophin levels to rise once more, and the cycle repeats.

Female sex hormones Oestrogens and progesterone are female sex hormones secreted by the ovaries. These steroids have a range of effects, among the most important of which are their actions on the endometrial lining of the uterus. It is the cyclical changes in oestrogen and progesterone levels which control the uterine cycle, as considered below. Some other actions are listed here.

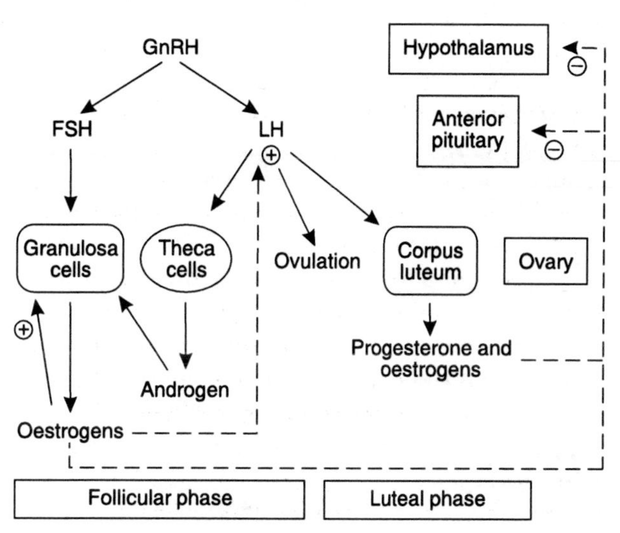

Fig. 176 Endocrine control of ovarian function.

Nine

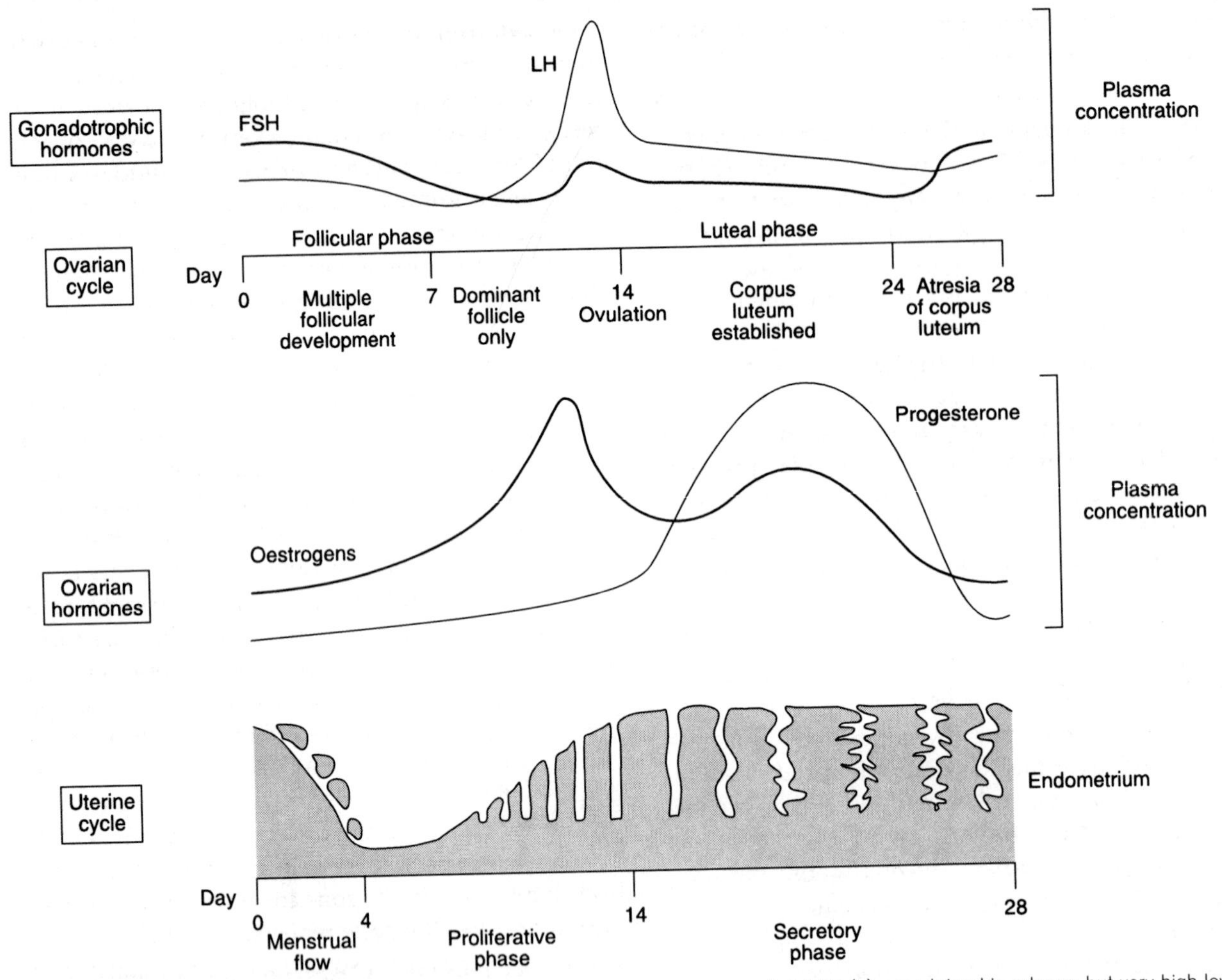

Fig. 177 Endocrine control of ovarian function. Oestrogens and progesterone normally inhibit (–) gonadotrophin release, but very high levels of oestrogen may also stimulate (+) LH production just prior to ovulation.

Oestrogens are a group of related chemicals. Oestradiol and oestrone make up most of the circulating oestrogen in the non-pregnant state during the reproductive years, but oestrone levels exceed estradiol postmenopausally. Placental oestriol becomes important during pregnancy.

Oestrogens:

- promote granulosa cell proliferation in response to FSH (Fig. 176)
- are responsible for the development of some female secondary sex characteristics at puberty, particularly fat accumulation in the subcutaneous tissues, round the hips and in the breasts
- promote conditions necessary for the maintenance of pregnancy after implantation
- promote further breast development during pregnancy by stimulating duct growth.

Progesterone:

- promotes conditions necessary for the maintenance of pregnancy
- promotes breast development, both at puberty and during pregnancy, by stimulating the formation of the secretory lobules.

The uterine cycle

The *endometrium* consists of a mixture of glands and blood vessels within a connective tissue stroma. At the beginning of each cycle, approximately two-thirds of the endometrial thickness sloughs off from the underlying endometrium and myometrium and is lost during *menstruation* (Fig. 177). This lasts 2–5 days. Rising oestrogen levels then stimulate hyperplasia of the endometrium, with the growth of many simple tube-like glands. This is the *proliferative phase* of the uterine cycle. Following ovulation, progesterone becomes the dominant sex hormone and this causes the glands to become tortuous, or saw-toothed, in outline. This is referred to as the *secretory phase* because the glands also fill up with secretions rich in glycogen. The purpose of this is to provide a highly vascular, fluid- and nutrient-rich environment suitable for implantation of a developing embryo. If no pregnancy occurs, the rapid fall in oestrogen and progesterone levels following atresia of the corpus luteum around day 24 causes vascular constriction, ischaemia and infarction of the endometrium, which is shed in the next menstrual flow.

Menopause

With increasing age, menstruation decreases in frequency and then finally ceases at about age 50. This is referred to as the menopause and reflects depletion of the follicular reserves during reproductive life. Oestrogen levels fall as a result, giving rise to many of the associated changes, including atrophy of the breasts, skin and genitalia. Loss of bone tissue (*osteoporosis*) can eventually lead to bone fractures, but this may be limited by long-term oestrogen replacement therapy.

9.5 Pregnancy, labour and lactation

Learning objectives

At the end of this section you should be able to:

- describe the process of fertilization
- outline the early steps in embryo development, and the necessary conditions for implantation
- outline how the placenta develops and list its functions
- describe how pregnancy affects maternal hormones and physiology
- define the stages of labour and describe the mechanisms responsible for parturition
- describe the hormonal control of lactation
- explain how fertility may be decreased pharmacologically or pathologically.

Pregnancy lasts from fertilization to the birth of the baby. This is normally 38 weeks in humans but, since the date of conception is not easily identifiable, calculations are usually based on the first day of the last menstrual period prior to conception, giving a gestation of 40 weeks from that date. This is divided into three *trimesters*, each lasting 3 months.

Fertilization

For fertilization to occur, spermatozoa and ovum must first come into contact with each other. Following ovulation, the ovum is propelled by the movement of the cilia lining the fimbriae and Fallopian tube. Progress is rather slow relative to the upward migration of the sperm through the female genital tract, and fertilization usually occurs in the outer third of the Fallopian tube. The spermatozoa are deposited in the vagina during sexual intercourse and first have to pass through the mucus which plugs the cervical canal (Fig. 173). This is only penetrable towards the end of the follicular phase of the ovarian cycle, around ovulation, when the ratio of oestrogen to progesterone is at its peak (Fig. 177). For a few days the mucus plug becomes much less dense, allowing sperm to swim through it, but even then less than 5% actually enter the uterus. Further transport depends on the contractile activity of the uterus, which randomly distributes sperm throughout the uterine cavity. Some enter the Fallopian tube and are propelled towards the ovum by retrograde peristalsis. Of as many as 300×10^6 spermatozoa released during ejaculation, perhaps only a few thousand are likely to reach the ovum. This explains the need for high sperm counts if a male is to be fertile. It should also be noted that the fertilizing ability of sperm is enhanced following several hours exposure to the female reproductive tract, a process referred to as *capacitation*.

Fertilization requires that a spermatozoon must first penetrate the corona radiata and the zona pellucida which surround the ovum (Fig. 178). The motility of the sperm orients their heads towards these barriers and forces them between the granulosa cells of the corona radiata. They then bind to receptors on the outer surface of the zona pellucida and digest a tunnel through it by releasing acrosomal enzymes. When the first spermatozoon contacts the plasma membrane of the secondary oocyte, the zona pellucida changes chemically, becoming impenetrable to other sperm. This prevents multiple fertilizations of the same ovum. The secondary oocyte completes the second meiotic division and the genetic

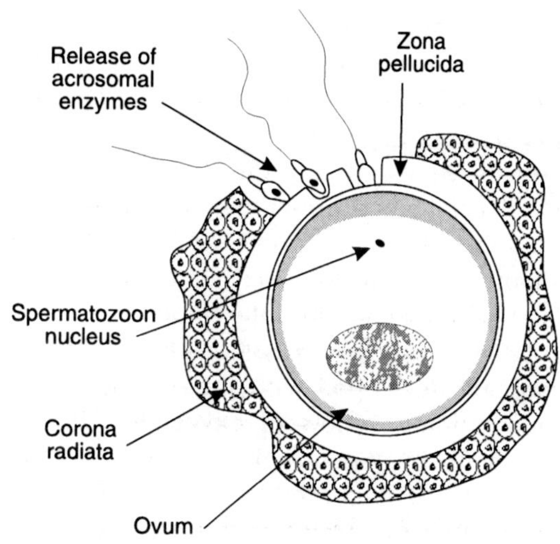

Fig. 178 Fertilization of the ovum.

material from the haploid gametes fuses to produce a diploid *zygote*. This then divides, initiating embryonic development.

Early development and implantation

Mitotic division leads to the formation of a solid ball of cells, the *morula*, which is transported towards the uterus by peristalsis and the action of cilia in the Fallopian tube (Fig. 179). In 3–4 days, it enters the uterine cavity and over the next 3–4 days develops into the *blastocyst*. This consists of:

- an outer shell of *trophoblast* cells, which will eventually form the fetal part of the placenta
- the fluid-filled *blastocoele*, which forms the yolk sac
- the *inner cell mass*, from which all embryonic structures develop.

During the first 7–10 days, the morula and blastocyst are nourished by glycogen released into the uterine cavity from the secretory endometrium. Further development, however, requires implantation of the blastocyst. The trophoblast layer adheres to the endometrial surface and then digests it, extending finger-like processes into the cavities formed. This digestion also releases nutrients for the developing embryo. The endometrium releases prostaglandins and becomes more vascular, oedematous and nutrient enriched, forming the *decidua*. Within 10–12 days after fertilization, the blastocyst has buried itself within this decidual endometrium (Fig. 179).

Placental development

After the first few weeks, the nutritional needs of the embryo can no longer be met simply by cannibalizing maternal tissue. Development of the placenta, which fulfils a nutritional role for the rest of pregnancy, begins with the erosion of maternal capillaries in the decidua by the *chorion*, a membrane derived from the trophoblast. Blood-filled spaces are formed in the decidua as a result and chorionic fingers project into these, eventually developing into the *placental villi*. Fetal blood from the umbilical arteries is circulated through the chorionic capillaries within these villi and is then recirculated via the umbilical vein. Maternal blood in the intervillous spaces is anticoagulated by chorionic secretions, and vascular connections are developed with the uterine blood vessels, maintaining a flow of oxygenated blood around the villi. By week 5, the placenta is well established with its fetal (chorionic) and maternal (decidual) components. Maternal and fetal blood are separated by chorionic tissue and capillary endothelium. This provides a protective barrier while allowing necessary exchange to occur.

Placental functions include:

- nutrient and gas exchange: fetal oxygen uptake is favoured by the high oxygen affinity of fetal haemoglobin (Section 4.4)
- removal of fetal waste products by diffusion into the maternal blood
- partial protection from potentially damaging agents in the maternal circulation; these include chemical toxins and infectious agents which might interfere with fetal development (*teratogens*)
- secretion of several endocrine hormones, including oestrogens and progesterone, human chorionic gonadotrophin and human chorionic somatomammotrophin.

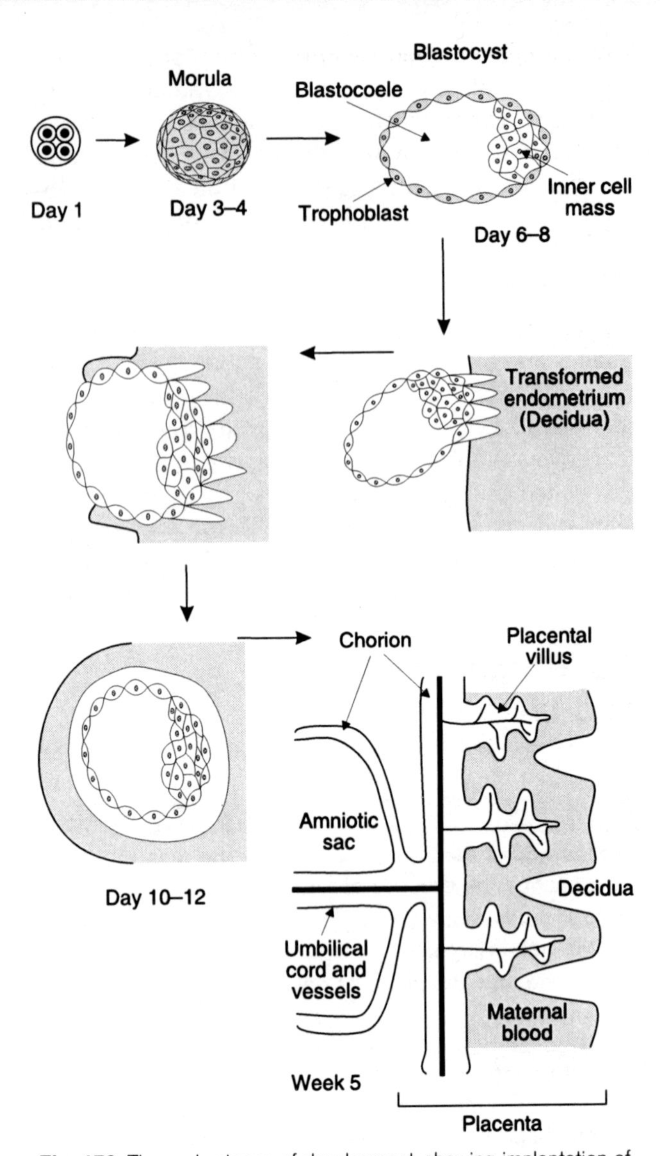

Fig. 179 The early stages of development, showing implantation of the blastocyst and development of the placenta.

Hormones of pregnancy

For a pregnancy to be maintained, high levels of oestrogen and progesterone must be produced. Soon after implantation, the chorion of the developing placenta starts to secrete *human chorionic gonadotrophin* (hCG). Like LH, hCG acts on the corpus luteum so that, instead of regressing 10 days after ovulation as normal, it is stimulated to develop further when a pregnancy becomes established. This prevents the fall in progesterone and oestrogen that usually precipitates menstruation, leading instead to continuously rising levels of sex steroids (Fig. 180). hCG can be detected in the urine within 2–3 weeks of fertilization and this is used for pregnancy testing.

Levels of hCG peak about 6 weeks after fertilization and then fall back to a much lower plateau, which is maintained from 10 weeks to the end of pregnancy. With the fall in hCG, the corpus luteum regresses partially, but *oestrogen* and *progesterone* continue to rise since the placenta itself becomes the chief source of sex hormones from about 3 months on, i.e., from the end of the first trimester.

Pregnancy-induced changes in maternal physiology

Pregnancy affects most body systems.

- *Morning sickness* is an early feature, which tends to diminish after the first trimester. This may be caused by the initial peak in hCG.
- *Cardiac output* rises to a maximum some 30–40% above normal at about 20 weeks.
- *Maternal blood volume* is increased by up to 20% at term, most of the expansion occurring in the second half of pregnancy.

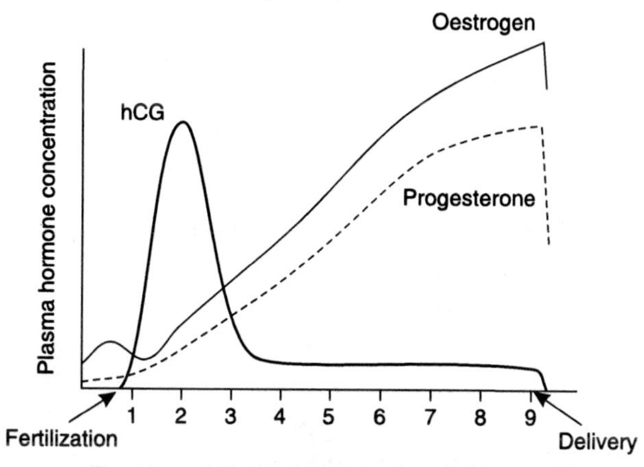

Fig. 180 Changes in the maternal concentrations of some of the hormones of pregnancy. The early rise in oestrogen and progesterone is caused by stimulation of the corpus luteum by hCG; the continuing rise during the second and third trimesters reflects placental production.

- *Oxygen consumption and CO_2 production* rise as a result of the contribution from fetal metabolism. This results in a 20% increase in respiratory minute volume.
- *Basal metabolic rate* is increased by up to 15% at term.
- *Renal reabsorption* of NaCl and water is increased by the elevated sex steroids, which have mild mineralocorticoid effects. Glomerular filtration is also increased, however, helping to limit the resulting fluid retention. In late pregnancy, mechanical compression of the bladder by the expanding uterus leads to increased frequency of micturition.
- *Nutritional needs* are increased because of the needs of the developing fetus. If there is any deficiency, the mother rather than the fetus is preferentially deprived. For example, if the calcium intake is low, the fetal supply may be maintained by demineralizing maternal bones. The energy and protein intake in the average diet is usually adequate, but mineral and vitamin intake may not be. Iron supplements are often required since an additional 1 g of iron is needed during the course of a pregnancy, over and above normal daily needs. This exceeds normal body iron stores (100–700 mg) and anaemia may result.
- *Weight gain* totals about 10 kg, mostly acquired during the third trimester. This gain consists of the fetus (3.5 kg), placenta and amniotic fluid (2 kg), breast growth (1 kg) and increased body weight (3.5 kg). About two-thirds of the change in body weight is retained fluid.

Parturition

Parturition refers to the mechanisms leading to the birth of a baby. Contraction of the *myometrium* provides the propulsive force and the muscle bulk of the uterus increases some 20 fold during pregnancy, largely under the influence of oestrogens. At about 40 weeks, *labour* begins spontaneously.

Labour is divided into three stages.

First stage of labour

During the first stage (which lasts 8–24 hours), the canal through the cervix becomes fully dilated up to 10 cm diameter. This occurs as the baby's head is forced downwards by myometrial contraction. As the cervix is stretched, a nervous reflex leads to *oxytocin* secretion from the posterior pituitary, and this further stimulates the uterine contractions (Fig. 181). Such positive feedback helps explain the steady increase in force, frequency and duration of contraction as labour progresses.

Oxytocin may also play a role in triggering the *onset of labour*. The high oestrogen concentration of late pregnancy increases the density of oxytocin receptors on the smooth muscle of the uterus, increasing the myometrial sensitivity to oxytocin, and this may eventually lead to spontaneous labour. *Prostaglandins* also stimulate uterine contraction, while the peptide hormone *relaxin*, secreted by the placenta and the corpus luteum of pregnancy, softens the connective tissue of the cervix and the pelvic ligaments. This eases cervical dilatation and reduces the resistance to the baby's passage through the pelvis.

Second stage of labour

The second stage (lasting 30–90 min) begins with full dilatation of the cervix and ends with the birth of the baby. Vaginal stretch stimulates reflex abdominal contractions, which can be reinforced voluntarily, this effort being most effective if timed to coincide with uterine contraction.

Third stage of labour

The third stage (complete in 15–30 min) involves the delivery of the placenta, which shears off the wall of the contracted uterus. Myometrial contraction usually limits blood loss to about 300–400 ml.

Post delivery

Following delivery, there is a dramatic drop in maternal oestrogen and progesterone levels caused by the loss of placental steroids and the regression of the corpus luteum (which relies on placental hCG). This leads to regression of the decidual endometrium, which is shed as a discharge referred to as *lochia*. The myometrium involutes and the pre-pregnant state is regained in 4–6 weeks.

Lactation

During pregnancy, the breasts become milk-secreting organs, with the high levels of sex steroids leading to structural development. Oestrogens stimulate duct growth while progesterone promotes the enlargement of the alveolar structures with their milk-secreting epithelium. Two other hormones of pregnancy, *prolactin* from the anterior pituitary and *human chorionic somato-mammotrophin* from the placenta, contribute to functional development by stimulating the synthesis of the enzymes necessary for milk production. By the second half of pregnancy, the breast is fully capable of milk production but this is inhibited by the high steroid levels.

After delivery, steroid levels fall and lactation can proceed. Both production and ejection of milk, however, depend on the reflex hormonal peaks stimulated by suckling (Fig. 182). Mechanical stimulation of the nipple promotes secretion of prolactin from the anterior pituitary by decreasing hypothalamic secretion of a *prolactin-inhibiting hormone* (probably dopamine), although prolactin-releasing hormone may also be involved (Section 8.2). The resulting peaks in prolactin levels stimulate milk production by the alveolar epithelium. Suckling also leads to release of *oxytocin*, which is manufactured within the hypothalamus and secreted from the posterior pituitary. This causes contraction of the myoepithelial cells surrounding the alveoli in the breast, leading to ejection of milk into the ducts for ingestion by the feeding baby.

Breast feeding can fulfil a contraceptive role, since prolactin inhibits GnRH secretion from the hypothalamus and can thus prevent ovulation and menstruation.

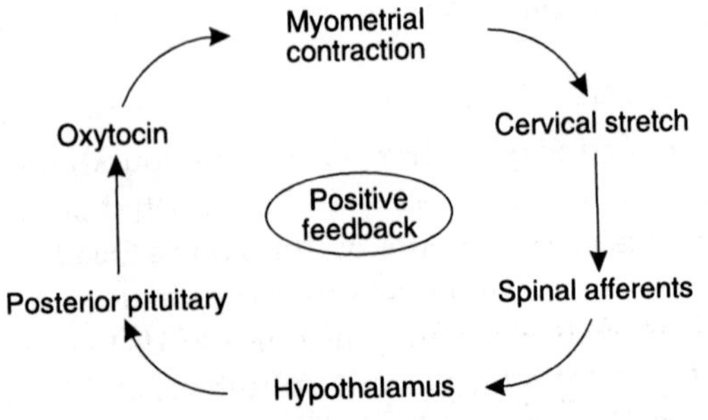

Fig. 181 Positive feedback mechanism promoting myometrial contraction through oxytocin release during labour.

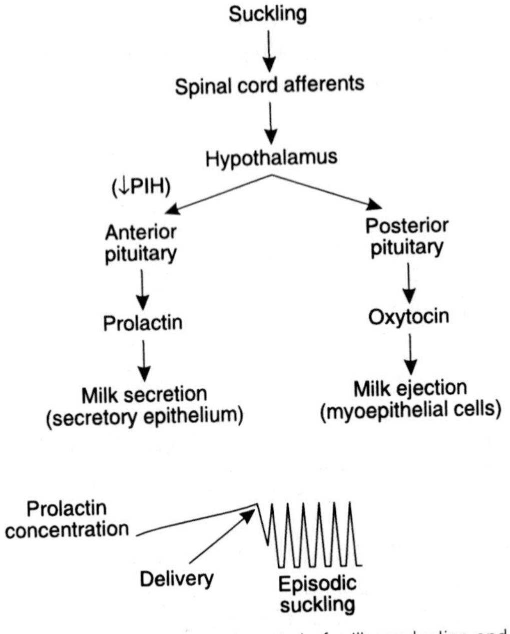

Fig. 182 Reflex hormonal control of milk production and release depends on secretion of prolactin and oxytocin in response to suckling by the infant.

Contents of breast milk

Normal breast milk contains carbohydrate in the form of lactose, triglycerides, protein, vitamins, calcium and phosphate. In the first 4–5 days, however, the secretion contains more protein with less fat and lactose. This *colostrum* provides protection against infection, since it contains lactoferrin, which has antibacterial actions, and maternal antibodies of the IgA type. These can be absorbed intact by the neonate's intestine and help provide passive immunity in the first few months of life when the baby's own immune system is immature.

Box 41 Clinical note: Contraception and infertility

These themes are closely related since fertility is reduced in both cases.

Contraception

Deliberate control of fertility relies on interrupting some of the processes necessary for conception and implantation. The main methods can be classified as hormonal, mechanical, chemical and natural.

Hormonal methods

These use artificial steroids to interfere with the normal feedback regulation of the ovarian and uterine cycles. The *combined pill* contains oestrogen and progestogen (artificial progesterone) and prevents both ovulation and the secretory phase of endometrial development. It also inhibits the change in cervical mucus seen in the middle of a normal cycle, so that sperm migration is hindered. The *progestogen-only pill* has less effect on ovulation and mainly acts on the endometrium and cervical mucus. Progestogens may also be injected or implanted to give longer-term contraception.

Mechanical and chemical methods

Simple barrier methods prevent fertilization by blocking sperm migration, e.g., male and female condoms, vaginal diaphragms, etc. These are often combined with spermicidal chemicals. Alternatively, intrauterine devices may be placed inside the body of the uterus for longer-term contraception. These interfere with ovum transport and implantation. Male and female sterilization can also be thought of as irreversible, mechanical methods of contraception, since they involve surgically tying off (ligating) the vasa deferentia or Fallopian tubes, respectively.

Natural methods

These rely on identifying the days in each month when a woman is fertile and then avoiding intercourse during that period. Sperm remain viable for 2–3 days after ejaculation, while the ovum survives 1–2 days after ovulation. Therefore, the fertile period extends from about 3 days before ovulation to 2 days after. The key is identification of the time of ovulation. Several indicators can be used, including the decrease in the viscosity of the cervical mucus around the time of ovulation and the small rise in body temperature (0.2–0.4°C) immediately after ovulation. Both of these features reflect the rise in progesterone around that time. In cycles of variable length, most of the variation occurs in the follicular phase, so the most secure approach is to wait until there is evidence that ovulation has taken place and restrict intercourse to the time between then and the next menstrual period.

Infertility

An involuntary loss of fertility may result from problems involving gamete production or transport, conditions which prevent fertilization, or failure of implantation following fertilization.

Failure of gamete production

This may reflect abnormal development of the gonads in utero, whether because of chromosomal defects or the damaging influence of some external agent (teratogen) during pregnancy. In adult life hormonal defects may arise, interfering with spermatogenesis or ovulation. Examples include gonadotrophin deficiencies and cases of inadequate production of testosterone or oestrogen, all of these hormones being necessary for normal gonadal function. Overproduction of prolactin by the anterior pituitary may also prevent ovulation or spermatogenesis by inhibiting gonadotrophin production. In males, viral infections may damage the testes, e.g., mumps orchitis. Usually, however, no underlying cause can be identified in such cases.

Failure of gamete transport

The Fallopian tubes may become mechanically blocked following pelvic infections, particularly in the case of sexually transmitted diseases. This prevents fertilization. Sperm motility may also be inhibited by antibodies, produced either by the man himself (auto-antibodies in the semen) or by his partner (female antisperm antibodies in the cervical mucus).

Failure of implantation

Successful fertilization cannot lead to an established pregnancy if implantation fails. This may occur because of a progesterone deficiency, leading to a poor endometrial secretory phase, or pathology of the uterus, e.g., endometrial infection or abnormal endometrial proliferation (endometriosis).

9.6 Physiology of the newborn

Learning objectives

At the end of this section you should be able to:

- describe the differences between the fetal and neonatal circulation
- outline the changes which occur in fetal physiology at birth
- identify areas in which neonatal homeostasis may be at risk and explain why.

At birth, a baby (termed a *neonate* for the first month of life) must take over the homeostatic functions previously carried out by placental exchange with the maternal circulation. This requires major adjustments in most organ systems following the transition from intrauterine to extrauterine life.

Cardiovascular adjustments

The newborn baby must rapidly shift from a fetal to an 'adult' pattern of circulation.

The fetal circulation

Several *vascular shunts* ensure the efficient circulation of fetal blood through the placenta and the delivery of a good supply of well-oxygenated blood to the developing brain, while minimizing blood flow through the functionally immature liver and lungs.

Oxygenated blood returns from the placenta in the single (left) umbilical vein (Fig. 183). Some of this blood circulates through the liver and enters the inferior vena cava via the hepatic vein. The rest is shunted through the *ductus venosus*, which links the umbilical, portal and hepatic veins, bypassing the liver itself. The inferior vena cava also receives deoxygenated blood from the legs and abdominal viscera, so it returns mixed oxygenated blood to the right atrium. This oxygenated blood does not circulate through the right ventricle to the pulmonary circulation but is shunted directly to the left atrium via the *foramen ovale*, an opening in the atrial septum. From there it passes to the left ventricle and is pumped into the aorta, which supplies the head and upper limbs in the usual way.

Deoxygenated blood is returned from the upper regions to the right atrium by the superior vena cava. The angle of flow of this blood directs it into the right ventricle and it is then pumped into the pulmonary trunk (Fig. 183). Since there is no alveolar gas exchange, pulmonary blood flow represents wasted cardiac effort and most of this deoxygenated output from the right ventricle is shunted into the descending aorta through the *ductus arteriosus*. The umbilical arteries carry this deoxygenated blood back to the placenta, completing the circulation.

Changes after birth

At birth, the main shunt pathways must close to achieve a normal pattern of circulation. In utero, the placental vasculature acts as a resistance in parallel with the rest of the systemic circulation and, therefore, reduces the total peripheral resistance to the systemic blood flow. The sudden interruption of the placental circulation at birth leads to an increase in peripheral resistance which causes pressure to rise in the aorta, left ventricle and left atrium. At the same time, the lungs are expanded as the neonate begins to breathe, and increased pulmonary oxygenation leads to pulmonary vasodilation and decreased pulmonary vascular resistance. As a result, the pressures in the pulmonary trunk, right ventricle and right atrium all fall. These changes reverse the pressure gradient from the right to the left atrium which existed in utero. Reverse flow does not occur, however, because a valve-like flap closes the foramen ovale, preventing blood flow from left to right. This functional closure is followed some months later by anatomical closure, with fibrosis of the potential opening.

The elevated aortic pressure also drives oxygenated blood into the pulmonary circulation through the ductus arteriosus. This shunt closes by active constriction after 1–8 days, possibly as a response to the high levels of P_{CO_2}. Anatomical closure, with fibrosis, follows in 1–4 months. The ductus venosus also closes functionally 1–3 hours after birth, with subsequent fibrosis, thus ensuring that all portal venous blood is passed through the liver.

Box 42 Clinical note: Persistent left to right shunts

There are a variety of cardiovascular defects which may result from abnormal embryonic development in utero. One group of abnormalities result from failure of the normal fetal shunts to close after birth. If either the foramen ovale or the ductus arteriosus remain open, blood will be shunted from the left side of the circulation to the right. This does not affect the oxygenation of the blood, which has already passed through the pulmonary circulation before the shunting occurs. It does increase the left ventricular work necessary to maintain a normal systemic blood flow, however, and may result in cardiac failure if not surgically corrected.

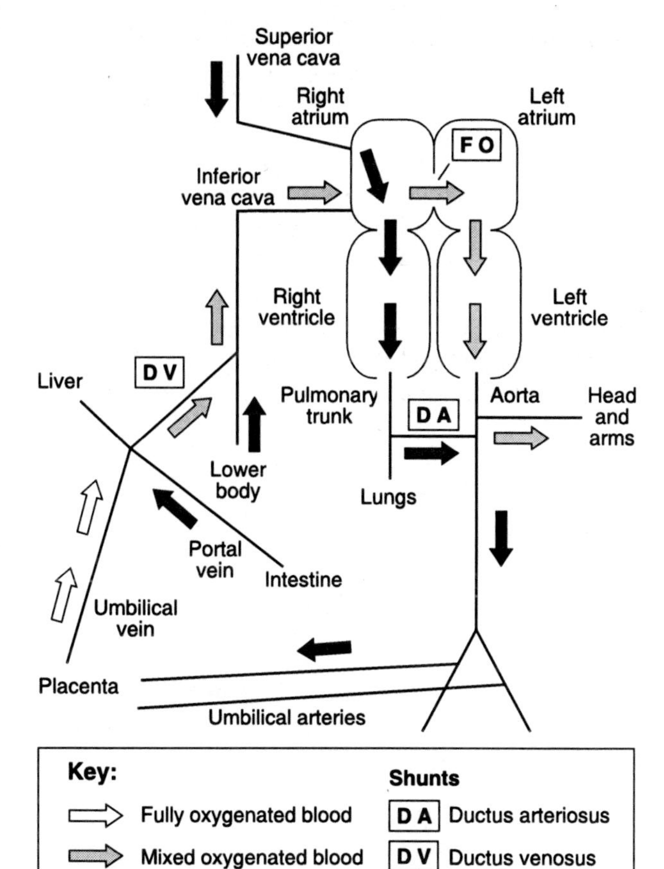

Fig. 183 Blood flow in the fetal circulation. Three major shunts bypass the immature liver (ductus venosus, DV) and the nonfunctioning pulmonary circulation (foramen ovale, FO, and ductus arteriosus, DA).

Respiratory adjustments

Respiratory movements are a normal feature of fetal behaviour in utero. These have no significance for fetal gas exchange, which is a placental function, but may prepare ventilatory muscles for extrauterine life. Adequate ventilation must be established immediately after birth if brain damage from anoxia is to be avoided. Lung expansion is initially resisted by the surface tension of the fluid which fills the alveoli at birth, so that large negative intrapleural pressures, around 60 cmH_2O, have to be generated during the first inspiratory phase. Inspiratory effort decreases rapidly over the first few breaths, as a stable functional residual capacity (FRC) is established, although the overall lung compliance remains lower than that in an adult. Expiration is also active rather than passive during early ventilation, since the airways are partly obstructed by fluid. Positive intrapleural pressures have to be generated to overcome this resistance.

Once breathing is established, the respiratory rate in neonates is high, at about 40 min^{-1}, with a respiratory minute volume of around 650 $ml\,min^{-1}$. This is about twice the adult rate when differences in body size are taken into account.

Premature babies often produce less than the normal amount of alveolar surfactant. Surfactant reduces the surface tension of intra-alveolar fluid and any deficiency further increases inspiratory work. This may lead to *respiratory distress syndrome* following preterm delivery.

Weight change and nutrition

Breast milk may be in short supply for the first 3–4 days following birth and the baby may take some time to master the art of suckling. This leads to an initial decline in weight, which may fall by as much as 10% mostly as a result of fluid loss. Birth weight is regained by about day 10, however, and growth continues at a dramatic rate, with a tripling of body mass in the first year of life. The nutritional demands for calcium, vitamin D and iron are particularly high during this period and infant stores are limited. Maternal supplements during breast feeding can help avoid deficiency states.

Metabolism and temperature control

An infant's metabolic rate is about twice that of an adult when scaled for body mass. Despite this, there is an initial drop in body temperature of 1–2°C after birth, and this only rises back to normal after 8–12 hours. Temperature regulation remains inefficient, so that the fluctuations in core temperature may be quite large for several weeks. Babies also have a high surface area to body volume ratio, favouring rapid heat loss to the environment. In order to counter this, infants can use sympathetic nerves to stimulate metabolic activity in *brown fat*. Mitochondrial function is uncoupled from ATP production in this tissue, so all the metabolic energy released by oxidation of glucose is converted to heat instead. This mechanism is unimportant in adult temperature control but can double heat production in infants.

Liver function

Several aspects of liver function are poorly developed at birth.

- Plasma bilirubin levels rise after birth, reaching a peak some five times greater than normal after 1–2 weeks. Concentrations then fall again over weeks or months as hepatic conjugation of bilirubin increases (Section 6.5)

- There is only limited glycogen storage and gluconeogenesis in the liver, so that blood glucose levels may fall as low as $2\,mmol\,L^{-1}$ in the first 24 hours of life, especially since lactose levels are low in colostrum.
- Plasma protein levels are initially lower than normal and the diminished concentration of clotting factors may increase the risk of bleeding.

Immune function

Active immunity is poorly developed in the newborn. Maternal immunoglobulins cross the placenta in utero (IgG class) and more can be absorbed from breast milk by the neonatal intestine, especially during the first few days of life (IgA class). This provides early passive immune protection, which declines over the first months of life while the infant's own Ig-producing mechanisms mature. Immunization programmes normally commence after 2–3 months when active immune responses can be evoked.

Self-assessment: questions

Multiple true/false questions

Each of the following statements consists of a stem followed by a number of possible endings. State whether each statement is True or False. For each stem, all, several or none of the statements may be true.

1. Meiosis:
   - **a.** does not involve replication of DNA
   - **b.** starts during intrauterine development of the fetal ovary
   - **c.** occurs only in the gonads
   - **d.** is responsible for multiplication of granulosa cells in the follicular phase of the ovarian cycle
   - **e.** is only completed in spermatozoa following fertilization of an ovum

2. Puberty:
   - **a.** is associated with increased plasma gonadotrophin levels
   - **b.** leads to appearance of the primary sex characteristics
   - **c.** is associated with increased rates of sex steroid secretion
   - **d.** generally begins earlier in males than in females
   - **e.** may be delayed in girls of very low body weight

3. Male sex hormones:
   - **a.** are only produced in the testis
   - **b.** lead to pubic hair growth in females
   - **c.** affect cellular transcription of DNA
   - **d.** stimulate gonadotrophin release
   - **e.** are produced by Sertoli cells
   - **f.** can be used as substrates for oestrogen synthesis in ovarian granulosa cells

4. Ejaculation of semen:
   - **a.** might be inhibited by drugs which block sympathetic neurotransmission
   - **b.** is independent of somatic nerves
   - **c.** is more likely to release 300×10^6 than 30×10^6 spermatozoa
   - **d.** involves contraction of skeletal muscle
   - **e.** involves contraction of smooth muscle

5. The ovarian cycle:
   - **a.** consists of proliferative and secretory phases
   - **b.** is controlled by FSH and LH from the posterior pituitary
   - **c.** involves development of only one primary follicle in the early part of each cycle
   - **d.** is associated with production of estradiol and progesterone
   - **e.** leads to the release of a secondary oocyte at ovulation

6. The menopause:
   - **a.** is followed by a rise in FSH concentration
   - **b.** is followed by a fall in oestrogen concentration
   - **c.** is caused by a decline in hypothalamic function
   - **d.** is followed by an increasing risk of osteoporosis
   - **e.** is followed by a rise in progesterone concentration

7. During pregnancy:
   - **a.** the placenta acts as an important endocrine structure
   - **b.** maternal extracellular fluid volume increases
   - **c.** the corpus luteum is sustained by luteinizing hormone
   - **d.** the mother may find warm conditions unusually uncomfortable (heat intolerance)
   - **e.** maternal prolactin concentrations rise

8. In a neonate (newborn baby):
   - **a.** parasympathetic stimulation of brown fat metabolism plays a significant role in temperature regulation
   - **b.** the intrapleural pressure during inspiration is likely to be more negative than normal in a premature baby
   - **c.** body weight usually falls immediately after birth
   - **d.** the levels of circulating bilirubin are higher than normal
   - **e.** active immune protection is provided by maternal immunoglobulins

Single best answer questions

For each of the following questions choose the single best answer.

1. Puberty:
   - **a.** is first indicated by a rise in sex hormones
   - **b.** is first indicated by a rise in gonadotrophins
   - **c.** Is first indicated by the development of the secondary sex characteristics
   - **d.** is first indicated by the adolescent growth spurt
   - **e.** is first indicated by the development of the secondary sex characteristics

2. During meiosis:
   - **a.** exchange of genetic material occurs during the 2nd meiotic division
   - **b.** cells formed after the 1st meiotic division are diploid
   - **c.** cells formed after the 2nd meiotic division are diploid
   - **d.** cells formed after the 2nd meiotic division have the normal number of genes
   - **e.** DNA replication occurs prior to the 1st meiotic division

3. In the testis, which of the following are *not* Sertoli cell functions?
   - **a.** phagocytosis
   - **b.** blood–testis barrier
   - **c.** secretion of testosterone
   - **d.** secretion of inhibin
   - **e.** production of seminiferous fluid

4. During oogenesis:
   - **a.** new primary oocytes are formed during each ovarian cycle
   - **b.** the primary oocytes complete the 1st meiotic division during fetal life
   - **c.** the polar bodies both contain the same quantity of DNA as a mature ovum
   - **d.** the second meiotic division is only completed on fertilization
   - **e.** the second meiotic division is only completed on ovulation

5. Which of the following are *not* a major source of oestrogens?
   - **a.** the theca cells in the follicular phase of the ovarian cycle
   - **b.** the granulosa cells in the follicular phase of the ovarian cycle
   - **c.** the corpus luteum following ovulation
   - **d.** the placenta
   - **e.** conversion of androgen precursors in peripheral tissues

6. Which of the following does not occur during pregnancy?
   - **a.** maternal cardiac output increases by about 40%
   - **b.** maternal blood volume increases by about 20%.
   - **c.** there is net maternal fluid retention
   - **d.** approximately 1g of additional Fe^{2+} is used for haemoglobin manufacture
   - **e.** maternal O_2 consumption doubles

7. During labour:
   - **a.** oxytocin is released from the anterior pituitary
   - **b.** relaxin inhibits uterine contraction
   - **c.** prostaglandins stimulate uterine contractions
   - **d.** cervical stretch promotes oxytocin release via sensory pathways which terminate in the medulla oblongata
   - **e.** vaginal stretch inhibits uterine contraction

8. Immediately following birth:
   - **a.** neonatal lung compliance is at its highest
   - **b.** bilirubin levels rise over the first 7–14 days
   - **c.** plasma protein levels are higher than in later infancy
   - **d.** mean variations in body core temperature are smaller than in later infancy
   - **e.** the foramen ovale closes due to increased levels of P_{O_2}.

Matching item questions

Theme: Hormones and reproduction

Options

A. Oestradiol
B. Testosterone
C. Progesterone
D. Luteinizing hormone
E. Follicle-stimulating hormone
F. Oxytocin

For each of the descriptions below choose the most appropriate option from the list above. Each option may be used once, more than once or not at all.

1. A peptide which stimulates Sertoli cell function.
2. The dominant steroid hormone during the luteal phase of the ovarian cycle.
3. Released from the posterior pituitary.
4. Stimulates ovulation.
5. Secreted by granulosa cells.
6. Secreted by interstitial cells of Leydig.

Theme: Phases of the menstrual cycle (assume a 28-day cycle)

Options

A. Day 14
B. Day 12
C. Day 27
D. Day 21
E. Day 0
F. Day 7

For each of the descriptions below choose the most appropriate option from the list above. Each option may be used once, more than once or not at all.

1. Progesterone levels are at their peak.
2. Luteinizing hormone is at its peak.
3. A single follicle becomes dominant.
4. Follicle-stimulating hormone is at its peak.
5. Corpus luteum is undergoing atresia.
6. Ovulation takes place.

Short notes

Write short notes on the following:

a. Sertoli cell function
b. erection of the penis
c. male sex hormones
d. development of the placenta

Modified essay

Male and female reproductive physiology shares many similarities, especially in relation to the control mechanisms involved. Understanding these controls is important, especially when considering contraception and infertility problems.

Questions

1. Name the main hormones involved in reproductive control in each sex and where they are produced.
2. What hormonal feedback loops can be identified in the control of reproduction?
3. Outline the main ovarian actions of relevant hormones named in you answer to Question 1.
4. Plot how the concentrations of these hormones change during a standard, 28-day, menstrual cycle.
5. Explain how feedback control mechanisms help explain the changes seen.
6. Describe the main changes in the uterus during the menstrual cycle.
7. Which hormones stimulate these changes? Relate the uterine responses seen to the changes in hormone concentration described in your answer to Question 4.

Data interpretation

A newborn baby is found to be apparently healthy at birth but after some weeks he does not seem to be thriving. When examined, the baby is found to be centrally cyanosed, i.e., his tongue and lips appear rather blue in colour. This indicates that the oxygen saturation of arterial haemoglobin is abnormally low. It is suspected that the cause is a congenital abnormality of the heart, leading to abnormal shunting of the blood from one side of the circulation to the other. Catheters are inserted allowing blood samples to be taken from a number of sites. The partial pressure of oxygen (P_{O_2}) is measured for each sample and the results obtained are given in Table 12.

Table 12 Measurement of P_{O_2} in blood samples from various sites around the circulation in a cyanosed baby (see text for details); normal values are quoted for comparison

	P_{O_2} (mmHg)	
Sample site	**Observed**	**Normal**
Brachial artery	35	97
Left ventricle	35	97
Pulmonary vein	103	103
Pulmonary trunk	10	40

Questions

1. Do these results support the suggestion that the arterial blood is deoxygenated?
2. Are these results more consistent with shunting of blood from the pulmonary to the systemic side of the circulation (right to left) or from the systemic to the pulmonary side (left to right)?
3. Based on the P_{O_2} values in Table 12, at what anatomical level does the mixing of oxygenated and deoxygenated blood appear to occur?
4. For such mixing to occur, would the hydrostatic pressure need to be higher on the right side of the circulation or the left? Is this the normal direction for the pressure gradient between the two sides of the circulation?
5. In the listed normal values, the P_{O_2} for the pulmonary veins is higher than that for the left ventricle. How might this be explained?
6. Assuming that the diagnosis of a congenital heart defect is correct, why did the baby not suffer damage in utero?

Clinical scenario

A married couple attend an infertility clinic. Preliminary investigations indicate that the husband has a very low sperm count and reduced plasma levels of testosterone.

Questions

1. What other hormone concentrations might one measure and how could the result suggest the possible anatomical site of the problem? Explain your answer.
2. Suitable treatment is found for the man's low fertility and the woman becomes pregnant. What changes would this lead to in the maternal concentrations of circulating gonadotrophins and sex steroids over the following 9 months? What is the source of each hormone?
3. The developing fetus carries XY sex chromosomes. Outline how this will affect the development of the fetal sex organs in utero.

After an uneventful pregnancy the woman goes into spontaneous labour at full term and gives birth to a healthy baby.

Question

4. Which hormones are believed to be involved in promoting labour and parturition, and what are their main actions?

The mother decides to breastfeed her baby and after a few days he is suckling strongly. He is soon gaining weight rapidly.

Questions

5. Which hormones control the supply of breast milk? What are their actions?
6. What are the main contents of breast milk?

Viva questions

1. Explain how the early embryo/fetus receives its nutrition.
2. Tell me about the uterine/menstrual cycle.
3. What role does the hypothalamus play in reproductive physiology?

Self-assessment: answers

Multiple true/false answers

1. a. **False.** DNA replication occurs prior to the 1st meiotic division.
   b. **True.** Leads to formation of primary oocytes.
   c. **True.** Meiosis gives rise to reproductive cells.
   d. **False.** This is normal mitotic division, only the oocyte undergoes meiosis.
   e. **False.** It is the secondary oocyte which completes the second meiotic division after fertilization.

2. a. **True.** This stimulates the onset of full reproductive activity in the gonads.
   b. **False.** Primary sex characteristics refer to gonads and genitalia which are present at birth; it is the secondary sex characteristics which develop during puberty.
   c. **True.** Through gonadotrophic stimulation of the gonads.
   d. **False.** About 1–2 years later in males.
   e. **True.** This may be seen in young athletes and those with eating disorders such as anorexia nervosa. A relative deficiency of sex hormone may explain this observation since adipose tissue is a source of oestrogens.

3. a. **False.** Androgens are also produced in the adrenal cortex.
   b. **True.** An effect of adrenal androgens.
   c. **True.** The general mode of action of steroid hormones.
   d. **False.** Testosterone inhibits gonadotrophin release as part of the negative feedback control of reproduction.
   e. **False.** Interstitial cells of Leydig.
   f. **True.** The androgens are produced by the thecal cells.

4. a. **True.** The emission phase is stimulated by sympathetic nerves.
   b. **False.** Reflex activity in somatic nerves is important for the expulsion phase.
   c. **True.** About 3 ml of semen at 100–120 $\times 10^6$ sperm ml^{-1}.
   d. **True.** Responsible for expulsion of semen from the urethra.
   e. **True.** Contracts the male reproductive glands and ducts during the emission phase.

5. a. **False.** These apply to the uterine cycle. The ovarian equivalents are the follicular and luteal phases.
   b. **False.** Gonadotrophins control the cycle but are anterior pituitary hormones.
   c. **False.** Several follicles begin development each cycle but only one achieves ovulation.
   d. **True.** From the developing follicles and the corpus luteum, respectively.
   e. **True.** The secondary oocyte is formed during ovulation.

6. a. **True.** Because of a loss of feedback inhibition by ovarian sex steroids.
   b. **True.** Since follicular development no longer occurs.
   c. **False.** GnRH levels rise initially with the reduced negative feedback.
   d. **True.** As a result of oestrogen deficiency.
   e. **False.** This falls in the absence of episodic corpus luteum development.

7. a. **True.** It produces hCG, human chorionic somatomammotropin and female sex steroids.
   b. **True.** As a result of mineralocorticoid effects of oestrogen and progesterone.
   c. **False.** hCG sustains the corpus luteum during pregnancy.
   d. **True.** This is a symptom of an elevated basal metabolic rate.
   e. **True.** This stimulates synthesis of milk-producing enzymes in the alveolar epithelium of the breast.

8. a. **False.** Sympathetic nerves stimulate heat production in brown fat.
   b. **True.** This reflects surfactant deficiency.
   c. **True.** Because of a poor initial supply of breast milk and limited suckling ability.
   d. **True.** The rates of bilirubin conjugation are low in the immature liver.

e. **False.** These provide passive immune protection.

Single best answers

1. b. A rise in FSH and LH levels precedes the other changes seen during puberty.
2. e. DNA replication and genetic crossover are features of the 1st meiotic division. This results in haploid cells, each of which has half the normal number of chromosomes. These cells still have two chromatids for each chromosome, and these separate during the 2nd meiotic division to produce haploid cells with one copy of each chromosome; i.e., with half the normal number of genes.
3. c. This is secreted by the interstitial cells of Leydig.
4. d. Primary oocytes are all formed in utero. The 1st meiotic division is completed at ovulation and the 2nd meiotic division on fertilization. The polar bodies contain the same amount of genetic material as the oocyte formed at that time. For the 1st meiotic division this is the same as that in the secondary oocyte, i.e., twice that in the mature ovum.
5. a. Peripheral conversion of androgen precursors to oestrones is an important oestrogen source postmenopausally, when ovarian secretion has ceased.
6. e. O_2 consumption increases but only by about 20–30%.
7. c. Oxytocin is released by the posterior pituitary, and its release is promoted by cervical stretch which sends sensory signals to the hypothalamus, the site of oxytocin manufacture. Vaginal stretch promotes additional uterine and abdominal contraction.
8. b. This is one aspect of immature liver function, which also leads to low plasma protein concentrations. Lung compliance is low, due to increased surface tension effects in the fluid filled lungs. Temperature regulation is initially poor. Closure of the foramen ovale reflects a change in the pressure gradient between left and right ventricles, and not the change in P_{O_2}, which is more important for the closure of the ductus arteriosus.

Matching item answers

Theme: Hormones and reproduction

1. E. FSH promotes spermatogenesis in this way. Testosterone also stimulates Sertoli cells but is a steroid hormone.
2. C. Progesterone is secreted from the corpus luteum in greater amounts than are the oestrogens.
3. F. Stimulates uterine contractions during labour and myoepithelial contraction during suckling.
4. D.
5. A. This is the dominant sex hormone during the follicular phase of the ovarian cycle.
6. B.

Theme: Phases of the menstrual cycle (assume a 28-day cycle)

1. D. Mid-luteal phase.
2. B. This occurs 1–2 days before ovulation (acts as signal for ovulation).
3. F. This occurs somewhere around day 7–10.
4. E. Loss of negative feedback by oestrogen and progesterone at the end of the luteal phase. Follicle stimulating hormone stimulates multiple follicles to start developing in the ovary.
5. C.
6. A. Ovulation occurs mid-cycle in a 'classical' 28 day cycle. Follicular phase may be quite variable in length, leading to variation in cycle length.

Short note answers

a. Sertoli cells provide nutritional support to the developing male gametes within the seminiferous tubules during spermatogenesis. They act as a protective diffusion barrier between the developing spermatocytes and blood-borne toxins, phagocytose the debris from dead or damaged cells and secrete the seminiferous fluid which washes spermatozoa into the epididymis. Sertoli cells also secrete inhibin, which is involved in feedback inhibition of FSH secretion.

b. Penile erection may be stimulated by erotic thoughts or images and is reflexly reinforced by mechanical stimulation of the glans penis. Sympathetic constrictor nerves are inhibited, while sacral parasympathetic dilator nerves are activated, leading to dilatation of the spiral arterioles which supply the vascular spaces in the corpora cavernosa and the corpus spongiosus. The increased blood flow congests these erectile tissues, and as the internal pressure rises, mechanical obstruction of the venous drainage may also contribute to the process.

c. Male sex hormones are steroids secreted from the adrenal cortex in both sexes but particularly from the interstitial cells of Leydig in the testis of the male. The production of testosterone in the developing XY fetus leads to the differentiation of the male genitalia. Testosterone levels rise at puberty as a result of stimulation of the testis by LH and this stimulates the onset of spermatogenesis. The secondary sex characteristics are also promoted, e.g., growth of pubic, facial and body hair and a drop in pitch of the voice. Testosterone also stimulates metabolic rate and the growth of muscle and long bones, leading to the pubertal growth spurt in adolescent boys. Male sex hormones have behavioural influences as well, stimulating the sex drive and aggressive traits.

d. Following implantation, the outer trophoblastic layer of the blastocyst develops into the chorion. Chorionic projections digest spaces in the surrounding decidual endometrium and these become filled with maternal blood anticoagulated by chorionic secretions. Fingers of chorionic tissue develop into placental villi, containing a core of fetal capillaries perfused from the umbilical arteries and draining to the umbilical vein. The thin placental membrane, which separates fetal and maternal blood, is formed from the capillary endothelium and the overlying chorionic epithelium. Uterine vessels anastomose with the intervillous spaces on the maternal side of the placenta, providing a continuous supply of blood to meet the gas exchange, nutritional and homeostatic needs of the fetus from about 4–5 weeks after conception until parturition.

Modified essay answers

1. The hypothalamus releases GnRH, and this stimulates release of FSH and LH from the anterior pituitary in both sexes. The male sex hormone testosterone is produced by the interstitial cells of the testis, while the female sex steroids oestrogen and progesterone are produced by developing ovarian follicles and the corpus luteum, respectively.

2. In general, secretion of regulatory hormones is inhibited by the gonadal hormones (negative feedback). Thus GnRH release from the hypothalamus and FSH/LH release from the anterior pituitary are inhibited by testosterone and inhibin in the male and by progesterone, inhibin and low concentrations of oestrogen in the female. FSH/LH may also inhibit GnRH secretion directly. Prolonged, high concentrations of oestrogen may stimulate LH secretion immediately prior to ovulation (positive feedback).

3. FSH stimulates follicular development during the follicular phase of the ovarian cycle (day 1–14 of the menstrual cycle for a 28 day cycle). This also requires adequate levels of oestrogens, which are both secreted by, and stimulate, the granulosa cells. Androgens from the thecal cells act as precursors for oestrogen production by the granulose cells. A rapid rise in LH (the LH-surge) promotes ovulation at mid-cycle, and the subsequent development and maintenance of the corpus luteum from the follicular remnants.

4. Plot should be similar to the upper two panels in Fig. 177.

5. At the start of the cycle FSH levels are high. This stimulates oestrogen production and negative feedback reduces FSH and LH up to about day 10. Positive feedback by high oestrogen levels then takes over and promotes the surge in LH. LH falls back again but remains relatively high during the luteal phase, promoting progesterone and oestrogen secretion. Negative feedback by these sex steroids again leads to a steady fall in gonadotrophin levels throughout the luteal phase. Programmed atresia of the corpus luteum about day 24 is followed by a rapid drop in progesterone and oestrogen levels. The loss of inhibition allows FSH and LH to rise again, stimulating the start of the next cycle.

6. These mainly affect the endometrium. Ischaemic tissue, blot and clots are shed as the menstrual flow for the first 3–4 days of the cycle This reduces the endometrial thickness by about two-thirds. There then follows the proliferative phase, when the endometrium undergoes hyperplasia and tubular glands reform. This phase is more or less complete by mid-cycle, after which there is little further hyperplasia. The glands become more tortuous in outline and start secreting glycogen rich fluid into the endometrial tissue and uterine lumen. Endometrial vascularity also increases. This secretory phase terminates after another 14 days with vascular constriction, haemorrhage and ischaemia, leading to the start of the next menstrual flow.
7. Oestrogen (mainly estradiol) stimulates the proliferative phase, while progesterone stimulates the secretory phase. The rapid fall in the levels of these hormones at the end of the luteal phase of the ovarian cycle acts as the signal for menstruation.

Data interpretation answers

1. Yes, since the P_{O_2} in the brachial artery is reduced considerably below normal. (Note: cyanosis is usually only detectable when there is more than $5\,g\,dl^{-1}$ of haemoglobin in the deoxygenated state. If we assume a haemoglobin concentration of $15\,g\,dl^{-1}$, $5\,g\,dl^{-1}$ of deoxygenated blood represents a saturation level of less than 70%. This occurs at a P_{O_2} of 40–50 mmHg (Section 4.4).)
2. Right to left shunts are more likely to produce deoxygenation of systemic blood since gas exchange in the pulmonary circulation is bypassed.
3. The drop in P_{O_2} between the pulmonary vein and the left ventricle suggests admixture of a large volume of deoxygenated blood with the oxygenated blood in the left atrium or left ventricle.
4. Deoxygenated blood appears to be flowing from right to left so the pressure must be higher on the right in this case. This is the reverse of the normal situation in which the pressure is higher in the aorta and on the left side of the heart. (Note: such a reversal of pressure gradient can occur, e.g., if there is an abnormal obstruction or narrowing of the pulmonary trunk leading to elevated pressure in the right ventricle and atrium.)
5. This suggests that some deoxygenated blood is mixed with the blood returning from the lungs under normal conditions. This occurs in the left atrium where deoxygenated blood from the bronchial blood vessels is returned to the circulation. This is a physiological shunt.
6. The placenta is responsible for gas exchange in utero and so the blood arriving at the right side of the heart is mixed, oxygenated blood. Right to left shunting is beneficial under these circumstances since it bypasses the redundant pulmonary circulation and increases blood flow to the systemic circulation. (Clinical note: the fetal circulation normally contains two such shunts, the foramen ovale and the ductus arteriosus, both of which close after birth. If the pressure remains high on the right side, however, e.g., because of a stenosis of the pulmonary trunk, closure may not take place and right to left shunting may persist. This is one cause of congenital cyanotic heart disease.)

Clinical scenario answers

1. Measurement of gonadotrophin levels might be useful. If the low testosterone is caused by a problem in the testis itself, FSH and LH levels should be elevated because of reduced negative feedback. Reduced gonadotrophin concentrations, however, would suggest a problem affecting the anterior pituitary or hypothalamus. Theoretically, GnRH measurement would distinguish between these, but concentrations in the systemic circulation are too low for easy analysis (local concentrations in the hypophyseal portal system are much higher). Prolactin measurements might also be useful, since high levels caused by an anterior pituitary tumour can inhibit gonadotrophin secretion in men and women.
2. Oestrogen and progesterone levels rise throughout pregnancy (Fig. 180). This inhibits FSH and LH levels, which are very low. Initially the sex steroids come from the corpus luteum, which is sustained by hCG from the placenta. hCG concentrations drop to a low plateau after 3 months, but steroid levels continue to rise because of placental production.

3. The fetus is a chromosomal male and the primitive gonads are genetically programmed to develop into testes. These secrete testosterone, which stimulates development of the male sex organs, and antimüllerian hormone, which inhibits development of the female reproductive tract.
4. Oxytocin from the posterior pituitary and prostaglandins released locally stimulate myometrial contraction. Stretching the uterine cervix reflexly increases oxytocin release, providing positive feedback. Oestrogen increases myometrial sensitivity to oxytocin towards term and this may help trigger labour. Relaxin from the placenta and corpus luteum increases elasticity in connective tissues, promoting dilatation of the cervix and stretching of pelvic ligaments to accommodate passage of the baby.
5. Prolactin from the anterior pituitary stimulates milk production by the alveolar epithelium while oxytocin from the posterior pituitary stimulates milk ejection into the ducts through myoepithelial cell contraction. Both are reflexly secreted in response to the baby's suckling at the nipple.
6. Water and electrolytes plus basic nutrients, i.e., carbohydrate (as lactose), lipids and protein, in descending order of concentration. Vitamins and minerals are also present, vitamin D, iron, calcium and phosphate being particularly important. Maternal IgA, along with the antibacterial lactoferrin, provides early protection against infection.

Viva answers

1. For the first week after fertilization first the morula and then the blastocyst is nourished by glycogen released into the uterus from the glands of the secretory endometrium. Further development requires implantation, during which digestion of the endometrium by the trophoblast layer of the blastocyst releases further nutrient substrates from the endometrial cells and stroma. Over the next month the placenta develops, in which fetal blood in the placental villi of the chorion is only separated from maternal blood by a thin trophoblastic layer. This allows diffusion of O_2 and nutrients from mother to fetus (and diffusion of CO_2 and waste products in the reverse direction) for the rest of the pregnancy.
2. Begin with a brief description of the endometrial changes during the three phases of the cycle (menstruation, proliferation, secretion). This will almost certainly lead to a discussion of the hormonal changes which cause these alterations. It is easiest to begin with the proliferative phase which follows menstruation. Rising oestrogen levels (released from the developing follicles) stimulate glandular and stromal growth in the endometrium. The secretory phase follows ovulation since it reflects progesterone dominance and this progesterone is produced by the corpus luteum. When the corpus luteum degenerates (about 10–12 days after ovulation), however, progesterone and oestrogen levels fall sharply. This causes endometrial ischaemia and infarction and leads to menstruation and the start of the next uterine (menstrual). cycle.
3. The hypothalamus plays a key role in reproductive physiology since it produces GnRH which is carried in the hypophyseal portal system and stimulates both FSH and LH secretion from the anterior pituitary. FSH stimulates spermatogenesis in the male and follicular development in the female. LH promotes testosterone production (necessary for spermatogenesis and maintenance of sex characteristics in men) and ovulation and formation of the corpus luteum in women. Thus, gonadal function is indirectly dependent on GnRH release. Puberty seems to be related to increased hypothalamic secretion of GnRH in the early teenage years, possibly due to genetically programmed development of relevant neurones.

Index

C

D

E

F

G

H

J

K

L

M

Q

R

S

T

W

X

Y

Z

Lightning Source UK Ltd.
Milton Keynes UK
11 December 2010

164237UK00001B/3/P

9 780443 102929